AF608697

NONLINEAR PHYSICAL SCIENCE

NONLINEAR PHYSICAL SCIENCE

Nonlinear Physical Science focuses on recent advances of fundamental theories and principles, analytical and symbolic approaches, as well as computational techniques in nonlinear physical science and nonlinear mathematics with engineering applications.

Topics of interest in *Nonlinear Physical Science* include but are not limited to:

- New findings and discoveries in nonlinear physics and mathematics
- Nonlinearity, complexity and mathematical structures in nonlinear physics
- Nonlinear phenomena and observations in nature and engineering
- Computational methods and theories in complex systems
- Lie group analysis, new theories and principles in mathematical modeling
- Stability, bifurcation, chaos and fractals in physical science and engineering
- Nonlinear chemical and biological physics
- Discontinuity, synchronization and natural complexity in the physical sciences

SERIES EDITORS

INTERNATIONAL ADVISORY BOARD

Albert C.J. Luo
Jian-Qiao Sun

Complex Systems

Fractionality, Time-delay and Synchronization

With 154 figures

Editors
Albert C.J. Luo
Department of Mechanical and Industrial Engineering
Southern Illinois University, Edwardsville
Edwardsville, IL 62026-1805, USA
Email: aluo@siue.edu

Jian-Qiao Sun
University of California, Merced
5200 N. Lake Road
P.O. Box 2039, Merced, CA 95344, USA
Email: jqsun@ucmerced.edu

ISSN 1867-8440 e-ISSN 1867-8459
Nonlinear Physical Science

ISBN 978-7-04-029710-2
Higher Education Press, Beijing

ISBN 978-3-642-17592-3 e-ISBN 978-3-642-17593-0
Springer Heidelberg Dordrecht London New York

Printed on acid-free paper

Springer is part of Springer Science+Business Media (www.springer.com)

Preface

This edited book covers recent developments on fractional dynamics, time-delay systems, system synchronization, and neuron dynamics.

Fractional calculus is extensively used as a powerful tool to investigate complex phenomena in engineering and science, and has received renewed attention recently. Chapter 1 of the book investigates fractional dynamics of complex systems. Some recent results and applications in fractional dynamics are presented. In Chapter 2, the synthesis and application of fractional-order controllers are presented. This is an active area of research. The fractional-order *PID* controllers are designed for the velocity control of an experimental modular servo system. The system consists of a digital servomechanism and open-architecture software environment for real-time implementation. Experimental results of fractional-order controllers are presented and analyzed. The effectiveness and superior performance of the fractional-order controls are compared with classical integer-order *PID* controllers.

Time delay is a common phenomenon in engineering, economical and biological systems, and has become a popular research topic in recent years. In Chapter 3, equilibrium stability, Lindsedt's method and Hopf bifurcation, and transient behaviors in differential-delay equations are presented. Multiple-scale and the center manifold analysis are addressed. These methods are applied to investigate dynamical behaviors of a differential-delay system modeling a section of the DNA molecule. Chapter 4 focuses on the methodologies for time-domain solutions and control design of time-delayed systems. Method of semi-discretization and continuous time approximation are discussed. The spectral properties of the methods will be investigated. A comparative study of stability of time-delayed linear time invariant systems is carried out by the Lyapunov method, Pad approximation and semi-discretization. The methods of solution for stochastic dynamical systems with time delay are also discussed, and a number of control examples and an experimental validation are presented.

Chapter 5 develops a theory for synchronization of multiple dynamical systems under constraints. The metric functionals based on the constraints are introduced to describe the synchronicity of two or more dynamical systems. The chapter provides a theoretic framework for designing controllers of slave systems which can be synchronized with master systems.

Finally in Chapter 6, complex dynamics of neurons with time-delay, stochasticity and impulsive discontinuity are presented. Complex dynamical behaviors include periodic spiking, chaotic spiking, periodic and chaotic bursting, and synchronization. In this chapter, a comprehensive review on recent developments and new results in nonlinear neural dynamics are presented.

It is our hope that the book presents a reasonably broad view of the state-of-the-art of complex systems, and provides a useful reference volume to scientists, engineers and students. Furthermore, we hope that the book will stimulate more researches in the rapidly evolving and interesting field of complex systems.

Edwardsville, Illinois — Albert C.J. Luo
Merced, California — Jian-Qiao Sun

June, 2010

Contents

Contributors

Dumitru Baleanu Department of Mathematics and Computer Science, Cankaya University, 06530 Ankara, Turkey; Institute of Space Sciences, P.O. BOX, MG-23, R 76900, Magurele-Bucharest, Romania, Email: dumitru@cankaya.edu.tr,baleanu @venus.nipne.ro

Ramiro S. Barbosa Dipartimento di Fisica, Università di Firenze, and INFN, Via Sansone 1, 50019 Sesto F.no (Firenze), Italy, Email: antoniaciani@gmail.com

M. Courbage Institute of Engineering of Porto, Dept. of Electrical Engineering, Rua Dr. Antonio Bernardino de Almeida, 431, 4200-072 Porto, Portugal, Email: rsb@isep.ipp.pt

Albert C.J. Luo Department of Mechanical and Industrial Engineering, Southern Illinois University Edwardsville, IL 62026-1805, USA, Email: aluo@siue.edu

J.A. Tenreiro Machado Institute of Engineering of Porto, Dept. of Electrical Engineering, Rua Dr. Antonio Bernardino de Almeida, 431, 4200-072 Porto, Portugal, Email: jtm@isep.ipp.pt

Richard Rand: Cornell University, Ithaca, NY 14853, USA, Email: rrand@cornell. edu

Manuel F. Silva Institute of Engineering of Porto, Dept. of Electrical Engineering, Rua Dr. Antonio Bernardino de Almeida, 431, 4200-072 Porto, Portugal, Email: mss@isep.ipp.pt

Bo Song School of Engineering, University of California, Merced, CA 95344, USA, Email: bosong1979@gmail.com

Jian-Qiao Sun School of Engineering, University of California, Merced, CA 95344, USA, Email: jqsun@ucmerced.edu

Yong Xie MOE Key Laboratory of Strength and Vibration, School of Aerospace, Xi'an Jiaotong University, Xi'an 710049, China, Email: yxie@mail.xjtv.edv.cn

Jian-Xue Xu MOE Key Laboratory of Strength and Vibration, School of Aerospace, Xi'an Jiaotong University, Xi'an 710049, China, Email: jxxu@mail.xjtu.edv.cn

Chapter 1
New Treatise in Fractional Dynamics

Dumitru Baleanu

Abstract Fractional calculus becomes a powerful tool used to investigate complex phenomena from various fields of science and engineering. In this context, the researchers paid a lot of attention for the fractional dynamics. However, the fractional modeling is still at the beginning of its developing. The aim of this chapter is to present some new results in the area of fractional dynamics and its applications.

1.1 Introduction

Fractional calculus deals with the generalization of differentiation and integration to non-integer orders. Fractional calculus is as old as the classical one and it has gained importance during the last few decades in various fields of science and engineering (Oldham and Spanier, 1974; Miller and Ross, 1993; Samko et al., 1993; Podlubny, 1999; Hilfer, 2000; Zaslavsky, 2005; Magin, 2000; Kilbas et al., 2006; West et al., 2003; Uchaikin, 2008; Lakshmikantham et al., 2009).

The fractional derivatives are the infinitesimal generators of a class of translation invariant convolution semigroups which appear universally as attractors. The fractional derivative at a point x is a local property only when α is an integer. Since the fractional derivatives represent the generalization of the classical ones, some of the classical properties are lost, e.g. the fractional Leibniz rule and the chain rule become more complicated than the classical counterparts (Oldham and Spanier, 1974; Miller and Ross, 1993; Samko et al., 1993; Podlubny, 1999; Kilbas et al., 2006).

Several applications of fractional calculus were simply based on replacing the time derivative in an evolution equation with a given derivative of fractional order.

Dumitru Baleanu
Department of Mathematics and Computer Science, Çankaya University, 06530 Ankara, Turkey;
Institute of Space Sciences, P.O. BOX, MG-23, R 76900, Magurele-Bucharest, Romania.
Emails: dumitru@cankaya.edu.tr, baleanu@venus.nipne.ro

Various recent results confirm that fractional derivatives seem to arise for important mathematical reasons (Podlubny, 1999; Hilfer, 2000; Zaslavsky, 2005; Magin, 2000; Kilbas et al., 2006; West et al., 2003; Uchaikin, 2008; Lakshmikantham et al., 2009; Gorenflo and Mainardi, 1997; Heymans and Podlubny, 2006; Mainardi et al., 2001; Scalas et al., 2004; Jesus and Machado, 2008; Chen et al., 2004; Barbosa et al., 2004; Carpinteri and Mainardi, 1997; Solomon et al., 1993; Fogleman et al., 2001; Nigmatullin and Mehaute, 2005; Momani, 2006; Tarasov, 2006, 2005; Zaslavsky, 2002; Lorenzo and Hartley, 2004; Baleanu et al., 2009a,b; Magin et al., 2008; Magin, 2009; Silva et al., 2008; Trujillo, 1999; Maraaba et al., 2008a,b; Baleanu et al., 2008a; Baleanu and Muslih, 2005a; Caputo, 2001; Lim and Muniandy, 2004; Mainardi et al., 2001; Mainardi, 1996; Tenreiro Machado, 2003, 2001; Metzler et al., 1995).

Based on the fact that the diffusion can be described by fractional differential equations, we ask the following questions:

Are mathematical models with fractional space and/or time derivatives consistent with the fundamental laws of nature? How can the fractional order of differentiation be observed experimentally?

Recently the fractional order differential equations started to play an important role in modeling the anomalous dynamics of various processes related to complex systems in the most diverse areas of science and engineering. However, only a few steps have been taken toward what may be called a coherent theory of these equations in the applied sciences (Oldham and S'panier, 1974; Miller and Ross, 1993; Samko et al., 1993; Podlubny, 1999; Hilfer, 2000; Kilbas, 2006; Uchaikin, 2008; Lakshmikantham, 2009).

The fractional Lagrangian and Hamiltonian are typical examples of non-local theories which were investigated in several physical problems (Pais and Uhlenbeck, 1950; Gomis et al., 2004, 2001; Gomis and Mehen, 2000; Llosa and Vives, 1994; Bering, online). Besides, a Hamilton formalism for nonlocal Lagrangian was proposed in Llosa and Vives (1994) and Bering (online), an equivalent singular first order Lagrangian was obtained and the corresponding Hamiltonian was pulled back on the phase space by making use of the corresponding constraints (Llosa and Vives, 1994). It was shown the space-time non-commutative field theories are acausal and the unitarity is lost (Seiberg et al., 2000; Alvarez-Gaume and Barbon, 2001).

The fractional variational principles represent an important part of fractional calculus and it is connected to the fractional quantization procedure (Riewe, 1996, 1997; Klimek, 2001; Klimek, 2002; Agrawal, 2002; Tarasov and Zaslavsky, 2006; Agrawal and Baleanu, 2007; Agrawal, 2006, 2007; Baleanu and Agrawal, 2006; Baleanu and Avkar, 2004; Baleanu, 2009; Rabei et al., 2009; Baleanu et al., 2008b; Baleanu, 2006, 2008; Baleanu and Trujillo, 2008). There are several proposed methods to obtain the fractional Euler-Lagrange equations and the corresponding Hamiltonian (Baleanu et al., 2008a,b; Rabei et al., 2007; Baleanu and Muslih, 2005b; Muslih and Baleanu, 2005a; Baleanu et al., 2006). However, this issue has not yet completely clarified and it requires more further analysis.

Quantization of systems with fractional derivatives is a novel area in the application of fractional differential and integral calculus. The interest in fractional quantization appears simply because it describes both conservative systems and non-conservative systems (Muslih and Baleanu, 2005b; Lim and Teo, 2009).

Schrödinger equation was considered with the first-order time derivative modified to Caputo fractional ones in Naber (2004), Dong and Xu (2008) and Jumarie (2009). However, in this case the obtained Hamiltonian was found to be non-Hermitian and non-local in time and the obtained wave functions are not invariant under the time reversal. The quantization of fractional Klein-Gordon field and fractional electromagnetic potential in the Coulomb gauge and the temporal gauge were subjected of intense debate (Lim and Teo, 2009).

The necessary conditions for the optimality in optimal control problems with dynamics described by differential equations of fractional order were obtained (Agrawal and Baleanu, 2007; Agrawal, 2004; Baleanu et al., 2009). By making use of an expansion formula for fractional derivative, optimality conditions and a new solution scheme is proposed.

It was proved that the fractional calculus models with differential equations can describe more complex biological systems by extending the scales (time and space) over which the models are effective and thus expand the range of phenomena under study.

The fractional wavelet transform (Unser and Blu, 2000b, 2002, 1999, 2000a) represents a new and important mathematical tool for signal and image analysis. The fractional wavelet analysis (Dinç and Baleanu, 2006, 2010) and the combination of this method with some other standard ones (Walczak, 2000; Dinç and Baleanu, 2007, 2004b,a; Dinç et al., 2003) were proposed very recently in order to investigate the composite signals of the components in complex drug mixtures.

This chapter is based mainly on the results obtained by the author and his collaborators in various fields of fractional calculus and its application. The chapter is organized as follows:

In Section 1.2 the main definitions and the properties of the fractional calculus are presented.

Section 1.3 is dedicated to the fractional variational principles and their applications.

Section 1.4 contains a brief review of the fractional optimal control formulation.

Section 1.5 is devoted to the application of the fractional calculus in nuclear magnetic resonance.

Fractional wavelet method and its applications in drug analysis are illustrated briefly in Section 1.6.

1.2 Basic definitions and properties of fractional derivatives and integrals

In this section we present the basic definitions of the fractional derivatives and integrals (Oldham and Spanier, 1974; Miller and Ross, 1993; Samko et al., 1993; Podlubny, 1999; Kilbas et al., 2006).

Definition 1. *Riemann-Liouville left-sided fractional integral of order* α is given by

$$ {}_a\mathbf{D}_x^{-\alpha}\phi(x) = \frac{1}{\Gamma(\alpha)}\int_a^x (x-t)^{\alpha-1}\phi(t)dt, \quad x > a. \tag{1.1}$$

Definition 2. *Riemann-Liouville right-sided fractional integral of order* α has the form

$$ {}_x\mathbf{D}_b^{-\alpha}\phi(x) = \frac{1}{\Gamma(\alpha)}\int_x^b (t-x)^{\alpha-1}\phi(t)dt, \quad x < b. \tag{1.2}$$

Here $\alpha > 0$ and $\Gamma(\alpha) = \int_0^\infty s^{\alpha-1}e^{-s}ds$ denotes the Gamma function.

Having defined the fractional integral, define the fractional derivative as the inverse operation, namely

$$ {}_a\mathbf{D}_x^{\alpha}\,{}_aD_x^{-\alpha}\phi = \phi, \tag{1.3}$$

$$ {}_a\mathbf{D}_x^{\mu}\phi = \frac{d^N}{dx^N}[{}_aD_x^{-\alpha}], \tag{1.4}$$

where $\alpha = N - \mu$, N is the smallest integer bigger than μ and $\mu > 0$.

Definition 3. *Left Riemann-Liouville fractional derivative of order* α is defined as

$$ {}_a\mathbf{D}_x^{\alpha}\phi(x) = \frac{1}{\Gamma(n-\alpha)}\left(\frac{d}{dx}\right)^n \int_a^x \frac{\phi(t)}{(x-t)^{\alpha+1-n}}dt, x > a. \tag{1.5}$$

Definition 4. *Right Riemann-Liouville fractional derivative of order* α becomes

$$ {}_x\mathbf{D}_b^{\alpha}\phi(x) = \frac{1}{\Gamma(n-\alpha)}\left(-\frac{d}{dx}\right)^n \int_x^b \frac{\phi(t)}{(t-x)^{\alpha+1-n}}dt, \quad x < b, \tag{1.6}$$

$n - 1 \leq \alpha < n$, and $\alpha > 0$.

Note that, for $\alpha = 1, 2, ...$ we have

$$ {}_a\mathbf{D}_x^{\alpha} = \left(\frac{d}{dx}\right)^\alpha, \tag{1.7}$$

$$ {}_x\mathbf{D}_b^{\alpha} = \left(-\frac{d}{dx}\right)^\alpha. \tag{1.8}$$

Definition 5. *The left Caputo fractional derivative* is defined as

$$ {}_a^CD_x^{\alpha}\phi(x) = {}_aD_x^{\alpha}\left(\phi(x) - \sum_{k=0}^{n-1}\frac{\phi^{(k)}(a)}{k!}(x-a)^k\right), \tag{1.9}$$

or

$$ {}_a^CD_x^{\alpha}\phi(x) = \frac{1}{\Gamma(n-\alpha)}\int_a^x \frac{\phi^n(t)}{(x-t)^{\alpha+1-n}}dt, \tag{1.10}$$

Definition 6. *The right Caputo fractional derivative* is defined as

$$ {}_{x}^{C}D_{b}^{\alpha}\phi(x) = \frac{(-1)^n}{\Gamma(n-\alpha)} \int_x^b \frac{\phi^n(t)}{(t-x)^{\alpha+1-n}} dt, \tag{1.11}$$

where $0 \leq n-1 < \alpha < n$ and $\phi(x)$ has n+1 continuous and bounded derivatives in $[a, b]$.

We notice that

$$ {}_{a}^{C}D_{x}^{\alpha} A = 0, \quad (A = constant) \tag{1.12}$$

and

$$ \lim_{x \to a} {}_{a}^{C}D_{x}^{\alpha}\phi = 0. \tag{1.13}$$

In an infinite domain we have the following results

$$ {}_{-\infty}^{C}D_{x}^{\alpha}\phi = {}_{-\infty}\mathbf{D}_{x}^{\alpha}\phi, {}_{x}^{C}D_{\infty}^{\alpha}\phi = {}_{x}\mathbf{D}_{\infty}^{\alpha}\phi. \tag{1.14}$$

Let us consider a function $f(x_1, x_2)$.

Definition 7. *A partial left Riemann-Liouville fractional derivative* of order α_2, $0 < \alpha_2 < 1$, in the second variable is defined as

$$ (\mathbf{D}_{a_2+}^{\alpha_2} f)(x) = \frac{1}{\Gamma(1-\alpha)} \frac{\partial}{\partial x_2} \int_{a_2}^{x_2} \frac{f(x_1, u)}{(x_2 - u)^{\alpha_2}} du. \tag{1.15}$$

Definition 8. *A partial right Riemann-Liouville fractional derivative* of order α_k has the form

$$ (\mathbf{D}_{a_2-}^{\alpha_2} f)(x) = \frac{1}{\Gamma(1-\alpha)} \frac{\partial}{\partial x_2} \int_{x_2}^{a_2} \frac{f(x_1, x_2)}{(-x_2 + u)^{\alpha_2}} du. \tag{1.16}$$

If the function f is differentiable we obtain

$$ (\mathbf{D}_{a_k+}^{\alpha_k} f)(x) = \frac{1}{\Gamma(1-\alpha_k)} \left[\frac{f(x_1, \ldots, x_{k-1}, a_k, x_{k+1}, \ldots, x_n)}{(x_k - a_k)^{\alpha_k}} \right. \\ \left. + \int_{a_k}^{x_k} \frac{\frac{\partial f}{\partial u}(x_1, \ldots, x_{k-1}, u, x_{k+1}, \ldots, x_n)}{(x_k - u)^{\alpha_k}} du. \right] \tag{1.17}$$

Definition 9. *Reflection operator* Θ is defined as follows

$$ \Theta\phi(x) = \phi(a + b - x) \tag{1.18}$$

and it has the following properties

$$ \Theta[{}_{a}\mathbf{D}_{x}^{-\alpha}\phi] = {}_{x}\mathbf{D}_{b}^{-\alpha}[\Theta\phi], \tag{1.19}$$

$$ \Theta[{}_{x}\mathbf{D}_{b}^{-\alpha}\phi] = {}_{a}\mathbf{D}_{x}^{-\alpha}[\Theta\phi]. \tag{1.20}$$

In addition, we have other properties of fractional derivatives and integrals, namely:

Semi-group property

$$ {}_a\mathbf{D}_x^{-\alpha}\,{}_a\mathbf{D}_x^{-\beta}\phi = {}_a\mathbf{D}_x^{-\alpha-\beta}\phi, \tag{1.21}$$

$$ {}_x\mathbf{D}_b^{-\alpha}\,{}_x\mathbf{D}_b^{-\beta}\phi = {}_x\mathbf{D}_b^{-\alpha-\beta}\phi, \tag{1.22}$$

where $\alpha > 0, \beta > 0$.

Reciprocity

$$ {}_a\mathbf{D}_x^{\alpha}\,{}_a\mathbf{D}_x^{-\alpha}\phi = \phi, \qquad \alpha > 0, t > a, \tag{1.23}$$

In the more general case we have

$$ {}_a\mathbf{D}_x^{\alpha}\,{}_a\mathbf{D}_x^{-\beta}\phi = {}_a\mathbf{D}_x^{\alpha-\beta}\phi. \tag{1.24}$$

Composition Rules

$$ \frac{d^n}{dx^n}[{}_a\mathbf{D}_x^{\alpha}\phi] = {}_a\mathbf{D}_x^{n+\alpha}\phi, \qquad \alpha \geq 0, \tag{1.25}$$

$$ {}_a\mathbf{D}_x^{\alpha}[\frac{d^n}{dx^n}\phi(x)] = {}_a\mathbf{D}_x^{n+\alpha}\phi(x) - \sum_{j=0}^{n-1}\frac{\phi^{(j)}(a)(x-a)^{j-\alpha-n}}{\Gamma(1+j-\alpha-n)}, \tag{1.26}$$

$$ n-1 \leq \alpha < n. $$

$\frac{d^n}{dx^n}$ and ${}_aD_x^{\alpha}$ are commutative only if $\phi^{(j)}(a) = 0$ for $j = 0, 1, ..., n-1$.

$$ {}_a\mathbf{D}_x^{\alpha}[{}_a\mathbf{D}_x^{\beta}\phi(x)] = {}_a\mathbf{D}_x^{\alpha+\beta}\phi(x) - \sum_{j=1}^{m}[{}_a\mathbf{D}_x^{\beta-j}\phi(x)]_{x=a}\frac{(x-t)^{-j-\alpha}}{\Gamma(1-j-\alpha)}, \tag{1.27}$$

$$ m-1 \leq \beta < m. $$

In particular, the fractional derivatives commute

$$ {}_a\mathbf{D}_x^{\alpha}[{}_a\mathbf{D}_x^{\beta}\phi] = {}_aD_x^{\beta}[{}_a\mathbf{D}_x^{\alpha}\phi] \tag{1.28}$$

if

$$ [{}_a\mathbf{D}_x^{\alpha-j}\phi(x)]_{x=a} = 0, \qquad j = 1, ..., n \tag{1.29}$$

and

$$ [{}_a\mathbf{D}_x^{\beta-j}\phi(x)]_{x=a} = 0, \qquad j = 1, ..., m. \tag{1.30}$$

Chain Rule has the following form:

For $\phi(x) = F(h(x))$ we obtain

$$ {}_a\mathbf{D}_x^{\alpha}F(h(x)) = \frac{(x-a)^{-\alpha}}{\Gamma(1-\alpha)}\phi(x) + \sum_{k=1}^{\infty}\binom{\alpha}{k}\frac{k!(x-a)^{k-\alpha}}{\Gamma(k-\alpha+1)}\frac{d^k}{dx^k}F(h(x)), \tag{1.31}$$

and it is obtained with the help of Faá di Bruno formula given below:

$$\frac{d^k}{dt^k}F(h(x)) = k! \sum_{m=1}^{k} F^{(m)}(h(x)) \sum \prod_{r=1}^{k} \frac{1}{a_r!} \left(\frac{h^{(r)}(x)}{r!}\right)^{a_r}, \tag{1.32}$$

and

$$\sum_{r=1}^{k} r a_r = k, \quad \sum_{r=1}^{k} a_r = m. \tag{1.33}$$

Leibniz Rule has a complicated formula as given below

$$_a\mathbf{D}_x^\alpha(\phi(x)\psi(x)) = \sum_{k=0}^{\infty} \binom{\alpha}{k} \phi^{(k)}(x)\, _a\mathbf{D}_x^{\alpha-k}\psi(x), \tag{1.34}$$

where $\phi(x)$ and $\psi(x)$ have continuous derivatives in $[a, t]$.

Definition 10. *Mittag-Leffler function* is defined as

$$E_\alpha = \sum_{k=0}^{\infty} \frac{t^k}{\Gamma(\alpha k + 1)}, \quad \alpha > 0, \ \alpha \in \mathbb{R}. \tag{1.35}$$

By using Eq. (1.35) we observe that

$$E_1(t) = \exp(t), \tag{1.36}$$

and

$$E_2(t) = \cosh\sqrt{t}. \tag{1.37}$$

Definition 11. *The two-parameter Mittag-Lefler function* has the form:

$$E_{\alpha,\beta}(t) = \sum_{0}^{\infty} \frac{t^k}{\Gamma(\alpha k + \beta)}, \quad \alpha > 0; \ \beta > 0, \ \alpha, \ \beta \in \mathbb{R}. \tag{1.38}$$

From (1.38) we conclude that

$$E_{1,2}(t) = \frac{e^t - 1}{t}, \tag{1.39}$$

$$E_{2,2}(t) = \frac{\sinh\sqrt{t}}{\sqrt{t}}, \tag{1.40}$$

$$E_{2,1}(t) = \cosh\sqrt{t}. \tag{1.41}$$

The Laplace transformations for several Mittag-Leffler functions are summarized in the following:

$$L(E_\alpha(-\lambda t^\alpha)) = \frac{s^{\alpha-1}}{s^\alpha + \lambda}, \tag{1.42}$$

$$L(t^{\alpha-1}E_{\alpha,\alpha}(-\lambda t^\alpha)) = \frac{1}{s^\alpha + \lambda}, \tag{1.43}$$

$$L(t^{\beta-1}E_{\alpha,\beta}(-\lambda t^{\alpha})) = \frac{s^{\alpha-\beta}}{s^{\alpha}+\lambda}. \tag{1.44}$$

In the following we are going to present a brief introduction of the generalized functions and they connect with the fractional derivatives.

Generalized functions have many interesting applications in science and engineering (Gelfand and Shilov, 1964).

Let us consider the Cauchy's integral formula as given below:

$$f^{(-n)} = \frac{1}{\Gamma(n)}\int_a^t f(\tau)(t-\tau)^{n-1}du. \tag{1.45}$$

Here n is a positive integer, $\Gamma(n)$ denotes the Gamma function and $a < t$. Let us consider $\Phi_n^+(t)$ as given below:

$$\Phi_n^+(t) = \frac{1}{\Gamma(n)}t^{n-1}, \quad t > 0 \tag{1.46}$$

and zero for $t \leq 0$.

Letting f be zero for $t < a$ and by making use of (1.46), we obtain that (1.45) becomes

$${}_a\mathbf{I}_t^n f(t) = f(t) * \Phi_n^+(t), \tag{1.47}$$

where $*$ denotes the convolution operation and it is given by

$$g(t) * f(t) = \int_{-\infty}^{+\infty} g(\tau)h(t-\tau)d\tau. \tag{1.48}$$

Equation (1.47) can be generalized for any $\alpha > 0$ as (Gelfand and Shilov, 1964)

$${}_a\mathbf{I}_t^\alpha f(t) = f(t) * \Phi_\alpha^+(t), \tag{1.49}$$

where $\Phi_\alpha^+(t)$ is a generalized function or distribution (Gelfand and Shilov, 1964).

Having in mind to define the convolution of two generalized functions we have to defined first the test functions (Gelfand and Shilov, 1964). For these reasons we choose the set K of all real functions $\phi(x)$ with continuous derivatives of all orders and with bounded support. We denote these functions the test functions. We can add and multiply by a scalar a test function in order to get new test functions, as a result, K becomes a linear space. Another interesting property of these test functions is that the sequence $\phi_1(x), \ldots, \phi_\nu(x)$ of test functions converges to zero in K if all above mentioned functions vanish outside a given fixed common bounded region and converge uniformly to zero together with their derivatives of any order.

We claim that f is a continuous linear functional on K if there exists some rule according to which we can associate with every $\phi(x)$ in K a real number (f, ϕ) such that (Gelfand and Shilov, 1964)

a) $(f, \alpha_1\phi_1 + \alpha_2\phi_2) = \alpha_1(f, \phi_1) + \alpha_2(f, \phi_2)$, for any real numbers α_1 and α_2 and for any two functions $\phi_1(x)$ and $\phi_2(x)$.

b) If the sequence $\phi_1, \phi_2, \ldots, \phi_\nu, \ldots$ converges to zero in K, then the sequence $(f, \phi_1), (f, \phi_2), \ldots, (f, \phi_\nu), \ldots$ converges to zero (Gelfand and Shilov, 1964).

The next step is to consider $k(t) = g(t) * h(t)$ and a test function $\phi(x)$. Therefore, we obtain the following (Podlubny, 1999; Gelfand and Shilov, 1964):

$$\begin{aligned} \langle k, \phi \rangle &= \int k(t)\phi(t)dt = it \int \left\{ \int g(\xi)h(t-\xi)d\xi \right\} \phi(t)dt \\ &= \int \int g(\xi)h(\eta)\phi(\xi+\eta)d\xi d\eta. \end{aligned} \tag{1.50}$$

Here the limits of the integrals are $-\infty$ and $+\infty$ respectively.

By making use of (1.50) we obtain the generalization of a convolution of two functions as mentioned in the following:

$$\langle g * h, \phi \rangle = \langle g(t), \langle h(\tau), \phi(t+\tau) \rangle \rangle. \tag{1.51}$$

From (1.51) we obtain the following properties of the convolution operation:

$$g * h = h * g, \tag{1.52}$$

$$f * (g * h) = (f * g) * h, \tag{1.53}$$

$$\mathbf{D}(g * h) = (\mathbf{D}g) * h = g * (\mathbf{D}h), \tag{1.54}$$

where $\mathbf{D}(\cdot)$ denotes the generalized derivative. The relation between the generalized derivative and the classical derivative becomes (Podlubny, 1999; Gelfand and Shilov, 1964)

$$\mathbf{D}^n f = f^{(n)} + \sum_{k=0}^{n-1} [\mathbf{D}^{n-k-1}\delta(t-a)] f^{(k)}(a). \tag{1.55}$$

For $\alpha < 0$ and $\Phi_\alpha^+(t)$ as a generalized function we introduce the notion of left fractional derivative as given below:

$${}_a\mathbf{D}_t^{-\alpha} = {}_a\mathbf{I}_t^\alpha[f], \tag{1.56}$$

or

$${}_a\mathbf{D}_t^{-\alpha}[f] = f(t) * \Phi_\alpha^+(t). \tag{1.57}$$

In the same way, we define

$${}_a\mathbf{D}_t^\alpha[f] = f(t) * \Phi_{-\alpha}^+(t), \; \alpha > 0. \tag{1.58}$$

The most interesting properties of the distributions $\Phi_\alpha^+(t)$ are (Podlubny, 1999; Gelfand and Shilov, 1964)

$$\mathbf{D}^{-n} = \Phi_n^+(t), n \in Z, \tag{1.59}$$

and

$$\Phi_\alpha^+(t-a) * \Phi_\beta^+(t) = \Phi_{\alpha+\beta}^+(t-a). \tag{1.60}$$

From (1.60) we obtain

$${}_a\mathbf{D}_t^\alpha({}_a\mathbf{D}_t^\beta(f)) = {}_a\mathbf{D}_t^{\alpha+\beta}(f). \tag{1.61}$$

For $0 \leq n-1 \leq \alpha < n$ we obtain the following important relations for the generalized fractional derivative

$$\begin{aligned} {}_a\mathbf{D}_t^\alpha(f) &= f(t) * \Phi_\alpha^+(t) = f(t) * (D^n \Phi_{n-\alpha}^+)(t) \\ &= (D^n f(t) * \Phi_{n-\alpha}^+(t)) = D^n(f(t) * \Phi_{n-\alpha}^+(t)). \end{aligned} \tag{1.62}$$

From (1.62) we observe that the distributional forms of Caputo and the Riemann-Liouville are the same.

The right fractional derivative can be define as follows

$${}_t\mathbf{D}_b^\alpha(f) = f(t) * \Phi_{-\alpha}^-, \tag{1.63}$$

where $\Phi_{-\alpha}^-$ is defined as

$$\begin{aligned} \Phi_{-\alpha}^-(t) &= \frac{(-t)^{\alpha-1}}{\Gamma(\alpha)}, \quad t < 0, \\ \Phi_{-\alpha}^-(t) &= 0, \qquad\qquad t \geqslant 0. \end{aligned} \tag{1.64}$$

In addition, we have

$$\Phi_n^-(t) = (-1)^n D^{-n}\delta(t^-), \tag{1.65}$$

for n being integer.

The integration by parts formula is valid for the generalized fractional derivatives, namely

$$\int {}_a\mathbf{D}_t^\beta[f]g(t)dt = \int {}_t\mathbf{D}_b^\beta[g]f(t)dt. \tag{1.66}$$

1.3 Fractional variational principles and their applications

The Lagrangian formulation of dynamical systems represents one of the most important principle in physics.

The corresponding Lagrangian for dissipative systems depends explicitly on time, therefore the Hamiltonian depends explicitly on time too.

One still open and important issue in this area is the fractional quantization procedure. The main obstacle for fractional calculus quantization is represented by its non-locality of the fractional derivatives.

The main advantage of this theory is that it incorporates, under certain limits, both the conservative and nonconservative systems.

1.3.1 *Fractional Euler-Lagrange equations for discrete systems*

The classical Euler-Lagrange differential equation is the fundamental equation of calculus of variations.

It states that J if is defined by an integral of the form

$$I = \int f(t, y, \dot{y}) dt, \tag{1.67}$$

where

$$\dot{y} = \frac{dy}{dt}, \tag{1.68}$$

then J has a stationary value if the Euler-Lagrange differential equation

$$\frac{\partial f}{\partial y} - \frac{d}{dt}\left(\frac{\partial f}{\partial y}\right) = 0. \tag{1.69}$$

Let us consider the following Lagrangian

$$L = \frac{1}{2}\dot{q}^2 - V(q). \tag{1.70}$$

As a result, the corresponding Euler-Lagrange equation becomes

$$\frac{\partial V(q)}{\partial q} + \ddot{q} = 0. \tag{1.71}$$

In the following we are giving the fractional generalization of the above results.

Let us assume that α_j $(j = 1, \ldots, n_1)$ and β_k $(k = 1, \ldots, n_2)$ are two sets of real numbers all greater than 0, $\alpha_{max} = \max(\alpha_1, \ldots, \alpha_{n1}, \beta_1, \ldots, \beta_{n2})$, and M is an integer such that $M - 1 \leq \alpha_{max} \leq M$. Let $J[q^\rho]$ be a functional of the type

$$\int_a^b L\left(t, q^\rho, {}_a\mathbf{D}_t^{\alpha_1} q^\rho, \ldots, {}_a\mathbf{D}_t^{\alpha_{n_1}} q^\rho,\ {}_t\mathbf{D}_b^{\beta_1} q^\rho, \ldots, {}_t\mathbf{D}_b^{\beta_{n_2}} q^\rho\right) dt, \tag{1.72}$$

defined on the set of n functions q^ρ, $\rho = 1, \ldots, n$ which have continuous left Riemann-Liouville fractional derivative of order α_j, $j = 1, \ldots, n_1$ and right Riemann-Liouville fractional derivative of order β_j, $j = 1, \ldots, n_2$ in $[a, b]$ and satisfy the boundary conditions $(q^\rho(a))^{(j)} = q^\rho_{aj}$ and $(q^\rho(b))^{(j)} = q^\rho_{bj}$, $j = 1, \ldots, M-1$. A necessary condition for $J[q^\rho]$ to admit an extremum for given functions $q^\rho(t)$, $\rho = 1, \ldots, n$ is that $q^\rho(t)$ satisfy Euler-Lagrange equations (Agrawal, 2002)

$$\frac{\partial L}{\partial q^{\rho}} + \sum_{j=1}^{n1} {}_t\mathbf{D}_b^{\alpha_j} \frac{\partial L}{\partial_a \mathbf{D}_t^{\alpha_j} q^{\rho}} + \sum_{j=1}^{n2} {}_a\mathbf{D}_t^{\beta_j} \frac{\partial L}{\partial_t \mathbf{D}_b^{\beta_j} q^{\rho}} = 0. \tag{1.73}$$

Here, if α_j is an integer, then ${}_a\mathbf{D}_t^{\alpha_j}$ and ${}_t\mathbf{D}_b^{\alpha_j}$ must be replaced with the ordinary derivatives $(d/dt)^{\alpha_j}$ and $(-d/dt)^{\alpha_j}$, respectively. The method initiated by Agrawal (Agrawal, 2002) was generalized and improved by Baleanu and coworkers (Baleanu et al., 2006; Baleanu and Avkar, 2004; Baleanu, 2004; Muslih and Baleanu, 2005c; Baleanu and Muslih, 2005b).

Let us start with the following classical Lagrangian

$$L(x, y, z) = \dot{x}\dot{z} + yz^3. \tag{1.74}$$

The classical solutions of Euler-Lagrange equations are given below

$$x(t) = at + b, z(t) = 0. \tag{1.75}$$

We notice that $y(t)$ has an undetermined evolution and a and b are constants to be determined from the initial conditions.

The fractional generalization of (1.74) becomes

$$L_f = ({}_a\mathbf{D}_t^{\alpha} x)\,{}_a\mathbf{D}_t^{\alpha} z + yz^3. \tag{1.76}$$

As a result, the Euler-Lagrange equations of (1.76) are given below

$${}_t\mathbf{D}_b^{\alpha}({}_a\mathbf{D}_t^{\alpha} z) = 0, z^3 = 0, \quad {}_t\mathbf{D}_b^{\alpha}({}_a\mathbf{D}_t^{\alpha} x) + 3yz^2 = 0. \tag{1.77}$$

From (1.77) we notice that $z = 0$, y is not determined. We note that x fulfills the following equation

$${}_t\mathbf{D}_b^{\alpha}({}_a\mathbf{D}_t^{\alpha} x) = 0. \tag{1.78}$$

The solution of (1.78), under the assumption of $1 < \alpha < 2$, is given by

$$\begin{aligned} x(t) &= A(t-a)^{\alpha-1} + B(t-a)^{\alpha-2} \\ &\quad + C(t-a)^{\alpha}\,{}_2F_1\left(1, 1-\alpha, 1+\alpha, \frac{t-a}{b-a}\right) \\ &\quad + D(t-a)^{\alpha}\,{}_2F_1\left(1, 2-\alpha, 1+\alpha, \frac{t-a}{b-a}\right). \end{aligned} \tag{1.79}$$

Here ${}_2F_1$ represents Gauss hypergeometric function and A, B, C, and D are real constants. When $\alpha \to 1^+$ and $a = 0$, the classical linear solution of one-dimensional space is recovered, namely

$$x(t) = A + Ct. \tag{1.80}$$

1.3.2 Fractional Hamiltonian formulation

1.3.2.1 A direct method with Riemann-Liouville fractional derivatives

In the following we introduce the meaning of fractional Hamiltonian. For simplicity, in the following we consider the following form of the fractional Euler-Lagrange equations (Agrawal, 2002)

$$\frac{\partial L}{\partial q^{\rho}(t)} + {}_t\mathbf{D}_b^{\alpha} \frac{\partial L}{\partial {}_a\mathbf{D}_t^{\alpha} q^{\rho}(t)} = 0, \quad 0 < \alpha < 1\,, \rho = 1, \ldots, N. \tag{1.81}$$

In the following by using (1.81) we define the generalized momenta as (see, for example, Ref.(Rabei et al., 2007) for more details)

$$p_{\alpha_\rho} = \frac{\partial L}{\partial {}_a\mathbf{D}_t^{\alpha} q^{\rho}(t)}, \quad \rho = 1, \ldots, N. \tag{1.82}$$

As a consequence of (1.81) and (1.82) a Hamiltonian function is defined as

$$H = p_{\alpha_\rho}{}_a\mathbf{D}_t^{\alpha} q^{\rho}(t) - L. \tag{1.83}$$

The canonical equations corresponding to (1.83) are given below

$$\frac{\partial H}{\partial t} = -\frac{\partial L}{\partial t}, \tag{1.84}$$

$$\frac{\partial H}{\partial p_{\alpha_\rho}} = {}_a\mathbf{D}_t^{\alpha} q^{\rho}, \tag{1.85}$$

$$\frac{\partial H}{\partial q^{\rho}} = {}_t\mathbf{D}_b^{\alpha} p_{\alpha_\rho}, \; 0 < \alpha < 1, \; \rho = 1, \ldots, N. \tag{1.86}$$

1.3.2.2 A direct method within Caputo fractional derivatives

In the following we present briefly the Hamiltonian formulation within Caputo's fractional derivatives (Baleanu and Agrawal, 2006). Let us consider the fractional Lagrangian as given below

$$L(q, {}_a^C D_t^{\alpha} q, t), \;\; 0 < \alpha < 1. \tag{1.87}$$

By using (1.87) we define the canonical momenta p_α as follows

$$p_\alpha = \frac{\partial L}{\partial {}_a^C D_t^{\alpha} q}. \tag{1.88}$$

We define the fractional canonical Hamiltonian as

$$H = p_\alpha \left({}_a^C D_t^{\alpha} q\right) - L. \tag{1.89}$$

Taking total differential of (1.89) and by using (1.88), we obtain

$$dH = dp_\alpha \left({}_a^C D_t^\alpha q\right) - \frac{\partial L}{\partial q} dq - \frac{\partial L}{\partial t} dt. \tag{1.90}$$

Taking into account the fractional Euler-Lagrange equations we obtain

$$dH = \left({}_a^C D_t^\alpha q\right) dp_\alpha + \left({}_t\mathbf{D}_b^\alpha p_\alpha\right) dq - \frac{\partial L}{\partial t} dt. \tag{1.91}$$

Finally, after some simple manipulations, the fractional Hamilton equations are obtained as follows

$$\frac{\partial H}{\partial t} = -\frac{\partial L}{\partial t}, \tag{1.92}$$

$$\frac{\partial H}{\partial p_\alpha} = {}_a^C D_t^\alpha q, \tag{1.93}$$

$$\frac{\partial H}{\partial q} = {}_t\mathbf{D}_b^\alpha p_\alpha. \tag{1.94}$$

1.3.2.3 Fractional Ostrogradski's formulation

The higher-derivatives theories (Gitman and Tytin, 1990; Nesterenko, 1989) appear naturally as corrections to general relativity and cosmic strings as well (Birell and Davies, 1982). The unconstrained higher-order derivatives have more degree of freedom than lower-derivative theories, as a result a lack a lower-energy bound was reported. A method how to remove all these problems was suggested in (Simon, 1990). It was reported that the non-local formulation can be written as an infinite order Ostrogradski's formulation (Gitman and Tytin, 1990; Nesterenko, 1989). On the other hand the fractional derivatives are non-local objects and we have a decomposition formula for them. In conclusion, a natural question is how to formulate a theory corresponding to the fractional case.

Let us consider an ordinary local Lagrangian depending on a finite number of derivatives at a given time as (Bering, online)

$$L\left(q(t), \dot{q}(t), ..., q^{(n)}(t)\right). \tag{1.95}$$

Let us consider a Lagrangian depending on a piece of the trajectory $q(t, \lambda)$ for $\forall \lambda$ belonging to an interval $[a, b]$, namely

$$L^{non}(t) = L(q(t + \lambda)). \tag{1.96}$$

Here a, b represent real numbers. Therefore, we have created a non-local Lagrangian and the corresponding action function is given by

$$S(q) = \int dt L^{non}(t). \tag{1.97}$$

We are able to write the Euler-Lagrange equation corresponding to (1.97) as (Bering, online)

$$\int dt \frac{\delta L^{non}(t)}{\delta(q(t))} = 0. \tag{1.98}$$

We observed that Equations (1.98) are functional relations to be satisfied by a Lagrangian constraint. Another observation is that there is no dynamics except the displacement inside the trajectory

$$q(t) \to q(t+\lambda). \tag{1.99}$$

The following step is to introduce the dynamical variable $Q(t,\lambda)$ as

$$Q(t,\lambda) = q(t+\lambda). \tag{1.100}$$

Let us consider a field $Q(t,\lambda)$ instead of a trajectory $q(t)$, namely

$$\dot{Q}(t,\lambda) = Q'(t,\lambda), \tag{1.101}$$

where $\dot{Q} = \frac{\partial Q(t,\lambda)}{\partial t}$ and $Q'(t,\lambda) = \frac{\partial Q(t,\lambda)}{\partial \lambda}$. In such a way we obtain a $1+1$ dimensional formulation of non-local Lagrangian (Bering, online).

The coordinates and momenta are suppose to have the following forms

$$Q(t,\lambda) = \sum_{m=0}^{\infty} e_m(\lambda) q^{(m)}(t), P(t,\lambda) = \sum_{m=0}^{\infty} e^m(\lambda) p_{(m)}(t), \tag{1.102}$$

where

$$\{q^{(n)}(t), p_{(m)}(t)\} = \delta^n_m \tag{1.103}$$

and

$$e_m(\lambda) = \frac{\lambda^m}{m!}, \quad e^m(\lambda) = (-\partial_\lambda)^m \delta(\lambda). \tag{1.104}$$

In conclusion, the Hamiltonian for $1+1$ dimensional field has the form

$$H(t,[Q,P]) = \int d\lambda P(t,\lambda) Q'(t,\lambda) - \tilde{L}(t,[Q]), \tag{1.105}$$

where P denotes the canonical momentum of Q. The phase space is $T*J$ equipped with the fundamental Poisson brackets

$$\{Q(t,\lambda), P(t,\lambda')\} = \delta(\lambda - \lambda'). \tag{1.106}$$

The functional $\tilde{L}(t,[Q])$ is defined as below

$$\tilde{L}(t,[Q]) = \int d\lambda \delta(\lambda) L(t,\lambda). \tag{1.107}$$

By analyzing (1.107), the primary constraint becomes

$$\phi(t, \lambda, [q, P]) = P(t, \lambda) - \int d\sigma \chi(\lambda, -\sigma)\varepsilon(t; \sigma, \lambda) \approx 0. \quad (1.108)$$

$\varepsilon(t; \sigma, \lambda)$ and $\chi(\lambda, -\sigma)$ have the following definitions:

$$\varepsilon(t; \sigma, \lambda) = \frac{\partial L(t, \sigma)}{\partial Q(t, \lambda)}, \quad \chi(\lambda, -\sigma) = \frac{\varepsilon(\lambda) - \varepsilon(\sigma)}{2}, \quad (1.109)$$

where $\varepsilon(\lambda)$ denotes the sigma distribution. By using this construction the Euler-Lagrange equation is guaranteed by itself

$$\dot{\phi} \sim \psi = \int d\sigma \xi(t; \sigma, \lambda). \quad (1.110)$$

In the following we would like to derive both the Lagrangian and the Hamiltonian formalisms for non singular Lagrangian with fractional order derivatives starting from the Hamiltonian formalism of non local-theories (Baleanu et al., 2006). Let us consider the following Lagrangian to start with

$$L(q, t) = L(t, q^{\alpha_m}), \quad (1.111)$$

where the generalized coordinates are defined as

$$q^{\alpha_m} = {}_a\mathbf{D}_t^{\alpha_m} x(t), \quad (1.112)$$

where m is a natural number.

To obtain the reduced phase space quantization, we start with the infinite dimensional phase space $T * J(t) = \{Q(t, \lambda), P(t, \lambda)\}$.

The key issue is to find an appropriate generalization of (1.104) for the fractional case (Baleanu et al., 2006). As it was pointed out in (Bering, online) the coordinates and the momenta are considered as a Taylor series. Therefore, the first step is to generalize the classical series to the fractional case. A natural extension is to use factorial instead of the Gamma function. In this way we introduce naturally the generalized functions instead of $e_m(\lambda)$ and $e^m(\lambda)$ given by (1.104).

As it is already known several fractional Taylor's series expansions were developed (Trujillo, 1999; Hardy, 1945), therefore we have to decide which one is appropriate for our generalization. Since we are dealing with fractional Riemann-Liouville derivatives we choose the following generalization proposed, namely

$$Q(t, \lambda) = \sum_{m=-\infty}^{\infty} e_{\alpha_m}(\lambda) q^{(\alpha_m)}(t),$$

$$P(t, \lambda) = \sum_{m=-\infty}^{\infty} e^{\alpha_m}(\lambda) p_{(\alpha_m)}(t), \quad (1.113)$$

where

$$e_{\alpha_m}(\lambda) = \frac{(\lambda-\lambda_0)^{\alpha_m}}{\Gamma(\alpha_m+1)}, \quad e^{\alpha_m}(\lambda) = \mathbf{D}_{\lambda}^{\alpha_m}\delta(\lambda-\lambda_0), \tag{1.114}$$

and $\alpha_m = m+\alpha$, with $0 \le \alpha < 1$. Here λ_0 is a constant. The coefficients in (1.113) are new canonical variables:

$$\{q^{(\alpha_m)}, p_{(\alpha_{m'})}\} = \delta_{\alpha_m}^{\alpha_{m'}}. \tag{1.115}$$

By using (1.115) we obtain that

$$\sum_{m=-\infty}^{\infty} e^{\alpha_m}(\lambda)e_{\alpha_m}(\lambda') = \delta(\lambda-\lambda'), \tag{1.116}$$

and

$$\int_{-\infty}^{+\infty} d\lambda e^{\alpha_m}(\lambda)e_{\alpha_{m'}}(\lambda) = \delta_{\alpha_{m'}}^{\alpha_m}. \tag{1.117}$$

Therefore, $e^{\alpha_m}(\lambda)$ and $e_{\alpha_m}(\lambda)$ form an orthonormal basis. We stress on the fact that (1.116) and (1.117) involve the generalized functions and the relations have the meaning in the sense of generalized functions approach (Gelfand and Shilov, 1964; Hardy, 1945).

The fractional Hamiltonian is now given by

$$H = \sum_{m=-\infty}^{\infty} p^{\alpha_m} q^{\alpha_{m+1}} - L(q^0, q^{\alpha_m}). \tag{1.118}$$

The momenta constraints become an infinite set of constraints

$$\phi_n = p_{\alpha_n}(t) - \sum_{m=n}^{\infty} {}_t\mathbf{D}_b^{\alpha_{m-n}} \frac{\partial L}{\partial q^{(\alpha_{m+1})}(t)} = 0. \tag{1.119}$$

The fractional Euler-Lagrange equations are as follows

$$\sum_{l=-\infty}^{\infty} {}_t\mathbf{D}_b^{\alpha_l} \frac{\partial L(t)}{\partial q^{\alpha_l}(t)} = 0. \tag{1.120}$$

An interesting property of the fractional series proposed by Riemann and discussed by Hardy in (Hardy, 1945) is that when α_m becomes integers the usual form of Taylor series is obtained. Therefore one should notice that for integer values of α_m we have

$$p_{\alpha_m}(t) - \sum_{l=0}^{n-m-1} \left(-\frac{d}{dt}\right)^l \frac{\partial L(t)}{\partial(\partial_t^{l+m+1} q(t))} = 0, \tag{1.121}$$

which is the definition of Ostrogradski's momenta (Gitman and Tytin, 1990).

In this case the Euler-Lagrange equation for original fractional derivative Lagrangian is given below

$$\sum_{l=0}^{n} {}_t\mathbf{D}_b^{\alpha_l} \frac{\partial L(t)}{\partial q^{\alpha_l}(t)} = 0. \tag{1.122}$$

Now, from this equation, for integer values of α_m we obtain the Euler-Lagrange equation for higher derivative Lagrangian, namely

$$\sum_{l=0}^{n} \left(-\frac{d}{dt}\right)^l \frac{\partial L(t)}{\partial(\partial_t^l q(t))} = 0, \tag{1.123}$$

The constraints (1.121) and (1.123) lead us to eliminate canonical pairs $\{q^{\alpha_l}, p_{\alpha_l}\}(l \geq n)$.

In this case the infinite dimensional phase space is reduced to a finite dimensional one. The reduced space is coordinated by $T * J^n = \{q^{\alpha_l}, p_{\alpha_l}\}$ with $l = 0, 1, ..., n-1$. The Hamiltonian in the reduced space is given by

$$H = \sum_{m=0}^{n-1} p^{\alpha_m} q^{\alpha_{m+1}} - L(q^0, q^{\alpha_m}). \tag{1.124}$$

One should notice that the canonical reduced phase space Hamiltonian (1.124) is obtained in terms of the reduce canonical phase space coordinates $\{q^{\alpha_l}, p_{\alpha_l}\}$ with $l = 0, 1, ..., n-1$. In this case the path integral quantization of filed system is given by

$$K = \int \prod_{m=0}^{n-1} dq^{\alpha_m} dp^{\alpha_m} e^{i\{\int dt(\sum_{m=0}^{n-1} p^{\alpha_m} q^{\alpha_{m+1}} - H)\}}. \tag{1.125}$$

We observe that when α are integers, we obtain the path integral for systems with higher order Lagrangian (Gitman and Tytin, 1990; Nesterenko, 1989).

1.3.2.4 Example

The classical Pais-Uhlenbeck fourth order oscillator is described by the following equation (Pais and Uhlenbeck, 1950)

$$\frac{d^4x}{dt^4} + (\omega_1^2 + \omega_2^2)\frac{d^2x}{dt^2} + \omega_1^2\omega_2^2 x = 0, \tag{1.126}$$

where γ, ω_1, and ω_2 are all positive constants. As it can be seen from (1.126), the model represents two oscillators coupled by a fourth-order equation of motion. The fractional generalization of (1.126) can be obtained by replacing the classical derivatives to the fractional one. It is easy to see that the fractional Lagrangian counterpart corresponding to the fractional generalization of (1.126) becomes

$$L_f = \frac{\gamma'}{2}[({}_a\mathbf{D}_t^{2\alpha}q)^2 - (\omega_1^2+\omega_2^2)({}_a\mathbf{D}_t^{\alpha}q)^2 + \omega_1^2\omega_2^2q^2] = 0, \tag{1.127}$$

or

$$L_f = \frac{\gamma'}{2}\left[\left(\frac{(t-a)^{-2\alpha}}{\Gamma(1-p)}q(t) + \sum_{k=1}^{\infty}\binom{2\alpha}{k}\frac{(t-a)^{k-2\alpha}}{\Gamma(k-2\alpha+1)}q^{(k)}(t)\right)^2 - (\omega_1^2+\omega_2^2)\left(\frac{(t-a)^{-\alpha}}{\Gamma(1-p)}q(t) + \sum_{k=1}^{\infty}\binom{\alpha}{k}\frac{(t-a)^{k-\alpha}}{\Gamma(k-\alpha+1)}q^{(k)}(t)\right)^2 + \omega_1^2\omega_2^2q^2(t)\right]. \tag{1.128}$$

Finally, we obtain the following Lagrangian

$$L_f[Q][t] = \frac{\gamma'}{2}\left[\left(\frac{(t-a)^{-2\alpha}}{\Gamma(1-2\alpha)}Q(t) + \sum_{k=1}^{\infty}\binom{2\alpha}{k}\frac{(t-a)^{k-2\alpha}}{\Gamma(k-2\alpha+1)}Q^{(k)}(t)\right)^2 - (\omega_1^2+\omega_2^2)\left(\frac{(t-a)^{-\alpha}}{\Gamma(1-\alpha)}Q(t) + \sum_{k=1}^{\infty}\binom{\alpha}{k}\frac{(t-a)^{k-\alpha}}{\Gamma(k-\alpha+1)}Q^{(k)}(t)\right)^2 + \omega_1^2\omega_2^2Q^2(t)\right]. \tag{1.129}$$

In our investigated case, we obtain the following fractional Euler-Lagrange and Hamilton equations

$$\dot{P}_{(n)}(t) + P_{(n-1)}(t) = \gamma'\binom{2\alpha}{n}\frac{(t-a)^{n-2\alpha}}{\Gamma(n-2\alpha+1)}{}_a\mathbf{D}_t^{2\alpha}Q(t) - \gamma'(\omega_1^2+\omega_2^2)\binom{\alpha}{n}\frac{(t-a)^{n-\alpha}}{\Gamma(n-\alpha+1)}{}_a\mathbf{D}_t^{\alpha}Q(t), \tag{1.130}$$

$$\dot{P}_{(0)}(t) = \gamma'\frac{(t-a)^{-2\alpha}}{\Gamma(1-2\alpha)}{}_a\mathbf{D}_t^{2\alpha}Q(t) - \gamma'(\omega_1^2+\omega_2^2)\frac{(t-a)^{-\alpha}}{\Gamma(1-\alpha)}{}_a\mathbf{D}_t^{\alpha}Q(t) + \gamma'\omega_1^2\omega_2^2Q(t). \tag{1.131}$$

1.3.2.5 Fractional path integral quantization

The classical Lagrangian to start with is given by

$$L = \frac{1}{2}(1+\epsilon^2\omega^2)\dot{x}^2 - \frac{1}{2}\omega^2 x^2 - \frac{1}{2}\epsilon^2\ddot{x}^2, \tag{1.132}$$

where ω and ϵ are constants.

The fractional generalization of (1.132) has the following form:

$$L' = \frac{1}{2}(1+\epsilon^2\omega^2)({}_t\mathbf{D}_a^\alpha x(t))^2 - \frac{1}{2}\omega^2 x^2 - \frac{1}{2}\epsilon^2[{}_t\mathbf{D}_a^\alpha({}_t\mathbf{D}_a^\alpha x(t))]^2. \tag{1.133}$$

The independent coordinates are $x(t)$ and ${}_t\mathbf{D}_a^\alpha x(t)$ respectively. Let us denote $p_1^\alpha = p_x$ and $p_2^\alpha = p_{({}_t\mathbf{D}_a^\alpha x(t))}$. The fractional canonical momenta are (Rabei et al., 2007)

$$p_1^\alpha = \frac{\partial L}{\partial_t \mathbf{D}_a^\alpha x(t)} - {}_t\mathbf{D}_a^\alpha(\frac{\partial L}{\partial_t \mathbf{D}_a^{2\alpha} x(t)}), p_2^\alpha = \frac{\partial L}{\partial_t \mathbf{D}_a^{2\alpha} x(t)}. \tag{1.134}$$

By making use of (1.133) we obtain the forms of the fractional canonical momenta as given below:

$$p_1^\alpha = (1+\epsilon^2\omega^2){}_t\mathbf{D}_a^\alpha x(t) + \epsilon^2{}_t\mathbf{D}_a^{3\alpha} x(t), \tag{1.135}$$

$$p_2^\alpha = -\epsilon^2{}_t\mathbf{D}_a^{2\alpha} x(t). \tag{1.136}$$

Taking into account (1.135) the fractional canonical Hamiltonian becomes

$$H = p_1^\alpha{}_t\mathbf{D}_a^\alpha x(t) + p_2^\alpha{}_t\mathbf{D}_a^{2\alpha} x(t) - L, \tag{1.137}$$

and after taking into account (1.133), (1.135) and (1.135), the fractional Hamiltonian has the form:

$$H = \frac{1}{2}[2p_1^\alpha{}_t\mathbf{D}_a^\alpha x(t) - \frac{(p_2^\alpha)^2}{\epsilon^2} + \omega^2 x^2(t) - (1+\epsilon^2\omega^2)({}_t\mathbf{D}_a^\alpha x(t))^2]. \tag{1.138}$$

By making use of (1.138) the fractional path integral is given by

$$K = \int dx d({}_t\mathbf{D}_a^\alpha x(t)) dp_1^\alpha dp_2^\alpha e^{i\{\int dt(p_1^\alpha x(t) + p_2^\alpha{}_t\mathbf{D}_a^\alpha x(t) - H)\}}. \tag{1.139}$$

1.3.3 Lagrangian formulation of field systems with fractional derivatives

A covariant form of the action would involve a Lagrangian density $\mathcal{L}$ via $S = \int \mathcal{L} d^4x = \int \mathcal{L} d^3x dt$ where $\mathcal{L} = \mathcal{L}(\phi, \partial_\mu\phi)$.

The corresponding classical covariant Euler-Lagrange equation is

$$\frac{\partial \mathcal{L}}{\partial \phi} - \partial_\mu \frac{\partial \mathcal{L}}{\partial(\partial_\mu \phi)} = 0, \tag{1.140}$$

where ϕ is the field variable. In the following we present the fractional generalization of the above mentioned result. Let us consider an action as given below:

$$S = \int \mathcal{L}\left(\phi(x), (\mathbf{D}_{a_k-}^{\alpha_k})\phi(x), (\mathbf{D}_{a_k+}^{\alpha_k})\phi(x), x\right) d^3x dt. \tag{1.141}$$

Here $0 < \alpha_k \leq 1$ and a_k correspond to x_1, x_2, x_3 and t respectively.

The fractional Euler-Lagrange equations are as follows (Baleanu and Muslih, 2005b):

$$\frac{\partial S}{\partial \phi} + \sum_{k}^{1,4} \left\{ (\mathbf{D}_{a_k+}^{\alpha_k}) \frac{\partial S}{\partial (\mathbf{D}_{a_k-}^{\alpha_k})\phi} + (\mathbf{D}_{a_k-}^{\alpha_k}) \frac{\partial S}{\partial (\mathbf{D}_{a_k+}^{\alpha_k})\phi} \right\} = 0. \tag{1.142}$$

For $\alpha_k \to 1$, Equation (1.142) is the usual Euler-Lagrange equations for classical fields.

1.3.3.1 Application 1: Fractional Dirac field

Lagrangian density for Dirac fields of order $\frac{2}{3}$ is proposed as follows (Muslih and Baleanu, 2005b; Raspini, 2001)

$$\mathcal{L} = \bar{\psi}\left(\gamma^{\alpha}\mathbf{D}_{\alpha-}^{2/3}\psi(x) + (m)^{2/3}\psi(x)\right). \tag{1.143}$$

By using (1.143) the generalized momenta become

$$(\pi_{t_-})_{\psi} = \bar{\psi}\gamma^0, \quad (\pi_{t_-})_{\bar{\psi}} = 0. \tag{1.144}$$

From (1.143) and (1.144) we construct the canonical Hamiltonian density as

$$\mathcal{H}_T = -\bar{\psi}\left(\gamma^k \mathbf{D}_{k_-}^{2/3}\psi(x) + (m)^{2/3}\psi(x)\right) + \lambda_1[(\pi_{t_-})_{\psi} - \bar{\psi}\gamma^0] \\ + \lambda_2[(\pi_{t_-})_{\bar{\psi}}]. \tag{1.145}$$

Making use of (1.145), the canonical equations of motion have the following forms

$$\mathbf{D}_{t_+}^{2/3}(\pi_{t_-})_{\psi} = -(m)^{2/3}\bar{\psi}(x) - \mathbf{D}_{k_-}^{2/3}\gamma^k\bar{\psi}(x), \tag{1.146}$$

$$\mathbf{D}_{t_+}^{2/3}(\pi_{t_-})_{\bar{\psi}} = -(m)^{2/3}\psi(x) - \gamma^k\mathbf{D}_{k_-}^{2/3}\psi(x) - \gamma^0\lambda_1 = 0, \tag{1.147}$$

$$\mathbf{D}_{t_+}^{2/3}\psi = \frac{\partial \mathcal{H}_T}{\partial(\pi_{t_-})_{\psi}} = \lambda_1, \tag{1.148}$$

$$\mathbf{D}_{t_+}^{2/3}\bar{\psi} = \frac{\partial \mathcal{H}_T}{\partial(\pi_{t_-})_{\bar{\psi}}} = \lambda_2, \tag{1.149}$$

which lead us to the following equations of motion:

$$\mathbf{D}_{+}^{2/3}\gamma^{\alpha}\bar{\psi}(x) + (m)^{2/3}\bar{\psi}(x) = 0, \tag{1.150}$$

$$\gamma^{\alpha}\mathbf{D}_{+}^{2/3}\psi(x) + (m)^{2/3}\psi(x) = 0. \tag{1.151}$$

The path integral for this system is given by

$$K = \int d(\pi_{t_-})_{\psi}\, d(\pi_{t_-})_{\bar{\psi}}\, d\psi\, d\bar{\psi}\delta[(\pi_{t_-})_{\psi} - \bar{\psi}\gamma^0]\delta[(\pi_{t_-})_{\bar{\psi}}]$$
$$\times \exp i\left[\int d^4x \left\{(\pi_{t_-})_{\psi}\mathbf{D}_{t_-}^{2/3}\psi + (\pi_{t_-})_{\bar{\psi}}\mathbf{D}_{t_-}^{2/3}\bar{\psi} - \mathcal{H}\right\}\right]. \tag{1.152}$$

Integrating over $(\pi_{\alpha_-})_{\psi}$ and $(\pi_{\alpha_-})_{\bar{\psi}}$, we arrive at the result

$$K = \int d\psi\, d\bar{\psi} \exp i\left[\int d^4x\mathcal{L}\right]. \tag{1.153}$$

1.3.3.2 Application 2: Fractional Schrödinger equation, a Lagrangian approach

The classic Schrödinger equation is given by

$$i\hbar\frac{d\psi}{dt} + \frac{\hbar^2}{2m}\triangle\psi - V(x)\psi = 0. \tag{1.154}$$

The classical Lagrangian corresponding to (1.154) is

$$\mathcal{L} = \frac{i\hbar}{2}\left(\psi^{\dagger}\frac{d\psi}{dt} - \psi\frac{d\psi^{\dagger}}{dt}\right) - \frac{\hbar^2}{2m}\left(\nabla\psi\nabla\psi^{\dagger}\right) - V(x)\psi\psi^{\dagger}. \tag{1.155}$$

We proposed the following Lagrangian density for the fractional Schrödinger equation:

$$\mathcal{L} = \frac{i\hbar}{2}(\psi^{\dagger}\mathbf{D}_{a_{t_+}}^{\alpha_t}\psi - \psi\mathbf{D}_{a_{t_+}}^{\alpha_t}\psi^{\dagger}) - \frac{\hbar^2}{2m}(\mathbf{D}_{a_{x_+}}^{\alpha_x}\psi\mathbf{D}_{a_{x_+}}^{\alpha_x}\psi^{\dagger}) - V(x)\psi\psi^{\dagger}. \tag{1.156}$$

The generalized momenta are

$$(\pi_+)_{\psi} = \frac{i\hbar}{2}\psi^{\dagger}, \quad (\pi_+)_{\psi^{\dagger}} = -\frac{i\hbar}{2}\psi. \tag{1.157}$$

The total canonical Hamiltonian density reads as

$$\mathcal{H}_T = \frac{\hbar^2}{2m}\mathbf{D}_{+}^{\alpha_x}\psi\mathbf{D}_{+}^{\alpha_x}\psi^{\dagger} + V(x)\psi\psi^{\dagger}$$
$$+\lambda_1[(\pi_+)_{\psi} - \frac{i\hbar}{2}\psi^{\dagger}] + \lambda_2[(\pi_+)_{\psi^{\dagger}} + \frac{i\hbar}{2}\psi]. \tag{1.158}$$

The corresponding fractional canonical equations of motion are

$$\mathbf{D}_{-}^{\alpha_t}(\pi_+)_\psi = \frac{\hbar^2}{2m}\mathbf{D}_{-}^{\alpha_x}\mathbf{D}_{+}^{\alpha_x}\psi^\dagger + V(x)\psi^\dagger + \frac{i\hbar}{2}\lambda_2, \tag{1.159}$$

$$\mathbf{D}_{-}^{\alpha_t}(\pi_+)_{\psi^\dagger} = \frac{\hbar^2}{2m}\mathbf{D}_{-}^{\alpha_x}\mathbf{D}_{+}^{\alpha_x}\psi + V(x)\psi - \frac{i\hbar}{2}\lambda_1, \tag{1.160}$$

$$\mathbf{D}_{+}^{\alpha_t}\psi^\dagger = \frac{\partial\mathcal{H}_T}{\partial(\pi_+)_{\psi^\dagger}} = \lambda_2, \tag{1.161}$$

$$\mathbf{D}_{+}^{\alpha_t}\psi = \frac{\partial\mathcal{H}_T}{\partial(\pi_+)_\psi} = \lambda_1, \tag{1.162}$$

and after some calculations we arrive at the following fractional equations of motion

$$\frac{i\hbar}{2}(\mathbf{D}_{-}^{\alpha_t} - \mathbf{D}_{+}^{\alpha_t})\psi^\dagger = \frac{\hbar^2}{2m}\mathbf{D}_{-}^{\alpha_x}\mathbf{D}_{+}^{\alpha_x}\psi^\dagger + V(x)\psi^\dagger, \tag{1.163}$$

$$-\frac{i\hbar}{2}(\mathbf{D}_{-}^{\alpha_t} - \mathbf{D}_{+}^{\alpha_t})\psi = \frac{\hbar^2}{2m}\mathbf{D}_{-}^{\alpha_x}\mathbf{D}_{+}^{\alpha_x}\psi + V(x)\psi. \tag{1.164}$$

The path integral for fractional Schrödinger equation is found to be as

$$K = \int d\psi\, d\psi^\dagger \exp i\left[\int d^4x\mathcal{L}\right]. \tag{1.165}$$

1.4 Fractional optimal control formulation

Fractional optimal control problem gained a lot of importance during the last decades.

In the following we present briefly how this technique works on a given example. The aim is to minimize the following performance index:

$$J(u) = \int_a^b f(x,u,t)dt, \tag{1.166}$$

such that

$$_a\mathbf{D}_t^\alpha x = g(x,u,t), \tag{1.167}$$

and the terminal conditions $x(a) = c$ and $x(b) = d$. Here t denotes the time, $x(t)$ and $u(t)$ are a $n_x \times 1$ state and $n_u \times 1$ control vectors, f and g are a scalar and a $n_x \times 1$ vector functions. The dimensions n_x and n_u fulfill the relation $n_u \leq n_x$.

A fractional order formulation for this problem for the case of scalar variables and functions was developed in (Agrawal and Baleanu, 2007). The same formulation is applicable for the vector case.

A modified performance index is defined as

$$\bar{J}(u) = \int_a^b [H(x,u,t) - \lambda^{\mathrm{T}}\, {}_a\mathbf{D}_t^\alpha x]dt, \tag{1.168}$$

where $H(x,u,\lambda,t)$ is the following Hamiltonian:

$$H(x,u,\lambda,t) = f(x,u,t) + \lambda^{\mathrm{T}} g(x,u,t), \tag{1.169}$$

and λ is a $n_x \times 1$ vector of Lagrange multipliers. Here the superscript T represents the transpose of the vector. Using (1.168) and (1.169) and the techniques from fractional variational principles, the necessary equations for the fractional control problem can be written as given below (Agrawal, 2004):

$${}_t\mathbf{D}_b^\alpha \lambda = \frac{\partial H}{\partial x}, \tag{1.170}$$

$$\frac{\partial H}{\partial u} = 0, \tag{1.171}$$

$${}_a\mathbf{D}_t^\alpha x = \frac{\partial H}{\partial \lambda}. \tag{1.172}$$

We notice that if $x_i(b)$ is not specified, then we require $\lambda_i(b)$ to be 0. Here x_i and λ_i are the ith components of the vectors x and λ. Equations (1.170)–(1.72) along with the above condition on λ represent the necessary conditions in terms of a Hamiltonian for the fractional optimal control problem. To solve the obtained set of differential equations we used the *Grünwald-Letnikov* definition. For the discretized time interval $[a,b]$ with $(N+1)$ equally-spaced grid points, at node i we have

$$\begin{aligned} {}_a\mathbf{D}_t^\alpha y_i &= \frac{1}{h^\alpha}\sum_{j=0}^{i} w_j^{(\alpha)} y_{i-j}, \qquad i = 0,\ldots,N, \\ {}_t\mathbf{D}_b^\alpha y_i &= \frac{1}{h^\alpha}\sum_{j=0}^{N-i} w_j^{(\alpha)} y_{i+j}, \qquad i = N,\ldots,0, \end{aligned} \tag{1.173}$$

respectively, where $y_i = y(t_i)$, $h = 1/N$, $t_i = ih$, and

$$w_j^{(\alpha)} = (-1)^j \binom{\alpha}{j}. \tag{1.174}$$

1.4.1 Example

To demonstrate the above proposed formulation, let us minimize (Baleanu et al., 2009)

$$J = \frac{1}{2}\int_0^2 [{}_0\mathbf{D}_t^\alpha\, {}_0\mathbf{D}_t^\alpha \theta]^2 dt, \tag{1.175}$$

subjected to the following dynamic constraint, ${}_0\mathbf{D}_t^\alpha {}_0\mathbf{D}_t^\alpha \theta(t) = u(t)$. Here we have $a = 0$, $b = 2$, and ${}_0\mathbf{D}_t^\alpha {}_0\mathbf{D}_t^\alpha \theta(t)$ is the sequential derivative of θ. By making use of $\theta(t) = x_1(t)$, ${}_0\mathbf{D}_t^\alpha \theta(t) = x_2(t)$ the modified performance index in (1.175) becomes

$$J = \int_0^2 [H(\mathbf{x}, u, \lambda) - \lambda^{\mathrm{T}} {}_0\mathbf{D}_t^\alpha \mathbf{x}(t))]dt, \tag{1.176}$$

where

$$H(\mathbf{x}, u, \lambda) = \frac{1}{2}u^2(t) + \lambda^{\mathrm{T}}(A\mathbf{x}(t) + \mathbf{b}\, u(t)) \tag{1.177}$$

denotes the Hamiltonian of the system, and

$$\mathbf{x}(t) = \begin{pmatrix} x_1(t) \\ x_2(t) \end{pmatrix}, \quad \lambda(t) = \begin{pmatrix} \lambda_1(t) \\ \lambda_2(t) \end{pmatrix}, \qquad \mathbf{b} = \begin{pmatrix} 0 \\ 1 \end{pmatrix}, \quad A = \begin{pmatrix} 0 & 1 \\ 0 & 0 \end{pmatrix}. \tag{1.178}$$

Using (1.170)-(1.72), we obtain

$$\begin{aligned} &{}_t\mathbf{D}_2^\alpha \lambda_1 = 0, \quad {}_t\mathbf{D}_2^\alpha \lambda_2 - \lambda_1 = 0, \quad u + \lambda_2 = 0, \\ &{}_0\mathbf{D}_t^\alpha x_1 - x_2 = 0, \quad {}_0\mathbf{D}_t^\alpha x_2 - u = 0. \end{aligned} \tag{1.179}$$

We observed that the variable u from the above mentioned equations can be eliminated using the third equality shown in (1.179). In addition, we use the following terminal conditions, $\theta(0) = {}_0\mathbf{D}_t^\alpha \theta(0) = 0$ and $\theta(2) = {}_0\mathbf{D}_t^\alpha \theta(2) = 1$ which can be translated into $x_1(0) = x_2(0) = 0$ and $x_1(2) = x_2(2) = 1$.

The presentation of the numerical method which is used to solve the equations in (1.179) is given below.

This method uses the scheme developed in (Agrawal and Baleanu, 2007) for scalar case. Briefly, we do the followings (Baleanu et al., 2009):

a) we divide the time domain into N sub-domains, where N represents an integer,

b) approximate the fractional derivatives in (1.179) at each node using the Grünwald-Letnikov definitions given in (1.173),

c) apply the terminal conditions,

d) solve the resulting equations.

For $N = 128$ and $\alpha = 0.75, 0.85, 0.95$ and 1, the values of the state variables x_1 and x_2 and the control variable u are presented in the following Figs.1.1, 1.2 and 1.3. These figures also include the analytical solutions for $\alpha = 1$. The results show that as α approaches 1, the analytical solution is recovered.

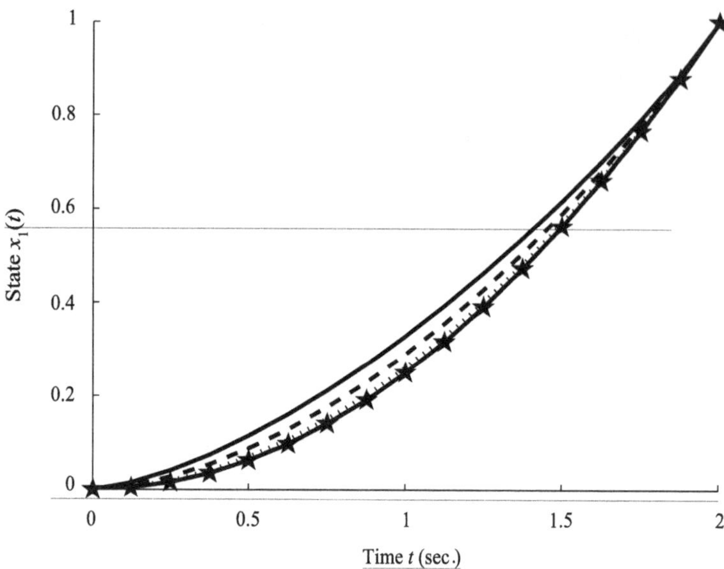

Fig. 1.1 State $x_1(t)$ as a function of t for different α ($-$: $\alpha = 0.75$, --- : $\alpha = 0.85$, $\cdots$: $\alpha = 0.95$, -·-·- : $\alpha = 1$, $\star$: $\alpha = 1$ (Analytical))

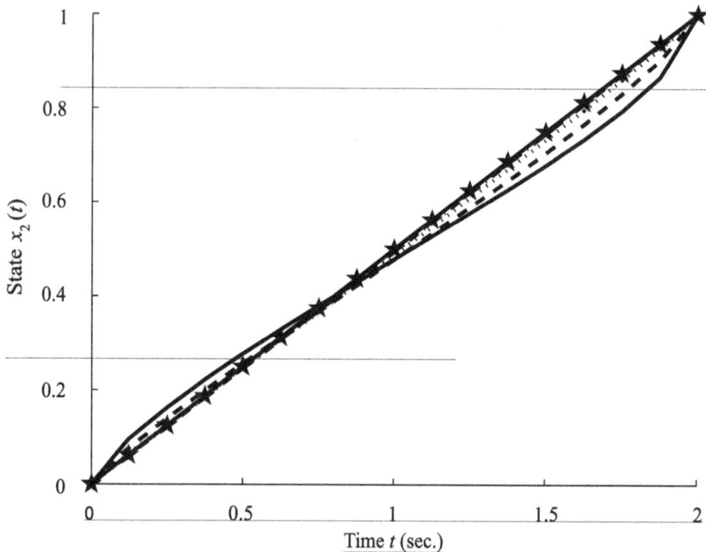

Fig. 1.2 State $x_2(t)$ as a function of t for different α ($-$: $\alpha = 0.75$, --- : $\alpha = 0.85$, $\cdots$: $\alpha = 0.95$, -·-·- : $\alpha = 1$, $\star$: $\alpha = 1$ (Analytical))

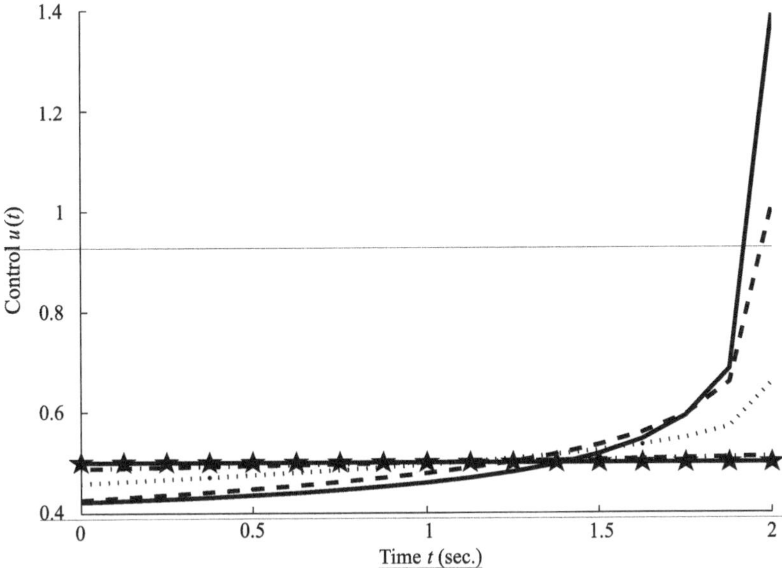

Fig. 1.3 Control $u(t)$ as a function of t for different α (– : $\alpha = 0.75$, --- : $\alpha = 0.85$, ··· : $\alpha = 0.95$, -·-·- : $\alpha = 1$, ⋆: $\alpha = 1$ (Analytical))

1.5 Fractional calculus in nuclear magnetic resonance

As it was proved in (Magin et al., 2008; Magin, 2009), the generalization of the Bloch equation through extension of the time derivative to fractional order offers a number of interesting possibilities concerning spin dynamics and magnetization relaxation.

Let us denote M_x, M_y, and M_z as the system magnetizations (x, y and z components). M_0 will denote as the equilibrium magnetization, T_1 is the spin-lattice relaxation time, T_2 represents the spin-spin relaxation time. Let us denote ω_0 the resonant frequency given by the Larmor relationship $\omega_0 = \gamma B_0$, where B_0 is the static magnetic field (z-component) and γ is the gyromagnetic ratio for spin $\frac{1}{2}$ particles.

As a result the classical equations are given in the following (see Refs. (Magin et al., 2008; Magin, 2009) and the references therein)

$$\frac{dM_z}{dt} = \frac{M_0 - M_z}{T_1}, \tag{1.180}$$

$$\frac{dM_x}{dt} = \omega_0 M_y - \frac{M_x}{T_2}, \tag{1.181}$$

$$\frac{dM_y}{dt} = -\omega_0 M_x - \frac{M_y}{T_2}. \tag{1.182}$$

Several approaches were presented to follow in fractional generalization; we ultimately should employ the form best suited for fitting experimental data. The assumption of a time domain fractional derivative suggests a modulation — or weighting — of system memory, an assumption that plays an important role in affecting the spin dynamics described by the Bloch equations.

In addition, the fractional order systems of differential equations (Kilbas et al., 2006) depend on the initial conditions; as a result, we should choose the fractional derivative most appropriate for handling the initial conditions of our problem.

In Nuclear Magnetic Resonance, the initial state of the system is specified by the components of the magnetization, hence these need to be clearly identified (Magin et al., 2008; Magin, 2009). Another more general, and still open issue, is the physical meaning of the fractional Bloch equations and this question ultimately goes back to the basic formulation-the Schrödinger equation as a fractional order partial differential equation in quantum mechanics - a topic beyond the scope of this presentation (Rabei et al., 2007; Naber, 2004; Baleanu and Muslih, 2005b).

$$
{}_0^C D_t^{\alpha} M_+(t) = -i\bar{\omega} M_+(t) - \frac{1}{T_2'} M_+(t). \tag{1.183}
$$

In order to preserve the meaning of the initial conditions for the magnetization for our problem,namely $M_x(0)$, $M_y(0)$, and $M_z(0)$, we will use the fractional order Caputo derivative.

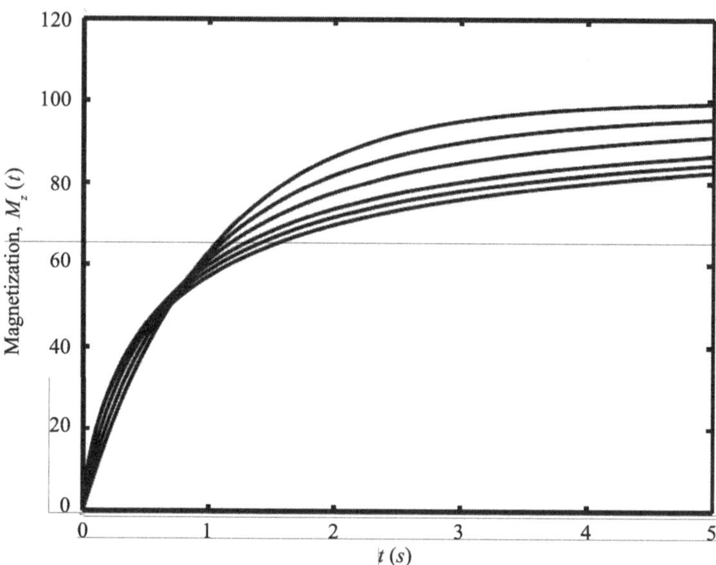

Fig. 1.4 Plots of $M_z(t)$ for different values of α in the range of α= 0:5 (bottom curve) to α = 1:0 (top curve) in steps of 0.1. For these plots, equation (186) was used with $M_z(0) = 0$, T_1=1 sec and $M_0 = 100$. The Mittag-Leffler function was evaluated using the Matlab function m-file mlf.m found at http://www.mathworks.com/matlabcentral.

As a result, we obtain a set of fractional order Bloch equations as given below

$$ {}_a^C D_x^\alpha M_z = \frac{M_0 - M_z}{T_1}, \tag{1.184} $$

$$ {}_a^C D_x^\alpha M_x = \bar{\omega}_0 M_y - \frac{M_x}{T_2}, \tag{1.185} $$

$$ {}_a^C D_x^\alpha M_y = -\bar{\omega}_0 M_x - \frac{M_y}{T_2}. \tag{1.186} $$

Here $\bar{\omega}_0$, $\frac{1}{T_1'}$, and $\frac{1}{T_2'}$ each have the units of $(sec)^{-\alpha}$.

Using either fractional calculus or the Laplace transformation, the solution for $M_z(t)$ is given as follows

$$ M_z(t) = M_z(0) E_\alpha \left(\frac{-t^\alpha}{T_1'} \right) + \frac{M_0}{T_1'} t^\alpha E_{\alpha,\alpha+1} \left(\frac{-t^\alpha}{T_1'} \right). \tag{1.187} $$

The solutions for $M_x(t)$ and $M_y(t)$ can be found by solving the corresponding fractional order differential equations. If we suppose that

$$ M_+(t) = M_x(t) + iM_y(t), \tag{1.188} $$

and use that

$$ M_+(0) = M_x(0) + iM_y(0), \tag{1.189} $$

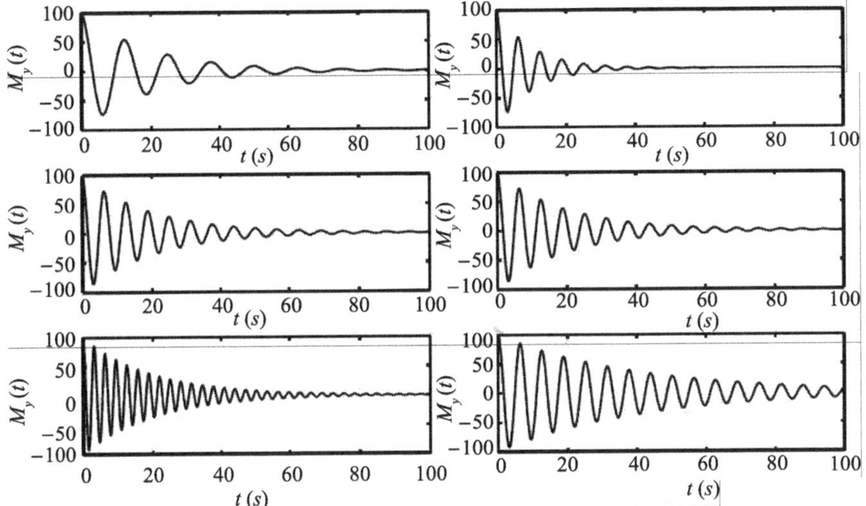

Fig. 1.5 Plots of $M_y(t)$ for T_2 = 20 ms and $f_0 = (\frac{2\pi}{\omega_0})$ = 80, 160, and 320 Hz, top to bottom, respectively. In the right hand side we have the plots of $M_y(t)$ for f_0 = 160 Hz and T_2 = 10, 20, and 40 ms, top to bottom, respectively. For all plots, equation (195) was used with $M_x(0) = 0$ and $M_y(0) = 100$.

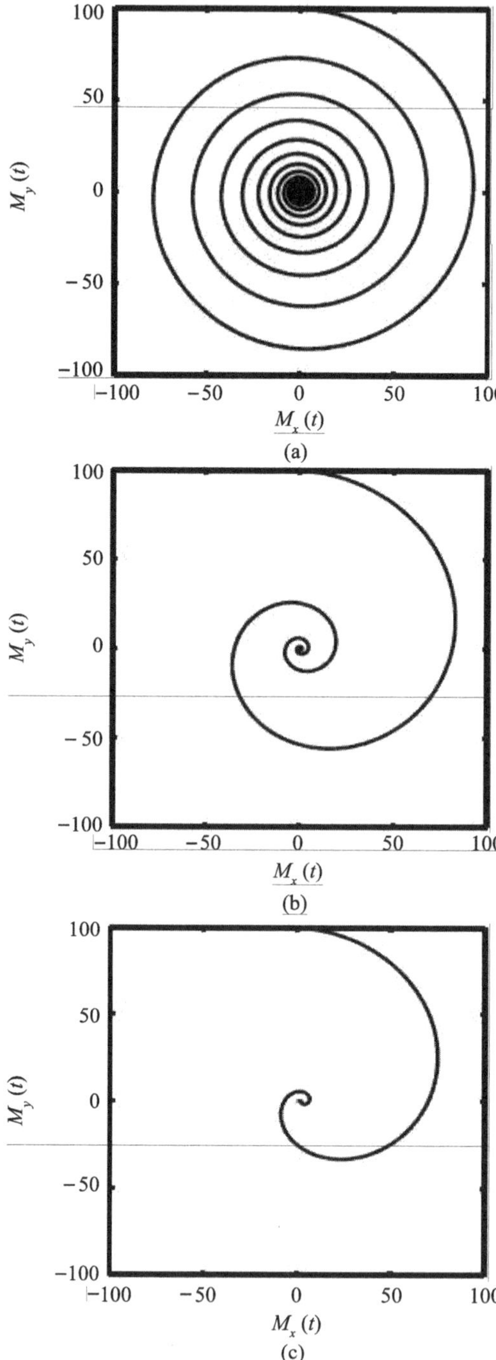

Fig. 1.6 Plots of $M_+(t)$ in the complex plane with $\alpha = 1$ (a, classical model), $\alpha = 0.9$ (b) and $\alpha = 0.8$ (c). For these plots, Equations (187) and (190) were used with $M_x(0) = 0$, $M_y(0) = 100$, $T_2 = 20$ ms and $f_0 = 160$ Hz.

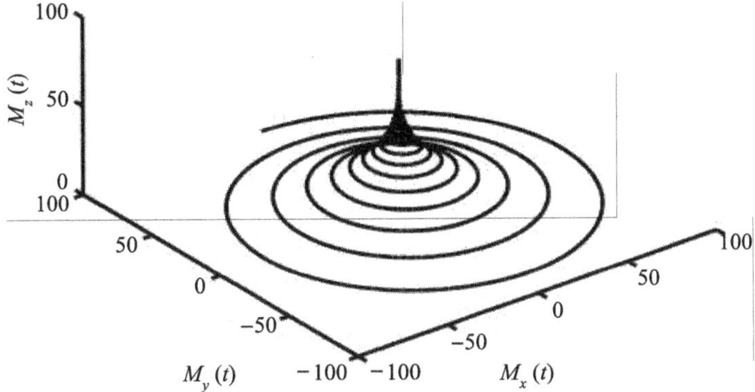

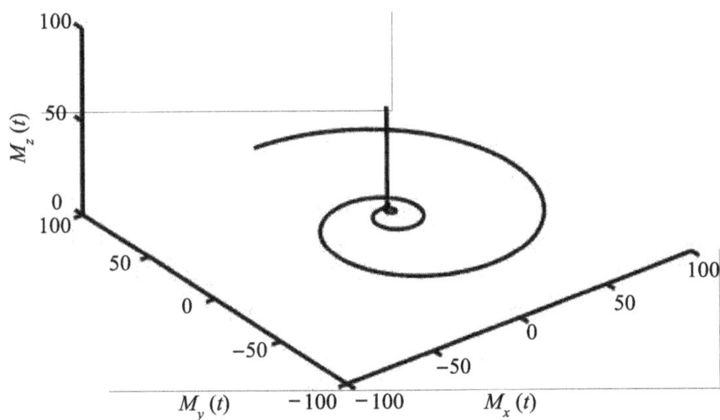

Fig. 1.7 A plot of fractional order solution to the Bloch equations with $\alpha = 1$ (classical model). In the right hand side we have a plot of fractional order solution to the Bloch equations with $\alpha = 0.9$. For these plots, Equations (1.186), (1.187) and (1.189) were used with $M_x(0) = 0$, $M_y(0) = 100$, $M_z(0) = 0$, $T_1 = 1$ s, $T_2 = 20$ ms and $f_0 = 160$ Hz.

then we observe that we can combine the two equations for the x and the y components of magnetization given above to yield.

The next step is to assume that

$$M_+(t) = M_+(0)E_\alpha(-\lambda t^\alpha), \tag{1.190}$$

and to introduce (1.190) into (1.183). By making use of

$${}_0^C D_t^\alpha E_\alpha(-\lambda t^\alpha) = -\lambda E_\alpha(-\lambda t^\alpha), \tag{1.191}$$

we obtain the values of λ as given below:

$$\lambda = \frac{1}{T_2'} + i\bar{\lambda}_0. \tag{1.192}$$

Also, we observe that

$$M_+(t=0) = M_+(0), \quad M_+(t=\infty) = 0. \tag{1.193}$$

In fractional calculus one of the questions is to verify the classical results. In this case we get for $\alpha = 1$ the following expressions:

$$M_x(t) = exp\left(-\frac{t}{T_2}\right)(M_x(0)\cos(\omega_0 t) + M_y(0)\sin(\omega_0 t)), \tag{1.194}$$

$$M_y(t) = exp\left(-\frac{t}{T_2}\right)(M_x(0)\cos(\omega_0 t) - M_y(0)\sin(\omega_0 t)), \tag{1.195}$$

which both agree with the classical results.

1.6 Fractional wavelet method and its applications in drug analysis

Various spectrophotometric methods such as graphical an numerical approaches have been used for the quantitative analysis of samples containing multicomponents.

In some complex mixture analysis the classical graphical and numerical spectroscopic methods do not provide always desirable and reliable results. Namely, the derivative spectrophotometry and its modified versions have been used extensively in fast quantitative analysis of mixtures. However, these spectral methods may not lead to desirable analysis results due to the strong spectral overlapping characteristics of compounds, decreasing signal intensity with worsening signal-to-noise ratio in higher derivative orders.

The developments of wavelet transform and its applications in the field of analytical chemistry has significantly amplified the potential power of various spectral analysis techniques. It was shown that the continuous wavelet transform (CWT) approach represents an important signal processing method for data reduction, denoising, baseline correction and resolution of multi-component overlapping spectra (Walczak, 2000). CWT methods have been successfully used for the quantitative resolution of multicomponent mixtures without using any priori separation procedure (see for example Ref.(Dinç and Baleanu, 2007) and the references therein). The application of the fractional wavelet transform for the simultaneous determination of the compounds in a binary mixture was applied for the first time in the literature by Dinc and Baleanu (Dinç and Baleanu, 2006). A new wavelet transform that is based on a recently defined family of scaling functions so called the fractional B-splines was introduced (Unser and Blu, 2000b, 2002, 1999, 2000a). The interest of this family is that they interpolate between the integer degrees of polynomial B-splines and that they allow a fractional order of approximation. In the following the notion of B-spline is introduced (Unser and Blu, 2000b, 2002, 1999, 2000a). A B-spline represents a generalization of the Bezier curve.

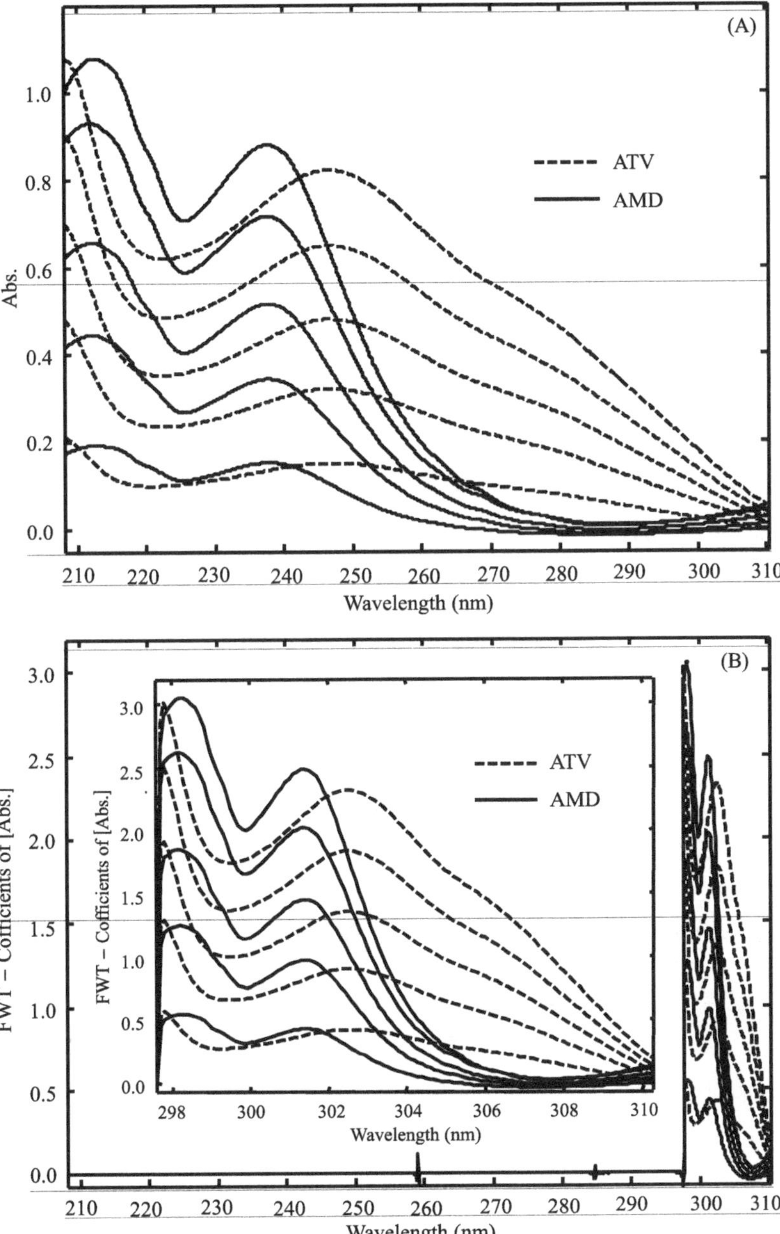

Fig. 1.8 Absorption spectra (A) and FWT spectra (B) of ATV (— — —) and AMD (⟼) in the calibration range of 4-20 $\mu g/mL$. Due to the large wavelength interval the FWT spectra was described on a small figure. Here ATV denotes the atorvastatin and AMT represents amlodipine in their mixture.

Let a vector known as the knot be defined by $T = \{t_0, t_1, \ldots, t_m\}$ where T is a non-decreasing sequence with $t_i \in [0,1]$, and define the control points P_0, P_n. The knots $t_{p+1}, ..., t_{m-p-1}$ are called internal knots.
We define the basis functional as

$$N_{i,0}(t) = \begin{cases} 1, \text{ if } t_i \leq t \prec t_{i+1} \text{ and } t_i \prec t_{i+1}, \\ 0, \text{ otherwise}, \end{cases} \tag{1.196}$$

and

$$N_{i,p}(t) = \frac{t - t_i}{t_{i+p} - t_i} N_{i,p-1}(t) + \frac{t_{i+p+1} - t}{t_{i+p+1} - t_{i+1}} N_{i+1,p-1}(t).$$

As a result the spline is a curve defined as below:

$$C(t) = \sum_{i=0}^{n} P_i N_{i,p}(t). \tag{1.197}$$

The next step is to introduce the fractional B spline as (Unser and Blu, 2000a)

$$\beta_+^\alpha(x) = \frac{\Delta_+^{\alpha+1} x_+^\alpha}{\Gamma(\alpha+1)} = \frac{\sum_{k=0}^{+\infty} (-1)^k \binom{\alpha+1}{k} (x-k)_+^\alpha}{\Gamma(\alpha+1)}, \tag{1.198}$$

where the Gamma function is defined as

$$\Gamma(\alpha+1) = \int_0^{+\infty} x^\alpha e^{-x} dx, \tag{1.199}$$

and $(x-k)_\alpha^+ = \max(x-k,0)^\alpha$.

The forward fractional finite difference operator of order α is given below:

$$\Delta_+^\alpha f(x) = \sum_{k=0}^{+\infty} (-1)^k \binom{\alpha}{k} f(x-k), \tag{1.200}$$

where the binomial coefficient has the form:

$$\binom{\alpha}{k} = \frac{\Gamma(\alpha+1)}{\Gamma(k+1)\Gamma(\alpha-k+1)}. \tag{1.201}$$

The centered fractional B-splines of degree α is defined by

$$\beta_\star^\alpha(x) = \frac{1}{\Gamma(\alpha+1)} \sum_{k \in Z} (-1)^k \binom{\alpha+1}{k} |x-k|_\star^\alpha. \tag{1.202}$$

We mention that

$$|x|_\star^\alpha = \begin{cases} \dfrac{|x|^\alpha}{-2\sin\left(\frac{\frac{\pi}{2}}{\alpha}\right)}, & \alpha \text{ odd}, \\ \dfrac{x^{2n}\log(x)}{(-1)^{1+n}\pi}, & \alpha \text{ even}. \end{cases} \tag{1.203}$$

The fractional B-spline wavelets is defined as

$$\psi_+^\alpha\left(\frac{x}{2}\right) = \sum_{k\in Z} \frac{(-1)^k}{2^\alpha} \sum_{l\in Z} \binom{\alpha+1}{1} \beta_\star^{2\alpha+1}(l+k-1)\beta_+^\alpha(x-k). \tag{1.204}$$

We observe that

$$\int_{-\infty}^{+\infty} x^n \psi_+^\alpha(x)dx = 0. \tag{1.205}$$

The Fourier transform fulfills the following relation

$$\hat{\psi}_+^\alpha(\omega) = C(j\omega)^{\alpha+1},\ as\ \omega \to 0, \tag{1.206}$$

and

$$\hat{\psi}_\star^\alpha(\omega) = C(j\omega)^{\alpha+1},\ as\ \omega \to 0, \tag{1.207}$$

where $\hat{\psi}_\star^\alpha(\omega)$ is symmetric.

We notice that fractional spline wavelet looks like a fractional derivative operator in *Grünwald*-Letnikov representation. To illustrate the advantages of the combined CWT and fractional wavelet transform for a given mixture we show below a picture showing the action of the fractional wavelet transform on a CWT coefficients. The implications of the obtained results from the chemical point of view were described in details in (Dinç and Baleanu, 2010). Namely, by making use of the fractional wavelet transform we obtain higher peak amplitude, less noise and sharper peaks (Dinç and Baleanu, 2010). Fractional wavelet transform is a powerful tools for the data reduction, de-noising, compressing and baseline correction of the analytical signals and resolution of multicomponent overlapping signals (se for example Ref.(Dinç and Baleanu, 2006) and references therein).

References

Agrawal O.P., 2002, Formulation of Euler-Lagrange equations for fractional variational problems, *Journal of Mathematical Analysis and Applications*, **272**, 368-379.

Agrawal O.P., 2004, A general formulation and solution scheme for fractional optimal control problems, *Nonlinear Dynamics*, **38**, 191-206.

Agrawal O.P., 2006, Fractional variational calculus and the transversality conditions, *Journal of Physics A:Mathematical and General*, **39**, 10375-10384.

Agrawal O.P., 2007, Generalized Euler-Lagrange equations and transversality conditions for FVPs in terms of the Caputo derivative, *Journal of Vibration and Control*, **13**, 1217-1237.

Agrawal O.P. and Baleanu D., 2007, Hamiltonian formulation and a direct numerical scheme for fractional optimal control problems, *Journal of Vibration and Control*, **13**, 1269-1281.

Alvarez-Gaume L. and Barbon J.L.F., 2001, Gauge theory on a quantum phase space, *International Journal of Modern Physics A*, **16**, 1123-1134.

Baleanu D., 2004, Constrained systems and Riemann-Liouville fractional derivative, *Proceedings of 1st IFAC Workshop on Fractional Differentiation and its Applications*, Bordeaux, France, July 19-21, 597-601.

Baleanu D., 2006, Fractional Hamiltonian analysis of irregular systems, *Signal Processing*, **86**, 2632-2636.

Baleanu D., 2008, New applications of fractional variational principles, *Reports on Mathematical Physics*, **61**, 199-206.

Baleanu D., 2009, About fractional quantization and fractional variational principles, *Communications in Nonlinear Science and Numerical Simulation*, **14**, 2520-2523.

Baleanu D. and Agrawal O.P., 2006, Fractional Hamilton formalism within Caputo's derivative, *Czechoslovak Journal of Physics*, **56**, 1087-1092.

Baleanu D. and Avkar T., 2004, Lagrangians with linear velocities within Riemann-Liouville fractional derivatives, *Nuovo Cimento B*, **119**, 73-79.

Baleanu D. and Trujillo J.J., 2008, On exact solutions of a class of fractional Euler-Lagrange equations, *Nonlinear Dynamics*, **52**, 331-335.

Baleanu D. and Muslih S.I., 2005a, About fractional supersymmetric quantum mechanics, *Czechoslovak Journal of Physics*, **55**, 1063-1066.

Baleanu D. and Muslih S.I., 2005b, Lagrangian formulation of classical fields within Riemann-Liouville fractional derivatives, *Physica Scripta*, **72**, 119-121.

Baleanu D., Defterli O. and Agrawal O.P., 2009, A central difference numerical scheme for fractional optimal control problems, *Journal of Vibration and Control*, **15**, 583-597.

Baleanu D., Golmankhaneh A.K. and Golmankhaneh A.K., 2009a, The dual action of fractional multi time Hamilton equations, *International Journal of Theoretical Physics*, **48**, 2558-2569.

Baleanu D., Golmankhaneh A.K. and Golmankhaneh A.K., 2009b, Fractional Nambu mechanics, *International Journal of Theoretical Physics*, **48**, 1044-1052.

Baleanu D., Maaraba T. and Jarad F., 2008a, Fractional variational principles with delay, *Journal of Physics A:Mathematical and Theoretical*, **41**, Art. No. 315403.

Baleanu D., Muslih S.I. and Rabei E.M., 2008b, On fractional Euler-Lagrange and Hamilton equations and the fractional generalization of total time derivative, *Nonlinear Dynamics*, **53**, 67-74.

Baleanu D., Muslih S.I. and Tas K., 2006, Fractional Hamiltonian analysis of higher order derivatives systems, *Journal of Mathematical Physics*, **47**, Art. No. 103503.

Barbosa R.S., Machado J.A.T. and Ferreira I.M., 2004, Tuning of PID controllers based on Bode's ideal transfer function, *Nonlinear Dynamics*, **38**, 305-321.

Bering K., online, A Note on Non-locality and Ostrogradski's Construction, hep-th/0007192.

Birell N.D. and Davies P.C.W., 1982, *Quantum Field in Curved Space*, Cambridge University Press, Cambridge.

Caputo M., 2001, Distributed order differential equations modelling dielectric induction and diffusion, *Fractional Calculus and Applied Analysis*, **4**, 421-442.

Carpinteri A. and Mainardi F. (Eds), 1997, *Fractals and Fractional Calculus in Continuum Mechanics*, Springer, Berlin.

Chen Y.Q., Vinagre B.M. and Podlubny I., 2004, Continued fraction expansion approaches to discretizing fractional order derivatives – an expository review, *Nonlinear Dynamics*, **38**, 155-170.

Dinç E. and Baleanu D., 2004a, Application of the wavelet method for the simultaneous quantitative determination of benazepril and hydrochlorothiazide in their mixtures, *Journal of AOAC International*, **87**, 834-841.

Dinç E. and Baleanu D., 2004b, Multicomponent quantitative resolution of binary mixtures by using continuous wavelet transform, *Journal of AOAC International*, **87**, 360-365.

Dinç E. and Baleanu D., 2006, A new fractional wavelet approach for simultaneous determination of sodium and sulbactam sodium in a binary mixture, *Spectrochimica Acta Part A*, **63**, 631-638.

Dinç, E. and Baleanu D., 2007, A review on the wavelet transform applications in analytical chemistry, *Mathematical Methods in Engineering*, Springer, Eds. K. Tas, J.A. Tenreiro Machado, D. Baleanu, pp. 265-285.

Dinç E. and Baleanu D., 2010, Fractional wavelet transform for the quantitative spectral resolution of the composite signals of the active compounds in a two-component mixture, *Communications in Nonlinear Science and Numerical Simulation*,**15**, 812-818.

Dinç E., Baleanu D. and Ustundag O., 2003, An approach to quantitative two-component analysis of a mixture containing hydrochlorothiazide and spironolactone in tablets by one-dimensional continuous Daubechies and biorthogonal wavelet analysis of UV-spectra, *Spectroscopy Letters*, **36**, 341-355.

Dong J.P. and Xu M.Y., 2008, Space-time fractional Schrödinger equation with time-independent potentials, *Journal of Mathematical Analysis and Applications*, **344**, 1005-1017.

Fogleman M.A., Fawcett M.J. and Solomon T.H., 2001, Lagrangian chaos and correlated Lévy flights in a non-Beltrami flow: Transient versus-long term transport, *Physical Review E*, **63**, 020101-1.

Gelfand I.M. and Shilov G.E., 1964, *Generalized Functions, Vol.I, Properties and Operators*, Accademic Press, New York.

Gitman D.M. and Tytin I.V., 1990, Quantization of fields with constraints, Springer, Berlin.

Gomis J. and Mehen T., 2000, Space-time noncommutative field theories and unitarity, *Nuclear Physics B*, **591**, 265-276.

Gomis J., Kamimura K. and Llosa J., 2001, Hamiltonian formalism for space-time noncommutative theories, *Physical Review D*, **63**, 045003.

Gomis J., Kamimura K., Ramirez T. and GomisJ., 2004, Physical degrees of freedom of non-local theories, *Nuclear Physics B*, **696**, 263-275.

Gorenflo R. and Mainardi F., 1997, *Fractional Calculus: Integral and Differential Equations of Fractional Orders*, Fractals and Fractional Calculus in Continuum Mechanics, Springer, Wien and New York.

Hardy G.H., 1945, Riemann's form of Taylor's series, *Journal of the London Mathematical Society*, **20**, 48-56.

Heymans N. and Podlubny I., 2006, Physical interpretation of initial conditions for fractional differential equations with Riemann-Liouville fractional derivatives, *Rheologica Acta*, **45**, 765-771.

Hilfer R., 2000, *Applications of Fractional Calculus in Physics*,World Scientific Publishing Company, Singapore.

Jesus I.S. and Machado J.A.T., 2008, Fractional control of heat diffusion systems, *Nonlinear Dynamics*, **54**, 263-282.

Jumarie G., 2009, From Lagrangian mechanics fractal in space to space fractal Schrödinger's equation via fractional Taylor's series, *Chaos Solitons and Fractals*, **41**, 1590-1604.

Kilbas A.A., Srivastava H.H. and Trujillo J.J., 2006, *Theory and Applications of Fractional Differential Equations, Volume* 204 (North-Holland Mathematics Studies) Elsevier Science, Amsterdam.

Klimek M., 2001, Fractional sequential mechanics — models with symmetric fractional derivative,*Czechoslovak Journal of Physics*, **51**, 1348-1354.

Klimek M., 2002, Lagrangean and Hamiltonian fractional sequential mechanics, *Czechoslovak Journal of Physics*, **52**, 1247-1253.

Lakshmikantham V., Leela S. and Vasundhara Devi J., 2009, *Theory of Fractional Dynamic Systems*, Cambridge Scientific Publishers Ltd., Cambridge.

Lim S.C. and Teo L.P., 2009, The fractional oscillator process with two indices, *Journal of Physics A:Mathematical and Theoretical*, **42**, Art. No. 065208.

Lim S.C. and Muniandy S.V., 2004, Stochastic quantization of nonlocal fields, *Physics Letters A*, **324**, 396-405.

Llosa J. and Vives J., 1994, Hamiltonian- formalism for nonlocal Lagrangians, *Journal of Mathematical Physics*, **35**, 2856-2877.

Lorenzo C.F. and Hartley T.T., 2004, Fractional trigonometry and the spiral functions, *Nonlinear Dynamics*, **38**, 23-34.

Magin R.L., 2000, *Fractional Calculus in Bioengineering*, Begell House Publisher, Connecticut.

Magin R., Feng X. and Baleanu D., 2009, Solving the fractional order bloch equation, *Concepts in Magnetic Resonance Part A*, **34A**, 16-23.

Magin R.L., Abdullah O., Baleanu D. and Zhou X.H.J., 2008, Anomalous diffusion expressed through fractional order differential operators in the Bloch-Torrey equation, *Journal of Magnetic Resonance*, **190**, 255-270.

Mainardi F., 1996, Fractional relaxation-oscillation and fractional diffusion-wave phenomena,*Chaos*, **7**, 1461-1476.

Mainardi F., Luchko Yu. and Pagnini G., 2001, The fundamental solution of the space-time fractional diffusion equation, *Fractional Calculus and Applied Analysis*, **4**, 153-192.

Maraaba T.A., Baleanu D. and Jarad F., 2008a, Existence and uniqueness theorem for a class of delay differential equations with left and right Caputo fractional derivatives, *Journal of Mathematical Physics*, **49**, Art. No. 083507.

Maraaba T.A., Jarad F. and Baleanu D., 2008b, On the existence and the uniqueness theorem for fractional differential equations with bounded delay within Caputo derivatives, *Science in China Series A-Mathematics*, **51**, 1775-1786.

Metzler R., Schick W., Kilian H.G. and Nonennmacher T.F., 1995 , Relaxation in filled polymers: A fractional calculus approach, *Journal of Chemical Physics*, **103**, 160137-1.

Miller K.S. and Ross B., 1993, *An Introduction to the Fractional Integrals and Derivatives-Theory and Applications*, John Wiley and Sons Inc., New York.

Momani S., 2006, A numerical scheme for the solution of multi-order fractional differential equations, *Applied Mathematics and Computation*, **182**, 761-786.

Muslih S.I. and Baleanu D., 2005a, Hamiltonian formulation of systems with linear velocities within Riemann-Liouville fractional derivatives, *Journal of Mathematical Analysis and Applications*, **304**, 599-606.

Muslih S.I. and Baleanu D., 2005b, Quantization of classical fields with fractional derivatives, *Nuovo Cimento B*, **120**, 507-512.

Muslih S.I. and Baleanu D., 2005c, Hamiltonian formulation of systems with linear velocities within Riemann-Liouville fractional derivatives, *Journal of Mathematical Analysis and Applications*, **304**, 599-606.

Naber M., 2004, Time fractional Schrödinger equation, *Journal of Mathematical Physics*, **45**, 3339-3352.

Nesterenko V.V., 1989, Singular Lagrangians with higher deratives, *J. Phy. A- Math. Gen.*, **22**, 1673-1687.

Nigmatullin R.R. and Mehaute A.L., 2005, Is there geometrical/physical meaning of the fractional integral with complex exponent? *Journal of Non-Crystalline Solids*, **351**, 2888-2899.

Oldham K.B. and S'panier, J., 1974, *The Fractional Calculus*, Academic Press, New York.

Pais A. and Uhlenbeck G.E., 1950, On field theories with non-localized action, *Physical Review*, **79**, 145-165.

Podlubny I., 1999, *Fractional Differential Equations*, Academic Press, San Diego CA.

Rabei E.M., Nawafleh K.I., Hijjawi R.S., Muslih S.I. and Baleanu D., 2007, The Hamilton formalism with fractional derivatives,*Journal of Mathematical Analysis and Applications*, **327**, 891-897.

Rabei E.M., Altarazi I.M.A., Muslih S.I. and Baleanu D., 2009, Fractional WKB approximation, *Nonlinear Dynamics*, **57**,171-175.

Raspini A., 2001, Simple solutions of the fractional Dirac equation of order 2/3, *Physica Scripta*, **64**, 20-22.

Riewe F., 1996, Nonconservative Lagrangian and Hamiltonian mechanics, *Physical Review E*, **53**, 1890-1899.

Riewe F., 1997, Mechanics with fractional derivatives,*Physical Review E*, **55**, 3581-3592.

Samko S.G., Kilbas A.A. and Marichev O.I., 1993, *Fractional Integrals and Derivatives – Theory and Applications*, Gordon and Breach, Linghorne.

Scalas E., Gorenflo R. and Mainardi F., 2004, Uncoupled continuous-time random walks: Solution and limiting behavior of the master equation,*Physical Review E*, **69**, 011107-1.

Seiberg N., Susskind L. and Toumbas T., 2000, Strings in background electric field, space/time noncommutativity and a new noncritical string theory, *Journal of High Energy Physics*, **6**, Art. No. 021.

Silva M.F., Tenreiro-Machado J.A.T. and Barbosa R.S., 2008, Using fractional derivatives in joint control of hexapod robots, *Journal of Vibration and Control*, **14**, 1473-1485.

Simon J.Z., 1990, Higher-derivative Lagrangians, nonlocality, problems and solutions,*Physical Review D*, **41**, 3720-3733.

Solomon T.H., Weeks E.R. and Swinney H.L., 1993, Observation of anomalous diffusion and Lévy flights in a two-dimensional rotating flow, *Physical Review Letters*, **71**, 3975-3978.

Tarasov V.E., 2005, Continuous medium model for fractal media, *Physics Letters A*, **336**, 167-174.

Tarasov V.E., 2006, Fractional statistical mechanics, *Chaos*, **16**, 033108-033115.

Tarasov V.E. and Zaslavsky G.M., 2006, Nonholonomic constraints with fractional derivatives, *Journal of Physics A:Mathematical and General*, **39**, 9797-9815.

Tenreiro Machado, J.A., 2001, Discrete-time fractional-order controllers, *Fractional Calculus and Applied Analysis*, **4**, 47-66.

Tenreiro Machado J. A., 2003, A probabilistic interpretation of the fractional order differentiation, *Fractional Calculus and Applied Analysis*, **6**, 73-81.

Trujillo J.J., 1999, On a Riemann-Liouville generalized Taylor's formula, *Journal of Mathematical Analysis and Applications*, **231**, 255-264.

Uchaikin, V.V., 2008, *The Method of the Fractional Derivatives*, Art-i-Shock Publisher, Ulianovsk (in Russian).

Unser M. and Blu T., 1999, Construction of fractional spline wavelet bases, in *proc. SPIE Wavelets Applications in Signal and Image Processing VII*, Denver, CO, **3813**, 422-431.

Unser M. and Blu T., 2000a, Fractional splines and wavelets, *SIAM Review*, **42**, 43-67.

Unser M. and Blu T., 2000b, The fractional spline wavelet transform: definition and implementation, *Proceedings of the Twenty-Fifth IEEE International Conference on Acoustics, Speech, and Signal Processing (ICASSP'00)*, Istanbul, Turkey, June 5-9, vol. I, pp. 512-515.

Unser M. and Blu T., 2002, Wavelets, fractals, and radial basis functions, *IEEE Transactions on Signal Processing*, **50**, 543-553.

Walczak B., 2000, *Wavelets in Chemistry*, Elsevier Press, Amsterdam.

West B.J., Bologna M. and Grigolini P., 2003, *Physics of Fractal Operators*, Springer, New York.

Zaslavsky G.M., 2002, Chaos, fractional kinetics, and anomalous transport, *Physics Reports*, **371**, 461-580.

Zaslavsky G., 2005, *Hamiltonian Chaos and Fractional Dynamics*, Oxford University Press, Oxford.

Chapter 2
Realization of Fractional-Order Controllers: Analysis, Synthesis and Application to the Velocity Control of a Servo System

Ramiro S. Barbosa, Isabel S. Jesus, Manuel F. Silva, J. A. Tenreiro Machado

Abstract The synthesis and application of fractional-order controllers is now an active research field. This article investigates the use of fractional-order *PID* controllers in the velocity control of an experimental modular servo system. The system consists of a digital servomechanism and open-architecture software environment for real-time control experiments using MATLAB/Simulink. Different tuning methods will be employed, such as heuristics based on the well-known Ziegler-Nichols rules, techniques based on Bode's ideal transfer function and optimization tuning methods. Experimental responses obtained from the application of the several fractional-order controllers are presented and analyzed. The effectiveness and superior performance of the proposed algorithms are also compared with classical integer-order *PID* controllers.

2.1 Introduction

Fractional calculus (FC) is an area of mathematics that extends derivatives and integrals to an arbitrary order (real or, even, complex order) and emerged at the same time as the classical differential calculus. FC generalizes the classical differential operator $D_t^n \equiv d^n/dt^n$ to a fractional operator D_t^α, where α can be a non-integer value (Oldham and Spanier, 1974; Oustaloup, 1991, 1995; Podlubny, 1999a; Samko et al., 1993; Miller and Ross, 1993; Carpinteri and Mainardi, 1997). However, its inherent complexity delayed the application of the associated concepts.

Nowadays, FC is extensively applied in science and engineering (Oldham and Spanier, 1974; Oustaloup, 1991, 1995; Podlubny, 1999a; Carpinteri and Mainardi, 1997; Hilfer, 2000; Westerlund, 2002), being recognized its ability to yield a su-

Ramiro S. Barbosa
ISEP-Institute of Engineering of Porto, Dept. of Electrotechnical Engineering, Rua Dr. António Bernardino de Almeida, 431, 4200-072 Porto, Portugal.
Email: rsb@isep.ipp.pt

perior modelling and control in many dynamical systems. We may cite its adoption in areas such as viscoelasticity and damping, diffusion and wave propagation, electromagnetism, chaos and fractals, heat transfer, biology, electronics, signal processing, robotics, system identification, traffic systems, genetic algorithms, percolation, modelling and identification, telecommunications, chemistry, irreversibility, physics, control, economy and finance. We may say that all areas of science are touched by this fascinating new (and, at the same time, old) topic.

In what concerns the area of automatic control, the fractional-order controllers (FOCs) are now an active field of research (Oustaloup, 1991, 1995; Podlubny, 1999a,b; Barbosa et al., 2004b; Machado, 1997; Ma and Hori, 2003; Feliu-Batlle et al., 2007; Valério and Sá da Costa, 2004; Sabatier et al., 2004; Silva and Machado, 2006, 2008; Silva et al., 2008; Machado et al., 2007; Jesus et. al., 2006a; Jesus et al., 2007c, 2008a; Jesus and Machado, 2008a, 2009). Ma and Hori (2003) used a $PI^{\alpha}D$ controller for the speed control of two-inertia system. The superior robustness performance against input torque saturation and load inertia variation are shown by comparison with integer *PID* control. Feliu-Batlle et al. (2007) applied FOCs in the control of main irrigation canals, which revealed to be robust to changes in the time delay and gain. Valério and Sá da Costa (2004) introduced a FOC in a two degree of freedom flexible robot, achieving a stable response for the position of its tip. Sabatier et al. (2004) applied the CRONE (French acronym for "Commande Robuste d'Ordre Non Entier") algorithm to a robust speed control of a low damped electromechanical system with backlash. The CRONE scheme ensures robust speed control of the load in spite of plant parametric variations and speed observation errors. Silva et al. (2006, 2008) compared the performance of legged robots locomotion when controlled using integer and fractional *PD* control algorithms. Through a simulation study, they showed the advantages of using a PD^{α} algorithm in the joint control of hexapod robots with two and three degrees of freedom per leg. This controller allows the minimization of two global measures of the overall performance of the mechanism (in an average sense), being one index inspired on the system dynamics and the other based on the trajectory tracking errors (Silva and Machado, 2008). It was also verified that the superior performance of the fractional PD^{α} joint leg controller, for the fractional order $\alpha \approx 0.5$, is kept for different ground properties. In (Machado et al., 2007; Jesus and Machado, 2009), It is studied the skin effect phenomena and the electrical potential in the FC perspective. It was verified that the classical electromagnetism and the Maxwell equation, with integer order derivatives, lead to models requiring a FC viewpoint to be fully interpreted. Jesus et al. (2006a, 2007c) analyzed the electrical impedance in botanical elements. The fractional order behavior as well as its relation with the electrical impedance was established and an equivalent fractional order circuit model was presented. Also, in (Jesus et al., 2008a; Jesus and Machado, 2008a), they studied an electrolytic process for developing fractional order capacitors. The results revealed capacitances of fractional order that can constitute an alternative to the classical integer order capacitors. It was verified that it is possible to get fractional order elements by adopting non-classical electrodes and dielectrics. For more application examples, see the therein references of the above cited works.

In spite of the recent progresses, the truth is that simple and effective tuning rules, such as those for classical *PID* controllers, are still lacking. In this chapter, we apply several types of fractional *PID* controllers in the velocity control of an experimental servo system. Also, several tuning methods will be employed in order to assess the performance of such algorithms. Firstly, we use the well-known Ziegler-Nichols (Z-N) heuristic rules (Ziegler and Nichols, 1942) and analyze the effect of fractional (derivative and integral) orders upon the system's performance. After, a simple analytical method for tuning FOCs based on Bode's ideal transfer function is outlined. Finally, several optimal fractional *PID* will be applied to the servo system. Comparison with conventional integer *PID* is also pursued. The experiments show that the extra parameters (derivative and integral orders) provided by the FOCs can effectively enhance the system performance and help to adjust more carefully the dynamics of an automatic control system.

Bearing these ideas in mind, this chapter is organized as follows. Section 2.2 presents the fundamentals of FOCs while Section 2.3 outlines the Oustaloup's frequency approximation method used in this work for the realization of the fractional algorithms. Section 2.4 describes the experimental servo system setup and Section 2.5 makes the mathematical modelling and identification of the servo system. Section 2.6 describes the FOC scheme used in the experiments. Here, we will present the several FOCs employed. Section 2.7 shows the experimental results obtained from the application of the well-known Ziegler-Nichols (Z-N) tuning rules. Responses from the quarter decay ratio and oscillatory behavior Z-N heuristics will be given. In Section 2.8 we propose a simple analytical method for tuning FOCs and in Section 2.9 we investigate the use of optimal FOCs in the control of the servo system. Finally, Section 2.10 draws the main conclusions.

2.2 Fractional-order control systems

FOCs may be defined as systems that possess fractional derivatives or integrals in the system to be controlled or in the controller. The fundamentals for the analysis and synthesis of fractional-order systems are given in next subsections.

2.2.1 Basic theory

In general, a fractional system can be described by a Linear Time Invariant (LTI) fractional-order differential equation of the form:

$$\begin{aligned} & a_n D_t^{\beta_n} y(t) + a_{n-1} D_t^{\beta_{n-1}} y(t) + \cdots + a_0 D_t^{\beta_0} y(t) \\ & = b_m D_t^{\alpha_m} u(t) + b_{m-1} D_t^{\alpha_{m-1}} u(t) + \cdots + b_0 D_t^{\alpha_0} u(t), \end{aligned} \tag{2.1}$$

where β_k, α_k $(k = 0,1,2,\ldots)$ are real numbers, $\beta_k > \cdots > \beta_1 > \beta_0, \alpha_k > \cdots > \alpha_1 > \alpha_0$ and a_k, b_k $(k = 0,1,2,\ldots)$ are arbitrary constants.

The corresponding continuous transfer function has the form:

$$G(s) = \frac{Y(s)}{U(s)} = \frac{b_m s^{\alpha_m} + b_{m-1} s^{\alpha_{m-1}} + \cdots + b_0 s^{\alpha_0}}{a_n s^{\beta_n} + a_{n-1} s^{\beta_{n-1}} + \cdots + a_0 s^{\beta_0}}. \tag{2.2}$$

A z-transfer function of (2.2) can be obtained by using a discrete approximation of the fractional-order operators, yielding:

$$G(z) = \frac{b_m [w(z^{-1})]^{\alpha_m} + b_{m-1} [w(z^{-1})]^{\alpha_{m-1}} + \cdots + b_0 [w(z^{-1})]^{\alpha_0}}{a_n [w(z^{-1})]^{\beta_n} + a_{n-1} [w(z^{-1})]^{\beta_{n-1}} + \cdots + a_0 [w(z^{-1})]^{\beta_0}}, \tag{2.3}$$

where $w(z^{-1})$ denotes the discrete equivalent of the Laplace operator s, expressed as a function of the complex variable z or the shift operator z^{-1}.

The generalized operator ${}_aD_t^\alpha$, where a and t are the limits and α the order of operation, is usually given by the Riemann-Liouville definition $(\alpha > 0)$:

$${}_aD_t^\alpha x(t) = \frac{1}{\Gamma(n-\alpha)} \frac{d^n}{dt^n} \int_a^t \frac{x(\tau)}{(t-\tau)^{\alpha-n+1}} d\tau, \quad n-1 < \alpha < n, \tag{2.4}$$

where $\Gamma(u)$ represents the Gamma function of u. Another common definition is that given by the Grünwald-Letnikov approach $(\alpha \in \Re)$:

$${}_aD_t^\alpha x(t) = \lim_{h \to 0} \frac{1}{h^\alpha} \sum_{k=0}^{\left[\frac{t-a}{h}\right]} (-1)^k \binom{\alpha}{k} x(t-kh), \tag{2.5}$$

where h is the time increment and $[v]$ means the integer part of v.

Several attempts have been made to give a meaningfully interpretation of fractional derivatives/integrals (Nigmatullin, 1992; Rutman, 1994, 1995; Adda, 1997; Monsrefi-Torbati and Hammond, 1998; Podlubny, 2002; Machado, 2003, 2009). Podlubny (2002) gives a geometric and physical interpretation to fractional differentiation and integration for various kinds of well-known operators. Machado (2003, 2009), introduces a probabilistic interpretation of the fractional derivative based on the Grünwald-Letnikov definition. Other different approaches have been proposed (Nigmatullin, 1992; Rutman, 1994, 1995; Adda, 1997; Monsrefi-Torbati and Hammond, 1998), but at author's knowledge, no consensual interpretation (geometric or physical) is yet been established for these operators.

For a wide class of functions, important for applications, definitions (2.4) and (2.5) are equivalent (Oldham and Spanier, 1974; Podlubny, 1999a; Samko et al., 1993; Miller and Ross, 1993). This allows one to use the Riemann-Liouville definition during problem formulation, and then turn to the Grünwald-Letnikov definition for obtaining the numerical solution. Moreover, an important fact revealed by both definitions is that the evaluation of fractional-order derivatives in any instant t requires the whole history of $y(t)$. This means that fractional-order derivatives are "global" operators having a memory of the entire past in opposition with the integer-order derivatives that are "local" operators. This property is being used to model hereditary and memory effects in most materials and systems. While this brings a new viewpoint over many areas of science and engineering, it poses, however, evaluation problems due to the unlimited memory imposed for their computation (*e.g.*, for large values of t). To overcome this difficulty, Podlubny (1999a) suggested the use of the so-called "short memory" principle, which takes into account the behavior of $y(t)$ only in the "recent past", i.e. in the interval $[t-L,t]$, where L is the "memory length" and, consequently, maximizing the amount of computation to L seconds. This method was applied successfully for the numerical solution of linear ordinary fractional-order differential equations with constant and non-constant coefficients and non-linear ordinary fractional-order differential equations (Podlubny, 1999a).

The fractional-order derivatives can also be defined in the transform domain. It is shown that the Laplace transform (L) of a fractional derivative of a signal $x(t)$ is given by:

$$L\{D^{\alpha}x(t)\} = s^{\alpha}X(s) - \sum_{k=0}^{n-1} s^{k}\, D^{\alpha-k-1}x(t)\Big|_{t=0}, \tag{2.6}$$

where $X(s) = L\{x(t)\}$. Considering null initial conditions, expression (2.6) reduces to the simple form ($\alpha \in \Re$):

$$L\{D^{\alpha}x(t)\} = s^{\alpha}X(s), \tag{2.7}$$

which is a direct generalization of the integer-order scheme with the multiplication of the signal transform $X(s)$ by the Laplace s-variable raised to a real power α. The Laplace transform reveals to be a valuable tool for the analysis and design of FOC systems.

2.2.2 *Fractional-Order controllers and their implementation*

The FOC concept was first introduced by Oustaloup (1991, 1995), who developed the so-called CRONE controller. Some earlier authors that produced important results in the control area include Bode (1945), Tustin et al. (1958) and Manabe (1961, 1963). Also, Machado (1997, 1999) discussed the design of fractional-order discrete-time controllers. More recently, Podlubny (1999b) proposed a generalization of the *PID* controller, the $PI^{\lambda}D^{\mu}$ controller, involving an integrator of order λ and a differentiator of order μ. The transfer function $G_c(s)$ of such a controller has the form:

$$G(s) = \frac{U(s)}{E(s)} = K_p + K_I s^{-\lambda} + K_D s^{\mu}, \quad \lambda,\, \mu > 0, \tag{2.8}$$

where $E(s)$ is the error signal and $U(s)$ the controller's output. The parameters (K_p, K_I, K_D) are the proportional, integral, and derivative gains of the controller, respectively.

The $PI^\lambda D^\mu$ controller is represented by a fractional integro-differential equation of type:

$$u(t) = K_P e(t) + K_I D^{-\lambda} e(t) + K_D D^\mu e(t). \tag{2.9}$$

Clearly, depending on the values of the orders λ and μ, we get an infinite number of choices for the controller's type (defined continuously on the (λ,μ)- plane). For example, taking $(\lambda,\mu) \equiv \{(1,1),(1,0),(0,1),(0,0)\}$ gives the $\{PID, PI, PD, P\}$ controllers. All these classical types of controllers are the particular cases of the fractional $PI^\lambda D^\mu$ algorithm. Thus, the $PI^\lambda D^\mu$ is more flexible and gives the possibility of adjusting more carefully the dynamical properties of a control system (Podlubny, 1999a,b).

As shown above, the fractional-order operators are characterized by having irrational continuous transfer functions in the Laplace domain or infinite dimensional discrete transfer functions in time domain. These facts preclude their direct utilization both in time and frequency domains. Therefore, the usual approach for the analysis and synthesis of fractional-order systems is the development of continuous and discrete integer-order approximations of these operators (Machado, 1997, 1999; Vinagre et al., 2000; Chen and Moore, 2002; Vinagre et al., 2003; Charef et al., 1992; Barbosa et al., 2006; Chen et al., 2004; Carlson and Halijak, 1964).

For instance, the usual approach for obtaining discrete equivalents of continuous operators of type s^α ($\alpha \in \Re$) adopts the Euler, Tustin and Al-Alaoui generating functions (Machado, 1997; Vinagre et al., 2000; Chen and Moore, 2002; Vinagre et al., 2003; Barbosa et al., 2006; Chen et al., 2004; Al-Alaoui, 1993, 1997). However, the fractional-order conversion schemes lead to non-rational z-formulae. In order to get rational expressions we may adopt two possibilities. One way is to perform a power series expansion (PSE) (Taylor series) over them and the final approximation corresponds to a truncated z-polynomial function (FIR filter) (Machado, 1997; Vinagre et al., 2000; Barbosa et al., 2006). For example, using the backward Euler rule, $H(z^{-1}) = (1-z^{-1})/T$, and performing a PSE of $[(1-z^{-1})/T]^\alpha$ gives the discretization formula corresponding to the Grünwald-Letnikov definition (2.5):

$$\begin{aligned} D^\alpha(z) &= \frac{Y(z)}{X(z)} = \left(\frac{1}{T}\right)^\alpha \text{PSE}\left\{\left(1-z^{-1}\right)^\alpha\right\}_N \\ &= \left(\frac{1}{T}\right)^\alpha P_N\left(z^{-1}\right) = \left(\frac{1}{T}\right)^\alpha \left(c_0^{(\alpha)} + c_1^{(\alpha)} z^{-1} + \cdots + c_N^{(\alpha)} z^{-N}\right), \end{aligned} \tag{2.10}$$

where P is a polynomial of degree N and $c_k^{(\alpha)}$ are binomial coefficients which may be calculated recursively as:

$$c_0^{(\alpha)} = 1, \quad c_k^{(\alpha)} = \left(1 - \frac{1+\alpha}{k}\right) c_{k-1}^{(\alpha)}, \quad k = 1, 2, \ldots \tag{2.11}$$

Another possible way is to obtain a discrete transfer function in the form of rational function (*i.e.*, as the ratio of two polynomials) (IIR filter) by application of the continued fraction expansion (CFE) method (Chen and Moore, 2002; Barbosa et al., 2006; Chen et al., 2004). It is well known that the CFE is a method of evaluation of functions that frequently converges much more rapidly than power series expansions, and converges in a much larger domain in the complex plane. A method for obtaining discrete equivalents of the fractional-order operators, which combines the well known advantages of the trapezoidal rule (commonly designated as the Tustin method in the control community) and the advantages of the CFE uses the generating function:

$$\left(w\left(z^{-1}\right)\right)^{\alpha} = \left(\frac{2}{T}\frac{1-z^{-1}}{1+z^{-1}}\right)^{\alpha}. \tag{2.12}$$

By doing so over expression (2.12), results in the discrete transfer function, approximating continuous fractional-order operators, expressed as:

$$\begin{aligned} D^{\alpha}(z) &= \frac{Y(z)}{X(z)} = \left(\frac{2}{T}\right)^{\alpha} \mathrm{CFE}\left\{\left(\frac{1-z^{-1}}{1+z^{-1}}\right)^{\alpha}\right\}_{m,n} \\ &= \left(\frac{2}{T}\right)^{\alpha} \frac{P_m\left(z^{-1}\right)}{Q_n\left(z^{-1}\right)} = \left(\frac{2}{T}\right)^{\alpha} \frac{p_0 + p_1 z^{-1} + \cdots + p_m z^{-m}}{q_0 + q_1 z^{-1} + \cdots + q_n z^{-n}}, \end{aligned} \tag{2.13}$$

where T is the sampling period, $\mathrm{CFE}\{u\}$ denotes the function from applying the continued fraction expansion to the function u, $Y(z)$ is the z-transform of the output sequence $y(nT)$, $X(z)$ is the z-transform of the input sequence $x(nT)$, m and n are the orders of the approximation, and P and Q are polynomials of degrees m and n, correspondingly, in the variable z^{-1}.

2.3 Oustaloup's frequency approximation method

In this study we adopt integer rational transfer functions of the continuous fractional operators introduced by Oustaloup (Oustaloup, 1991; Oustaloup et al., 2000), and hence commonly named the Oustaloup-recursive-approximation method. So, in order to implement the term s^{α} $(\alpha \in \Re)$ of the FOC, a frequency-band limited approximation is used by cutting out both high and low frequencies of transfer function $(s/\omega_u)^{\alpha}$ to a given frequency range $[\omega_b, \omega_h]$, distributed geometrically around the unit gain frequency $\omega_u = (\omega_b \omega_h)^{1/2}$ (Oustaloup et al., 2000), yielding:

$$D(s) = \left(\frac{\omega_u}{\omega_h}\right)^{\alpha} \left(\frac{1+s/\omega_b}{1+s/\omega_h}\right)^{\alpha}. \tag{2.14}$$

The synthesis of such transfer function (2.14) results in a recursive distribution of poles and zeros, giving:

$$D(s) = \lim_{N\to\infty} D_N(s), \tag{2.15}$$

where

$$D_N(s) = \left(\frac{\omega_u}{\omega_h}\right)^{\alpha} \prod_{k=-N}^{N} \frac{1+s/\omega'_k}{1+s/\omega_k} \tag{2.16}$$

and

$$\omega'_k = \left(\frac{\omega_h}{\omega_b}\right)^{\frac{k+N+\frac{1}{2}-\frac{\alpha}{2}}{2N+1}} \omega_b, \quad \omega_k = \left(\frac{\omega_h}{\omega_b}\right)^{\frac{k+N+\frac{1}{2}+\frac{\alpha}{2}}{2N+1}} \omega_b. \tag{2.17}$$

Taking N, ω_b, ω_h and α, permits the determination of the values of the set of zeros and poles of (2.17) and consequently, the synthesis of the desired transfer function (2.16).

2.4 The experimental modular servo system

The Modular Servo System (MSS) consists of the INTECO (Inteco, 2006) digital servomechanism and open-architecture software environment for real-time control experiments. The MSS supports the real-time design and implementation of advanced control methods using MATLAB/Simulink tools.

Fig. 2.1(a) illustrates the MSS setup, which consists of several modules mounted in a metal rail and coupled with small clutches. The modules are arranged in the chain such that the DC motor with the generator module is at the front and the gearbox with the output disk is at the end of the chain, see Fig. 2.1(b).

The DC motor can be coupled with the modules of inertia, magnetic brake, backlash and gearbox with the output disk. The angle of rotation of the DC motor shaft is measured using an incremental encoder. The generator is connected directly to the DC motor and generates a voltage proportional to the angular velocity.

The servomechanism is connected to a computer where a control algorithm is implemented based on the measurement of the angular position and/or velocity. The accuracy of measurement of the position is 0.1% while the accuracy of measured velocity is 5%. The armature voltage of the DC motor is controlled by a PWM signal $v(t)$ excited by a dimensionless control signal in the form $u(t) = v(t)/v_{max}$. The admissible controls satisfy $|u(t)| \leq 1$ and $v_{max} = 12$ V (Inteco, 2006).

2.5 Mathematical modelling and identification of the servo system

The experiments in the MSS include the modules of the DC motor with tachogenerator, inertia load, encoder module and gearbox module with output disk (see Fig. 2.1).

The linear model of the setup system is represented in Fig. 2.2. It is assumed that the armature inductance of the motor is negligible. Also, the static and dry frictions, as well the saturation are neglected. Based on these considerations, the electrical

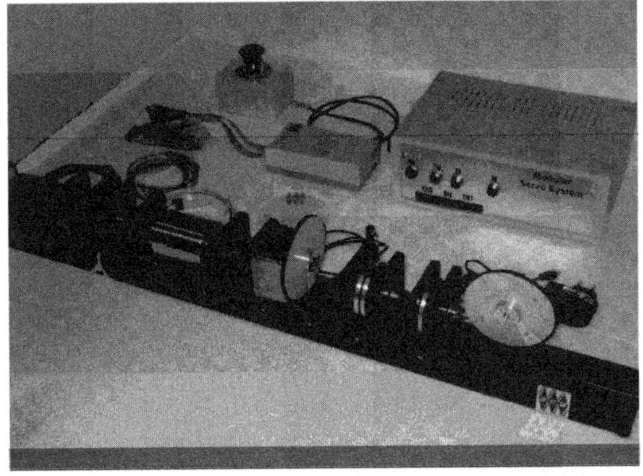

(a) Setup

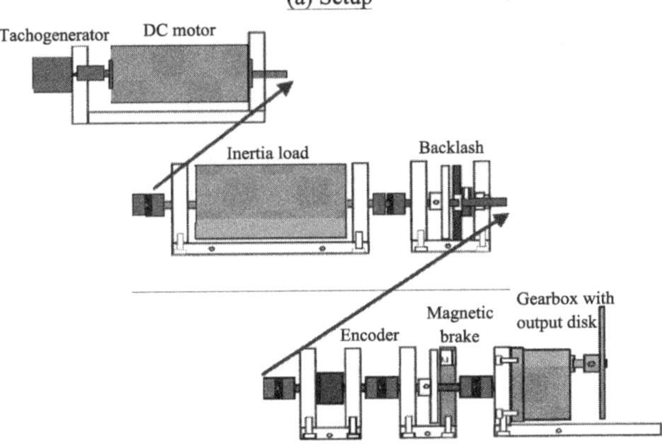

(b) Mechanical construction

Fig. 2.1 The modular servo system (MSS) (Inteco, 2006).

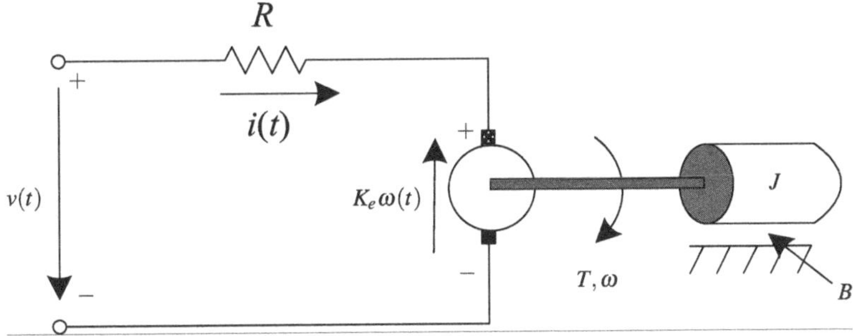

Fig. 2.2 Schematic of DC Motor.

and mechanical equations that model the system are, respectively, given by:

$$v(t) = Ri(t) + K_e\omega(t), \tag{2.18}$$

$$J\dot{\omega}(t) + B\omega(t) = K_t i(t), \tag{2.19}$$

where

$v(t)$ is the applied voltage,
$i(t)$ is the armature current,
$\omega(t)$ is the angular velocity of the rotor,
R is the resistance of armature winding,
J is the moment of inertia of the moving parts,
B is the damping coefficient due to viscous friction,
$K_e\omega(t)$ is the back electromotive force (EMF),
$T = K_t i(t)$ is the electromechanical torque.

By combining expressions (2.18) and (2.19) we get

$$T_s\dot{\omega}(t) + \omega(t) = K_{sm}v(t), \tag{2.20}$$

where the motor time constant T_s and motor gain K_{sm} are given by:

$$T_s = \frac{RJ}{BR + K_e K_t}, \quad K_{sm} = \frac{K_t}{BR + K_e K_t}. \tag{2.21}$$

The transfer function for the motor velocity $\omega(t)$ has the form:

$$G(s) = \frac{\omega(s)}{V(s)} = \frac{K_{sm}}{T_s s + 1}. \tag{2.22}$$

The control applied to the system has the form of a PWM signal. Thus, we assume the dimensionless control signal as the scaled input voltage, $u(t) = v(t)/v_{max}$. The admissible control satisfies $|u(t)| \leq 1$. With respect to $K_s = K_{sm}v_{max}$ we obtain the transfer function in the form:

$$G(s) = \frac{\omega(s)}{V(s)} = \frac{K_s}{T_s s + 1}, \tag{2.23}$$

where the parameters K_s and T_s must be identified by the user.

For the identification process, a unit step input signal, $u(t) = 1(t)$, is applied to the servo system and the angular velocity *versus* time is acquired. Next, the least squares method is used to find the system parameters of transfer function (2.23).

Figure 2.3 shows the velocity obtained from the measurements and the velocity calculated from the model (2.23). Clearly, the fitting of both curves is good. The calculated values of system parameters are $K_s = 203.5344$ rad/s and $T_s = 0.8258$ s. We must note that these values were obtained for the configuration described in the first paragraph of this section and that other setup of the system will generate different parameters.

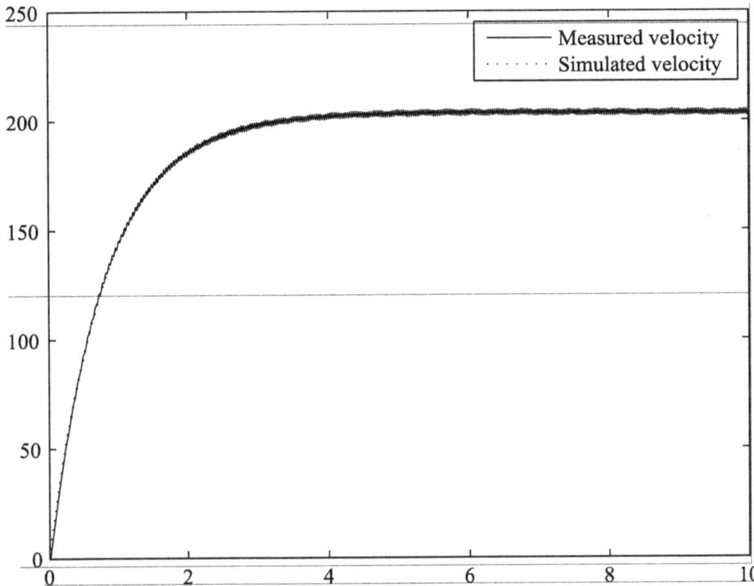

Fig. 2.3 Measured and simulated velocities.

2.6 Fractional-order real-time control system

The MSS setup for the experiments include the modules of DC motor with tacho-generator, inertia load, encoder and gearbox with output disk (see Fig. 2.1).

All real-time control experiments are performed using the MATLAB/Simulink real-time model of Fig. 2.4. A fixed-step solver (Euler's integration method) with a step size of 0.01 (sampling period of $T = 0.01$ s) is chosen.

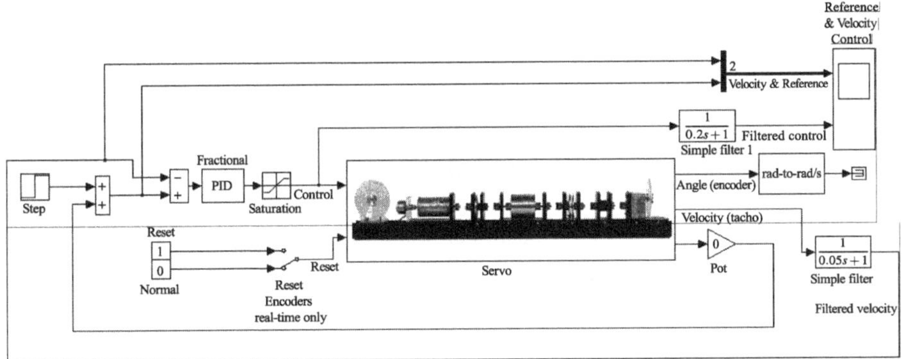

Fig. 2.4 Real-time model with the fractional *PID* controller (adapted from (Inteco, 2006)).

The fractional term s^α ($\alpha \in \Re$) in the fractional *PID* controller transfer function (2.8) is implemented by using the Oustaloup's frequency approximation method described in Section 2.3. The values used were $N = 5$, $\omega_b = 1$ rad/s and $\omega_h = 1000$ rad/s.

The fractional-order controllers are implemented in digital form by discretization of the continuous controller transfer functions. The discretization technique used consists in the bilinear (or Tustin's) approximation with a sampling period of $T = 0.01$ s.

This work investigates the application of several types of fractional *PID*s in the control of the angular velocity of the servo systems in which were was adopted the D^μ-controller, the I^λ-controller, the PI^λ-controller and the $PI^\lambda D$-controller, correspondingly given by the transfer functions ($\mu,\ \lambda > 0$):

$$G_c(s) = K_D s^\mu, \tag{2.24}$$

$$G_c(s) = \frac{K_I}{s^\lambda}, \tag{2.25}$$

$$G_c(s) = K_P + \frac{K_I}{s^\lambda}, \tag{2.26}$$

$$G_c(s) = K_P + \frac{K_I}{s^\lambda} + K_D s, \tag{2.27}$$

where the gains ($K_P,\ K_I,\ K_D$) and orders ($\mu,\ \lambda$) of the FOCs are the parameters to be tuned.

In order to assure a good steady state error, the term $1/s^\lambda$ in expressions (2.25)-(2.27) must be implemented by means of an integer integrator (Franklin et al., 2006; Axtell and Bise, 1990). The modified integral term of the mentioned FOCs is then given in the form:

$$I^\lambda = K_I \frac{s^{1-\lambda}}{s}, \quad 0 < \lambda < 1 \tag{2.28}$$

The steady-state behavior could be also improved by multiplying the FOC by a term of the form $(s+\eta)/s$, with η being a small value (Feliu-Batlle et al., 2007).

2.7 Ziegler-Nichols tuning rules

Ziegler and Nichols (1942) proposed two methods for tuning the controller parameters based on the transient response characteristics of a given plant. In the first method, the choice of controller parameters is designed to result in a closed-loop step response transient with a decay ratio of approximately 0.25 (that is, the transient decays to a quarter of its value after one period of oscillation). In the second method, the criterion for tuning the controller parameters consists in evaluating the system at the limit of stability (*ultimate sensitivity method*). Here, the proportional gain is increased until the system becomes marginally stable and we observe continuous oscillations. The corresponding gain K_u and the period of oscillation P_u (also called *ultimate gain* and *ultimate period*, respectively) are then determined (Franklin et al., 2006). In this section, we will apply both methods to the servo system.

Several works for tuning fractional/integer order *PID* controllers based on Ziegler-Nichols rules have been proposed (Barbosa et al., 2008a,b; Valério and da Costa, 2006a,b). Barbosa et al. (2008a,b) investigated the adoption of different fractional *PID* algorithms in the velocity control of an experimental servo system by using the Ziegler-Nichols rules. In (Barbosa et al., 2008a), they used the ultimate sensitivity method while in (Barbosa et al., 2008b) the quarter-decay ratio was adopted. The fractional orders, as well the constants of the controllers, were varied and their effect on system's performance was analyzed. It was shown that the FOCs can effectively enhance the system performance providing extra tuning parameters useful for the adjustment of the control system dynamics. Also, the Zeigler-Nichols rules revealed to be simple and effective in the final tuning of the fractional algorithms. Valério and da Costa (2006b) developed tuning rules for fractional *PID* controllers. These rules are quadratic and require the same plant time-response used by the quarter-decay ratio Ziegler-Nichols rules for integer *PID*. The fractional *PID* tuned with these rules compare well with integer *PID* tuned according to the Ziegler–Nichols rules, and provide a roughly constant overshoot when the gain of the plant undergoes variations. Also, in (Valério and da Costa, 2006a) the same authors have developed tuning rules for integer *PID*s that behave, to the possible extent, as fractional *PID*s, while keeping the simplicity of the Ziegler-Nichols rules. The results obtained showed that the rules lead to *PID* controllers that behave better than those tuned using Ziegler-Nichols rules, but worst than ruled-tuned fractional *PID*s (which follow more complex specifications with greater ease).

2.7.1 Ziegler-Nichols tuning rules: quarter decay ratio

Ziegler and Nichols (Z-N) recognized that the step responses of a large number of process control systems exhibit a process reaction curve like that shown in Fig. 2.5.

The S-shape of the curve is characteristic of many higher-order systems, and such plant transfer function may be approximated by a first-order system plus a time delay (FOPTD) of t_d seconds:

$$\frac{Y(s)}{U(s)} = \frac{Ae^{-t_d s}}{\tau s+1}, \tag{2.29}$$

where τ is the system time constant and A is the gain. The parameters (A, t_d, τ) are determined from the unit step response of the process. If a tangent is drawn at the inflection point of the S-shaped curve, then the slope of the line is $R = A/\tau$ and the intersections of the tangent line with the time axis and line $y(t) = A$ identify the time delay $L = t_d$ and time constant τ (Franklin et al., 2006).

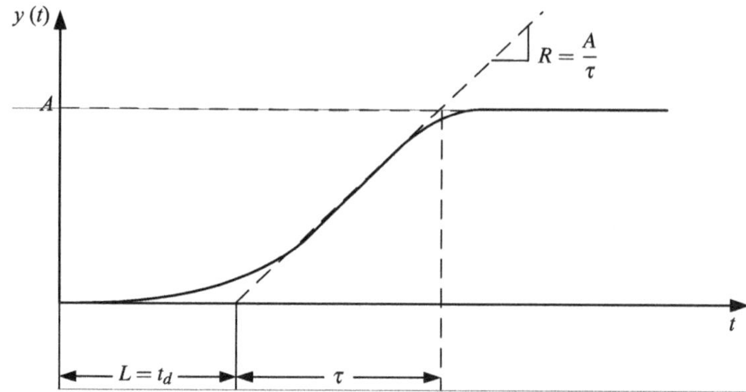

Fig. 2.5 Process reaction curve.

The choice of controller parameters is designed to result in a closed-loop step response transient with a decay ratio of approximately 0.25 in one period of oscillation. This corresponds to $\zeta = 0.21$ and is a good compromise between quick response and adequate stability margins. Table 2.1 lists the controller parameters suggested by Ziegler and Nichols to tune the proportional gain K_p, integral time T_I, and derivative time T_D.

Once the values of T_I and T_D have been obtained, the gains K_I and K_D, are computed as:

$$K_I = \frac{K_P}{T_I}, \quad K_D = K_P T_D. \tag{2.30}$$

Table 2.1 Ziegler-Nichols tuning for the controller $G_c(s) = K_P(1 + 1/T_I s + T_D s)$, for a decay ratio of 0.25.

Type of controller	K_p	T_I	T_D
P	$\frac{1}{RL}$	∞	0
PI	$\frac{0.9}{RL}$	$\frac{L}{0.3}$	0
PID	$\frac{1.2}{RL}$	$2L$	$0.5L$

For the identification of the FOPTD model parameters, a unit step input is applied to the system and the process reaction curve is acquired, as shown in Fig. 2.6. Note that the response is slightly different of that of Fig. 2.3, particularly in what concerns the gain of the system. This fact is a consequence of variation of the system dynamics over time, since these responses were obtained in different times.

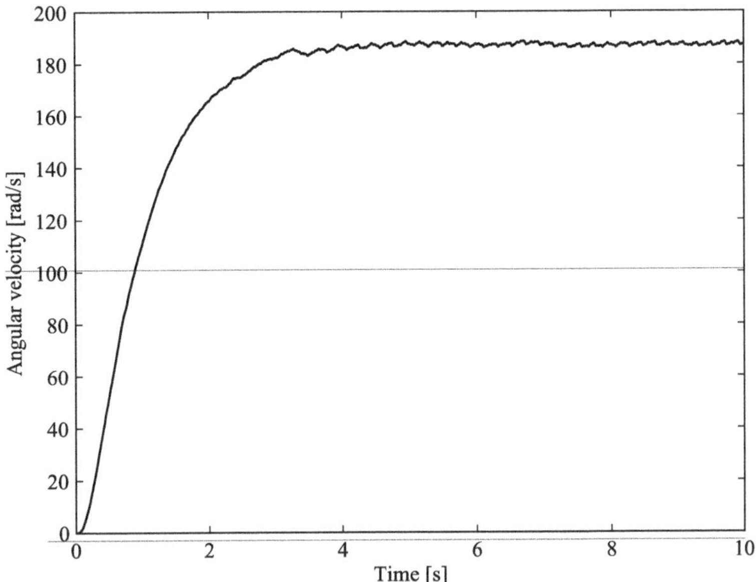

Fig. 2.6 Unit step response of the servo system.

Following the method of Ziegler-Nichols we identify the parameters $A = 187.2106$, $\tau = 1.1841$ and $L = 0.1753$. The integer *PID* parameters are then calculated according to the formulae given in Table 2.1.

In the next experiments, a velocity step input of amplitude 40 rad/s is applied to the closed-loop servo system (Fig. 2.4) and the angular velocity *versus* time is acquired for different types of fractional *PID* controllers. The obtained experimental responses are then presented and analyzed.

Figures 2.7—2.10 illustrate the velocity responses of the experimental system with the D^{μ}, I^{λ}, PI^{λ} and $PI^{\lambda}D$ controllers, respectively, for several values of derivative order μ and integrative order λ. In all the cases, the gains K_P, K_D and K_I are fixed and given by the following rules:

- For the D^{μ} and I^{λ} controllers we have used the proportional gain of the integer *P* controller obtained from application of the Z-N rules (see Table 2.1), that is, $K_D = K_I = 1/RL = 0.0361$.
- For the PI^{λ} we have used the parameters of the integer *PI* obtained from application of the Z-N rules (see Table 2.1), that is, $K_P = 0.9/RL = 0.0325$ and $K_I = 0.3K_p/L = 0.0556$.
- For the $PI^{\lambda}D$ we have used the parameters of the integer *PID* obtained from application of the Z-N rules (see Table 2.1), that is, $K_P = 1.2/RL = 0.0433$, $K_I = K_p/(2L) = 0.1235$ and $K_D = 0.5LK_P = 0.0038$.

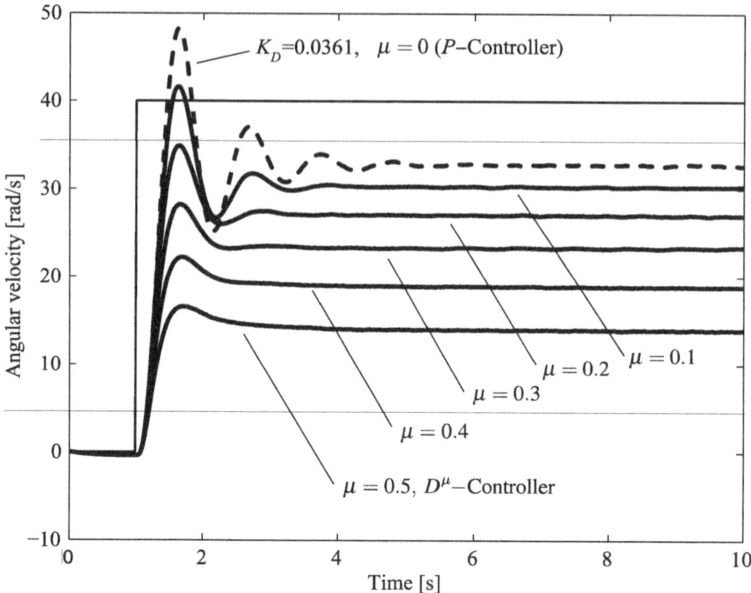

Fig. 2.7 Velocity response of the real-system with the D^{μ} controller and $\mu = \{0, 0.1, 0.2, 0.3, 0.4, 0.5\}$.

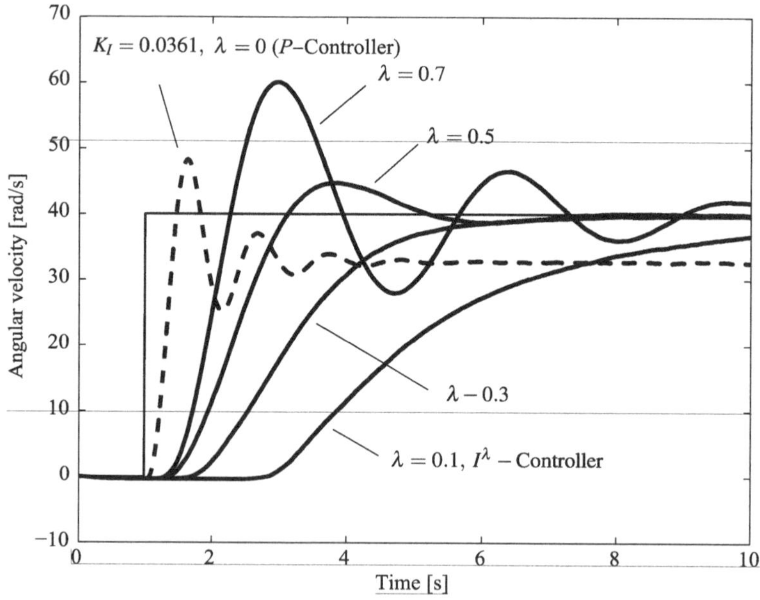

Fig. 2.8 Velocity response of the real-system with the I^{λ} controller and $\lambda = \{0, 0.1, 0.3, 0.5, 0.7\}$.

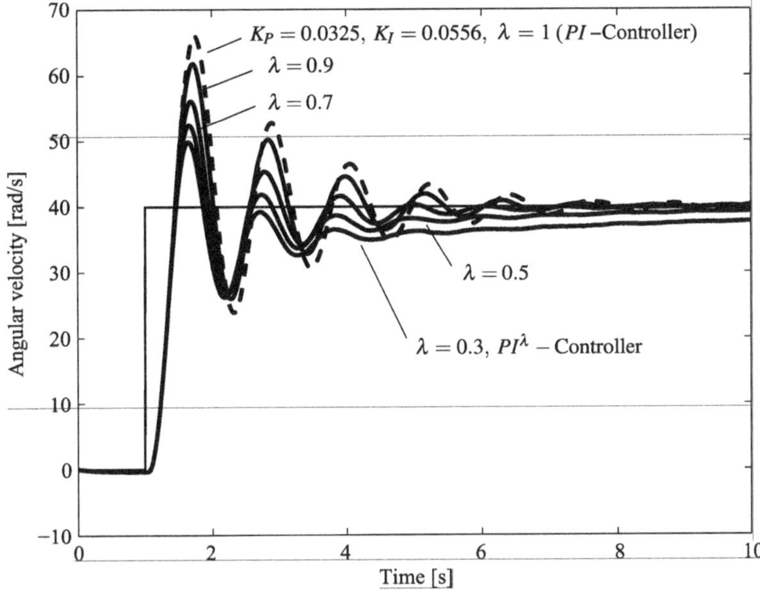

Fig. 2.9 Velocity response of the real-system with the PI^{λ} controller and λ={0.3,0.5,0.7,0.9,1}.

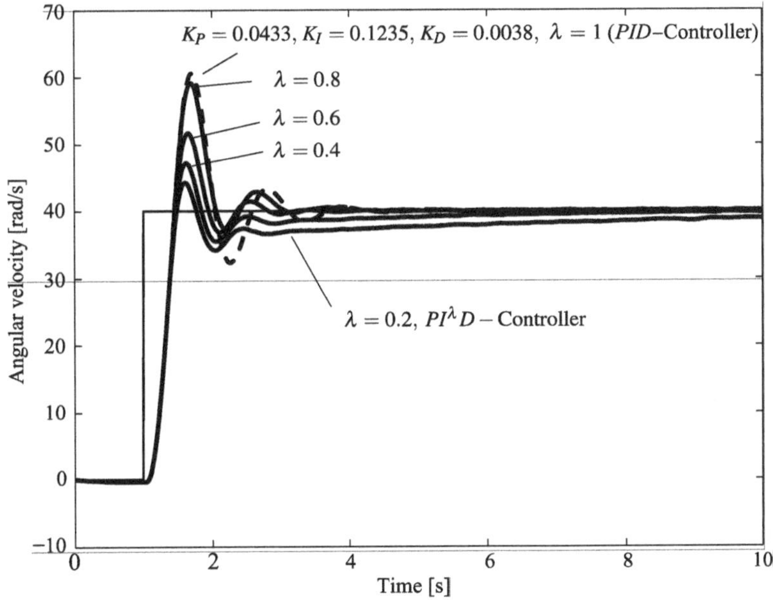

Fig. 2.10 Velocity response of the real-system with the $PI^{\lambda}D$ controller and $\lambda = \{0.2, 0.4, 0.6, 0.8, 1\}$.

2.7.2 Ziegler-Nichols tuning rules: oscillatory behavior

In the second method, a P controller is applied to the velocity servo system until the system shows non-decaying oscillations, as shown in Fig. 2.11. The ultimate gain and period yield $K_u = 0.08$ and $P_u = 0.74$ s, respectively. The controller parameters are then calculated according to the Ziegler-Nichols rules illustrated in Table 2.2. Once the values of T_I and T_D have been obtained, the gains K_I and K_D are computed by using formulae (2.30).

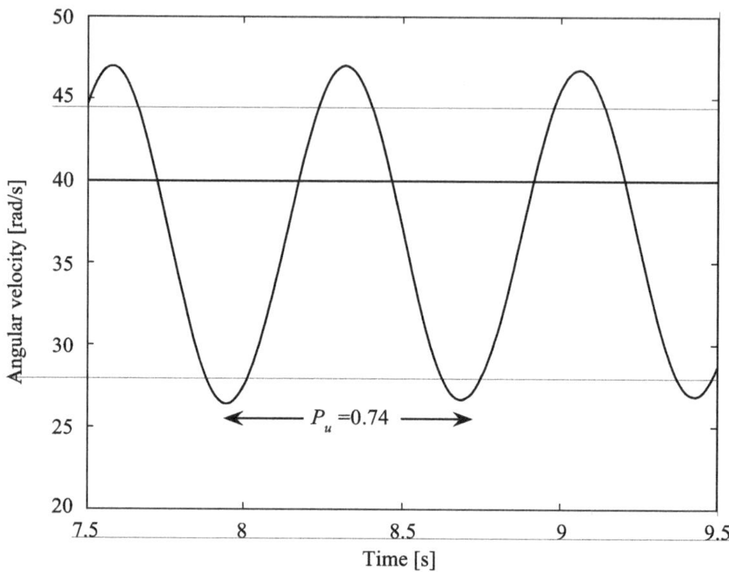

Fig. 2.11 Ultimate gain $K_u = 0.08$ and ultimate period $P_u = 0.74$s.

Table 2.2 Ziegler-Nichols tuning for controller $G_c(s) = K_P(1 + 1/T_I s + T_D s)$ based on oscillatory behavior.

Type of controller	K_P	T_I	T_D
P	$0.5K_u$	∞	0
PI	$0.45K_u$	$\frac{1}{1.2}P_u$	0
PID	$0.6K_u$	$\frac{1}{2}P_u$	$\frac{1}{8}P_u$

Once more, a velocity step input of amplitude 40 rad/s is applied to the closed-loop servo system (Fig. 2.4) and the angular velocity *versus* time is acquired for different types of fractional *PID* controllers. The obtained experimental responses are then presented and analyzed.

Figures 2.12—2.14 illustrate the velocity responses of the experimental system with the D^{μ}, I^{λ} and PI^{λ} controllers, respectively, for several values of derivative order μ and integrative order λ. In all the cases, the gains K_P, K_D and K_I are fixed and given by the following rules:

- For the D^{μ} and I^{λ} controllers we have used the proportional gain of the integer P controller obtained from application of the Z-N rules (see Table 2.2), that is, $K_D = K_I = 0.5K_u = 0.04$.
- For the PI^{λ} we have used the parameters of the integer PI obtained from application of the Z-N rules (see Table 2.2), that is, $K_P = 0.45K_u = 0.0364$ and $K_I = 1.2K_P/P_u = 0.0590$.

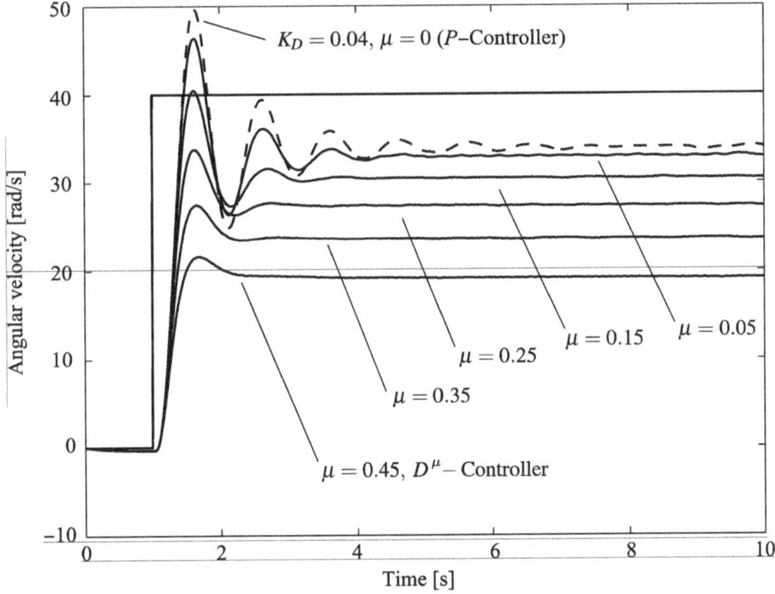

Fig. 2.12 Velocity response of the real-system with the D^{μ} controller and $\mu = \{0, 0.05, 0.15, 0.25, 0.35, 0.45\}$.

2.7.3 Comments on the results

From the analysis of previous Figs. 2.7—2.10 and 2.12—2.14, we conclude that the application of both Z-N tuning methods leads to similar results. However, the velocity responses differ from the application of different fractional PID controllers.

In fact, Figs. 2.7 and 2.12 reveal that the steady-sate error increases as the order μ of the D^{μ} controller increases. The variation of the gain K_D was also tested (with

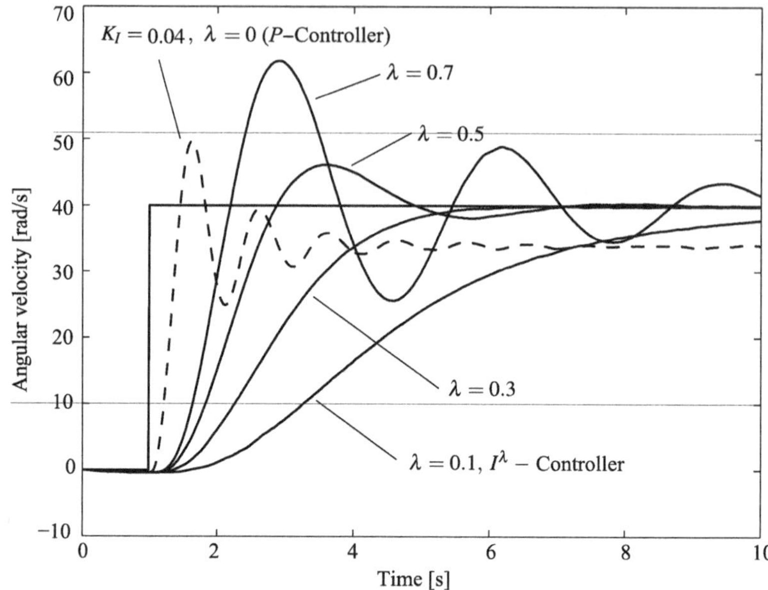

Fig. 2.13 Velocity response of the real-system with the I^λ controller and $\lambda = \{0.1, 0.3, 0.5, 0.7, 1\}$.

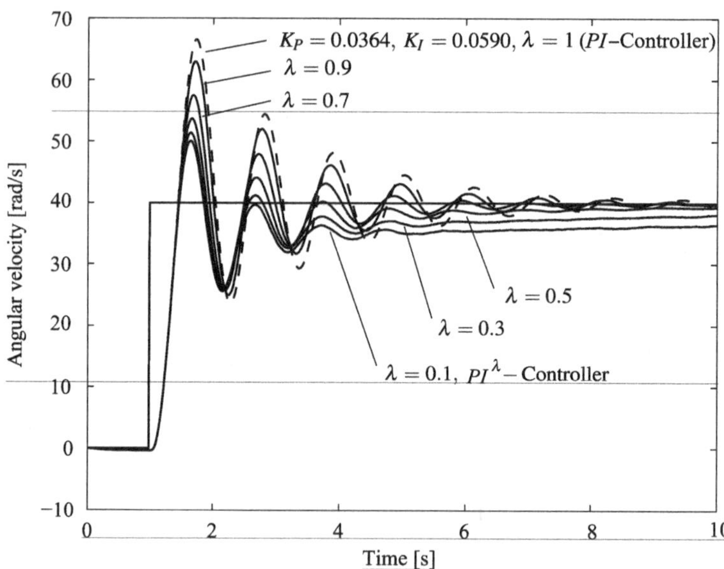

Fig. 2.14 Velocity response of the real-system with the PI^λ controller and $\lambda = \{0.1, 0.3, 0.5, 0.7, 0.9, 1\}$.

a fixed value of the derivative order) and, as expected, the system showed a diminishing steady-state error as K_D increases. However, the overshoot and settling time are more acceptable for the case where the order μ is changed. We verify that the extra degree of tuning provided by the fractional algorithm, in comparison to the classical P controller, may be useful to yield a satisfactory control. In Figs. 2.8 and 2.13, we observe that the steady-state error due to the use of an I^{λ} controller is very small. We must note that the real system is nonlinear and, therefore, the oscillations are damped very quickly. Once more, we verify that the fractional order λ is a very useful parameter for adjusting the dynamics of the control system. In fact, the order λ has a large influence upon the system dynamics, as illustrated in Figs. 2.8 and 2.13. As in previous case, Figs. 2.9, 2.10 and 2.14 show that the steady-state error is very small and that the order λ has a large influence in the overshoot and settling time of the system. An adequate phase margin can be easily established by a proper choice of λ. However, the output converges to its final value more slowly, as should be expected by a weak fractional integral term.

2.8 A simple analytical method for tuning fractional-order controllers

In his work on design of feedback amplifiers, Bode suggested an ideal shape of the loop transfer function of the form (Bode, 1945; Astrom, 2000):

$$L(s) = \left(\frac{\omega_c}{s}\right)^{\alpha} \tag{2.31}$$

where ω_c is the gain crossover frequency and α is an arbitrary non-integer value. Bode called (2.31) the ideal cut-off characteristic, but in the terminology of automatic control it is best known as the Bode's ideal loop transfer function. The slope α is typical positive. In this study we consider $1 < \alpha < 2$.

The interpretation of loop transfer function (2.31) in the frequency domain is very simple. The Bode diagram of amplitude is simply a straight line of slope -20α dB/dec, while the phase curve is a horizontal line positioned at $-\alpha\pi/2$ rad (see Fig. 2.15). The Nyquist curve is a straight line through the origin with $\arg L(j\omega) = -\alpha\pi/2$ rad.

This choice of $L(s)$ gives a closed-loop system with the desirable property of being insensitive to gain changes. If the gain changes, the crossover frequency ω_c will change, but the phase margin of the system remains PM $= \pi(1-\alpha/2)$ rad, independently of the value of the gain (see Fig. 2.15). The gain margin is infinite. The slopes $\alpha = 1.333$, 1.5 and 1.667 correspond to phase margins of PM $= 60°$, $45°$ and $30°$, respectively.

The transfer function (2.31) is irrational for non-integer values of α. Therefore, for its practical implementation, it will be approximated by a rational transfer function of type (2.16), that is, with a recursive set of poles and zeros over a specified

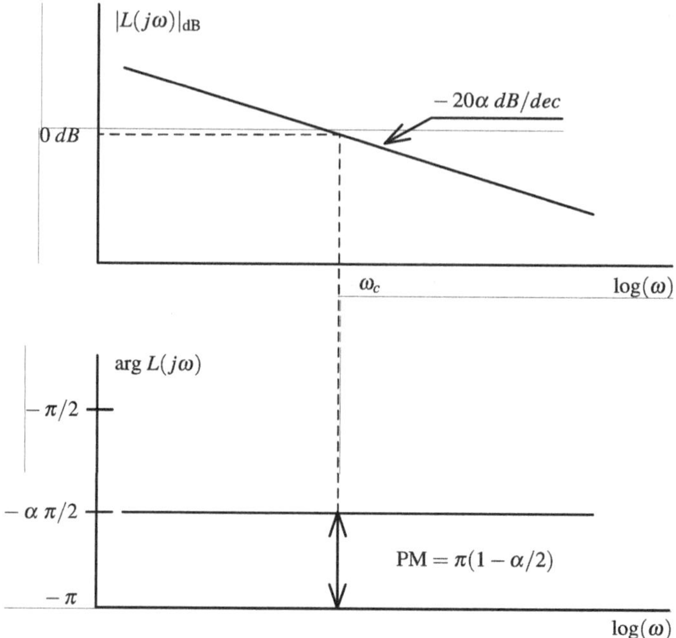

Fig. 2.15 Bode diagrams of amplitude and phase of $L(j\omega)$ for $1 < \alpha < 2$.

band of frequencies. Bode suggested that it is sufficient to approximate $L(s)$ over a frequency range of interest around the crossover frequency ω_c.

Several studies for tuning fractional, as well integer, controllers have been proposed based on Bode's ideal loop transfer function (Barbosa et al., 2004b; Karimi et al., 2002a,b; Barbosa et al., 2003, 2004a; Chen et al, 2003; Chen and More, 2005; Djouambi et al., 2008). Barbosa et al, (2003, 2004a,b) suggested the use of the Bode's ideal shape as reference function for the tuning of integer *PID* controllers. They verified that specifying a desired gain crossover frequency and the slope at that frequency (which is equivalent to defining a specific phase margin), and minimizing a performance criterion like the integral of square error (ISE), can ensure that the phase around the gain crossover frequency is nearly flat. Assuring this feature, we then obtain closed-loop systems more robust to gain variations and step responses exhibiting an almost iso-damping property. Adopting a similar approach, Y.Q. (Chen et al, 2003; Chen and More, 2005) proposed a *PID* tuning method for a class of unknown, stable and minimum phase plants. They designed a *PID* controller to ensure that the phase plot is flat at a given frequency called "tangent frequency" so that the closed-loop system is robust to gain variations and that the step responses exhibit an iso-damping property. With this method, no plant models are assumed during the *PID* controller design; only several relay tests are needed. More recently, Djouambi et al. (2008) proposed a method for tuning a controller that guarantees that the open-loop transfer function of a unity feedback control system will be the Bode's ideal transfer function. Like the case of the above cited works, the idea of the proposed controller is to assure a flat phase around the crossover frequency so that the closed-loop system is robust to gain variations and the step response exhibits an iso-damping property. The proposed method is based on the rational approximation of the fractional-order operators (Charef et al., 1992), which can somehow be manipulated to easily control the width and the position of the flat phase to achieve the desired performances.

2.8.1 The proposed analytical tuning method

Here we outline a simple analytical method for tuning fractional-order controllers based on Bode's ideal loop transfer function (Barbosa et al., 2004b; Astrom, 2000).

Let us consider again the mathematical model of the velocity servo system:

$$G(s) = \frac{K}{Ts+1}. \tag{2.32}$$

Since $L(s) = C(s)G(s)$, the controller transfer function that gives the Bode's ideal loop transfer function (2.31) has the form:

$$C(s) = K_P \frac{Ts+1}{s^{\alpha}}, \tag{2.33}$$

which can be expressed as

$$C(s) = K_P \left(\frac{1}{s^{\alpha}} + Ts^{1-\alpha} \right). \tag{2.34}$$

As can be seen by expression (2.34), the transfer function is, in fact, a $I^{\lambda}D^{\mu}$ controller of fractional integration λ and fractional derivation μ (where $\mu = 1-\lambda$).

The time-domain equation of the controller $C(s)$ is:

$$u(t) = K_P \left(D_t^{-\alpha} e(t) + TD_t^{1-\alpha} e(t) \right). \tag{2.35}$$

The open-loop transfer function $L(s) = C(s)G(s)$ is then given by:

$$L(s) = \frac{K_p K}{s^{\alpha}}, \qquad 1 < \alpha < 2. \tag{2.36}$$

The phase margin ϕ_m of the system and the controller's gain can be calculated through the relations:

$$\phi_m = \pi + \arg L(j\omega_c) = \pi \left(1 - \frac{\alpha}{2}\right), \tag{2.37}$$

$$|L(j\omega_c)| = 1 \;\Rightarrow\; K_p = \frac{\omega_c^{\alpha}}{K}, \tag{2.38}$$

where ω_c is the gain crossover frequency. So, given a specified phase margin, from (2.37) we get the fractional order α and, given the gain crossover frequency, we get from relation (2.38) the controller gain K_P.

Therefore, the parameters of the fractional controller (2.33) are tuned to attain two specifications: the desired phase margin ϕ_m, which provides the adequate overshoot of the system; and the gain crossover frequency ω_c, which defines the desired speed of response of the system.

The design procedure can be outlined as follows:

1. Find the fractional order α by using formula (2.37) from the desired phase margin ϕ_m;
2. Calculate the proportional gain K_P by using formula (2.38) from the gain crossover frequency ω_c and the nominal gain of the system process K.

As an example, the closed-loop system should satisfy the following specifications:

1. Phase margin $\phi_m = 60°$;
2. Gain crossover frequency $\omega_c = 1$ rad/s.

The parameters of the servo system obtained from the identification experiment of Section 2.5 are $K = 203.5344$ and $T = 0.8258$ s. Following the above design procedure, the parameters of the FOC are $\alpha = 1.333$ and $K_P = 0.0049$. So, the transfer function of the fractional controller is:

$$C(s) = 0.0049\frac{0.8258s+1}{s^{1.333}}. \tag{2.39}$$

Once again, a velocity step input of amplitude 40 rad/s is applied to the closed-loop servo system with fractional controller (2.39) and the angular velocity *versus* time is acquired for several different design specifications. Next, the obtained experimental responses are presented and analyzed.

Figure 2.16 shows the experimental velocity step responses of the servo system for the phase margins $\phi_m = \{30°,\ 45°,\ 60°,\ 80°\}$ and the same gain crossover frequency $\omega_c = 1$ rad/s while the corresponding control signals are shown in Fig. 2.17. In Fig. 2.18 we show the experimental velocity step responses of the servo system for $\phi_m = 45°$ and $\omega_c = \{0.5,\ 0.75,\ 1,\ 1.5\}$ rad/s. Figure 2.19 illustrates the corresponding control signals.

From the results, we verify that the system behaves like the desired specifications. That is, with the phase margin we change mainly the overshoot of the closed-loop response and with the crossover frequency we change the speed of response, although, in this case, the overshoot is also slightly changed. Therefore, we prove the effectiveness of the simple analytical method for tuning FOCs and its applicability in the velocity control of a servo system.

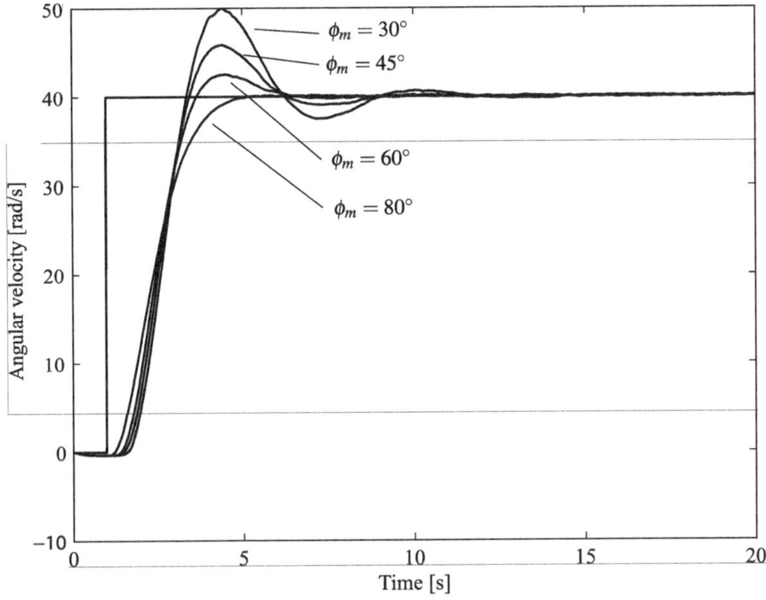

Fig. 2.16 Velocity response of the real-system with the fractional controller (2.39) and $\phi_m = \{30°, 45°, 60°, 80°\}$.

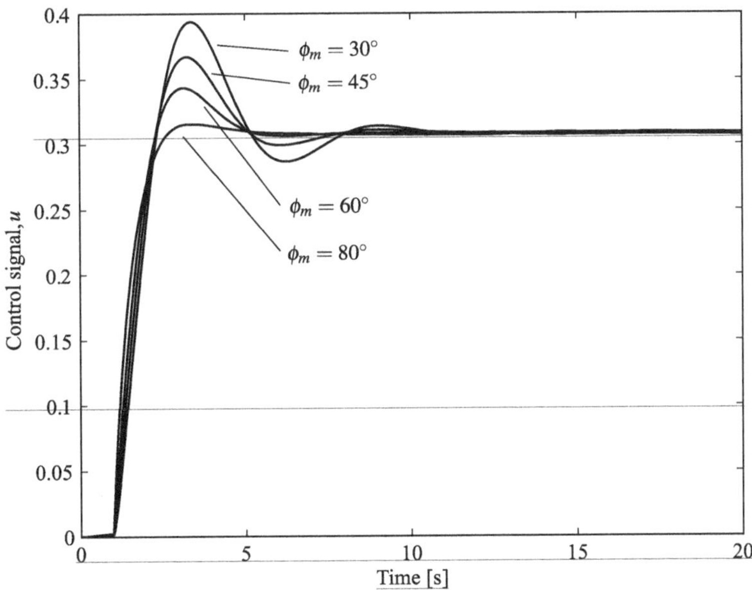

Fig. 2.17 Control signal of the real-system with the fractional controller (2.39) and $\phi_m = \{30°, 45°, 60°, 80°\}$.

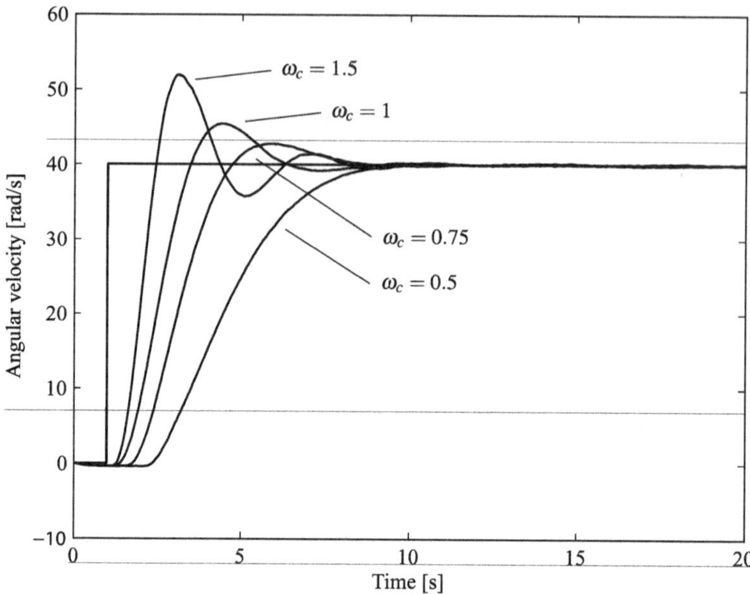

Fig. 2.18 Velocity response of the real-system with the fractional controller (2.39) and $\omega_c = \{0.5, 0.75, 1, 1.5\}$ rad/s.

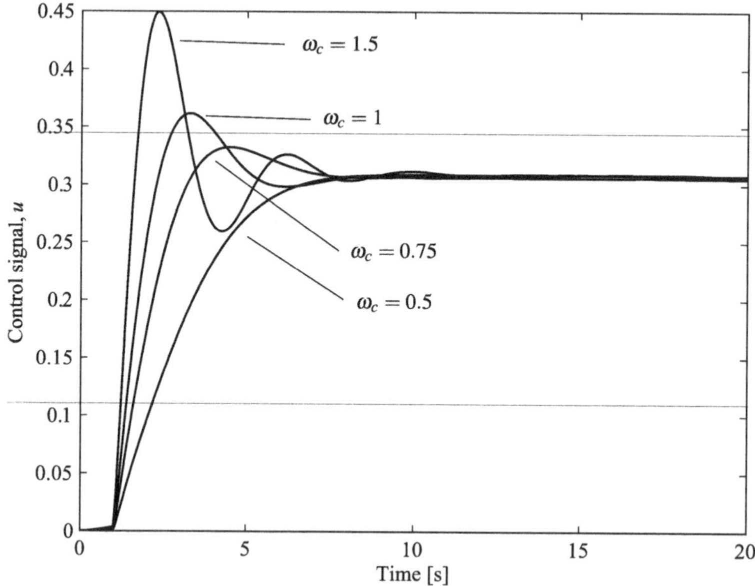

Fig. 2.19 Control signal of the real-system with the fractional controller (2.39) and $\omega_c = \{0.5, 0.75, 1, 1.5\}$ rad/s.

2.9 Application of optimal fractional-order controllers

In recent years, the number of works dedicated to the design of optimal fractional-order controllers has been increasing (Leu et al., 2002; Xue et al., 2006; Jesus et al., 2006b, 2007a; Machado et al., 2006; Jesus and Machado, 2007; Jesus et al., 2007b; Jesus and Machado, 2008b; Jesus et al., 2008b). Leu et al. (2002) proposed the design of a fractional *PID* controller by taking both time and frequency domain specifications. The controller parameters and the fractional orders of the fractional *PID* controller are determined to minimize an integral square error (ISE) performance index while satisfying the specified gain and phase margins. A comparison between optimal fractional *PID* and optimal integer *PID* algorithms is presented for controlling integer as well fractional processes. The design examples show that the performance of the control system can significantly be improved by using a fractional *PID*. Xue et al. (2006) investigated the use of a fractional *PID* for a position servomechanism control system considering actuator saturation and the shaft torsional flexibility. Extensive simulations of the position servomechanism controlled by optimal integer *PID*/*PI* and optimal fractional *PID*/*PI* algorithms are presented to illustrate the superior robustness of the fractional scheme. It was shown that the best fractional *PID* works better than the best integer *PID*. Jesus et al. (2006b, 2007a) analyzed the heat diffusion system in the perspective of fractional calculus by applying a conventional *PID* controller. The *PID* parameters are tuned by using the so-called Ziegler-Nichols open loop (ZNOL) method (Jesus and Machado, 2008a; Machado et al., 2006). However, the poor results indicated that the method of tuning might not be the most adequate for the control of the heat system. Therefore, in (Jesus et al., 2007a,b) they proposed the use of FOCs tuned through the minimization of the ISE, ITSE, IAE and ITAE indices. The results demonstrate the effectiveness of the FOCs when used for the control of fractional systems. More recently, in (Jesus et al., 2008b) a nonlinear controller (NLC) with a fractional model was presented and compared with other algorithms. The results reveal the superior performance of the NLC based on the fractional algorithm, namely in the dynamics of systems of non-integer order.

In this section we will apply several optimal fractional/integer *PID* controllers to the velocity control of the servo system. The presence of the phenomena of saturation and backlash in the servo system are analyzed also in this study. For that, we consider again the servo (real-time) feedback control system illustrated in Fig. 2.4. The controller is given by a *PID*/$PI^{\lambda}D$ controller and the nonlinearity (Figure 2.20) is described by equation:

$$n(m) = \begin{cases} m, & |m| < \delta, \\ \delta \operatorname{sign}(m), & |m| \geq \delta, \end{cases} \tag{2.40}$$

where sign(m) is the signal function.

The controller is tuned by the minimization of an integral performance index. For that purpose, we analyze the indices that measure the response error, namely the integral absolute error (IAE) and the integral time absolute error (ITAE) criteria,

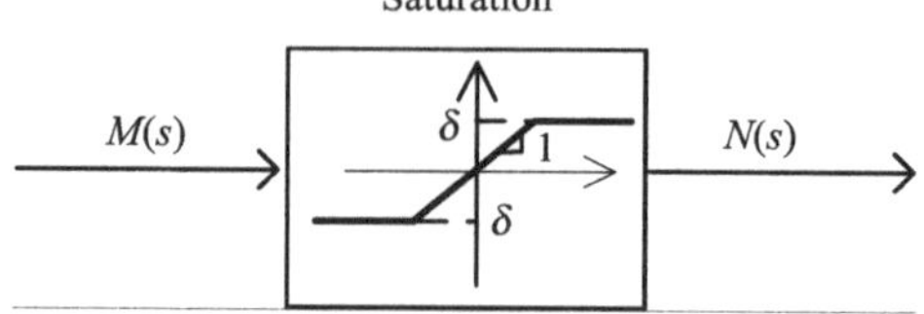

Fig. 2.20 Non-linearity of the saturation type ($\delta = [-1, 1]$), inserted in the closed-loop system.

defined as:

$$\mathrm{IAE} = \int_0^{\infty} |r(t) - c(t)|\, dt, \tag{2.41}$$

$$\mathrm{ITAE} = \int_0^{\infty} t|r(t) - c(t)|\, dt, \tag{2.42}$$

where $r(t)$ is the input reference to the system and $c(t)$ the corresponding response of the closed-loop system.

We can use other performance criteria, such as the integral square error (ISE) or the integral time square error (ITSE); however, in the present case, the IAE and the ITAE criterion produce the best results and are adopted in the study.

Another possible performance index consists on the energy E_n at the controller output $n(t)$, given by the expression:

$$E_n = \int_0^{T_e} n^2(t)dt, \tag{2.43}$$

where T_e is the time window needed to stabilize the system output $c(t)$.

2.9.1 Tuning of the PID and PI$^\lambda$ D controllers

In this work we compare the performances of two controller architectures, namely an integer *PID* and a FOC of type $PI^\lambda D$, when controlling the servo system without or with backlash for the IAE and ITAE indices.

Once more, a velocity step input of amplitude 40 rad/s is applied to the closed-loop servo system with the optimal integer/fractional *PID* controllers and the angular velocity *versus* time is acquired.

The first step in the work consisted in tuning both control schemes in the perspective of optimizing the IAE and ITAE indices.

Table 2.3 presents the results of tuning both control algorithms, when minimizing the IAE index, and using the servo system without the presence of backlash. In this table is depicted the optimum value of the fractional order λ for the integrative term of the $PI^\lambda D$ ($\lambda = 0.8$), the minimum values of the IAE and the controller energy and the corresponding parameters (K_P, K_I, K_D) for the $PI^\lambda D$ and *PID* controllers. Also, it is given some specifications that characterize the system time response, namely:

Table 2.3 Optimum values for the integer *PID* and fractional $PI^{\lambda}D$ controllers, and corresponding values of the IAE, controller parameters (K_P, K_I, K_D) and time specifications (PO, t_r, t_s) for the system without backlash.

λ	IAE	E_n	K_P	K_I	K_D	PO (%)	t_r	t_s
1.00	34.0668	2.8269	0.0998	0.0375	0.0027	35.1405	0.27	8.19
0.80	17.8305	1.8528	0.0397	0.0348	0.0041	24.7334	0.28	2.93

- The percentual overshoot, PO;
- The rise time, t_r;
- And the settling time, t_s.

Analysing this table we conclude that the $PI^{\lambda}D$ scheme leads not only to lower values of the index IAE and of the controller energy E_n, but also to a better transient response with lower values of PO and t_s.

Table 2.4 presents the results of tuning both control algorithms, when minimizing the IAE index, and using the servo system in the case of having backlash. In this table is presented the optimum value of λ of the $PI^{\lambda}D$, the minimum values of the IAE, the controller energy E_n, the controller parameters (K_P, K_I, K_D) and the specifications (PO, t_r, t_s) for the $PI^{\lambda}D$ and *PID* controllers.

Table 2.4 Optimum values for the integer *PID* and fractional $PI^{\lambda}D$ controllers, and corresponding values of the IAE, controller parameters (K_P, K_I, K_D) and time specifications (PO, t_r, t_s) for the system with backlash.

λ	IAE	E_n	K_P	K_I	K_D	PO (%)	t_r	t_s
1.00	30.4017	2.8335	0.0998	0.0375	0.0027	33.6425	0.28	7.01
0.80	18.1299	2.0250	0.0397	0.0348	0.0041	23.4787	0.29	2.88

It is possible to conclude that, as in the previous case of the system with backlash, the $PI^{\lambda}D$ scheme, with $\lambda = 0.8$, leads not only to lower values for the IAE and for the controller energy E_n, but also to a faster transient response having lower values of the PO and t_s.

The previous tuning procedure is now repeated considering the minimization of the ITAE index.

Table 2.5 presents the results of tuning both algorithms, when minimizing the ITAE, and using the servo system without the presence of backlash. In this table is presented the optimum value of λ of the $PI^{\lambda}D$, the minimum values of the ITAE, the controller energy E_n, the controller parameters (K_P, K_I, K_D) and the specifications (PO, t_r, t_s) for the $PI^\lambda D$ and PID controllers.

Table 2.5 Optimum values for the integer PID and fractional $PI^\lambda D$ controllers, and corresponding values of the ITAE, controller parameters (K_P, K_I, K_D) and time specifications (PO, t_r, t_s) for the system without backlash.

λ	ITAE	E_n	K_P	K_I	K_D	PO (%)	t_r	t_s
1.00	62.6920	2.2555	0.0671	0.0373	0.0015	40.4646	0.27	4.90
0.85	39.8773	1.6899	0.0243	0.0353	0.0062	16.5204	0.40	3.35

Analyzing the values presented in Table 2.5 we conclude that, like in the previous cases of the IAE with or without the backlash, the $PI^\lambda D$ scheme, with $\lambda = 0.85$, leads to lower values for the ITAE and for the controller energy E_n, and also gives a faster transient response with lower values for PO and t_s.

Finally, Table 2.6 presents the results of tuning both algorithms, when minimizing the ITAE, and using the servo system in the case of having backlash. In this table is presented the optimum value of λ of the $PI^\lambda D$, the minimum values of the ITAE, the controller energy E_n, the controller parameters (K_P, K_I, K_D) and the specifications (PO, t_r, t_s) for the $PI^\lambda D$ and PID controllers.

Table 2.6 Optimum values for the integer PID and fractional $PI^\lambda D$ controllers, and corresponding values of the ITAE, controller parameters (K_P, K_I, K_D) and time specifications (PO, t_r, t_s) for the system with backlash.

λ	ITAE	E_n	K_P	K_I	K_D	PO (%)	t_r	t_s
1.00	57.2717	2.3740	0.0671	0.0373	0.0015	37.7394	0.28	4.98
0.85	39.9524	1.9431	0.0243	0.0353	0.0062	16.0887	0.41	3.36

Observing the values presented in Table 2.6 we can conclude again that the $PI^\lambda D$ scheme, with $\lambda = 0.85$, leads to lower values for the ITAE and for the controller energy E_n, and (as previously) gives also a faster transient response with lower values for PO and t_s.

Comparing the results presented in Tables 2.3–2.6 we conclude that the $PI^\lambda D$ controller (with $\lambda \in [0.8, 0.85]$) minimizes the controller energy E_n when controlling the servo system with or without the presence of backlash. Furthermore, it also leads to a faster transient response with lower values of time specifications PO and t_s for both situations.

Comparing the results in the tables, we may conclude that when the control algorithms are tuned using the minimization of the ITAE, both the controller energy and the percentual overshoot are lower, when adopting the $PI^{\lambda}D$. In this case, it is then preferable to tune the control algorithm by minimizing the ITAE. However, these results are obtained at the cost of an increase in the values of t_r and t_s.

The same conclusions can be drawn analyzing Figs. 2.21 – 2.28. Figs. 2.21 and 2.22 show the velocity step responses of the closed-loop system, for the integer PID tuned in the IAE and the ITAE perspectives, respectively, without or with backlash. We note that the responses are very similar, if we look at specifications PO and t_s for both situations.

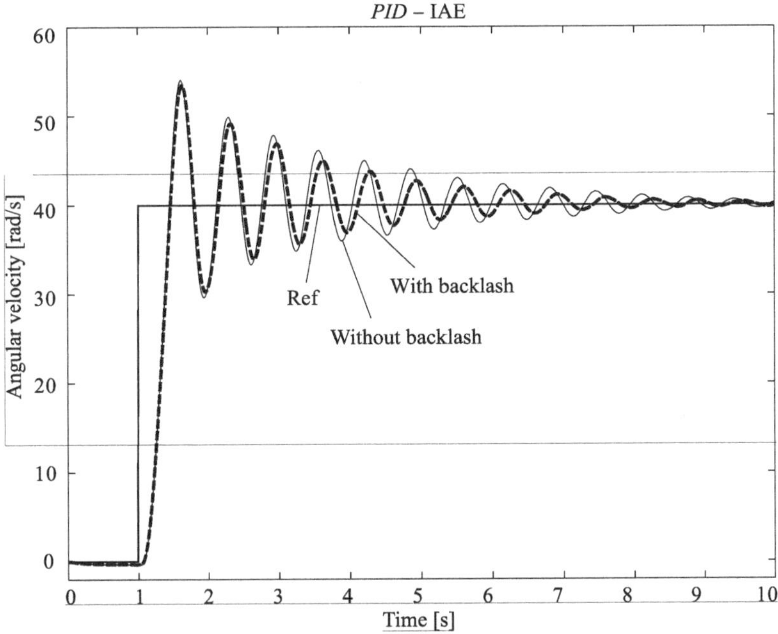

Fig. 2.21 Comparison of the velocity response without or with backlash, under the action of a *PID* controller, tuned in the viewpoint of the minimization of the IAE.

Figures 2.23 and 2.24 illustrate the velocity step responses of the closed-loop system, for the $PI^{\lambda}D$ tuned in the IAE and the ITAE perspectives, respectively, without or with backlash. From these figures, we observe that when the controller is tuned from the viewpoint of minimizing the IAE the step response show higher PO and longer t_s. Also, in both cases, we note that the responses are almost equal, independently of the fact of the system including backlash or not.

Figures 2.25 and 2.26 show the velocity step responses of the closed-loop system, for the integer PID and the fractional $PI^{\lambda}D$ controllers tuned in the IAE and

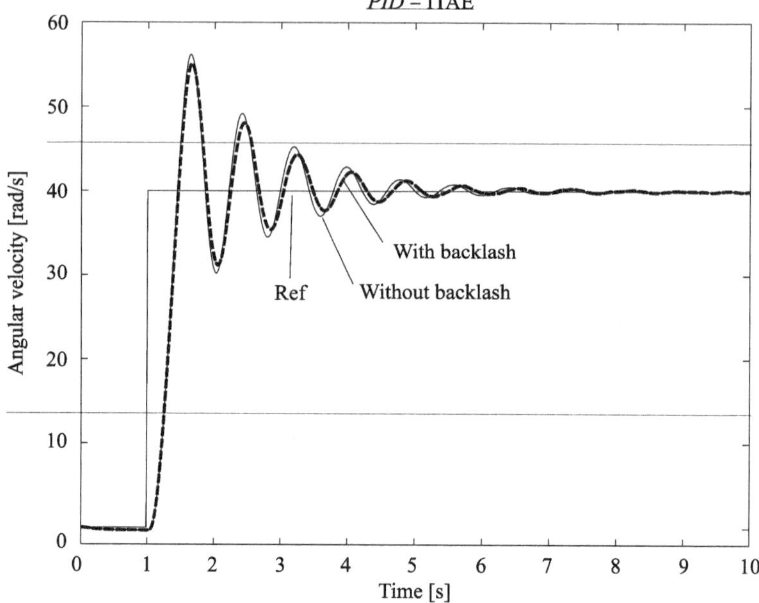

Fig. 2.22 Comparison of the velocity response without or with backlash, under the action of a *PID* controller, tuned in the viewpoint of the minimization of the ITAE.

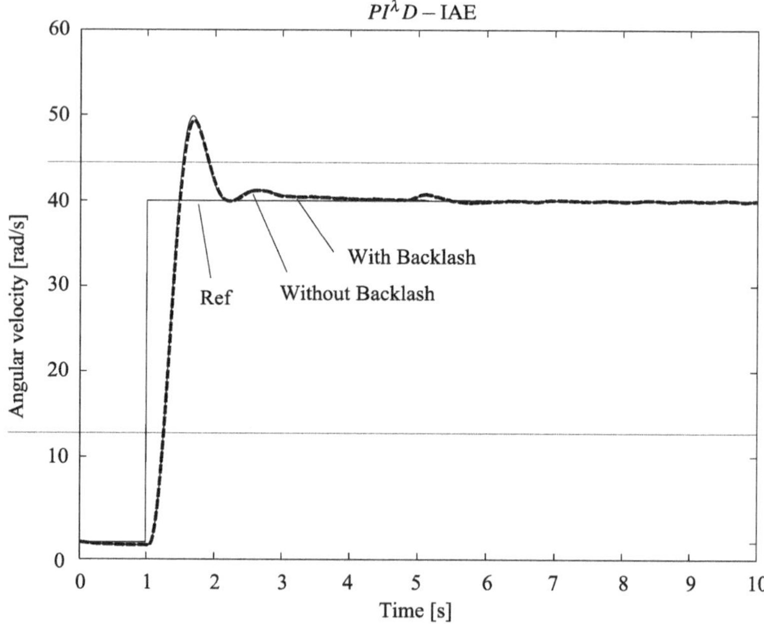

Fig. 2.23 Comparison of the velocity response without or with backlash, under the action of a $PI^{\lambda}D$ controller, tuned in the viewpoint of the minimization of the IAE.

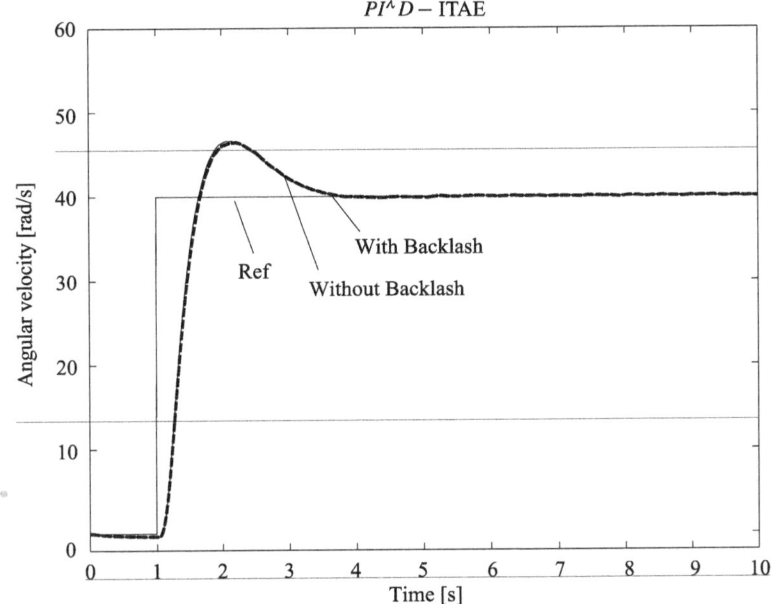

Fig. 2.24 Comparison of the velocity response without or with backlash, under the action of a $PI^{\lambda}D$ controller, tuned in the viewpoint of the minimization of the ITAE.

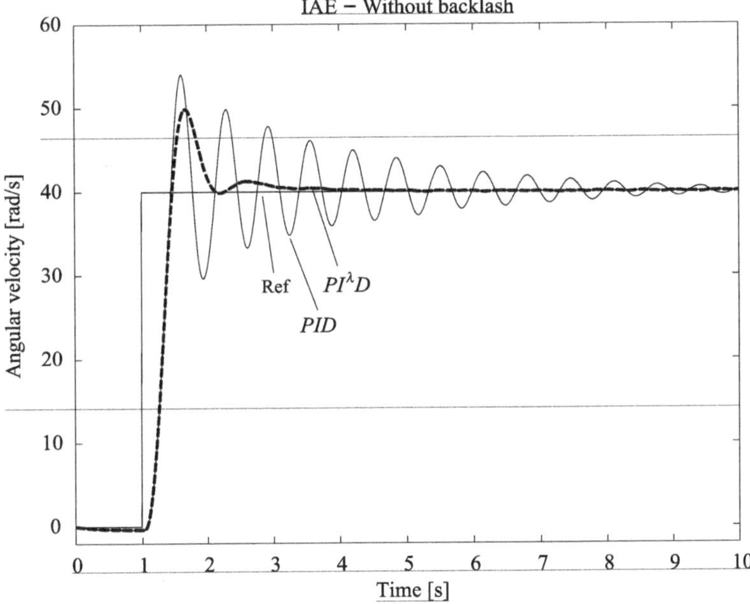

Fig. 2.25 Comparison of the velocity response, under the action of a *PID* and $PI^{\lambda}D$ controller, tuned in the viewpoint of the minimization of the IAE, when the system is backlash free.

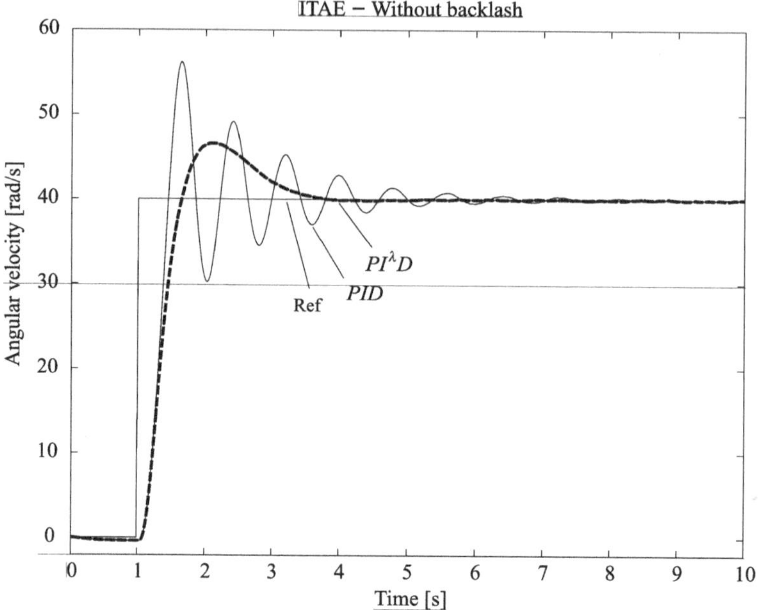

Fig. 2.26 Comparison of the velocity response, under the action of a *PID* and $PI^{\lambda}D$ controller, tuned in the viewpoint of the minimization of the ITAE, when the system is backlash free.

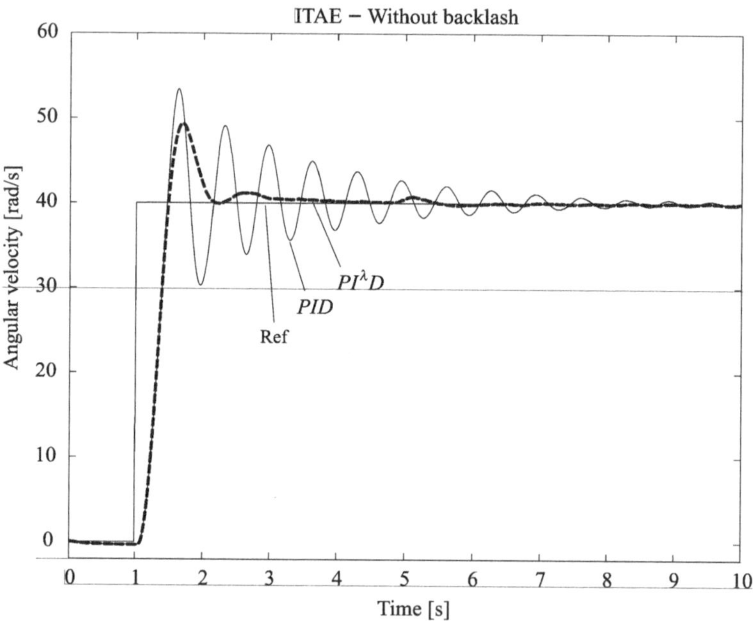

Fig. 2.27 Comparison of the velocity response, under the action of a *PID* and $PI^{\lambda}D$ controller, tuned in the viewpoint of the minimization of the IAE, when the system includes backlash.

the ITAE perspectives, respectively, without system backlash. We conclude that the $PI^{\lambda}D$ controller presents a better transient response, namely it reveals a lower PO and smaller t_s. These results are independent of the criteria followed to tune the control scheme.

Figures 2.27 and 2.28 illustrate the velocity step responses of the closed-loop system, for the integer PID and the fractional $PI^{\lambda}D$ controllers tuned in the IAE and the ITAE perspectives, respectively, when the system includes backlash. As previously observed in Figures 2.25 and 2.26, we conclude also that the $PI^{\lambda}D$ controller presents a better transient response, revealing a lower PO and smaller t_s. In the same way, these results are independent of the indices minimized to tune the control scheme.

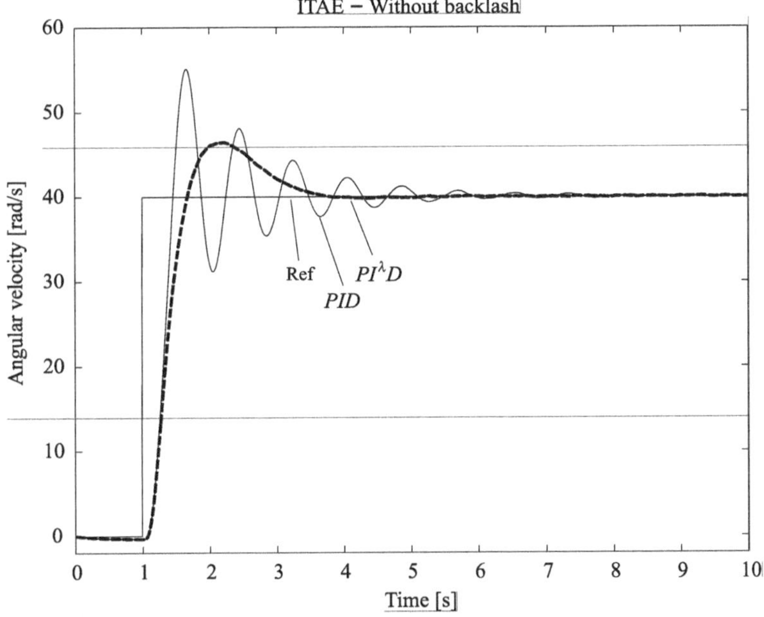

Fig. 2.28 Comparison of the velocity response, under the action of a PID and $PI^{\lambda}D$ controller, tuned in the viewpoint of the minimization of the ITAE, when the system includes backlash.

2.10 Conclusions

In this chapter we investigated the velocity control of a servo system by using several fractional PID controllers and tuning methods:

- First, we adapted the well-known Ziegler-Nichols rules for the tuning of fractional PID controllers. It was shown that the FOCs can effectively enhance the control system performance providing extra tuning parameters useful for the adjustment of the control system dynamics. The Ziegler-Nichols rules revealed to be simple and effective in the final tuning of the FOCs;

- Second, a simple analytical tuning method was devised and used to tune a FOC. With this method it is possible to attain two design specifications: one to establish the overshoot of the closed-loop system, and the other to define the speed of response of the system. The FOC tuned by this method was applied to the servo system proving to be effective;
- Third, several optimum fractional *PID* controllers were proposed for the control of the servo system. The controllers were tuned by using the IAE and ITAE performance indices. With the ITAE the advantage of using FOCs is more evident, revealing better transient response and smaller controller energy. Furthermore, it was shown that the system with the FOCs performs better than the integer *PID* without or with the presence of backlash.

References

Adda F.B., 1997, Geometric interpretation of the fractional derivative, *Journal of Fractional Calculus*, **11**, 21-52.

Astrom K.J., 2000, Limitations on control system performance, *European Journal of Control*, **6**, 1-19.

Al-Alaoui M.A., 1993, Novel digital integrator and diferentiator, *Electronics Letters*, **29**, 376–378.

Al-Alaoui M.A., 1997, Filling the gap between the bilinear and the backward-difference transforms: An interactive design approach, *Int. J. Elect. Eng. Educ.*, **34**, 331–337.

Axtell M. and Bise M.E., 1990, Fractional calculus applications in control systems, In: *Proceedings of the IEEE 1990 National Aerospace and Electronics Conference*, New York, 563-566.

Barbosa R.S., Machado J.A.T. and Ferreira I.M., 2003, A fractional calculus perspective of PID tuning, In: *Proceedings of the ASME International 19th Biennial Conference on Mechanical Vibration and Noise (VIB)*, Chicago, Illinois, USA.

Barbosa R.S., Machado J.A.T. and Ferreira I.M., 2004a, PID controller tuning using fractional calculus concepts, *FCAA-Journal of Fractional Calculus & Applied Analysis*, **7**, 119-134.

Barbosa R.S., Machado J.A.T. and Ferreira I.M., 2004b, Tuning of PID controllers based on Bode's ideal transfer function, *Nonlinear Dynamics*, **38**, 305-321.

Barbosa R.S., Machado J.A.T. and Silva M.F., 2006, Time domain design of fractional differintegrators using least-squares, *Signal Processing*, **86**, 2567-2581.

Barbosa R.S., Machado J.A.T. and Jesus I.S., 2008a, On the fractional PID control of a laboratory servo system, In: *Proceedings of the IFAC'08 – 17th IFAC World Congress*, Seoul, Korea, 15273-15278.

Barbosa R.S., Machado J.A.T. and Silva M.F., 2008b, Fractional PID control of an experimental servo system, In: *Proceedings of the FDA'08 – Third IFAC Workshop on Fractional Differentiation and Its Applications*, Ankara, Turkey.

Bode H.W., 1945, *Network Analysis and Feedback Amplifier Design*, Van Nostrand, New York.

Carlson G.E. and Halijak C.A., 1964, Approximation of fractional capacitors $(1/s)^{1/n}$ by a regular Newton process. *IRE Transactions on Circuit Theory*, **CT-11**, 210-213.

Carpinteri A. and Mainardi F., 1997, *Fractals and Fractional Calculus in Continuum Mechanics*, CISM Courses and Lectures No. 378, Springer, Wien-New York.

Charef A., Sun H.H., Tsao Y.Y. and Onaral B., 1992, Fractal system as represented by singularity function, *IEEE Transactions on Automatic Control*, **37**, 1465-1470.

Chen Y.Q. and Moore K.L., 2002, Discretization schemes for fractional-order differentiators and integrators, *IEEE Transactions on Circuits and Systems I: Fundamental Theory and Applications*, **49**, 363-367.

Chen Y.Q., Vinagre B.M. and Podlubny I., 2004, Continued fraction expansion approaches to discretizing fractional order derivatives: An expository review, *Nonlinear Dynamics*, **38**, 155-170.

Chen Y., Hu C.H. and More K.L., 2003, Relay feedback tuning of robust PID controllers with iso-damping property, In: *Proceedings of the 42nd IEEE Conference on Decision and Control*, Maui, Hawaii, USA, 2180-2185.

Chen Y. and More K.L., 2005, Relay Feedback Tuning of Robust PID Controllers with Iso-Damping Property, *IEEE Transactions on Systems, Man, and Cybernetics-Part B: Cybernetics*, **35**, 23-31.

Djouambi A., Charef A. and Besancon A., 2008, Fractional order robust controller based on Bode's ideal transfer function, *JESA - Journal Européen des Systèmes Automatisés*, **42**, 999-1014.

Feliu-Batlle V., Pérez R.R. and Rodríguez L.S., 2007, Fractional robust control of main irrigation canals with variable dynamic parameters, *Control Engineering Practice*, **15**, 673-686.

Franklin G.F., Powel J.D. and Emami-Naeini A., 2006, *Feedback Control of Dynamic Systems*, 5th edition, Addison-Wesley, Reading, MA.

Hilfer R., 2000, *Aplications of Fractional Calculus in Physics*, World Scientific, Singapore.

Inteco M., 2006, *Modular Servo System (MSS): User's Manual*, Krakow, Poland.

Jesus I.S. and Machado J.A.T., 2007, Application of fractional calculus in the control of heat systems, *Journal of Advanced Computational Intelligence and Intelligent Informatics*, **11**, 1086-1091.

Jesus I.S. and Machado J.A.T., 2008a, Development of fractional order capacitors based on electrolyte processes, *Journal Nonlinear Dynamics*, **56**, 45-55.

Jesus I.S. and Machado J.A.T., 2008b, Fractional control of heat diffusion systems, *Journal Nonlinear Dynamics*, **54**, 263-282.

Jesus I.S. and Machado J.A.T. , 2009, Implementation of fractional-order electromagnetic potential through a genetic algorithm, *Journal of Communications in Nonlinear Science and Numerical Simulation*, **14**, 1838-1843.

Jesus I.S., Machado J.A.T., Cunha J.B. and Silva M.F., 2006a, Fractional order electrical impedance of fruits and vegetables, In: *Proceedings of the MIC'06 - 25th IASTED International Conference on Modelling, Identification and Control*, Spain, 489-494.

Jesus I.S., Barbosa R.S., Machado J.A.T. and Cunha J.B., 2006b, Strategies for the control of heat diffusion systems based on fractional calculus, In: *IEEE Int. Conf. on Computational Cybernetics*, Estonia.

Jesus I.S., Machado J.A.T. and Cunha J.B., 2007a, Fractional dynamics and control of heat diffusion systems, In: *Proceedings of the MIC'07 – IASTED International Conference on Modelling, Identification and Control*, Austria.

Jesus I.S., Machado J.A.T., Barbosa R.S. and Pires E.S., 2007b, Towards the PID^{β} control of heat diffusion systems, In: *Seventh International Conference on Intelligent Systems Design and Applications*, Brazil.

Jesus I.S., Machado J.A.T. and Cunha J.B., 2007c, Fractional electrical impedances in botanical elements, *Journal of Vibration and Control - Special issue Fractional Differentiation and its Applications*, **14**, 1389-1402.

Jesus I.S., Machado J.A.T. and Silva M.F., 2008a, Fractional order capacitors, In: *Proceedings of the MIC'08 – 27th IASTED International Conference on Modelling, Identification and Control*, Austria.

Jesus I.S., Machado J.A.T. and Barbosa R.S., 2008b, Fractional order nonlinear control of heat system, In: *Proceedings of the FDA'08 – Third IFAC Workshop on Fractional Differentiation and Its Applications*, Ankara, Turkey.

Karimi A., Garcia D. and Longchamp R., 2002a, PID controller design using Bode's integrals, In: *Proceedings of the ACC'02 -American Control Conference*, Anchorage, AK, USA, 5007-5012.

Karimi A., Garcia D. and Longchamp R., 2002b, Iterative controller tuning using Bode's integrals, In: *Proceedings of the CDC'02 – 41st IEEE Conference on Decision and Control*, Las Vegas, NV, USA, 4227-4232.

Leu J.-F., Tsay S.-Y. and Hwang C., 2002, Design of optimal fractional-order PID controllers, *J. Chin. Inst. Chem. Engrs.*, **33**, 193-202.

Ma C. and Hori Y., 2003, Design of robust fractional order $PI^{\alpha}D$ speed control for two-inertia system, In: *Proceedings of the Japan Industry Applications Society Conference*, Tokyo, Japan, 1-4.

Machado J.A.T., 1997, Analysis and design of fractional-order digital control systems, *SAMS Journal of Systems Analysis, Modelling and Simulation*, **27**, 107-122.

Machado J.A.T., 1999, Fractional-order derivative approximations in discrete-time control systems, SAMS Journal of *Systems Analysis, Modelling and Simulation*, **34**, 419-434.

Machado J.A.T., 2003, A probabilistic interpretation of the fractional-order differentiation, *FCAA Fractional Calculus & Applied Analysis*, **6**, 73-80.

Machado J.A.T., 2009, Fractional derivatives: probability interpretation and frequency response of rational approximations, *Communications in Nonlinear Science and Numerical Simulations*, **14**, 3492-3497.

Machado, J.A.T., Jesus I.S., Cunha J.B. and Tar J.K., 2006, Fractional dynamics and control of distributed parameter systems, In: *Intelligent Systems at the Service of Mankind*, Vol. 2, 295-305.

Machado J.T., Jesus I., Galhano A., Cunha J.B. and Tar J.K., 2007, Electrical skin phenomena: a fractional calculus analysis, In: *Advances in Fractional Calculus: Theoretical Developments and Applications in Physics and Engineering*, 323-332.

Manabe S., 1961, The non-integer integral and its application to control systems, *ETJ of Japan*, **6**, 83-87.

Manabe S., 1963, The system design by the use of a model consisting of a saturation and non-integer integrals, *ETJ of Japan*, **8**, 147-150.

Miller K.S. and B. Ross, 1993, *An Introduction to the Fractional Calculus and Fractional Differential Equations*, Wiley & Sons, New York.

Monsrefi-Torbati M. and Hammond J.K., 1998, Physical and geometrical interpretation of fractional operators, *Journal of Franklin Institute*, **335B**, 1077-1086.

Nigmatullin R.R., 1992, A fractional integral and its physical interpretation, *Theoret. and Math. Phys.*, **90**, 242-251.

Oldham K.B. and Spanier J., 1974, *The Fractional Calculus*, Academic Press, New York.

Oustaloup A., 1991, *La Command CRONE: Commande Robuste d'Ordre Non Entier*, Editions Hermès, Paris.

Oustaloup A., 1995, *La Dérivation Non Entière: Théorie, Synthèse et Applications*, Editions Hermès, Paris.

Oustaloup A., Levron F., Mathieu B. and Nanot F.M., 2000, Frequency-band complex noninteger differentiator: characterization and synthesis, *IEEE Transactions on Circuits and Systems-I: Fundamental Theory and Applications*, **47**, 25-39.

Podlubny I., 1999a, *Fractional Differential Equations*, Academic Press, San Diego.

Podlubny I., 1999b, Fractional-Order systems and $PI^{\lambda}D^{\mu}$-controllers, *IEEE Transactions on Automatic Control*, **44**, 208-214.

Podlubny I., 2002, Geometric and physical interpretation of fractional integration and fractional differentiation, *FCAA Fractional Calculus & Applied Analysis*, **5**(4), 367-386.

Rutman R.S., 1994, On the paper by R.R. Nigmatullin: A fractional integral and its physical interpretation, *Theoret. Math. Phys.*, **100**, 1154-1156.

Rutman R.S., 1995, On physical interpretations of fractional integration and differentiation, *Theoret. Math. Phys.*, **105**, 393-404.

Sabatier J., Poullain S., Latteux P. and Oustaloup A., 2004, Robust speed control of a low damped electromechanical system based on CRONE control: application to a four mass experimental test bench, *Nonlinear Dynamics*, **38**, 383-400.

Samko S.G., Kilbas A.A. and Mariche O. I., 1993, *Fractional Integrals and Derivatives: Theory and Applications*, Gordon and Breach Science, Amsterdam.

Silva M.F. and Machado J.A.T., 2006, Fractional order PD^{α} joint control of legged robots, *Journal of Vibration and Control – Special Issue on Modeling and Control of Artificial Locomotion Systems*, **12**, 1483-1501.

Silva M.F. and Machado J.A.T., 2008, Kinematic and dynamic performance analysis of artificial legged systems, *ROBOTICA*, **26**, 19-39.

Silva M.F., Machado J.A.T. and Barbosa R.S., 2008, Using fractional derivatives in joint control of hexapod robots, *Journal of Vibration and Control - Special Issue on Fractional Differentiation and its Applications*, **14**, 1473-1485.

Tustin A., Allanson J.T., Layton J.M. and Jakeways R.J., 1958, Position control of massive objects, *The Proceedings of the Institution of Electrical Engineers*, **105**(1–suppl., part C), 1-57.

Valério D. and da Costa J.S., 2004, Non-integer order control of a flexible robot, In: *Proceedings of the First IFAC Workshop on Fractional Differentiation and Its Applications*, Bordeaux, France, 520-525.

Valério D. and da Costa J.S., 2006a, Tuning rules for integer PIDs that make them look like fractional, In: *Proceedings of the CONTROLO'06 – 7th Portuguese Conference on Automatic Control*, Lisbon, Portugal.

Valério J. and da Costa S., 2006b, Tuning of fractional PID controllers with Ziegler-Nichols-Type rules, *Signal Processing*, **86**, 2771-2784.

Vinagre B.M., Podlubny I., Hernández A. and Feliu V., 2000, Some approximations of fractional order operators used in control theory and applications, *FCAA Fractional Calculus and Applied Analysis*, **3**, 231-248.

Vinagre B. M., Chen Y. Q. and Petras I., 2003, Two direct tustin discretization methods for fractional-order differentiator/integrator, *Journal of the Franklin Institute* **340**, 349–362.

Westerlund S., 2002, *Dead Matter Has Memory*, Causal Consulting, Kalmar, Sweden.

Xue D., Zhao C. and Chen Y.Q., 2006, Fractional order PID control of a DC-motor with elastic shaft: A case study, In: *Proceedings of the ACC'06 American Control Conference*, Minneapolis, Minnesota, USA.

Ziegler J.G. and Nichols N.B., 1942, Optimum settings for automatic controllers, *Transactions of the ASME*, **64**, 759-765.

Chapter 3
Differential-Delay Equations

Richard Rand

Abstract Periodic motions in DDE (Differential-Delay Equations) are typically created in Hopf bifurcations. In this chapter we examine this process from several points of view. Firstly we use Lindstedt's perturbation method to derive the Hopf Bifurcation Formula, which determines the stability of the periodic motion. Then we use the Two Variable Expansion Method (also known as Multiple Scales) to investigate the transient behavior involved in the approach to the periodic motion. Next we use Center Manifold Analysis to reduce the DDE from an infinite dimensional evolution equation on a function space to a two dimensional ODE (Ordinary Differential Equation) on the center manifold, the latter being a surface tangent to the eigenspace associated with the Hopf bifurcation. Finally we provide an application to gene copying in which the delay is due to an observed time lag in the transcription process.

3.1 Introduction

Some dynamical processes are modeled as differential-delay equations (abbreviated DDE). An example is

$$\frac{dx(t)}{dt} = -x(t-T) - x(t)^3. \tag{3.1}$$

Here the rate of growth of x at time t is related both to the value of x at time t, and also to the value of x at a previous time, $t-T$.

Applications of DDE include laser dynamics (Wirkus and Rand, 2002), where the source of the delay is the time it takes light to travel from one point to another; machine tool vibrations (Kalmar-Nagy et al., 2001), where the delay is due to the

---

Richard Rand
Cornell University, Ithaca, NY 14853, USA,
Email: rrand@cornell.edu

dependence of the cutting force on thickness of the rotating workpiece; gene dynamics (Verdugo and Rand, 2008a), where the delay is due to the time required for messenger RNA to copy the genetic code and export it from the nucleus to the cytoplasm; investment analysis (Kot, 1979), where the delay is due to the time required by bookkeepers to determine the current state of the system; and physiological dynamics (Camacho et al., 2004), where the delay comes from the time it takes a substance to travel via the bloodstream from one organ to another.

A generalized version of Eq.(3.1) is

$$\frac{dx(t)}{dt} = \alpha x(t) + \beta x(t-T) + f(x(t), x(t-T)) \tag{3.2}$$

where α and β are coefficients and f is a strictly nonlinear function of $x(t)$ and $x(t-T)$. Here the linear terms $\alpha x(t)$ and $\beta x(t-T)$ have been separated from the strictly nonlinear terms, a step which facilitates stability analysis.

3.2 Stability of equilibrium

Equation (3.2) has the trivial equilibrium solution $x(t) = 0$. Is it stable? In order to find out, we linearize Eq.(3.2) about $x = 0$:

$$\frac{dx(t)}{dt} = \alpha x(t) + \beta x(t-T). \tag{3.3}$$

Since Eq.(3.3) has constant coefficients, we look for a solution in the form $x = e^{\lambda t}$, which gives the characteristic equation:

$$\lambda = \alpha + \beta e^{-\lambda T}. \tag{3.4}$$

Equation (3.4) is a transcendental equation and will in general have an infinite number of roots, which will either be real or will occur in complex conjugate pairs. The equilibrium $x = 0$ will be stable if all the real parts of the roots are negative, and unstable if any root has a positive real part. In the intermediate case in which no roots have positive real part, but some roots have zero real part, the linear stability analysis is inadequate, and nonlinear terms must be considered.

As an example, we consider Eq.(3.1), for which Eq.(3.4) becomes

$$\lambda = -e^{-\lambda T}. \tag{3.5}$$

Since λ will be complex in general, we set $\lambda = \nu + i\omega$, where ν and ω are the real and imaginary parts. Substitution into Eq.(3.5) gives two real equations:

$$\nu = -e^{-\nu T} \cos \omega T, \tag{3.6}$$

$$\omega = e^{-\nu T} \sin \omega T. \tag{3.7}$$

The question of stability will depend upon the value of the delay parameter T. Certainly when $T = 0$ the system is stable. By continuity, as T is increased from zero, there will come a first positive value of T for which $x = 0$ is not (linearly) stable. This can happen in one of two ways. Either a single real root will pass through the origin in the complex-λ plane, or a pair of complex conjugate roots will cross the imaginary axis. Since ν=ω=0 does not satisfy Eqs.(3.6) and (3.7), the first case cannot occur.

In order to consider the second case of a purely imaginary pair of roots, we set $\nu = 0$ in Eqs.(3.6) and (3.7), giving

$$0 = -\cos \omega T, \tag{3.8}$$

$$\omega = \sin \omega T. \tag{3.9}$$

Equation (3.8) gives ωT=$\pi/2$, whereupon Eq.(3.9) gives $\omega = 1$, from which we conclude that the critical value of delay T= T_{cr}=$\pi/2$. That is, **$\mathbf{x = 0}$ in Eq.(3.1) is stable for $\mathbf{T < \pi/2}$ and is unstable for $\mathbf{T > \pi/2}$.** Stability for T= T_{cr}=$\pi/2$ requires consideration of nonlinear terms.

In order to check these results we numerically integrate Eq.(3.1) using the MATLAB package DDE23. Note that this requires that the values of x be given on the entire interval $-T \leq t \leq 0$. Figures 3.1 and 3.2 show the results of numerical integration using the initial condition $x = 0.01$ on $-T \leq t \leq 0$. Figure 3.1 is for $T = \pi/2 - 0.01$ and shows stability, while Fig. 3.2 is for $T = \pi/2 + 0.01$ and shows instability, in agreement with the foregoing analysis.

3.3 Lindstedt's method

The change in stability observed in the preceding example will be accompanied by the birth of a limit cycle in a Hopf bifurcation. In order to obtain an approximation for the amplitude and frequency of the resulting periodic motion, we use Lindstedt's method (Rand and Armbruster, 1987; Rand, 2005).

We begin by stretching time,

$$\tau = \omega t. \tag{3.10}$$

Replacing t by τ as independent variable, Eq.(3.1) may be written in the form:

$$\omega \frac{dx(\tau)}{d\tau} = -x(\tau - \omega T) - x(\tau)^3. \tag{3.11}$$

Next we choose the delay T to be close to the critical value T_{cr}=$\pi/2$:

$$T = \frac{\pi}{2} + \Delta. \tag{3.12}$$

We introduce a perturbation parameter $\varepsilon \ll 1$ by scaling x:

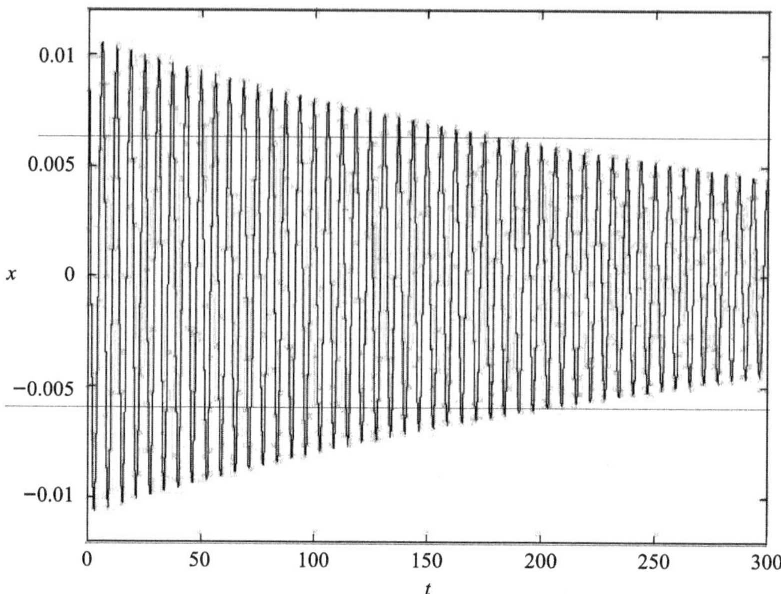

Fig. 3.1 Numerical integration of Eq.(3.1) for the initial condition $x = 0.01$ on $-T \leq t \leq 0$, for $T = \pi/2 - 0.01$.

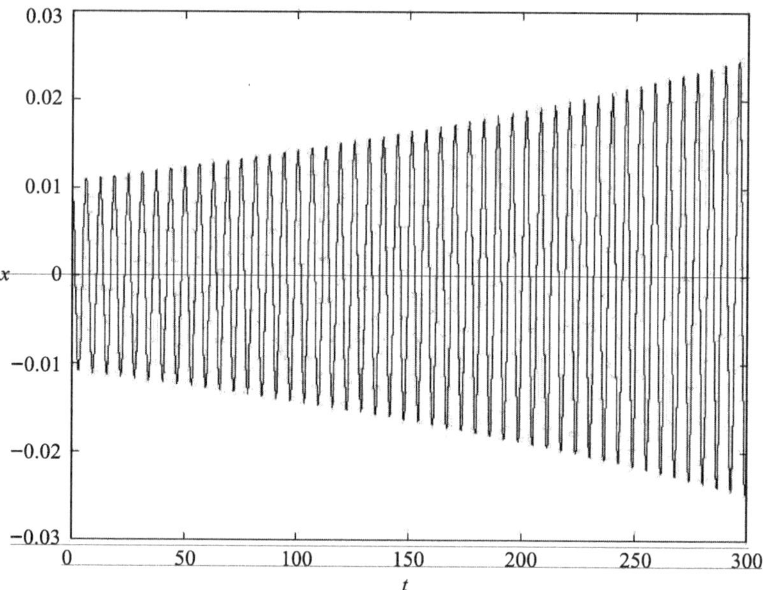

Fig. 3.2 Numerical integration of Eq.(3.1) for the initial condition $x = 0.01$ on $-T \leq t \leq 0$, for $T = \pi/2 + 0.01$.

$$x = \sqrt{\varepsilon} u. \tag{3.13}$$

Using Eq.(3.13), Eq.(3.11) becomes:

$$\omega \frac{du(\tau)}{d\tau} = -u(\tau - \omega T) - \varepsilon u(\tau)^3. \tag{3.14}$$

Next we scale Δ

$$\Delta = \varepsilon \mu, \tag{3.15}$$

and we expand u and ω in power series of ε:

$$u(\tau) = u_0(\tau) + \varepsilon u_1(\tau) + O(\varepsilon^2), \tag{3.16}$$
$$\omega = 1 + \varepsilon k_1 + O(\varepsilon^2), \tag{3.17}$$

where we have used the fact that ω=1 when T=T_{cr}.

The delay term $u(\tau - \omega T)$ is handled by expanding it in Taylor series about ε=0:

$$u(\tau - \omega T) = u\left(\tau - (1 + \varepsilon k_1 + O(\varepsilon^2))(\frac{\pi}{2} + \varepsilon \mu)\right) \tag{3.18}$$
$$= u\left(\tau - \frac{\pi}{2} - \varepsilon(k_1 \frac{\pi}{2} + \mu) + O(\varepsilon^2)\right) \tag{3.19}$$
$$= u(\tau - \frac{\pi}{2}) - \varepsilon(k_1 \frac{\pi}{2} + \mu)\frac{du}{d\tau}(\tau - \frac{\pi}{2}) + O(\varepsilon^2). \tag{3.20}$$

Substituting into Eq.(3.14) and collecting terms gives:

$$\varepsilon^0 : \quad \frac{du_0}{d\tau} + u_0(\tau - \frac{\pi}{2}) = 0, \tag{3.21}$$
$$\varepsilon^1 : \quad \frac{du_1}{d\tau} + u_1(\tau - \frac{\pi}{2}) = -k_1 \frac{du_0}{d\tau} + (k_1 \frac{\pi}{2} + \mu)\frac{du_0}{d\tau}(\tau - \frac{\pi}{2}) - u_0^3. \tag{3.22}$$

Since Eq.(3.1) is autonomous, we may choose the phase of the periodic motion arbitrarily. This permits us to take the solution to Eq.(3.21) as:

$$u_0(\tau) = A \cos \tau, \tag{3.23}$$

where A is the approximate amplitude of the periodic motion. Substituting Eq.(3.23) into (3.22), we obtain:

$$\frac{du_1}{d\tau} + u_1(\tau - \frac{\pi}{2}) = k_1 A \sin \tau + (k_1 \frac{\pi}{2} + \mu) A \cos \tau - A^3 \left(\frac{3}{4}\cos \tau + \frac{1}{4}\cos 3\tau\right), \tag{3.24}$$

For no secular terms, we equate to zero the coefficients of $\sin \tau$ and $\cos \tau$ on the RHS of Eq.(3.24). This gives:

$$k_1 = 0 \quad \text{and} \quad A = \frac{2}{\sqrt{3}}\sqrt{\mu}. \tag{3.25}$$

Now A is the amplitude in u. In order to obtain the amplitude in x, we multiply the second one of Eqs.(3.25) by $\sqrt{\varepsilon}$, which, together with Eqs.(3.13) and (3.15), gives

$$\text{The Amplitude of periodic motion in Eq.(3.1) is } \frac{2}{\sqrt{3}}\sqrt{\Delta}, \text{namely, } \frac{2}{\sqrt{3}}\sqrt{\left(T-\frac{\pi}{2}\right)}. \tag{3.26}$$

This predicts, for example, that when $T = \pi/2 + 0.01$, the limit cycle born in the Hopf will have approximate amplitude of 0.1155. For comparison, numerical integration gives a value of 0.1145, see Fig. 3.3.

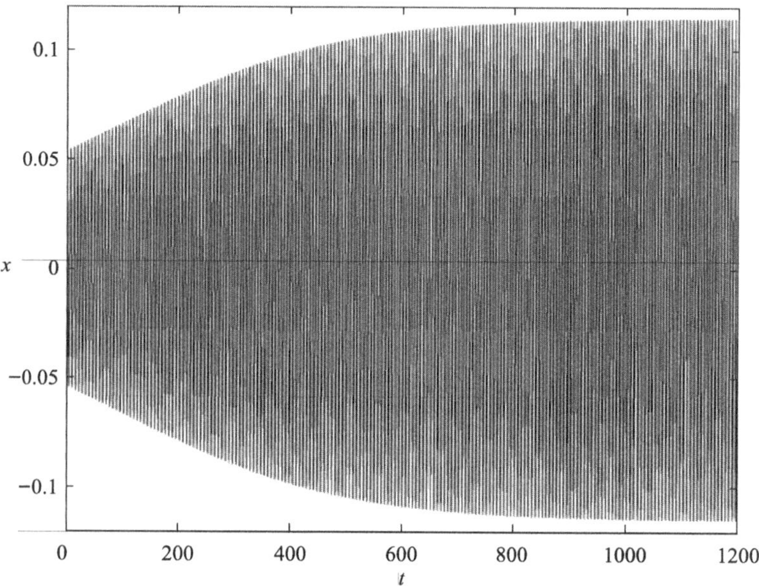

Fig. 3.3 Numerical integration of Eq.(3.1) for the initial condition $x = 0.05$ on $-T \le t \le 0$, for $T = \pi/2 + 0.01$.

3.4 Hopf bifurcation formula

The treatment of the Hopf bifurcation (Rand and Verdugo, 2007) in the previous section for Eq.(3.1) can be generalized to apply to a wide class of DDEs. In this section we present a formula for the amplitude of the resulting limit cycle for the DDE:

$$\frac{dx}{dt} = \alpha x + \beta x_d + a_1 x^2 + a_2 x x_d + a_3 x_d^2 + b_1 x^3 + b_2 x^2 x_d + b_3 x x_d^2 + b_4 x_d^3, \tag{3.27}$$

where $x = x(t)$ and $x_d = x(t-T)$. Here T is the delay. Associated with (3.27) is a linear DDE

$$\frac{dx}{dt} = \alpha x + \beta x_d. \tag{3.28}$$

We assume that (3.28) has a critical delay T_{cr} for which it exhibits a pair of pure imaginary eigenvalues $\pm\omega i$ corresponding to the solution

$$x = c_1 \cos \omega t + c_2 \sin \omega t \tag{3.29}$$

Then for values of delay T which lie close to T_{cr},

$$T = T_{cr} + \mu, \tag{3.30}$$

the nonlinear Eq.(3.27) may exhibit a periodic solution which can be written in the approximate form:

$$x = A \cos \omega t, \tag{3.31}$$

where the amplitude A can be obtained from the following expression for A^2:

$$A^2 = \frac{P}{Q}\,\mu, \tag{3.32}$$

where

$$P = 4\beta^3\,(\beta-\alpha)\,(\beta+\alpha)^2\,(-5\beta+4\alpha), \tag{3.33}$$

$$\begin{aligned}
Q = {} & 15 b_4 \beta^6 T_{cr} + 5 b_2 \beta^6 T_{cr} + 3\alpha b_4 \beta^5 T_{cr} - 15\alpha b_3 \beta^5 T_{cr} + \alpha b_2 \beta^5 T_{cr} \\
& - 15\alpha b_1 \beta^5 T_{cr} - 22 a_3{}^2 \beta^5 T_{cr} - 7 a_2 a_3 \beta^5 T_{cr} - 14 a_1 a_3 \beta^5 T_{cr} - 3 a_2{}^2 \beta^5 T_{cr} \\
& - 7 a_1 a_2 \beta^5 T_{cr} - 4 a_1{}^2 \beta^5 T_{cr} - 12\alpha^2 b_4 \beta^4 T_{cr} - 3\alpha^2 b_3 \beta^4 T_{cr} + 6\alpha^2 b_2 \beta^4 T_{cr} \\
& - 3\alpha^2 b_1 \beta^4 T_{cr} + 12 a_3{}^2 \alpha \beta^4 T_{cr} + 37 a_2 a_3 \alpha \beta^4 T_{cr} + 30 a_1 a_3 \alpha \beta^4 T_{cr} \\
& + 7 a_2{}^2 \alpha \beta^4 T_{cr} + 19 a_1 a_2 \alpha \beta^4 T_{cr} + 18 a_1{}^2 \alpha \beta^4 T_{cr} + 12\alpha^3 b_3 \beta^3 T_{cr} \\
& + 2\alpha^3 b_2 \beta^3 T_{cr} + 12\alpha^3 b_1 \beta^3 T_{cr} + 4 a_3{}^2 \alpha^2 \beta^3 T_{cr} - 20 a_2 a_3 \alpha^2 \beta^3 T_{cr} \\
& - 16 a_1 a_3 \alpha^2 \beta^3 T_{cr} - 12 a_2{}^2 \alpha^2 \beta^3 T_{cr} - 26 a_1 a_2 \alpha^2 \beta^3 T_{cr} - 8 a_1{}^2 \alpha^2 \beta^3 T_{cr} \\
& - 8\alpha^4 b_2 \beta^2 T_{cr} - 4 a_2 a_3 \alpha^3 \beta^2 T_{cr} + 8 a_2{}^2 \alpha^3 \beta^2 T_{cr} + 8 a_1 a_2 \alpha^3 \beta^2 T_{cr} \\
& + 5 b_3 \beta^5 + 15 b_1 \beta^5 - 15\alpha b_4 \beta^4 + \alpha b_3 \beta^4 - 15\alpha b_2 \beta^4 + 3\alpha b_1 \beta^4 - 4 a_3{}^2 \beta^4 \\
& - 9 a_2 a_3 \beta^4 - 18 a_1 a_3 \beta^4 - a_2{}^2 \beta^4 - 9 a_1 a_2 \beta^4 - 18 a_1{}^2 \beta^4 - 3\alpha^2 b_4 \beta^3 \\
& + 6\alpha^2 b_3 \beta^3 - 3\alpha^2 b_2 \beta^3 - 12\alpha^2 b_1 \beta^3 + 26 a_3{}^2 \alpha \beta^3 + 19 a_2 a_3 \alpha \beta^3 \\
& + 30 a_1 a_3 \alpha \beta^3 + 11 a_2{}^2 \alpha \beta^3 + 33 a_1 a_2 \alpha \beta^3 + 12 a_1{}^2 \alpha \beta^3 + 12\alpha^3 b_4 \beta^2 \\
& + 2\alpha^3 b_3 \beta^2 + 12\alpha^3 b_2 \beta^2 - 8 a_3{}^2 \alpha^2 \beta^2 - 32 a_2 a_3 \alpha^2 \beta^2 - 12 a_1 a_3 \alpha^2 \beta^2 \\
& - 14 a_2{}^2 \alpha^2 \beta^2 - 18 a_1 a_2 \alpha^2 \beta^2 - 8\alpha^4 b_3 \beta - 8 a_3{}^2 \alpha^3 \beta + 8 a_2 a_3 \alpha^3 \beta \\
& + 4 a_2{}^2 \alpha^3 \beta + 8 a_2 a_3 \alpha^4
\end{aligned} \tag{3.34}$$

In Eq.(3.32), A is real so that $A^2 > 0$, which means that μ must have the same sign as $\frac{P}{Q}$.

Eq.(3.34) depends on μ, α, β, a_i, b_i and T_{cr}. This equation may be alternately written with T_{cr} expressed as a function of α and β. This relationship may be obtained by considering the linear DDE (3.28). Substituting Eq.(3.31) into Eq.(3.28) and equating to zero coefficients of $\sin(\omega t)$ and $\cos(\omega t)$, we obtain the two equations:

$$\beta \sin(\omega \mathrm{T_{cr}}) = -\omega, \qquad \beta \cos(\omega \mathrm{T_{cr}}) = -\alpha. \tag{3.35}$$

Squaring and adding these we obtain

$$\omega = \sqrt{\beta^2 - \alpha^2}. \tag{3.36}$$

Substituting (3.36) into the second of (3.35), we obtain the desired relationship between T_{cr} and α and β:

$$T_{cr} = \frac{\arccos\left(\frac{-\alpha}{\beta}\right)}{\sqrt{\beta^2 - \alpha^2}}. \tag{3.37}$$

3.4.1 Example 1

As an example, we consider the following DDE:

$$\frac{dx}{dt} = -x - 2x_d - xx_d - x^3. \tag{3.38}$$

This corresponds to the following parameter assignment in Eq.(3.27):

$$\alpha = -1,\ \beta = -2,\ a_1 = a_3 = b_2 = b_3 = b_4 = 0,\ a_2 = b_1 = -1. \tag{3.39}$$

The associated linearized equation (3.28) is stable for zero delay. As the delay T is increased, the origin first becomes unstable when $T = T_{cr}$, where Eq.(3.37) gives that

$$T_{cr} = \frac{\arccos \frac{-1}{2}}{\sqrt{3}} = \frac{2\pi}{3\sqrt{3}}. \tag{3.40}$$

Substituting (3.39) and (3.40) into (3.32), (3.33), (3.34), we obtain:

$$A^2 = \frac{648\mu}{40\sqrt{3}\pi + 171} = 1.667\mu, \tag{3.41}$$

where we have set

$$T = T_{cr} + \mu = \frac{2\pi}{3\sqrt{3}} + \mu = 1.2092 + \mu. \tag{3.42}$$

Thus the origin is stable for $\mu < 0$ and unstable for $\mu > 0$. In order for A^2 in (3.41) to be positive, we require that $\mu > 0$. Therefore the limit cycle is born out of an unstable equilibrium point. Since the stability of the limit cycle must be the opposite of the stability of the equilibrium point from which it is born, we may conclude that the limit cycle is stable and that we have a *supercritical* Hopf. This result is in agreement with numerical integration of Eq.(3.38).

3.4.2 Derivation

In order to derive the result (3.32), (3.33), (3.34), we use Lindstedt's method. We begin by introducing a small parameter ε via the scaling

$$x = \varepsilon u. \tag{3.43}$$

The detuning μ of Eq.(3.30) is scaled like ε^2:

$$T = T_{cr} + \mu = T_{cr} + \varepsilon^2 \hat{\mu}. \tag{3.44}$$

Next we stretch time by replacing the independent variable t by τ, where

$$\tau = \Omega t. \tag{3.45}$$

This results in the following form of Eq.(3.27):

$$\Omega \frac{du}{d\tau} = \alpha u + \beta u_d + \varepsilon(a_1 u^2 + a_2 u u_d + a_3 u_d^2) + \varepsilon^2(b_1 u^3 + b_2 u^2 u_d + b_3 u u_d^2 + b_4 u_d^3), \tag{3.46}$$

where $u_d = u(\tau - \Omega T)$. We expand Ω in a power series in ε, omitting the $O(\varepsilon)$ term for convenience, since it turns out to be zero:

$$\Omega = \omega + \varepsilon^2 k_2 + \cdots \tag{3.47}$$

Next we expand the delay term u_d:

$$u_d = u(\tau - \Omega T) = u(\tau - (\omega + \varepsilon^2 k_2 + \cdots)(T_{cr} + \varepsilon^2 \hat{\mu})) \tag{3.48}$$

$$= u(\tau - \omega T_{cr} - \varepsilon^2(k_2 T_{cr} + \omega \hat{\mu}) + \cdots) \tag{3.49}$$

$$= u(\tau - \omega T_{cr}) - \varepsilon^2(k_2 T_{cr} + \omega \hat{\mu}) u'(\tau - \omega T_{cr}) + O(\varepsilon^3). \tag{3.50}$$

Finally we expand $u(\tau)$ in a power series in ε:

$$u(\tau) = u_0(\tau) + \varepsilon u_1(\tau) + \varepsilon^2 u_2(\tau) + \cdots \tag{3.51}$$

Substituting and collecting terms, we find:

$$\omega \frac{du_0}{d\tau} - \alpha u_0(\tau) - \beta u_0(\tau - \omega T_{cr}) = 0, \tag{3.52}$$

$$\omega \frac{du_1}{d\tau} - \alpha u_1(\tau) - \beta u_1(\tau - \omega T_{cr})$$
$$= a_1 u_0(\tau)^2 + a_2 u_0(\tau) u_0(\tau - \omega T_{cr}) + a_3 u_0(\tau - \omega T_{cr})^2, \tag{3.53}$$

$$\omega \frac{du_2}{d\tau} - \alpha u_2(\tau) - \beta u_2(\tau - \omega T_{cr}) \tag{3.54}$$
$$= \cdots$$

where $\cdots$ stands for terms in u_0 and u_1, omitted here for brevity. We take the solution of the u_0 equation as (cf. Eq.(3.29) above):

$$u_0(\tau) = \hat{A}\cos(\tau). \tag{3.55}$$

We substitute (3.55) into (3.53) and obtain the following expression for u_1:

$$u_1(\tau) = m_1 \sin(2\tau) + m_2 \cos(2\tau) + m_3, \tag{3.56}$$

where m_1 is given by the equation:

$$m_1 = -\frac{\hat{A}^2\,(2a_3\,\beta + a_2\,\beta - 2a_1\,\beta - 2a_3\,\alpha)\,\sqrt{\beta^2 - \alpha^2}}{2\beta\,(\beta + \alpha)\,(5\beta - 4\alpha)}. \tag{3.57}$$

and where m_2 and m_3 are given by similar equations, omitted here for brevity. In deriving (3.57), ω has been replaced by $\sqrt{\beta^2 - \alpha^2}$ as in Eq.(3.36).

Next the expressions for u_0 and u_1, Eqs.(3.55),(3.56), are substituted into the u_2 equation, Eq.(3.55), and, after trigonometric simplifications have been performed, the coefficients of the resonant terms, $\sin\tau$ and $\cos\tau$, are equated to zero. This results in Eq.(3.32) for A^2 as well as an expression for k_2 (cf. Eq.(3.47)) which does not concern us here. (Note that $A = \varepsilon\hat{A}$ from Eqs.(3.31),(3.43),(3.55), and $\mu = \varepsilon^2\hat{\mu}$ from (3.44). The perturbation method gives $\hat{A}^2$ as a function of $\hat{\mu}$, but multiplication by ε^2 converts to a relation between A^2 and μ.)

3.4.3 Example 2

As a second example, we consider the case in which the quantity Q in Eqs.(3.32),(3.34) is zero. To generate such an example for the DDE (3.27), we embed the previous example in a one-parameter family of DDE's:

$$\frac{dx}{dt} = -x - 2x_d - x x_d - \lambda x^3, \tag{3.58}$$

and we choose λ so that $Q = 0$ in Eq.(3.32). This leads to the following critical value of λ:

$$\lambda = \lambda_{cr} = \frac{4\pi + 3\sqrt{3}}{18(2\pi + 3\sqrt{3})} = 0.0859 \tag{3.59}$$

Since this choice for λ leads to $Q = 0$, Eq.(3.32) obviously cannot be used to find the limit cycle amplitude A. Instead we use Lindstedt's method, maintaining terms of $O(\varepsilon^4)$. The correct scalings in this case turn out to be (cf.Eqs.(3.44),(3.47)):

$$T = T_{cr} + \mu = \frac{2\pi}{3\sqrt{3}} + \varepsilon^4 \hat{\mu}, \tag{3.60}$$

$$\Omega = \omega + \varepsilon^2 k_2 + \varepsilon^4 k_4 + \cdots \tag{3.61}$$

We find that the limit cycle amplitude A satisfies the equation:

$$A^4 = -\Gamma\mu, \tag{3.62}$$

where we omit the closed form expression for Γ and give instead its approximate value, Γ=620.477.

The analysis of this example has assumed that the parameter λ exactly takes on the critical value given in Eq.(3.59). Let us consider a more robust model which allows λ to be detuned:

$$\lambda = \lambda_{cr} + \Lambda = \frac{4\pi + 3\sqrt{3}}{18(2\pi + 3\sqrt{3})} + \varepsilon^2 \hat{\Lambda}. \tag{3.63}$$

Using Lindstedt's method we obtain for this case the following equation on A:

$$A^4 + \sigma\Lambda A^2 + \Gamma\mu = 0, \tag{3.64}$$

where we omit the closed form expression for σ and give instead its approximate value, σ=342.689. Equation (3.64) can have 0,1, or 2 positive real roots for A, each of which is a limit cycle in the original system. Thus the number of limit cycles which are born in the Hopf bifurcation depends on the detuning coefficients Λ and μ. Elementary use of the quadratic formula and the requirement that A^2 be both real and positive gives the following results: If $\mu < 0$ then there is one limit cycle. If $\mu > 0$ and $\sigma\Lambda < -2\sqrt{\Gamma\mu}$ then there are two limit cycles. If $\mu > 0$ and $\sigma\Lambda > -2\sqrt{\Gamma\mu}$ then there are no limit cycles.

3.4.4 Discussion

Although Lindstedt's method is a formal perturbation method, i.e., lacking a proof of convergence, our experience is that it gives the same results as the center manifold approach, which has a rigorous mathematical foundation. The center manifold approach has been described in many places, for example (Hassard et al., 1981; Campbell et al., 1995; Stepan, 1989; Kalmar-Nagy et al., 2001; Rand, 2005), and will be treated later in this Chapter. Since the DDE (3.27) is infinite dimensional (for example the characteristic equation of the linear DDE (3.28) is transcendental rather than polynomial, and hence has an infinite number of complex roots), the center manifold approach involves decomposing the original function space into a two dimensional center manifold (in which the Hopf bifurcation takes place) and an infinite dimensional function space representing the rest of the original phase space. The center manifold procedure is much more complicated than the Hopf calculation. Stepan refers to the center manifold calculation as "long and tedious" ((Stepan, 1989), p.112), and Campbell et al. refer to it as "algebraically daunting" ((Campbell et al., 1995), p.642). Thus the main advantage of the Hopf calculation is that it is simpler to understand and easier to execute than the center manifold approach.

The idea of using Lindstedt's method on bifurcation problems in DDE goes back to a 1980 paper by Casal and Freedman (Casal and Freedman, 1980). That work provided the algorithm but not the Hopf bifurcation formula. We present the general expression for the Hopf bifurcation, as in Eqs.(3.32)—(3.34), as a convenience for researchers in DDE.

3.5 Transient behavior

We have seen that Lindstedt's method can be used to obtain an approximation for the periodic motion of a DDE. This section is concerned with the use of perturbation methods to obtain approximate expressions for the *transient* behavior of DDEs, e.g. for the approach to a steady state periodic motion. In the case of ordinary differential equations (ODEs), a very popular method for obtaining transient behavior is the two variable expansion method (also know as multiple scales) (Cole, 1968; Nayfeh, 1973; Rand and Armbruster, 1987; Rand, 2005). In this section we show how this method can be applied to a DDE. See also (Das and Chatterjee, 2002; Wang and Hu, 2003; Das and Chatterjee, 2005; Nayfeh, 2008). Although this approximate method is strictly formal, its use is justified by center manifold considerations. Although the DDE is an infinite dimensional system, a wide class of problems involves the presence of a two dimensional invariant manifold, and it is the approximation of the transient flow on this surface which is the goal of this perturbation method.

3.5.1 Example

In order to illustrate the manner in which this perturbation method may be applied to DDEs, we choose a simple DDE problem, one that has an exact solution, namely:

$$\frac{dx}{dt} = -x(t-T), \quad T = \frac{\pi}{2} + \varepsilon\mu. \tag{3.65}$$

3.5.2 Exact solution

In Eq.(3.65) we set

$$x(t) = \exp(\lambda t), \tag{3.66}$$

giving the characteristic equation

$$\lambda = -\exp(-\lambda T). \tag{3.67}$$

When $\varepsilon = 0$, that is when $T = \pi/2$, Eq.(3.67) has the exact solution $\lambda = i$. Thus for $\varepsilon > 0$ we set

$$\lambda = i + \varepsilon(a + ib). \tag{3.68}$$

Substituting Eq.(3.68) into (3.67) and equating real and imaginary parts to zero, we obtain the following two equations on a and b:

$$a\varepsilon = \exp\left(-a\varepsilon^2\mu - \frac{\pi a\varepsilon}{2}\right)\sin\left(b\varepsilon^2\mu + \varepsilon\mu + \frac{\pi b\varepsilon}{2}\right), \tag{3.69}$$

$$b\varepsilon + 1 = \exp\left(-a\varepsilon^2\mu - \frac{\pi a\varepsilon}{2}\right)\cos\left(b\varepsilon^2\mu + \varepsilon\mu + \frac{\pi b\varepsilon}{2}\right). \tag{3.70}$$

If Eqs.(3.69) and (3.70) were to be solved for a and b, we would obtain a solution to (3.65) in the form:

$$x(t) = \exp(\varepsilon a t)\begin{cases} \sin & (1+\varepsilon b)t, \\ \cos & (1+\varepsilon b)t. \end{cases} \tag{3.71}$$

In order to obtain a version of the exact solution (3.71) which will be useful for comparing solutions to Eq.(3.65) obtained by the perturbation method, we now derive approximate expressions for a and b. Taylor expanding Eqs.(3.69),(3.70) for small ε, we obtain

$$a\varepsilon + \cdots = \frac{(b\pi + 2\mu)\,\varepsilon}{2} + \cdots \tag{3.72}$$

$$1 + b\varepsilon + \cdots = 1 - \frac{\pi a\varepsilon}{2} + \cdots \tag{3.73}$$

Solving Eqs.(3.72),(3.73) for a and b, we obtain the approximate expressions:

$$a = \frac{4\mu}{\pi^2+4} + O(\varepsilon), \quad b = -\frac{2\pi\mu}{\pi^2+4} + O(\varepsilon) \tag{3.74}$$

3.5.3 Two variable expansion method (also known as multiple scales)

In applying this method to the example of Eq.(3.65), we replace time t by two time variables: regular time $\xi = t$, and slow time $\eta = \varepsilon t$. The dependent variable $x(t)$ is then replaced by $x(\xi,\eta)$, and Eq.(3.65) becomes

$$\frac{\partial x}{\partial \xi} + \varepsilon \frac{\partial x}{\partial \eta} = -x(\xi - T, \eta - \varepsilon T). \tag{3.75}$$

Note that since $T = \pi/2 + \varepsilon\mu$, the delayed term may be expanded for small ε as follows:

$$x(\xi - T, \eta - \varepsilon T) = x(\xi - \frac{\pi}{2}, \eta) - \varepsilon\mu \frac{\partial x_d}{\partial \xi} - \varepsilon \frac{\pi}{2}\frac{\partial x_d}{\partial \eta} + O(\varepsilon^2), \tag{3.76}$$

where x_d is an abbreviation for $x(\xi - \frac{\pi}{2}, \eta)$. Next we expand $x = x_0 + \varepsilon x_1 + O(\varepsilon^2)$ and collect terms in Eqs.(3.75),(3.76), giving

$$\frac{\partial x_0}{\partial \xi} + x_0(\xi - \frac{\pi}{2}, \eta) = 0, \tag{3.77}$$

$$\frac{\partial x_1}{\partial \xi} + x_1(\xi - \frac{\pi}{2}, \eta) = \mu \frac{\partial x_{0d}}{\partial \xi} + \frac{\pi}{2}\frac{\partial x_{0d}}{\partial \eta} - \frac{\partial x_0}{\partial \eta}. \tag{3.78}$$

Eq.(3.77) has the periodic solution

$$x_0 = R(\eta)\cos(\xi - \theta(\eta)). \tag{3.79}$$

where as usual in this method, $R(\eta)$ and $\theta(\eta)$ are as yet undetermined functions of slow time η. The next step is to substitute Eq.(3.79) into (3.78). Before doing so, we rewrite (3.78) by noting that (3.77) can be written in the form $x_{0d} = -\partial x_0/\partial \xi$:

$$\frac{\partial x_1}{\partial \xi} + x_1(\xi - \frac{\pi}{2}, \eta) = -\mu \frac{\partial^2 x_0}{\partial \xi^2} - \frac{\pi}{2}\frac{\partial^2}{\partial \eta \partial \xi} x_0 - \frac{\partial x_0}{\partial \eta}. \tag{3.80}$$

Now we substitute (3.79) into (3.80) and require the coefficients of both $\cos(\xi - \theta)$ and $\sin(\xi - \theta)$ to vanish, giving the following slow flow on R and θ:

$$R' + \frac{\pi}{2}R\theta' - \mu R = 0, \tag{3.81}$$

$$\frac{\pi}{2}R' - R\theta' = 0, \tag{3.82}$$

where primes represent differentiation with respect to slow time η. Solving for R' and θ', we get

$$R' = \frac{4\mu}{\pi^2 + 4}R, \tag{3.83}$$

$$\theta' = \frac{2\pi\mu}{\pi^2 + 4}, \tag{3.84}$$

from which we obtain

$$R(\eta) = R_0 \exp\left(\frac{4\mu\eta}{\pi^2+4}\right), \tag{3.85}$$

$$\theta(\eta) = \frac{2\pi\mu}{\pi^2+4}\eta + \theta_0. \tag{3.86}$$

Eq.(3.79) thus gives

$$x \approx x_0 = R_0 \exp\left(\frac{4\mu\varepsilon t}{\pi^2+4}\right) \cos\left(t - \frac{2\pi\mu}{\pi^2+4}\varepsilon t - \theta_0\right), \tag{3.87}$$

which agrees with the exact solution given by Eq.(3.71) with a and b given by (3.74).

3.5.4 Approach to limit cycle

Now let us use the two variable method on Eq.(3.1). We have seen by use of Lindstedt's method that this DDE has a limit cycle of amplitude $2\sqrt{\mu}/\sqrt{3}$, see Eq.(3.25). The question arises as to the stability of this limit cycle. This may be determined as follows: After scaling x as in Eq.(3.13), we may obtain Eq.(3.1) by adding the term $-\varepsilon x(t)^3$ to the RHS of Eq.(3.65). This results in the term $-x_0^3$ being added to the RHS of Eqs.(3.78) and (3.80). After trigonometric reduction, this new term causes the term $-3R^3/4$ to be added to the RHS of Eq.(3.81), resulting in the new slow flow

$$R' = \frac{4\mu R - 3R^3}{\pi^2+4}, \tag{3.88}$$

$$\theta' = \frac{2\pi\mu - (3\pi/2)R^2}{\pi^2+4}. \tag{3.89}$$

Here we see that Eq.(3.88) has two equilibria, $R = 0$ and $R = 2\sqrt{\mu}/\sqrt{3}$. Treating (3.88) as a flow on the R−line immediately shows that for $\mu > 0$ the $R = 0$ equilibrium is unstable, a fact which we have already observed via a different approach, since $R = 0$ corresponds to the trivial solution of Eq.(3.1), which was investigated in Eqs.(3.5)–(3.9). In addition (3.88) shows that for $\mu > 0$ the equilibrium $R = 2\sqrt{\mu}/\sqrt{3}$ is stable, from which we may conclude that the corresponding limit cycle is stable and that the Hopf bifurcation is supercritical. This conclusion agrees with the numerical integration displayed in Figure 3.3.

3.6 Center manifold analysis

We have seen earlier that the equilibrium solution $x = 0$ in Eq.(3.1) is stable for $T < \pi/2$ and is unstable for $T > \pi/2$. The question remains as to the stability of

$x=0$ when $T=\pi/2$. More generally, in order to determine the stability of the x=0 solution of a DDE in the form of Eq.(3.2),

$$\frac{dx(t)}{dt} = \alpha x(t) + \beta x(t-T) + f(x(t), x(t-T)), \tag{3.90}$$

in the case that the delay T takes on its critical value T_{cr}, it is necessary to take into account the effect of nonlinear terms. This may be accomplished by using a center manifold reduction. In order to accomplish this, the DDE is reformulated as an evolution equation on a function space. Our treatment closely follows that presented in (Kalmar-Nagy et al., 2001).

The idea here is that the initial condition for Eq.(3.90) consists of a function defined on $-T \le t \le 0$. As t increases from zero we may consider the piece of the solution lying in the time interval $[-T+t, t]$ as having evolved from the initial condition function. In order to avoid confusion, the variable θ is used to identify a point in the interval $[-T, 0]$, whereupon $x(t+\theta)$ will represent the piece of the solution which has evolved from the initial condition function at time t. From the point of view of the function space, t is a parameter, and it is θ which is the independent variable. To emphasize this, we write:

$$x_t(\theta) = x(t+\theta). \tag{3.91}$$

The evolution equation, which acts on a function space consisting of continuously differentiable functions on $[-T, 0]$, is written:

$$\frac{\partial x_t(\theta)}{\partial t} = \begin{cases} \dfrac{\partial x_t(\theta)}{\partial \theta}, & \text{for } \theta \in [-T, 0), \\ \alpha x_t(0) + \beta x_t(-T) + f(x_t(0), x_t(-T)), & \text{for } \theta = 0. \end{cases} \tag{3.92}$$

Here the DDE (3.90) appears as a boundary condition at $\theta = 0$. The rest of the interval goes along for the ride, which is to say that the equation $\frac{\partial x_t(\theta)}{\partial t} = \frac{\partial x_t(\theta)}{\partial \theta}$ is an identity due to Eq.(3.91).

The RHS of Eq.(3.92) may be written as the sum of a linear operator A and a nonlinear operator F:

$$A x_t(\theta) = \begin{cases} \dfrac{\partial x_t(\theta)}{\partial \theta}, & \text{for } \theta \in [-T, 0), \\ \alpha x_t(0) + \beta x_t(-T), & \text{for } \theta = 0. \end{cases} \tag{3.93}$$

$$F x_t(\theta) = \begin{cases} 0, & \text{for } \theta \in [-T, 0), \\ f(x_t(0), x_t(-T)), & \text{for } \theta = 0. \end{cases} \tag{3.94}$$

We now assume that the delay T is set at its critical value for a Hopf bifurcation, i.e. the characteristic equation has a pair of pure imaginary roots, $\lambda = \pm \omega i$. Corresponding to these eigenvalues are a pair of eigenfunctions $s_1(\theta)$ and $s_2(\theta)$ which satisfy the eigenequation:

$$A(s_1(\theta)+is_2(\theta)) = i\omega(s_1(\theta)+is_2(\theta)). \quad (3.95)$$

That is,

$$As_1(\theta) = -\omega s_2(\theta), \quad (3.96)$$

$$As_2(\theta) = \omega s_1(\theta). \quad (3.97)$$

From the definition (3.93) of the linear operator A we find that $s_1(\theta)$ and $s_2(\theta)$ must satisfy the following differential equations and boundary conditions:

$$\frac{d}{d\theta}s_1(\theta) = -\omega s_2(\theta), \quad (3.98)$$

$$\frac{d}{d\theta}s_2(\theta) = \omega s_1(\theta), \quad (3.99)$$

$$\alpha s_1(0)+\beta s_1(-T) = -\omega s_2(0), \quad (3.100)$$

$$\alpha s_2(0)+\beta s_2(-T) = \omega s_1(0). \quad (3.101)$$

Let's illustrate this process by using Eq.(3.1) as an example. We saw earlier that at T=T_{cr}=$\pi/2$, ω=1, which permits us to solve Eqs.(3.98), (3.99) as:

$$s_1(\theta) = c_1\cos\theta - c_2\sin\theta, \quad (3.102)$$

$$s_2(\theta) = c_1\sin\theta + c_2\cos\theta, \quad (3.103)$$

where c_1 and c_2 are arbitrary constants. In the case of Eq.(3.1), the boundary conditions (3.100), (3.101) become (α=0, β=−1):

$$-s_1(-\pi/2) = -s_2(0), \quad (3.104)$$

$$-s_2(-\pi/2) = s_1(0). \quad (3.105)$$

Equations (3.102), (3.103) identically satisfy Eqs.(3.104), (3.105) so that the constants of integration c_1 and c_2 remain arbitrary at this point.

In preparation for the center manifold analysis, we write the solution $x_t(\theta)$ as a sum of points lying in the center subspace (spanned by $s_1(\theta)$ and $s_2(\theta)$) and of points which do not lie in the center subspace, i.e., the rest of the solution, here called w:

$$x_t(\theta) = y_1(t)s_1(\theta)+y_2(t)s_2(\theta)+w(t)(\theta). \quad (3.106)$$

Here $y_1(t)$ and $y_2(t)$ are the components of $x_t(\theta)$ lying in the directions $s_1(\theta)$ and $s_2(\theta)$ respectively.

The idea of the center manifold reduction is to find w as an approximate function of y_1 and y_2 (the center manifold), and then to substitute $w(y_1,y_2)$

into the equations on $y_1(t)$ and $y_2(t)$. The result is that we will have replaced the original infinite dimensional system with a two dimensional approximation.

In order to find the equations on $y_1(t)$ and $y_2(t)$, we need to project $x_t(\theta)$ onto the center subspace. In this system, projections are accomplished by means of a bilinear form:

$$(v,u) = v(0)u(0) + \int_{-T}^{0} v(\theta+T)\beta u(\theta)d\theta, \tag{3.107}$$

where $u(\theta)$ lies in the original function space, i.e. the space of continuously differentiable functions defined on $[-T,0]$, and where $v(\theta)$ lies in the adjoint function space of continuously differentiable functions defined on $[0,T]$. See the Appendix to this chapter for a discussion of the adjoint operator A^*.

In order to project $x_t(\theta)$ onto the center subspace, we will need the adjoint eigenfunctions $n_1(\theta)$ and $n_2(\theta)$ which satisfy the adjoint eigenequation:

$$A^*(n_1(\theta)+in_2(\theta)) = -i\omega(n_1(\theta)+in_2(\theta)), \tag{3.108}$$

That is,

$$A^*n_1(\theta) = \omega n_2(\theta), \tag{3.109}$$
$$A^*n_2(\theta) = -\omega n_1(\theta), \tag{3.110}$$

where the adjoint operator A^* is defined by

$$A^*v(\theta) = \begin{cases} -\dfrac{dv(\theta)}{d\theta}, & \text{for } \theta \in (0,T], \\ \alpha v(0)+\beta v(T), & \text{for } \theta = 0. \end{cases} \tag{3.111}$$

In addition, the adjoint eigenfunctions n_i are defined to be orthonormal to the eigenfunctions s_i:

$$(n_i,s_j) = \begin{cases} 1, & \text{if } i=j, \\ 0, & \text{if } i \neq j. \end{cases} \tag{3.112}$$

From the definition (3.111) of the linear operator A^* we find that $n_1(\theta)$ and $n_2(\theta)$ must satisfy the following differential equations and boundary conditions:

$$-\frac{d}{d\theta}n_1(\theta) = \omega n_2(\theta), \tag{3.113}$$
$$-\frac{d}{d\theta}n_2(\theta) = -\omega n_1(\theta), \tag{3.114}$$
$$\alpha n_1(0)+\beta n_1(T) = \omega n_2(0), \tag{3.115}$$
$$\alpha n_2(0)+\beta n_2(T) = -\omega n_1(0). \tag{3.116}$$

We continue to illustrate by using Eq.(3.1) as an example. With $\omega=1$, Eqs.(3.113), (3.114) may be solved as:

$$n_1(\theta) = d_1\cos\theta - d_2\sin\theta, \tag{3.117}$$

$$n_2(\theta) = d_1 \sin\theta + d_2 \cos\theta, \tag{3.118}$$

where d_1 and d_2 are arbitrary constants. In the case of Eq.(3.1), the boundary conditions (3.115), (3.116) become (α=0, β=−1):

$$-n_1(\pi/2) = n_2(0), \tag{3.119}$$
$$-n_2(\pi/2) = -n_1(0). \tag{3.120}$$

As in the case of the s_i eigenfunctions, Eqs.(3.117), (3.118) identically satisfy Eqs.(3.119), (3.120) so that the constants of integration d_1 and d_2 remain arbitrary.

The four arbitrary constants c_1, c_2, d_1, d_2 of Eqs.(3.102),(3.103),(3.117),(3.118) will be related by the orthonormality conditions (3.112). Let's illustrate this by computing (n_1,s_1) for the example of Eq.(3.1). Using the definition of the bilinear form (3.107), we obtain:

$$(n_1,s_1) = n_1(0)s_1(0) + \int_{-\frac{\pi}{2}}^{0} n_1\left(\theta+\frac{\pi}{2}\right)(-1)s_1(\theta)d\theta, \tag{3.121}$$

$$(n_1,s_1) = \frac{(2c_2+\pi c_1)\,d_2 + (2c_1-\pi c_2)\,d_1}{4} = 1. \tag{3.122}$$

Similarly, we find:

$$(n_1,s_2) = \frac{(\pi c_2-2c_1)\,d_2 + (2c_2+\pi c_1)\,d_1}{4} = 0. \tag{3.123}$$

The other two orthonormality conditions give no new information since it turns out that $(n_2,s_1) = -(n_1,s_2)$ and $(n_2,s_2) = (n_1,s_1)$. Thus Eqs.(3.122) and (3.123) are two equations in four unknowns, c_1, c_2, d_1, d_2. Without loss of generality we take $d_1 = 1$ and $d_2 = 0$, giving $c_1 = \frac{8}{\pi^2+4}, c_2 = -\frac{4\pi}{\pi^2+4}$. Thus the eigenfunctions s_i and n_i for Eq.(3.1) become:

$$s_1(\theta) = \frac{4\pi\sin\theta + 8\cos\theta}{\pi^2+4}, \tag{3.124}$$
$$s_2(\theta) = \frac{8\sin\theta - 4\pi\cos\theta}{\pi^2+4}, \tag{3.125}$$
$$n_1(\theta) = \cos\theta, \tag{3.126}$$
$$n_2(\theta) = \sin\theta. \tag{3.127}$$

Recall that our purpose in solving for n_1 and n_2 was to obtain equations on $y_1(t)$ and $y_2(t)$, the components of $x_t(\theta)$ lying in the directions $s_1(\theta)$ and $s_2(\theta)$ respectively, see Eq.(3.106). We have:

$$y_1(t) = (n_1,x_t), \quad y_2(t) = (n_2,x_t). \tag{3.128}$$

Differentiating (3.128) with respect to t:

$$\dot{y}_1(t) = (n_1,\dot{x}_t), \quad \dot{y}_2(t) = (n_2,\dot{x}_t). \tag{3.129}$$

Let us consider the first of these:

$$\dot{y}_1(t) = (n_1, \dot{x}_t) = (n_1, Ax_t + Fx_t) = (n_1, Ax_t) + (n_1, Fx_t). \tag{3.130}$$

Now by definition of the adjoint operator A^*,

$$(n_1, Ax_t) = (A^* n_1, x_t) = (\omega n_2, x_t) = \omega(n_2, x_t) = \omega y_2. \tag{3.131}$$

So we have

$$\dot{y}_1 = \omega y_2 + (n_1, Fx_t),$$

and similarly (3.132)

$$\dot{y}_2 = -\omega y_1 + (n_2, Fx_t).$$

In Eqs.(3.132), the quantities (n_i, Fx_t) are given by (cf. Eq.(3.107)):

$$(n_i, Fx_t) = n_i(0)Fx_t(0) + \int_{-T}^{0} n_i(\theta + T)\beta Fx_t(\theta)d\theta \tag{3.133}$$

$$= n_i(0)f(x_t(0), x_t(-T)) \text{ since } Fx_t(\theta)\text{=0 unless } \theta\text{=0}, \tag{3.134}$$

in which $x_t = y_1(t)s_1(\theta) + y_2(t)s_2(\theta) + w(t)(\theta)$. Continuing with the example of Eq.(3.1), Eqs.(3.132) become, using $f = -x(t)^3$:

$$\dot{y}_1 = y_2 - \left(\frac{8y_1}{\pi^2+4} - \frac{4\pi y_2}{\pi^2+4} + w(\theta = 0)\right)^3 \quad \text{and} \quad \dot{y}_2 = -y_1, \tag{3.135}$$

where we have used $s_1(0)=\frac{8}{\pi^2+4}$, $s_2(0)=\frac{-4\pi}{\pi^2+4}$, $n_1(0)=1$ and $n_2(0)=0$ from Eqs.(3.124) –(3.127).

The next step is to look for an approximate expression for the center manifold, which is tangent to the y_1-y_2 plane at the origin, and which may be written in the form:

$$w(y_1, y_2)(\theta) = m_1(\theta)y_1^2 + m_2(\theta)y_1y_2 + m_3(\theta)y_2^2. \tag{3.136}$$

The procedure is to substitute (3.136) into the equations of motion, collect terms, and solve for the unknown functions $m_i(\theta)$. Then the resulting expression is to be substituted into the y_1-y_2 equations (3.132). Note that if this is done for the example of Eq.(3.1), i.e. for Eqs.(3.135), the contribution made by w will be of degree 4 and higher in the y_i. However, stability of the origin will be determined by terms of degree 2 and 3, according to the following formula (obtainable by averaging). Suppose the y_1-y_2 equations are of the form:

$$\dot{y}_1 = \omega y_2 + h(y_1, y_2) \quad \text{and} \quad \dot{y}_2 = -\omega y_1 + g(y_1, y_2). \tag{3.137}$$

Then the stability of the origin is determined by the sign of the quantity Q, where

$$16Q = h_{111} + h_{122} + g_{112} + g_{222} - \frac{1}{\omega}\left[h_{12}(h_{11} + h_{22}) - g_{12}(g_{11} + g_{22}) - h_{11}g_{11} + h_{22}g_{22}\right], \quad (3.138)$$

where subscripts represent partial derivatives, which are to be evaluated at $y_1 = y_2 = 0$. $Q > 0$ means unstable, $Q < 0$ means stable. See (Guckenheimer and Holmes, 1983) pp.154-156, where it is shown that the flow on the y_1-y_2 plane in the neighborhood of the origin can be approximately written in polar coordinates as:

$$\frac{dr}{dt} = Qr^3 + O(r^5). \quad (3.139)$$

Applying this criterion to Eqs.(3.135) (in which w is assumed to be of the form (3.136) and hence contributes terms of higher order in y_i), we find:

$$Q = -\frac{48}{(\pi^2 + 4)^2} = -0.2495. \quad (3.140)$$

The origin in Eq.(3.1) for $T = T_{cr} = \pi/2$ is therefore predicted to be stable. This result is in agreement with numerical integration, see Fig. 3.4.

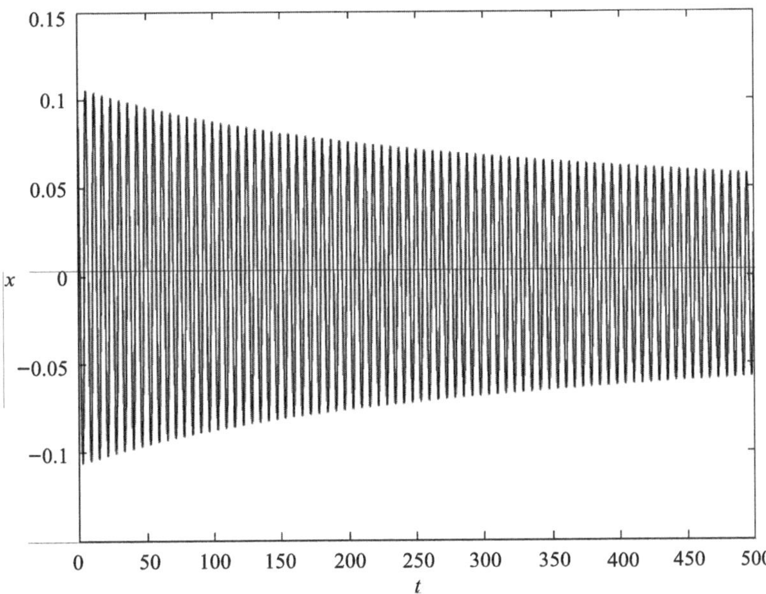

Fig. 3.4 Numerical integration of Eq.(3.1) for the initial condition $x = 0.1$ on $-T \leq t \leq 0$, for $T = \pi/2$.

Now let's change the example a little so that w plays a significant role in determining the stability of the origin:

$$\frac{dx(t)}{dt} = -x(t-T) - x(t)^2. \tag{3.141}$$

Since the linear parts of this example and of the previous example are the same, our previously derived expressions for s_1, s_2, n_1 and n_2, Eqs.(3.124)–(3.127), still apply. Eqs.(3.135) now become:

$$\dot{y}_1 = y_2 - \left(\frac{8y_1}{\pi^2+4} - \frac{4\pi y_2}{\pi^2+4} + w(\theta=0)\right)^2 \quad \text{and} \quad \dot{y}_2 = -y_1. \tag{3.142}$$

So our goal now is to find the functions $m_i(\theta)$ in the expression for the center manifold (3.136), and then to substitute the result into Eq.(3.142) and use the formula (3.138) to determine stability.

We begin by differentiating the expression for the center manifold (3.136) with respect to t:

$$\frac{\partial w(y_1,y_2)(\theta)}{\partial t} = 2m_1(\theta)y_1\dot{y}_1 + m_2(\theta)(y_1\dot{y}_2 + y_2\dot{y}_1) + 2m_3(\theta)y_2\dot{y}_2. \tag{3.143}$$

We substitute the equations (3.132) on $\dot{y}_1$ and $\dot{y}_2$ in (3.143) and neglect terms of degree higher than 2 in the y_i:

$$\begin{aligned}\frac{\partial w(y_1,y_2)(\theta)}{\partial t} &= 2m_1(\theta)y_1\omega y_2 + m_2(\theta)(-y_1\omega y_1 + y_2\omega y_2)\\ &\quad -2m_3(\theta)y_2\omega y_1 + \cdots &(3.144)\\ &= \omega[-m_2(\theta)y_1^2 + 2(m_1(\theta) - m_3(\theta))y_1y_2\\ &\quad +m_2(\theta)y_2^2] + \cdots &(3.145)\end{aligned}$$

We obtain conditions on the functions $m_i(\theta)$ by deriving another expression for $\dot{w}$ and equating them. Let us differentiate Eq.(3.106) with respect to t:

$$\frac{\partial x_t(\theta)}{\partial t} = \dot{y}_1(t)s_1(\theta) + \dot{y}_2(t)s_2(\theta) + \frac{\partial w(t)(\theta)}{\partial t}. \tag{3.146}$$

Using the functional DE (3.92)–(3.94), and rearranging terms, we get:

$$\begin{aligned}\frac{\partial w(t)(\theta)}{\partial t} &= \frac{\partial x_t(\theta)}{\partial t} - \dot{y}_1(t)s_1(\theta) - \dot{y}_2(t)s_2(\theta) &(3.147)\\ &= Ax_t(\theta) + Fx_t(\theta) - \dot{y}_1(t)s_1(\theta) - \dot{y}_2(t)s_2(\theta) &(3.148)\\ &= A(y_1(t)s_1(\theta) + y_2(t)s_2(\theta) + w(t)(\theta))\\ &\quad +Fx_t(\theta) - \dot{y}_1(t)s_1(\theta) - \dot{y}_2(t)s_2(\theta) &(3.149)\\ &= y_1As_1 + y_2As_2 + Aw + Fx_t - \dot{y}_1s_1 - \dot{y}_2s_2 &(3.150)\\ &= y_1(-\omega s_2) + y_2(\omega s_1) + Aw + Fx_t\\ &\quad -(\omega y_2 + (n_1,Fx_t))s_1 - (-\omega y_1 + (n_2,Fx_t))s_2 &(3.151)\\ &= Aw + Fx_t - (n_1,Fx_t)s_1 - (n_2,Fx_t)s_2 &(3.152)\end{aligned}$$

where we have used Eqs.(3.96), (3.97) and (3.132), and where the quantities (n_i, Fx_t) are given by Eq.(3.134).

Eq.(3.152) is an equation for the time evolution of w. Since the operator A is defined differently for $\theta \in [-T,0)$ and for $\theta = 0$, we consider each of these cases separately when we substitute Eq.(3.136) for the center manifold. In the $\theta \in [-T,0)$ case, Eq.(3.152) becomes:

$$\frac{\partial w(t)(\theta)}{\partial t} = m_1' y_1^2 + m_2' y_1 y_2 + m_3' y_2^2 - (n_1, Fx_t)s_1(\theta) - (n_2, Fx_t)s_2(\theta), \quad (3.153)$$

where primes denote differentiation with respect to θ. In the $\theta = 0$ case, Eq.(3.152) becomes:

$$\begin{aligned} \frac{\partial w(t)(\theta)}{\partial t} &= \alpha(m_1(0)y_1^2 + m_2(0)y_1y_2 + m_3(0)y_2^2) \\ &\quad + \beta(m_1(-T)y_1^2 + m_2(-T)y_1y_2 + m_3(-T)y_2^2) \\ &\quad + f(x_t(0), x_t(-T)) - (n_1, Fx_t)s_1(0) - (n_2, Fx_t)s_2(0). \quad (3.154) \end{aligned}$$

Now we equate Eqs.(3.153) and (3.154) to the previously derived expression for $\dot{w}$, Eq.(3.145). Equating (3.153) to (3.145), we get:

$$\begin{aligned} &m_1' y_1^2 + m_2' y_1 y_2 + m_3' y_2^2 - (n_1, Fx_t)s_1(\theta) - (n_2, Fx_t)s_2(\theta) = \\ &\omega[-m_2 y_1^2 + 2(m_1 - m_3)y_1 y_2 + m_2 y_2^2] + \cdots \quad (3.155) \end{aligned}$$

Equating (3.154) to (3.145), we get:

$$\begin{aligned} &\alpha(m_1(0)y_1^2 + m_2(0)y_1y_2 + m_3(0)y_2^2) \\ &+\beta(m_1(-T)y_1^2 + m_2(-T)y_1y_2 + m_3(-T)y_2^2) \\ &+f(x_t(0), x_t(-T)) - (n_1, Fx_t)s_1(0) - (n_2, Fx_t)s_2(0) = \\ &\omega[-m_2(0)y_1^2 + 2(m_1(0) - m_3(0))y_1y_2 + m_2(0)y_2^2] + \cdots \quad (3.156) \end{aligned}$$

Now we equate coefficients of y_1^2, $y_1 y_2$ and y_2^2 in Eqs.(3.155) and (3.156) and so obtain 3 first order ODE's on m_1, m_2 and m_3 and 3 boundary conditions. From Eq.(3.134), the nonlinear terms (n_i, Fx_t) become:

$$(n_i, Fx_t) = n_i(0) f(x_t(0), x_t(-T)), \quad (3.157)$$

in which $x_t = y_1(t)s_1(\theta) + y_2(t)s_2(\theta) + w(t)(\theta) \approx y_1(t)s_1(\theta) + y_2(t)s_2(\theta)$. In the case of the example system (3.141) we have $\alpha = 0$, $\beta = -1$, $\omega = 1$, $T = \pi/2$, $f(x(t), x(t-T)) = -x(t)^2$ and

$$f(x_t(0), x_t(-T)) = -x_t(0)^2 = -(y_1 s_1(0) + y_2 s_2(0))^2. \quad (3.158)$$

For this example, Eq.(3.155) becomes

$$m_1' y_1^2 + m_2' y_1 y_2 + m_3' y_2^2 + (y_1 s_1(0) + y_2 s_2(0))^2 s_1(\theta)$$

$$= -m_2 y_1^2 + 2(m_1 - m_3) y_1 y_2 + m_2 y_2^2 \tag{3.159}$$

which gives the following 3 ODE's on $m_i(\theta)$:

$$m_1' + s_1(0)^2 s_1(\theta) = -m_2, \tag{3.160}$$

$$m_2' + 2 s_1(0) s_2(0) s_1(\theta) = 2(m_1 - m_3), \tag{3.161}$$

$$m_3' + s_2(0)^2 s_1(\theta) = m_2. \tag{3.162}$$

For this example, Eq.(3.156) becomes

$$\begin{aligned}
&-(m_1(-\pi/2) y_1^2 + m_2(-\pi/2) y_1 y_2 + m_3(-\pi/2) y_2^2) \\
&-(y_1 s_1(0) + y_2 s_2(0))^2 + (y_1 s_1(0) + y_2 s_2(0))^2 s_1(0) = \\
&-m_2(0) y_1^2 + 2(m_1(0) - m_3(0)) y_1 y_2 + m_2(0) y_2^2,
\end{aligned} \tag{3.163}$$

which gives the following 3 boundary conditions on $m_i(\theta)$:

$$-m_1(-\pi/2) - s_1(0)^2 + s_1(0)^3 = -m_2(0), \tag{3.164}$$

$$-m_2(-\pi/2) - 2 s_1(0) s_2(0) + 2 s_1(0)^2 s_2(0) = 2(m_1(0) - m_3(0)), \tag{3.165}$$

$$-m_3(-\pi/2) - s_2(0)^2 + s_2(0)^2 s_1(0) = m_2(0). \tag{3.166}$$

So we have 3 linear ODE's (3.160)–(3.162) with 3 boundary conditions (3.164)–(3.166) for the $m_i(\theta)$. In these equations, s_1 and s_2 are given by Eqs.(3.124), (3.125). The solution of these equations is algebraically complicated. I used a computer algebra system to obtain a closed form solution for the $m_i(\theta)$. For brevity, a numerical version of the center manifold is given below:

$$\begin{aligned}
w = \\
&(0.20216 \sin 2\theta + 0.16022 \cos 2\theta - 0.6953 \sin\theta + 0.39537 \cos\theta - 0.5768)\, {y_1}^2 \\
&+ (0.32044 \sin 2\theta - 0.40432 \cos 2\theta + 0.09393 \sin\theta + 0.5034 \cos\theta)\, y_1\, y_2 \\
&+ (-0.20216 \sin 2\theta - 0.16022 \cos 2\theta + 0.0299 \sin\theta + 0.64984 \cos\theta - 0.5768)\, {y_2}^2.
\end{aligned} \tag{3.167}$$

Next we substitute the algebraic version of Eq.(3.167) into the y_1-y_2 Eqs.(3.142) and use Eq.(3.138) to compute the stability parameter Q:

$$Q = \frac{32\,(9 - \pi)}{5\,(\pi^2 + 4)^2} = 0.19491 > 0. \tag{3.168}$$

Thus the center manifold analysis predicts that origin of Eq.(3.141) is unstable. This result is in agreement with numerical integration, see Fig. 3.5.

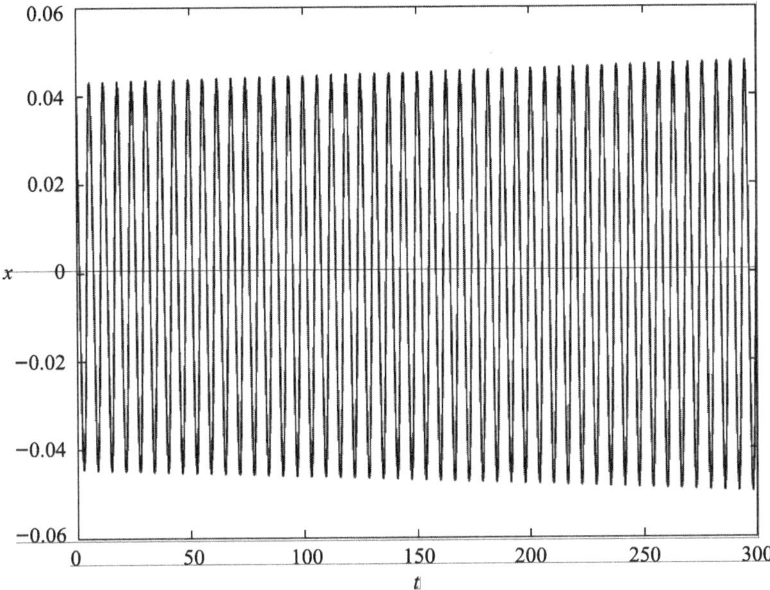

Fig. 3.5 Numerical integration of Eq.(3.141) for the initial condition $x = 0.04$ on $-T \leq t \leq 0$, for $T = \pi/2$.

3.6.1 Appendix: The adjoint operator A^*

The adjoint operator A^* is defined by the relation:

$$(v, Au) = (A^* v, u), \tag{3.169}$$

where the bilinear form (v, u) is given by Eq.(3.107):

$$(v, u) = v(0)u(0) + \int_{-T}^{0} v(\theta + T)\beta u(\theta) d\theta, \tag{3.170}$$

where $u(\theta)$ lies in the original function space, i.e. the space of continuously differentiable functions defined on $[-T, 0]$, and where $v(\theta)$ lies in the adjoint function space of continuously differentiable functions defined on $[0, T]$.

The linear operator A is given by Eq.(3.93):

$$Au(\theta) = \begin{cases} \dfrac{du(\theta)}{d\theta}, & \text{for } \theta \in [-T, 0), \\ \alpha u(0) + \beta u(-T), & \text{for } \theta = 0, \end{cases} \tag{3.171}$$

from which (v, Au) can be written as follows:

$$(v,Au) = v(0)Au(0) + \int_{-T}^{0} v(\theta+T)\beta Au(\theta)d\theta \tag{3.172}$$

$$= v(0)[\alpha u(0) + \beta u(-T)] + \int_{-T}^{0} v(\theta+T)\beta \frac{du(\theta)}{d\theta} d\theta. \tag{3.173}$$

Using integration by parts, the last term of Eq.(3.173) can be written:

$$\int_{-T}^{0} v(\theta+T)\beta \frac{du(\theta)}{d\theta} d\theta$$

$$= v(\theta+T)\beta u(\theta)|_{-T}^{0} - \int_{-T}^{0} \beta u(\theta) \frac{dv(\theta+T)}{d\theta} d\theta \tag{3.174}$$

$$= v(T)\beta u(0) - v(0)\beta u(-T) - \int_{\phi=0}^{T} \beta u(\phi - T) \frac{dv(\phi)}{d\phi} d\phi \tag{3.175}$$

where $\phi = \theta + T$. Substituting (3.175) into (3.173), we get

$$(v,Au) = [\alpha v(0) + \beta v(T)]u(0) + \int_{\phi=0}^{T} \left(-\frac{dv(\phi)}{d\phi}\right) \beta u(\phi - T) d\phi \tag{3.176}$$

$$= (A^* v, u) \tag{3.177}$$

from which we may conclude that the adjoint operator A^* is given by:

$$A^* v(\phi) = \begin{cases} -\dfrac{dv(\phi)}{d\phi} & \text{for } \phi \in (0,T], \\ \alpha v(0) + \beta v(T) & \text{for } \phi = 0. \end{cases} \tag{3.178}$$

3.7 Application to gene expression

This section offers a timely example showing how DDEs occur in a mathematical model of gene expression (Monk, 2003; Verdugo and Rand, 2008a). The biology of the problem may be described as follows: A gene, i.e. a section of the DNA molecule, is copied *(transcribed)* onto messenger RNA (mRNA), which diffuses out of the nucleus of the cell into the cytoplasm, where it enters a subcellular structure called a ribosome. In the ribosome the genetic code on the mRNA produces a protein (a process called *translation)*. The protein then diffuses back into the nucleus where it represses the transcription of its own gene.

Dynamically speaking, this process may result in a steady state equilibrium, in which the concentrations of mRNA and protein are constant, or it may result in an oscillation. In this section we analyze a simple model previously proposed in the biological literature (Monk, 2003), and we show that the transition between equilibrium and oscillation is a Hopf bifurcation. The model takes the form of two equations, one an ordinary differential equation (ODE) and the other a delayed differential equation (DDE). The delay is due to an observed time lag in the transcription process.

Oscillations in biological systems with delay have been dealt with previously in (Mahaffy, 1988; Mahaffy et al., 1992; Mocek et al., 2005).
The model equations investigated here involve the variables $M(t)$, the concentration of mRNA, and $P(t)$, the concentration of the associated protein (Monk, 2003):

$$\dot{M} = \alpha_m \left(\frac{1}{1+\left(\frac{P_d}{P_0}\right)^n} \right) - \mu_m M, \tag{3.179}$$

$$\dot{P} = \alpha_p M - \mu_p P \tag{3.180}$$

where dots represent differentiation with respect to time t, and where we use the subscript d to denote a variable which is delayed by time T, thus $P_d = P(t-T)$. The model constants are as given in (Monk, 2003): α_m is the rate at which mRNA is transcribed in the absence of the associated protein, α_p is the rate at which the protein is produced from mRNA in the ribosome, μ_m and μ_p are the rates of degradation of mRNA and of protein, respectively, P_0 is a reference concentration of protein, and n is a parameter. We assume μ_m=μ_p=μ.

3.7.1 Stability of equilibrium

We begin by rescaling Eqs. (3.179) and (3.180). We set $m = \frac{M}{\alpha_m}$, $p = \frac{P}{\alpha_m \alpha_p}$, and $p_0 = \frac{P_0}{\alpha_m \alpha_p}$, giving:

$$\dot{m} = \frac{1}{1+\left(\frac{p_d}{p_0}\right)^n} - \mu m, \tag{3.181}$$

$$\dot{p} = m - \mu p. \tag{3.182}$$

Equilibrium points, (m^*, p^*), for (3.181) and (3.182) are found by setting $\dot{m} = 0$ and $\dot{p} = 0$

$$\mu m^* = \frac{1}{1+\left(\frac{p^*}{p_0}\right)^n}, \tag{3.183}$$

$$m^* = \mu p^*. \tag{3.184}$$

Eliminating m^* from Eqs. (3.183) and (3.184), we obtain an equation on p^*:

$$(p^*)^{n+1} + p_0^n p^* - \frac{p_0^n}{\mu^2} = 0. \tag{3.185}$$

Next we define ξ and η to be deviations from equilibrium: $\xi = \xi(t) = m(t) - m^*$, $\eta = \eta(t) = p(t) - p^*$, and $\eta_d = \eta(t-T)$. This results in the nonlinear system:

$$\dot{\xi} = \frac{1}{1+\left(\frac{\eta_d+p^*}{p_0}\right)^n} - \mu(m^*+\xi), \tag{3.186}$$

$$\dot{\eta} = \xi - \mu\eta. \tag{3.187}$$

Expanding for small values of η_d, Eq.(3.186) becomes:

$$\dot{\xi} = -\mu\xi - K\eta_d + H_2\eta_d^2 + H_3\eta_d^3 + \cdots \tag{3.188}$$

where K, H_2 and H_3 depend on p^*, p_0, and n as follows:

$$K = \frac{n\beta}{p^*(1+\beta)^2}, \quad \text{where} \quad \beta = \left(\frac{p^*}{p_0}\right)^n, \tag{3.189}$$

$$H_2 = \frac{\beta n(\beta n - n + \beta + 1)}{2(\beta+1)^3 p^{*2}}, \tag{3.190}$$

$$H_3 = -\frac{\beta n(\beta^2 n^2 - 4\beta n^2 + n^2 + 3\beta^2 n - 3n + 2\beta^2 + 4\beta + 2)}{6(\beta+1)^4 p^{*3}}. \tag{3.191}$$

Next we analyze the linearized system coming from Eqs. (3.188) and (3.187):

$$\dot{\xi} = -\mu\xi - K\eta_d, \tag{3.192}$$

$$\dot{\eta} = \xi - \mu\eta. \tag{3.193}$$

Stability analysis of Eqs. (3.192) and (3.193) shows that for $T = 0$ (no delay), the equilibrium point (m^*, p^*) is a stable spiral. Increasing the delay, T, in the linear system (3.192)–(3.193), will yield a critical delay, T_{cr}, such that for $T > T_{cr}$, (m^*, p^*) will be unstable, giving rise to a Hopf bifurcation. For $T = T_{cr}$ the system (3.192)–(3.193) will exhibit a pair of pure imaginary eigenvalues $\pm\omega i$ corresponding to the solution

$$\xi(t) = B\cos(\omega t + \phi), \tag{3.194}$$

$$\eta(t) = A\cos\omega t, \tag{3.195}$$

where A and B are the amplitudes of the $\eta(t)$ and $\xi(t)$ oscillations, and where ϕ is a phase angle. Note that we have chosen the phase of $\eta(t)$ to be zero without loss of generality. Then for values of delay T close to T_{cr},

$$T = T_{cr} + \Delta, \tag{3.196}$$

the nonlinear system (3.181)–(3.182) is expected to exhibit a periodic solution (a limit cycle) which can be written in the approximate form of Eqs. (3.194), (3.195). Substituting Eqs. (3.194) and (3.195) into Eqs. (3.192) and (3.193) and solving for

ω and T_{cr} we obtain

$$\omega = \sqrt{K-\mu^2}, \tag{3.197}$$

$$T_{cr} = \frac{\arctan\left(\frac{2\mu\sqrt{K-\mu^2}}{K-2\mu^2}\right)}{\sqrt{K-\mu^2}}. \tag{3.198}$$

3.7.2 Lindstedt's method

We use Lindstedt's Method (Rand and Verdugo, 2007) on Eqs. (3.188) and (3.187). We begin by changing the first order system into a second order DDE. This results in the following form:

$$\ddot{\eta} + 2\mu\dot{\eta} + \mu^2\eta = -K\eta_d + H_2\eta_d^2 + H_3\eta_d^3 + \cdots \tag{3.199}$$

where K, H_2 and H_3 are defined by Eqs. (3.189)–(3.191). We introduce a small parameter ε via the scaling

$$\eta = \varepsilon u. \tag{3.200}$$

The detuning Δ of Eq. (3.196) is scaled like ε^2, $\Delta = \varepsilon^2\delta$:

$$T = T_{cr} + \Delta = T_{cr} + \varepsilon^2\delta. \tag{3.201}$$

Next we stretch time by replacing the independent variable t by τ, where

$$\tau = \Omega t. \tag{3.202}$$

This results in the following form of Eq. (3.199):

$$\Omega^2\frac{d^2u}{d\tau^2} + 2\mu\Omega\frac{du}{d\tau} + \mu^2 u = -K u_d + \varepsilon H_2 u_d^2 + \varepsilon^2 H_3 u_d^3, \tag{3.203}$$

where $u_d = u(\tau - \Omega T)$. We expand Ω in a power series in ε, omitting the $O(\varepsilon)$ for convenience, since it turns out to be zero:

$$\Omega = \omega + \varepsilon^2 k_2 + \cdots \tag{3.204}$$

Next we expand the delay term u_d:

$$u_d = u(\tau - \Omega T) = u(\tau - (\omega + \varepsilon^2 k_2 + \cdots)(T_{cr} + \varepsilon^2\delta)) \tag{3.205}$$

$$= u(\tau - \omega T_{cr} - \varepsilon^2(k_2 T_{cr} + \omega\delta) + \cdots) \tag{3.206}$$

$$= u(\tau - \omega T_{cr}) - \varepsilon^2(k_2 T_{cr} + \omega\delta)u'(\tau - \omega T_{cr}) + O(\varepsilon^3). \tag{3.207}$$

Now we expand $u(\tau)$ in a power series in ε:

$$u(\tau) = u_0(\tau) + \varepsilon u_1(\tau) + \varepsilon^2 u_2(\tau) + \cdots \tag{3.208}$$

Substituting and collecting terms, we find:

$$\omega^2 \frac{d^2 u_0}{d\tau^2} + 2\mu\omega \frac{du_0}{d\tau} + K u_0(\tau - \omega T_{cr}) + \mu^2 u_0 = 0, \tag{3.209}$$

$$\omega^2 \frac{d^2 u_1}{d\tau^2} + 2\mu\omega \frac{du_1}{d\tau} + K u_1(\tau - \omega T_{cr}) + \mu^2 u_1 = H_2 u_0^2(\tau - \omega T_{cr}), \tag{3.210}$$

$$\omega^2 \frac{d^2 u_2}{d\tau^2} + 2\mu\omega \frac{du_2}{d\tau} + K u_2(\tau - \omega T_{cr}) + \mu^2 u_2 = \cdots \tag{3.211}$$

where $\cdots$ stands for terms in u_0 and u_1, omitted here for brevity. We take the solution of the u_0 equation as:

$$u_0(\tau) = \hat{A}\cos\tau, \tag{3.212}$$

where from Eqs. (3.195) and (3.200) we know $A = \hat{A}\varepsilon$. Next we substitute (3.212) into (3.38) and obtain the following expression for u_1:

$$u_1(\tau) = m_1 \sin 2\tau + m_2 \cos 2\tau + m_3, \tag{3.213}$$

where m_1 is given by the equation:

$$m_1 = -\frac{2\hat{A}^2 H_2 \mu \sqrt{K - \mu^2}\,(\mu^2 - K)\,(2\mu^2 - 3K)}{K\,(16\mu^6 - 39K\mu^4 + 18K^2\mu^2 + 9K^3)}, \tag{3.214}$$

and where m_2 and m_3 are given by similar equations, omitted here for brevity. We substitute Eqs. (3.212) and (3.213) into (3.40), and after trigonometric simplifications have been performed, we equate to zero the coefficients of the resonant terms $\sin\tau$ and $\cos\tau$. This yields the amplitude, A, of the limit cycle that was born in the Hopf bifurcation:

$$A^2 = \frac{P}{Q}\Delta, \tag{3.215}$$

where

$$P = -8K^2(\mu^4 - K^2)(16\mu^6 - 39K\mu^4 + 18K^2\mu^2 + 9K^3), \tag{3.216}$$

$$Q = Q_0 T_{cr} + Q_1, \tag{3.217}$$

and

$$\begin{aligned} Q_0 = {} & 48\mathrm{H}_3 K^2 \mu^8 + 16\mathrm{H}_2{}^2 K \mu^8 - 69\mathrm{H}_3 K^3 \mu^6 + 32\mathrm{H}_2{}^2 K^2 \mu^6 \\ & - 63\mathrm{H}_3 K^4 \mu^4 - 162\mathrm{H}_2{}^2 K^3 \mu^4 + 81\mathrm{H}_3 K^5 \mu^2 + 108\mathrm{H}_2{}^2 K^4 \mu^2 \\ & + 27\mathrm{H}_3 K^6 + 30\mathrm{H}_2{}^2 K^5, \end{aligned} \tag{3.218}$$

$$\begin{aligned} Q_1 = {} & 96\mathrm{H}_3 K \mu^9 + 64\mathrm{H}_2{}^2 \mu^9 - 138\mathrm{H}_3 K^2 \mu^7 - 16\mathrm{H}_2{}^2 K \mu^7 \\ & - 126\mathrm{H}_3 K^3 \mu^5 - 308\mathrm{H}_2{}^2 K^2 \mu^5 + 162\mathrm{H}_3 K^4 \mu^3 + 296\mathrm{H}_2{}^2 K^3 \mu^3 \\ & + 54\mathrm{H}_3 K^5 \mu + 12\mathrm{H}_2{}^2 K^4 \mu. \end{aligned} \tag{3.219}$$

Eq. (3.217) depends on μ, K, H_2, H_3, and T_{cr}. By using Eq. (3.198) we may express Eq. (3.217) as a function of μ, K, H_2, and H_3 only. Removal of secular terms also yields a value for the frequency shift k_2, where, from Eq. (3.204), we have $\Omega = \omega + \varepsilon^2 k_2$:

$$k_2 = -\frac{R}{Q}\,\delta. \tag{3.220}$$

where Q is given by (3.217) and

$$R = \sqrt{K - \mu^2}\,Q_0. \tag{3.221}$$

An expression for the amplitude B of the periodic solution for $\xi(t)$ (see Eq. (3.194)) may be obtained directly from Eq. (3.187) by writing $\xi = \dot{\eta} + \mu\eta$, where $\eta \sim A\cos\omega t$. We find:

$$B = \sqrt{K}A, \tag{3.222}$$

where K and A are given as in (3.189) and (3.215) respectively.

3.7.3 Numerical example

Using the same parameter values as in (Monk, 2003)

$$\mu = 0.03/\text{min},\ p_0 = 100\,,\ n = 5, \tag{3.223}$$

we obtain

$$p^* = 145.9158\,,\ m^* = 4.3774, \tag{3.224}$$

$$K = 3.9089 \times 10^{-3}\,,\ H_2 = 6.2778 \times 10^{-5}\,,\ H_3 = -6.4101 \times 10^{-7}, \tag{3.225}$$

$$T_{cr} = 18.2470\,,\ w = 5.4854 \times 10^{-2}\,,\ \frac{2\pi}{w} = 114.5432. \tag{3.226}$$

Here the delay T_{cr} and the response period $2\pi/\omega$ are given in minutes. Substituting (3.224)–(3.226) into (3.215)–(3.222) yields the following equations:

$$A = 27.0215\,\sqrt{\Delta}, \tag{3.227}$$

$$k_2 = -2.4512 \times 10^{-3}\,\delta, \tag{3.228}$$

$$B = 1.6894\,\sqrt{\Delta}. \tag{3.229}$$

Note that since Eq. (3.227) requires $\Delta > 0$ for the limit cycle to exist, and since we saw in Eqs. (3.192) and (3.193) that the origin is unstable for $T > T_{cr}$, i.e. for $\Delta > 0$, we may conclude that the Hopf bifurcation is supercritical, i.e., the limit cycle is stable.

Multiplying (3.228) by ε^2 and substituting into (3.204) we obtain:

$$\Omega\ =\ 5.4854 \times 10^{-2} - 2.4512 \times 10^{-3}\,\Delta \tag{3.230}$$

where $\Delta = T - T_{cr} = T - 18.2470$. Plotting the period, $\frac{2\pi}{\Omega}$, against the delay, T, yields the graph shown in Fig. 3.6. These results are in agreement with those obtained by numerical integration of the original Eqs. (3.179) and (3.180) and with those presented in (Monk, 2003).

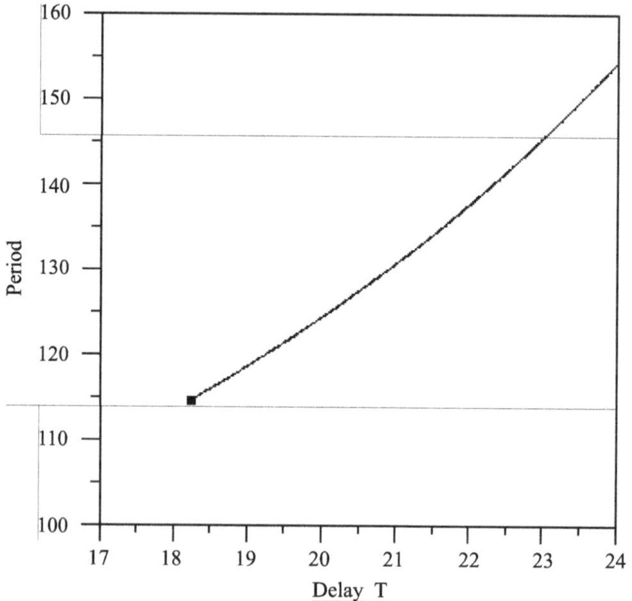

Fig. 3.6 Period of oscillation, $\frac{2\pi}{\Omega}$, plotted as a function of delay T, where Ω is given by Eq.(3.230). The initiation of oscillation at $T = T_{cr} = 18.2470$ is due to a supercritical Hopf bifurcation, and is marked in the Figure with a dot.

3.8 Exercises

Exercise 1

For which values of the delay $T > 0$ is the trivial solution in the following DDE stable?

$$\frac{dx(t)}{dt} = x(t) - 2x(t-T). \tag{3.231}$$

Exercise 2

Use Lindstedt's method to find an approximation for the amplitude of the limit cycle in the following DDE:

$$\frac{dx(t)}{dt} = -x(t-T) + x(t-T)^3. \tag{3.232}$$

Exercise 3
Use the center manifold approach to determine the stability of the x=0 solution in the following DDE:

$$\frac{dx(t)}{dt} = -x(t-\frac{\pi}{2})(1+x(t)). \tag{3.233}$$

Here is an outline of the steps involved in this complicated calculation:

1. Show that the parameters of the linearized equation

$$\frac{dx(t)}{dt} = -x\left(t-\frac{\pi}{2}\right), \tag{3.234}$$

have been chosen so that the delay is set at its critical value for a Hopf bifurcation, i.e. the characteristic equation has a pair of pure imaginary roots, $\lambda = \pm\omega i$. Find ω.
2. Find the eigenfunctions $s_1(\theta)$, $s_2(\theta)$ and the adjoint eigenfunctions $n_1(\theta)$, $n_2(\theta)$. These are determined by Eqs.(3.98)–(3.101),(3.113)–(3.116), where the constants c_i, d_i are related by the orthonomality conditions (3.112), in which the bilinear form (v,u) is given by Eq.(3.107).
3. By comparing Eq.(3.233) with the general form (3.90), identify α, β, and f for this system. This will permit you to write down Eqs.(3.155) and (3.156), in which (n_i, Fx_t) is given by Eq.(3.157) and $x_t = y_1(t)s_1(\theta) + y_2(t)s_2(\theta)$.
4. Equate coefficients of y_1^2, y_1y_2 and y_2^2 in Eqs.(3.155) and (3.156) and so obtain 3 first order linear ODE's on $m_1(\theta)$, $m_2(\theta)$ and $m_3(\theta)$, together with 3 boundary conditions.
5. Solve these for $m_i(\theta)$.
6. Substitute the resulting expressions for $m_i(\theta)$ into Eq.(3.136) for the center manifold.
7. Substitute your expression for the center manifold into the y_1-y_2 Eqs.(3.132). Here (n_i, Fx_t) is given by Eq.(3.134) and $x_t = y_1(t)s_1(\theta) + y_2(t)s_2(\theta) + w(t)(\theta)$.
8. Compute Q from Eq.(3.138).
   Answer: $Q = -\frac{4\pi}{5}\frac{(3\pi-2)}{(\pi^2+4)^2}$

Acknowledgements The author wishes to thank his associates Anael Verdugo, Si Mohamed Sah, Meghan Suchorsky, and Tamas Kalmar-Nagy for their help in preparing this work.

References

Camacho E., Rand R. and Howland, H., 2004, Dynamics of two van der Pol oscillators coupled via a bath, *International J. Solids and Structures,* **41**, 2133-2143.

Campbell S.A., Belair J., Ohira T. and Milton J., 1995, Complex dynamics and multistability in a damped harmonic oscillator with delayed negative feedback, *Chaos*, **5**, 640-645.

Casal A. and Freedman M., 1980, A Poincare-Lindstedt approach to bifurcation problems for differential-delay equations, *IEEE Transactions on Automatic Control*, **25**, 967-973.

Cole J.D., 1968, *Perturbation Methods in Applied Mathematics,* Blaisdell, Waltham, MA.

Das S.L. and Chatterjee A., 2002, Multiple scales without center manifold reductions for delay differential equations near Hopf bifurcations, *Nonlinear Dynamics*, **30**, 323-335.

Das S.L. and Chatterjee A., 2005, Second order multiple scales for oscillators with large delay, *Nonlinear Dynamics*, **39**, 375-394.

Guckenheimer J. and Holmes P., 1983, *Nonlinear Oscillations, Dynamical Systems, and Bifurcations of Vector Fields,* Springer, New York.

Hassard B.D., Kazarinoff N.D. and WanY-H., 1981, *Theory and Applications of Hopf Bifurcation*, Cambridge University Press, Cambridge.

Kalmar-Nagy T., Stepan G. and Moon F.C., 2001, Subcritical Hopf bifurcation in the delay equation model for machine tool vibrations, *Nonlinear Dynamics*, **26**, 121-142.

Kot M., 1979, *Delay-differential Models and the Effects of Economic Delays on the Stability of Open-access Fisheries,* M.S. thesis, Cornell University.

Mahaffy J.M., 1988, Genetic control models with diffusion and delays, *Mathematical Biosciences*, **90**, 519-533.

Mahaffy J.M., Jorgensen D.A. and Vanderheyden R.L., 1992, Oscillations in a model of repression with external control, *J. Math.Biology*, **30**, 669-691.

Mocek W.T., Rudbicki R. and Voit E.O., 2005, Approximation of delays in biochemical systems, *Mathematical Biosciences*, **198**, 190-216.

Monk N.A.M., 2003, Oscillatory expression of Hes1, p53, and NF-κB driven by transcriptional time delays, *Current Biology*, **13**, 1409-1413.

Nayfeh A.H., 1973, *Perturbation Methods,* Wiley, New York.

Nayfeh A.H., 2008, Order reduction of retarded nonlinear systems – the method of multiple scales versus center-manifold reduction, *Nonlinear Dynamics*, **51**, 483-500.

Rand R.H. and Armbruster D., 1987, *Perturbation Methods, Bifurcation Theory and Computer Algebra,* Springer, New York .

Rand R.H., 2005, *Lecture Notes on Nonlinear Vibrations (version 52)*, available online at http://audiophile.tam.cornell.edu/randdocs.

Rand R. and Verdugo A., 2007, Hopf Bifurcation formula for first order differential-delay equations, *Communications in Nonlinear Science and Numerical Simulation*, **12**, 859-864.

Sanders J.A. and Verhulst F., 1985, *Averaging Methods in Nonlinear Dynamical Systems* , Springer, New York.

Stepan G., 1989, *Retarded Dynamical Systems: Stability and Characteristic Functions,* Longman Scientific and Technical, Essex.

Verdugo A. and Rand R., 2008a, Hopf bifurcation in a DDE model of gene expression, *Communications in Nonlinear Science and Numerical Simulation*, **13**, 235-242.

Verdugo A. and Rand R., 2008b, Center manifold analysis of a DDE model of gene expression, *Communications in Nonlinear Science and Numerical Simulation*, **13**, 1112-11120.

Wang H. and Hu H., 2003, Remarks on the perturbation methods in solving the second-order delay differential equations, *Nonlinear Dynamics*, **33**, 379-398.

Wirkus S. and Rand R., 2002, Dynamics of two coupled van der Pol oscillators with delay coupling, *Nonlinear Dynamics*, **30**, 205-221.

Chapter 4
Analysis and Control of Deterministic and Stochastic Dynamical Systems with Time Delay

Jian-Qiao Sun, Bo Song

Abstract This chapter presents a comprehensive summary of recent advances in the analysis and control of time-delayed deterministic and stochastic systems. The studies of numerical methods for time-delayed systems in the mathematics literature are reviewed including a discussion of the abstract Cauchy problem for delayed differential equations. Several numerical methods for computing the response of and designing controls for time-delayed systems are presented. These include semi-discretization, continuous time approximation, lowpass filter based continuous time approximation, and continuous time approximation with Chebyshev nodes. A large number of examples are presented including optimal feedback gain design, stability domains in the feedback gain space of linear time-invariant and periodic systems, optimal control, Lyapunov stability, supervisory control of systems with uncertain time delay, moment stability, Fokker-Planck-Kolmogorov equation and reliability formulation of stochastic systems.

4.1 Introduction

Time delay is a common phenomenon in engineering, economical and biological systems. It is caused by signal transportation and communication lags, feedback delays and retarded hardware responses. It also arises when high order industry processes are approximated by low-order models with delay (Camacho and Bordons, 1999). Other than a few cases, time delay is undesirable. Control strategies to eliminate or minimize unwanted effects are often employed. Effects of time delay on the stability and performance of control systems have been a subject of many studies. For time-invariant linear systems with time delay, several methods are available

---

Jian-Qiao Sun, Bo Song
School of Engineering, University of California, Merced, CA 9533, USA.
Emails: jqsun@ucmerced.edu, bosong1979@gmail.com

for the design of PID controls and stability analysis. These methods including root locus and Nyquist criterion are quite mature.

4.1.1 Deterministic systems

Stability conditions of delayed time-varying systems have been extensively studied in the literature. The Lyapunov approach is a popular method to use (Wu and Mizukami, 1995; Kapila and Haddad, 1999). A non-Lyapunov based stability study of linear time-varying system by the Gauss-Seidel iteration is presented in (Xiao and Liu, 1994). An unconditional stability criterion is derived in (Li et al., 1989) for time-varying discrete systems. A study on stability and performance of feedback controls with multiple time delays is reported in (Ali et al., 1998) by considering the roots of the closed-loop characteristic equation. A survey of methods for stability analysis of deterministic delayed linear systems is presented in (Niculescu et al., 1998). Another excellent survey of stability and control of time-delayed systems can be found in (Gu and Niculescu, 2003).

Cao, Lin and Hu (2002), Fridman and Orlov (2009), and Kim (2008) analyzed stability of linear systems with time delay using the Lyapunov approach. Fan and Chan discussed asymptotic stability problem for a class of neutral systems with discrete and distributed delays via linear matrix inequality (Fan et al., 2002). de Oliveira and Geromel focused on synthesis of non-rational controllers for linear delay systems (de Oliveira and Geromel, 2004). Stabilization and performance design problems are expressed in terms of linear matrix inequalities. Gao, Chen and Lam studied stability and H_∞ controls of systems with two successive delay components (Gao et al., 2008). Han concerned with stability of linear time-delay systems of both retarded and neutral types by using time-independent and time-dependent Lyapunov-Krasovskii functional (Han, 2009). Ivanescu, Dion, Dugard and Niculecu (2000), Kolmanovskii and Richard (1999), Zhang, Tsiotras and Knospegave (2002) investigated delay-dependent and delay-independent stability conditions. Shao provided improved delay-dependent stability criteria for systems with a varying delay in a range (Shao, 2008). Xia and Jia considered the problem of robust stability and stabilization of linear systems with a constant time-delay in the state and subject to real convex polytopic uncertainty (Xia and Jia, 2003).

Although time delay is considered as a undesired characteristics which is frequently a source of instability and complicates the analysis and design in most of the applications, positive uses of time delay have also been investigated. From the early 1950s (Tallman and Smith, 1958) to the more recent days (Suh and Bien, 1979, 1980; Shanmugathasan and Johnston, 1988; Kwon et al., 1990), time delay has been used to improve system performance in various ways. The time-delayed feedback control is designed in (Fujii et al., 2000) to regulate the librational motion of gravity-gradient satellites in an elliptic orbit. Delayed feedback laws are investigated in (Atay, 2002) to control the amplitude of oscillations in planar systems with general nonlinearities. Olgac and his associates have published extensively on the use of delayed resonator for vibration suppression (see e.g., (Filipovic and Olgac, 1998)). Space teleoperation is notorious for time delays (Nohmi and Matsumoto, 2002). In reference (Singh, 1995), a time-delay filter is developed to design a fuel/time optimal control. A sampled-data control system is studied in (Ha and Ly, 1996) with a consideration of computation time delay. Yang and Wu (1998) and Stepan (1998) have studied structural systems with time delay. Over the years, researchers have come to a realization that the model predictive control offers a good tool to deal with time delay (Dumont et al., 1993; Normey-Rico and Camacho, 1999; Rawlings, 2000).

There have also been many studies of control systems with unknown and time-varying time delays. Chen et al. derived sufficient conditions for the existence of the guaranteed cost output-feedback controller in terms of matrix inequalities for uncertain dynamical systems with time delay (Chen et al., 2004). The Lyapunov method is used in (He et al., 2007) for the stability analysis of systems with time-varying delay with known lower and upper bounds. The Lyapunov function dependent on the known upper bound of uncertain state-delays is derived in the study of model predictive controls (MPC) for a constrained linear digital systems with uncertain state-delays (Hu and Chen, 2004). A class of iterative learning control systems with uncertain state delay and control delay is studied in (Ji and Luo, 2006). Robust stability of uncertain linear systems with interval time-varying delay is studied in (Jiang and Han, 2008). Stability of systems with bounded uncertain time-varying bounded delays in the feedback loop is studied in (Kao and Rantzer, 2007). The stability problem is treated in the integral quadratic constraint (IQC) framework. Kwon, Park and Lee investigated delay-dependent robust stability for neutral systems with the help of the Lyapunov method (Kwon et al., 2008). The system has time-varying structured uncertainties and interval time-varying delays. A compensation scheme consisting of a fuzzy-PID controller and a neural network compensator is proposed for real-time control over the network (Lin et al., 2008). This scheme reduces the influence of time delays on stability while maintaining the system performance. According to (Miller and Davison, 2005), given a finite-dimensional LTI plant and an upper bound on the admissible time delay, there is no general theory for designing a controller to handle an arbitrarily large uncertain delay. The authors show that given a finite-dimensional LTI plant and an upper bound on the admissible time delay, there exists a linear periodic controller which robustly stabilizes the plant. Robust stability of systems with random time-varying delay is studied in (Yue et al., 2009). The resulting system model has stochastic parameters. Sufficient conditions for the exponential mean square stability of the system are derived by using the Lyapunov functional method and the linear matrix inequality (LMI) technique.

When the uncertain time delay is bounded with known lower and upper bounds, the supervisory control (Morse, 1996, 1997; Hespanha et al., 1999, 2003) can be considered. The supervisory control proposes to use several estimates of uncertain parameters for the system model. For each estimate of the parameter, a control is designed to achieve the desired performance. A supervisor monitors the real-time response of the system, selects a plant model according to a switching criterion and implements the corresponding control. This chapter presents one such example of the supervisory control of systems with uncertain time delay.

4.1.2 Stochastic systems

There is a strong interest in the stochastic systems with time delay. An effective Monte Carlo simulation scheme that converges in a weak sense is presented by Kuchler and Platen (2002). Buckwar has studied numerical solutions of Itô type differential equations and their convergence where the system considered has time delay both in diffusion and drift terms (Buckwar, 2000). Guillouzic, L'Heureux and Longtin have studied first order delayed Itô differential equations using a small delay approximation and obtained probability density functions (PDF) as well as the second order statistics analytically (Guillouzic et al., 1999). Frank and Beek have obtained the PDFs using the Fokker-Planck-Kolmogorov (FPK) equation for linear delayed stochastic systems and studied the stability of fixed point solutions in biological systems (Frank and Beek, 2001). State feedback stabilization of nonlinear time delayed stochastic systems are investigated by Fu, Tian and Shi (2003) where a Lyapunov approach is used.

4.1.3 Methods of solution

Several methods of solution are available in the literature.

A method that fully discretizes the delayed control system in time domain has been extensively studied. Pinto and Goncalves (2002) have fully discretized a nonlinear SDOF system to study control problems with time delay. Klein and Ramirez (2001) have studied MDOF delayed optimal regulators with a hybrid discretization technique where the state equation is partitioned into discrete and continuous portions. Cai and Huang have studied optimal vibration controller with a delayed feedback where standard discretization techniques are used (Cai and Huang, 2002). Time-delayed systems have been studied using discretization techniques with an extended state vector. The Smith predictor is a well-known method (Smith, 1957) that proposes a compensator to stabilize the feedback control designed for the system without time delay.

A method using Chebyshev polynomials to approximate general nonlinear functions of time has been developed to handle linear and nonlinear time-delayed dynamical systems with periodic coefficients (Deshmukh et al., 2008, 2006; Ma et al., 2005, 2003). The method has also been applied to study optimal control problems. A temporal finite element method has been proposed in (Garg et al., 2007) to study the stability of time-delayed systems with parametric excitations. The work reported in (Kalmar-Nagy, 2005) makes use of the piece-wise exact solution of linear differential equations with a single time delay to create a map in order to study the stability of the system.

The semi-discretization (SD) is a well-established method in the literature and used widely in structural and fluid mechanics (Pfeiffer and Marquardt, 1996; Leugering, 2000). The method has been applied to delayed deterministic dynamical systems by Insperger and Stepan (Insperger and Stepan, 2001, 2002). The method has been extended to control systems with delayed feedback (Sheng et al., 2004; Sheng and Sun, 2005). The effect of various higher order approximations in semi-discretization on the computational efficiency and accuracy has been examined in (Elbeyli and Sun, 2004). The merit of the semi-discretization method as introduced by Insperger and Stepan lies in that it makes use of the exact solution of linear systems over a short time interval to construct the mapping of a finite dimensional state vector for the system with time delay. The disadvantage of the method is that it becomes difficult to handle multiple independent time delays with the mapping as well as nonlinear dynamical systems. The continuous time approximation (CTA) method is an extension of the method of semi-discretization and provides an alternative to handle systems with multiple independent time delays (Sun, 2009). The CTA method has been applied to study control problems of the time-delayed linear dynamical systems, and stochastic dynamical systems with time delay.

Most numerical methods for the solution and stability analysis of time-delayed systems focus on approximation of temporal responses of the system, and are not specifically developed to meet frequency domain requirements such as accurate representation of poles and zeros of the original system. Numerical methods based on the abstract Cauchy problem for computing the right-most characteristic roots of delay differential equations (DDEs) are presented in (Engelborghs and Roose, 2002; Breda et al., 2004, 2005). The convergence and stability of the method with the Chebyshev polynomial expansion of the delayed response are discussed. The abstract Cauchy problem can be stated in terms of a PDE, which is open to various numerical methods for solutions. A finite difference method to solve the differential-difference equation of the time-delayed system and the stability of the method are presented in (Bellen and Maset, 2000), and a method of lines for solving the PDE of the time-delayed system is investigated in (Maset, 2003; Koto, 2004). These methods are the same as the CTA method (Sun, 2009). The higher order Runge-Kutta methods and their convergence are studied in (Maset, 2003). The implicit-explicit (IMEX) linear multistep Runge-Kutta method for DDEs is studied in (Koto, 2009). The book (Bellen and Zennaro, 2003) presents a comprehensive discussion of studies of numerical methods for DDEs up to 2003. The Padé approximation of the transfer function is a method in frequency domain (Franklin et al., 1986; Vijta, 2000). This method provides a rational approximation of the transfer function of the time-delayed system, which contains the exponential term $e^{-\tau s}$ due to time delay. However, it does not focus on the accurate representation of the infinite number of poles and zeros of the transfer function, in particular, the dominant poles.

4.1.4 Outline of the chapter

The chapter is organized as follows. Section 4.2 reviews the abstract Cauchy problem, and points out that many numerical methods can be derived in this framework. Section 4.3 presents the method of semi-discretization. Section 4.4 discusses the method of continuous time approximation. Section 4.5 studies the spectral properties of these two methods. Section 4.6 presents a comparative study of stability of time-delayed linear time invariant systems by the Lyapunov method, Padé approximation and semi-discretization. A number of control examples and an experimental validation are presented in Sections 4.7 to 4.9. Section 4.10 presents a supervisory control of systems with unknown time delays.

We then switch our interest to stochastic systems with time delay. In Sections 4.11 and 4.12, we review the methods of solution for stochastic dynamical systems with time delay. Section 4.13 presents several examples of stability and response analysis of stochastic systems with time delay.

4.2 Abstract Cauchy problem for DDE

The discussion in this section follows closely the reference (Bellen and Zennaro, 2003). Consider an n-dimensional system with k discrete time delays

$$\begin{aligned}\dot{\mathbf{x}}(t) &= \mathbf{A}_0\mathbf{x}(t) + \sum_{l=1}^{k}\mathbf{A}_l\mathbf{x}(t-\tau_l) \equiv \mathbf{g}(\mathbf{x}(t)), \\ \mathbf{x}(\theta) &= \varphi(\theta), \quad \theta \in [-\tau, 0],\end{aligned} \tag{4.1}$$

where $\mathbf{x}(t) \in \Re^n$, $\mathbf{A}_0, \mathbf{A}_1, \ldots, \mathbf{A}_k \in \Re^{n\times n}$, $0 = \tau_0 < \tau_1 < \cdots < \tau_k = \tau$ and $\mathbf{g}(\mathbf{x}(t)) = \sum_{l=0}^{k}\mathbf{A}_l\mathbf{x}(t-\tau_l)$. The solution operator $\mathbf{T}(t)$ $(t \geq 0)$ of the system (4.1) is defined by

$$\mathbf{T}(t)\varphi(\theta) = \mathbf{x}(t+\theta), \quad \varphi(\theta) \in \mathbf{X}, \tag{4.2}$$

where the Banach space $\mathbf{X} = C([-\tau, 0], \Re^n)$ is endowed with the maximum norm

$$||\varphi(\theta)|| = \max_{\theta\in[-\tau,0]} |\varphi(\theta)|, \quad \varphi \in \mathbf{X}, \tag{4.3}$$

and $\mathbf{x}(t+\theta)$ denotes the solution of Eq. (4.1) with the initial condition $\varphi(\theta) \in \mathbf{X}$. The family $\{\mathbf{T}(t)\}_{t\geq 0}$ is a C_0-semigroup with an infinitesimal generator $\mathbf{A} : D(\mathbf{A}) \subseteq \mathbf{X} \rightarrow \mathbf{X}$ given by

$$\mathbf{A}\varphi = \frac{d\varphi(\theta)}{d\theta}, \quad \varphi \in D(\mathbf{A}), \tag{4.4}$$

where the domain $D(\mathbf{A})$ is defined as

$$D(\mathbf{A}) = \left\{ \varphi \in \mathbf{X} : \frac{d\varphi(\theta)}{d\theta} \in \mathbf{X} \text{ and } \frac{d\varphi}{d\theta}(0) = \sum_{l=0}^{k} \mathbf{A}_l \varphi(-\tau_l) \right\}. \tag{4.5}$$

The equation

$$\frac{d\varphi}{d\theta}(0) = \sum_{l=0}^{k} \mathbf{A}_l \varphi(-\tau_l) \tag{4.6}$$

is known as the *splicing condition*. When $\varphi(\theta)$ satisfies the splicing condition, the response $\mathbf{x}(t)$ of the DDE is C^1-continuous. When the splicing condition is not satisfied, $\mathbf{x}(t)$ is C^0-continuous (Bellen and Zennaro, 2003). The discontinuity in $\varphi(\theta)$ propagates to the solution $\mathbf{x}(t)$ making it nonsmooth.

The system (4.1) can be restated as an abstract Cauchy problem in terms of the infinitesimal generator

$$\begin{aligned} \frac{d\mathbf{x}(t+\theta)}{dt} &= \mathbf{A}(\theta)\mathbf{x}(t+\theta), \ t > 0, \\ \mathbf{x}(\theta) &= \varphi(\theta), \ \theta \in [-\tau, 0]. \end{aligned} \tag{4.7}$$

It should be pointed out that in general, the infinitessimal operator $\mathbf{A}$ is a function of time delay index θ. Hence, this simple looking linear system implicitly lives in an infinite dimensional state space.

Introduce a function

$$\mathbf{v}(t,\theta) = \mathbf{x}(t+\theta), \ \ t \geq 0, \ \ -\tau \leq \theta \leq 0. \tag{4.8}$$

Equations (4.4) and (4.7) lead to a hyperbolic PDE for $\mathbf{v}(t,\theta)$

$$\frac{\partial \mathbf{v}}{\partial t}(t,\theta) = \frac{\partial \mathbf{v}}{\partial \theta}(t,\theta), \tag{4.9}$$

with a boundary condition

$$\frac{\partial \mathbf{v}}{\partial \theta}(t,0) = \mathbf{g}(\mathbf{v}(t,0)), \ \ t \geq 0, \tag{4.10}$$

and an initial condition

$$\mathbf{v}(0,\theta) = \varphi(\theta), \ \ \theta \in [-\tau, 0]. \tag{4.11}$$

The abstract Cauchy problem (4.7) and the PDE (4.9) do not contain time delay explicitly and are amenable to various numerical methods of integration (Maset, 2003; Bellen and Maset, 2000; Koto, 2009; Bellen and Zennaro, 2003). The methods of semi-discretization and continuous time approximation can also be derived from the abstract Cauchy problem (4.7) and the PDE (4.9) (Elbeyli and Sun, 2004; Sun, 2009).

Next, we construct a discrete approximation of $\mathbf{A}$. Consider a mesh $\mathbf{\Omega}_N = \{\tau_{N,i}, i = 0, 1, \ldots, N\}$ of $N+1$ points in $[0, \tau]$ such that $0 = \tau_{N,0} < \tau_{N,1} <$

$\cdots < \tau_{N,N} = \tau$. The continuous space $\mathbf{X}$ is replaced by the space $\mathbf{X}_N$ of the discrete functions defined on the mesh Ω_N. That is, $\mathbf{x}(t+\theta)$ is discretized into a block-vector

$$\begin{aligned} \mathbf{y}(t) &= [\mathbf{x}(t), \mathbf{x}(t-\tau_{N,1}), \ldots, \mathbf{x}(t-\tau_{N,N})]^{\mathrm{T}} \\ &\equiv [\mathbf{y}_0(t), \mathbf{y}_1(t), \mathbf{y}_2(t), \ldots, \mathbf{y}_N(t)]^{\mathrm{T}}. \end{aligned} \tag{4.12}$$

Let $(\mathbf{L}_N\mathbf{y})(\theta)$ be the unique $\Re^{n(N+1)}$ valued interpolating polynomial of degree N with $(\mathbf{L}_N\mathbf{y})(\tau_{N,i}) = \mathbf{y}_i(t)$. In particular, $(\mathbf{L}_N\mathbf{y})(\tau_{N,0}) = \mathbf{y}_0(t) = \mathbf{x}(t)$.

The infinitesimal generator $\mathbf{A}$ is approximated by a spectral differentiation matrix $\mathbf{A}_N$ determined by the following equations

$$\begin{aligned} \dot{\mathbf{y}}_0 &= \dot{\mathbf{x}}(t) = \mathbf{f}(\mathbf{L}_N\mathbf{y}(\tau_{N,0})) = (\mathbf{A}_N\mathbf{y}(t))_0, \\ \dot{\mathbf{y}}_i &= -\frac{d(\mathbf{L}_N\mathbf{y})}{d\tau}(\tau_{N,i}) = (\mathbf{A}_N\mathbf{y}(t))_i, \quad i = 1, \ldots, N. \end{aligned} \tag{4.13}$$

It should be noted that the term $(\mathbf{A}_N\mathbf{y}(t))_0$ in the first equation should be interpreted in the sense of the operator as defined by $\mathbf{f}(\mathbf{L}_N\mathbf{y}(\tau_{N,0}))$. Other terms $(\mathbf{A}_N\mathbf{y}(t))_i$ $(i = 1, \ldots, N)$ are matrix multiplications. The system (4.7) now reads

$$\dot{\mathbf{y}}(t) = \mathbf{A}_N\mathbf{y}(t). \tag{4.14}$$

Note that $\mathbf{y}(t)$ is $n(N+1)\times 1$, and $\mathbf{A}_N$ is $n(N+1)\times n(N+1)$. The initial condition reads $\mathbf{y}(0) = [\varphi(0), \varphi(-\tau_{N,1}), \ldots, \varphi(-\tau_{N,N})]^{\mathrm{T}}$. In a nutshell, Eq. (4.14) is a continuous time approximation of the system (4.1), which is the same as the one presented in (Sun, 2009). A detailed matrix representation of $\mathbf{A}_N$ with the Chebyshev external nodes on $[0, -\tau]$ is presented in (Breda et al., 2004, 2005).

4.2.1 Convergence with Chebyshev nodes

Let $B(\lambda, \rho)$ be a closed ball in $\mathbf{C}$ centered at λ with radius ρ where λ is the eigenvalue of the original system. It is shown in (Breda et al., 2004, 2005) that when the Chebyshev external nodes on $[0, -\tau]$ are used for $(\mathbf{L}_N\mathbf{y})(\theta)$, the maximum error $e_{\max}$ of the collocation polynomial is bounded above by

$$e_{\max} \leq \frac{C_0}{\sqrt{N}}\left(\frac{C_1}{N}\right)^N ||\varphi||, \tag{4.15}$$

where C_0 and C_1 are constants determined by λ and ρ, but independent of N. Furthermore,

$$\max_{1\leq i\leq \nu} |\lambda - \lambda_i| \leq \rho_N, \; i = 1, \ldots, \nu, \tag{4.16}$$

where λ_i denotes the eigenvalue of the matrix $\mathbf{A}_N$ of multiplicity ν that matches the exact eigenvalue λ of the original system, and

$$\rho_N = \left(\frac{C_2}{C_3}\right)^{1/\nu} \left(\frac{1}{\sqrt{N}} \left(\frac{C_1}{N}\right)^N\right)^{1/\nu}, \tag{4.17}$$

with C_2 and $C_3 = C_3(\lambda)$ are constants. It should be noted that this convergence analysis does not apply to the extraneous eigenvalues introduced by the discretization, which don't match any true eigenvalues of the original system. It also does not address the accuracy of temporal response prediction. We shall numerically examine this issue in the example section, and further show that these extraneous eigenvalues are not negligible, and contribute to temporal responses.

For the convergence and stability analysis when the finite difference and Runge-Kutta methods are used to derive Eq. (4.14), the readers are referred to (Maset, 2003; Bellen and Maset, 2000; Koto, 2009; Bellen and Zennaro, 2003).

4.3 Method of semi-discretization

Consider a linear periodic system with time delay

$$\dot{\mathbf{x}}(t) = \mathbf{A}(t)\mathbf{x}(t) + \mathbf{A}_d(t)\mathbf{x}(t-\tau) + \mathbf{B}(t)\mathbf{u}(t), \tag{4.18}$$

where $\mathbf{x} \in \Re^n$ and $\mathbf{u} \in \Re^m$. $\mathbf{A}(t) \in \Re^{n\times n}$, $\mathbf{A}_d(t) \in \Re^{n\times n}$ and $\mathbf{B}(t) \in \Re^{n\times m}$ are periodic matrices with period T. We shall consider a feedback control with or without time delay in the following forms

$$\mathbf{u}(t) = -\mathbf{K}\mathbf{x}(t) \text{ or } \mathbf{u}(t) = -\mathbf{K}\mathbf{x}(t-\tau), \tag{4.19}$$

where $\mathbf{K} \in \mathbf{R}^{m\times n}$ is the gain matrix.

When we introduce the method of semi-discretization, we can focus on the system in Eq. (4.20) without loss of generality, because in the closed loop system, the control simply modifies the matrix $\mathbf{A}(t)$ or $\mathbf{A}_d(t)$.

$$\dot{\mathbf{x}}(t) = \mathbf{A}(t)\mathbf{x}(t) + \mathbf{A}_d(t)\mathbf{x}(t-\tau). \tag{4.20}$$

The time delay significantly complicates the solution process of the system, because the state vector of the system is no longer just $\mathbf{x}(t)$, but $(\mathbf{x}^{\mathrm{T}}(t), \mathbf{x}^{\mathrm{T}}(t-\tau_1))^{\mathrm{T}}$ for all $0 < \tau_1 \leq \tau$, which has an infinite dimension.

Let us discretize the period T into an integer k intervals of length Δt such that $T = k\Delta t$. For the sake of simplicity, we assume that the time delay $\tau = N\Delta t$ where $N < k$ is an integer. When an integer cannot be found, discretization of the time delay τ will be approximate. Details on how to treat this case can be found in (Insperger and Stepan, 2001). The CTA method introduced next can handle this issue naturally.

Consider Eq. (4.20) in a time interval $t \in [t_i, t_{i+1}]$ where $t_i = i\Delta t,\ i = 0, 1, 2, \ldots, k$. In each small time interval $[t_i, t_{i+1}]$, the delayed responses $\mathbf{x}(t-\tau)$ and the time dependent coefficients are assumed to be constant. We denote

$$\begin{aligned} &\mathbf{x}(t_i - \tau) = \mathbf{x}((i-N)\Delta t) = \mathbf{x}_{i-N}, \\ &\mathbf{A}(t_i) = \mathbf{A}_i, \ \ \mathbf{A}_d(t_i) = \mathbf{A}_{di}. \end{aligned} \tag{4.21}$$

Equation (4.20) becomes

$$\dot{\mathbf{x}}(t) - \mathbf{A}_i \mathbf{x}(t) = \mathbf{A}_d(t)\mathbf{x}(t-\tau)\ t \in [t_i, t_{i+1}], \ \ i = 0, 1, 2, \ldots, k. \tag{4.22}$$

The general solution of the equation is

$$\begin{aligned} \mathbf{x}(t) = e^{\mathbf{A}_i(t-t_i)}\mathbf{x}_i + \int_{t_i}^{t} e^{\mathbf{A}_i(t-t_i-\hat{t})}\mathbf{A}_d(\hat{t})\mathbf{x}(\hat{t}-\tau)d\hat{t}, \\ t \in [t_i, t_{i+1}], \ \ i = 0, 1, 2, \ldots, k. \end{aligned} \tag{4.23}$$

The integration on the RHS of the above equation can be computed by assuming that $\mathbf{A}_d(t)\mathbf{x}(t-\tau)$ is a constant or a linear function of time over the small interval $[t_i, t_{i+1}]$. When it is taken to be a constant, it can be either the value at the beginning of the interval or the mid-point average. The latter approximation has been shown as good as the linear approximation. The work in (Elbeyli and Sun, 2004) has studied the accuracy of these approximation schemes. In the numerical examples, the mid-point approximation has been used.

As an example, we show the case when $\mathbf{A}_d(t)\mathbf{x}(t-\tau)$ is assumed to be $\mathbf{A}_{di}\mathbf{x}_{i-N}$ over $[t_i, t_{i+1}]$. The response $\mathbf{x}_{i+1} = \mathbf{x}(t_{i+1})$ at time t_{i+1} can then be expressed in terms of the initial condition $\mathbf{x}_i$ and $\mathbf{x}_{i-N}$ in the following mapping

$$\mathbf{x}_{i+1} = \mathbf{Q}_i \mathbf{x}_i + \mathbf{P}_i \mathbf{x}_{i-N}, \tag{4.24}$$

where

$$\mathbf{P}_i = \int_{t_i}^{t_{i+1}} e^{\mathbf{A}_i(\Delta t - \tau)}\mathbf{A}_{di} d\tau, \ \ \mathbf{Q}_i = e^{\mathbf{A}_i \Delta t}. \tag{4.25}$$

Define an $(N+1) \times n$ dimensional state vector as

$$\mathbf{y}_i = \left[\mathbf{x}_i^{\mathrm{T}}\ \mathbf{x}_{i-1}^{\mathrm{T}}\ \mathbf{x}_{i-2}^{\mathrm{T}} \ldots \mathbf{x}_{i-N}^{\mathrm{T}}\right]^{\mathrm{T}}.$$

A mapping of the state vector over the interval $[t_i, t_{i+1}]$ can be found as

$$\mathbf{y}_{i+1} = \mathbf{H}_i \mathbf{y}_i, \tag{4.26}$$

where the transition matrix from time t_i to t_{i+1} is

$$\mathbf{H}_i = \begin{bmatrix} \mathbf{Q}_i & \mathbf{0}_{n\times(Nn-n)} & \mathbf{P}_i \\ \mathbf{I}_{n\times n} & \mathbf{0}_{n\times(Nn-n)} & \mathbf{0}_{n\times n} \\ \mathbf{0}_{(Nn-n)\times n} & \mathbf{I}_{(Nn-n)\times(nN-n)} & \mathbf{0}_{(Nn-n)\times n} \end{bmatrix}. \tag{4.27}$$

The mapping of the state vector over one period $T = k\Delta t$ is therefore

$$\mathbf{y}_{j+1} = \mathbf{\Phi}\mathbf{y}_j, \tag{4.28}$$

where the mapping matrix $\mathbf{\Phi}$ is given by

$$\mathbf{\Phi} = \mathbf{H}_{k-1}\mathbf{H}_{k-2}\cdots\mathbf{H}_1\mathbf{H}_0. \tag{4.29}$$

Note that the index j ($j = 0, 1, \ldots$) refers to the number of periods, i.e. $\mathbf{y}_j$ is the state vector at the beginning of the jth period.

The stability of the control system is determined by the eigenvalues of $\mathbf{\Phi}$. Let $|\lambda|_{\max}$ denote the largest absolute value of eigenvalues of the matrix $\mathbf{\Phi}$. Then,

$$|\mathbf{y}_{j+1}| \le |\lambda|_{\max}|\mathbf{y}_j|. \tag{4.30}$$

When $|\lambda|_{\max} < 1$, $\mathbf{\Phi}$ is a contraction, and the control system is asymptotically stable. The stability boundary is given by $|\lambda|_{\max} = 1$. Equation (4.30) indicates that the smaller $|\lambda|_{\max}$ is, the faster the system converges to zero. $|\lambda|_{\max}$ therefore also provides a measure of the control performance.

In the method of semi-discretization, delayed portion of the response is discretized and other part is kept continuous. And we use finite dimensional map to approximate an infinite dimensional system. The method of semi-discretization is an efficient and accurate method for analysis of time delayed periodic systems. Minimization of the largest eigenvalue of the mapping leads to an optimal feedback controller in the sensor of response decay over one mapping step. However, it is difficult to handle multiple independent delays or when the period of the system is not a multiple of the time delay. The method of continuous time approximation presented in Section 4.4 can deal with this problem.

4.3.1 General time-varying systems

In principle, the SD method can be extended to general time-varying linear systems. In this case, the mapping of the state vector over one mapping interval becomes

$$\mathbf{y}_{j+1} = \mathbf{\Phi}(j)\mathbf{y}_j, \tag{4.31}$$

where the matrix $\mathbf{\Phi}(j)$ is now a function of the mapping step j. The asymptotic stability of the system requires

$$|\lambda|_{\max}\left(\prod_{j=0}^{\infty}\mathbf{\Phi}(j)\right) < 1. \tag{4.32}$$

The computation to delineate the stability boundary is far more intensive. A stringent sufficient condition for asymptotic stability is that there exists a $J < \infty$ such that

$$|\lambda|_{\max}(\mathbf{\Phi}(j)) < 1 \text{ for all } j \geq J. \tag{4.33}$$

The extension of the method to nonlinear systems is nontrivial, and will lead to a nonlinear mapping $\mathbf{y}_{j+1} = \mathbf{F}(j, \mathbf{y}_j)$ with a high dimension. It would be very difficult just to locate all the equilibrium points of the mapping in the high dimensional state space.

4.3.2 *Feedback controls*

4.3.2.1 Optimal feedback gains

If we restrict our interest in a finite and compact region $\mathbf{\Omega} \subset \Re^{m \times n}$ in the parametric space $\mathbf{K}$, we can find the regions of stability and optimal control gains in the region to minimize $|\lambda|_{\max}$. This leads to the following optimization problem

$$\min_{\mathbf{K}\in\mathbf{\Omega}} [\max|\lambda(\mathbf{\Phi})|] \text{ subject to } |\lambda|_{\max} < 1. \tag{4.34}$$

This optimization formulation offers a different approach to the design of feedback controls for linear systems with time delay. The control performance criterion is the decay rate of the mapping $\mathbf{\Phi}$ over one period.

The implication of the optimal control gains obtained from Eq. (4.34) is studied next by examining the root locus of PID controls of the linear time-invariant system.

4.3.2.2 Implication of optimal feedback gains

Consider a delayed PID control of a linear time-invariant second-order system.

$$\dot{\mathbf{x}}(t) = \begin{bmatrix} 0 & 1 & 0 \\ 0 & 0 & 1 \\ 0 & -\omega^2 & -2\zeta\omega \end{bmatrix} \mathbf{x}(t) - \begin{bmatrix} 0 & 0 & 0 \\ 0 & 0 & 0 \\ k_i & k_p & k_d \end{bmatrix} \mathbf{x}(t-\tau), \tag{4.35}$$

where $\mathbf{x} = (x, \dot{x}, \ddot{x})^{\mathrm{T}}$, ζ is the damping ratio, and ω is the natural frequency. Because the system is autonomous, we can arbitrarily select a period $T > \tau$ to construct the mapping. For convenience, we choose the undamped natural period of the system as T. Note that the matrix $\mathbf{H}_i$ is independent of i in this case. The stability of the system is also determined by the eigenvalues of a single matrix $\mathbf{H}_i$.

We first consider PI controls. The system parameters are chosen as $\zeta = 0.05$, $\omega = 2$ and $N = 50$. The undamped period of the system is $T = \pi$. Figures 4.1 and 4.2 show the root locus of the closed loop system for varying feedback gain k_i when $k_p = -1.1217$. The closed loop poles corresponding to the optimal gains (k_p, k_i) are marked on the loci in the figure. It appears that the optimization problem stated in Eq. (4.34) leads to the feedback gains that stabilize all branches of the root

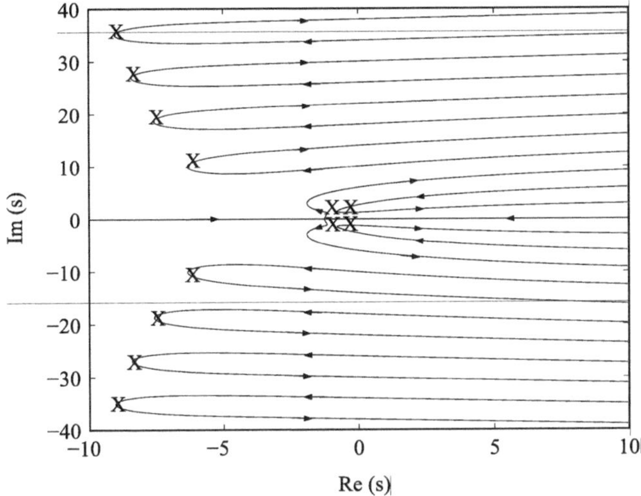

Fig. 4.1 Root locus of the second order system with PI control with respect to k_i when $k_p = -1.1217$. "**x**" indicates the optimal control gains on the root locus. Arrows show direction of increasing k_i gain.

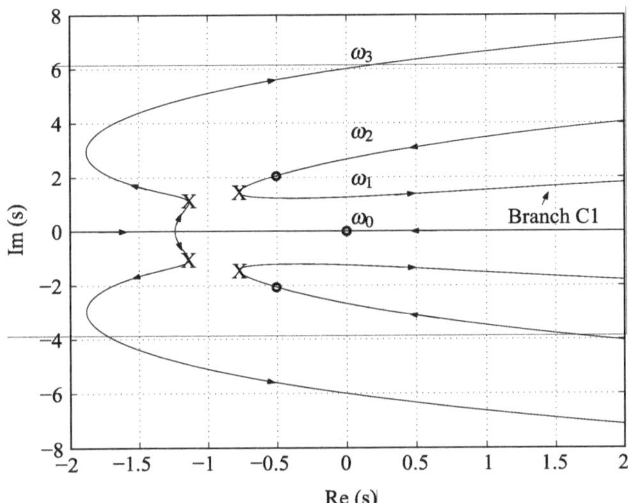

Fig. 4.2 Root locus of the second order system with PI control with respect to k_i when $k_p = -1.1217$ in the magnified region of interest. "**x**" indicates the optimal control gains on the root locus. "o" indicates the zero k_i gains. Arrows show direction of increasing k_i gain. ω_i $(i = 0, 1, 2, 3)$ correspond to the crossing frequency when the system is marginally stable. The branch C1 is the most "vulnerable" branch for the system.

locus. The branch $C1$ marked in Fig. 4.2 is the most vulnerable branch that could cause instability. The optimal control gains place the pole to the leftmost tip of the branch $C1$ to ensure the best convergence rate. The present design method of optimal feedback controls leads to a multi-variable optimization problem, and offers a complementary approach to the classic design methods such as root locus.

From Fig. 4.2, stability bounds of k_i for the fixed $k_p = -1.1217$ can be found as $[0, 2.713]$.

The case for varying k_p with a fixed optimal k_i gain has a similar result and is omitted here.

4.3.2.3 Tracking control

We now consider the tracking control problem. The control defined by Eq. (4.19) is given by $\mathbf{u}(t) = -\mathbf{K}\mathbf{e}(t)$ or $\mathbf{u}(t) = -\mathbf{K}\mathbf{e}(t-\tau)$ where $\mathbf{e}(t) = \mathbf{r}(t) - \mathbf{x}(t)$ is the tracking error and $\mathbf{r}(t)$ is the reference vector. The state equation now reads

$$\dot{\mathbf{x}}(t) = \mathbf{A}(t)\mathbf{x}(t) + \mathbf{A}_d(t)\mathbf{x}(t-\tau) + \mathbf{g}(\mathbf{r}(t)), \tag{4.36}$$

where $\mathbf{g}(\mathbf{r}(t))$ is a function of the reference input $\mathbf{r}(t)$ evaluated at t or $t - \tau$.

The one step mapping of the state vector $\mathbf{y}_i$ can be expressed as

$$\mathbf{y}_{i+1} = \mathbf{H}_i\mathbf{y}_i + \mathbf{b}_i^j, \tag{4.37}$$

where $\mathbf{b}_i^j$ is due to the contribution of $\mathbf{g}(\mathbf{r}(t))$ over the interval $[t_i, t_{i+1}]$ in the jth period. Consequently the mapping of the state vector over one period $T = k\Delta t$ is

$$\mathbf{y}_{j+1} = \mathbf{\Phi}\mathbf{y}_j + \mathbf{d}_j, \tag{4.38}$$

where

$$\mathbf{d}_j = \sum_{i=0}^{k-1}\left(\prod_{n=i+1}^{k-1}\mathbf{H}_n\right)\mathbf{b}_i^j. \tag{4.39}$$

Note that the index j ($j = 0, 1, ...$) still refers to the number of periods. This result suggests that the tracking control of linear periodic systems has the same stability region as that of the regulator.

The above control design and stability study are in terms of the extended state vector $\mathbf{y}_i$. Recall that $\mathbf{y}_i$ is a finite dimensional approximation of the original infinite dimensional state of the time delay system. Therefore, the stability of $\mathbf{y}_i$ implies that of the original system. Furthermore, the condition $|\lambda|_{\max} < 1$ is necessary and sufficient for $\mathbf{\Phi}$ to be a contraction. It is also necessary and sufficient for the stability of the system.

When $|\lambda|_{\max} < 1$, Equation (4.30) indicates that the magnitude of $\mathbf{y}_i$ decays in every mapping step. This does not guarantee the same decay of the magnitude of $\mathbf{x}_i$ unless the system is in steady state. For this reason, the optimal feedback control gains may not guarantee the transient performance of the closed-loop system in some cases. The numerical results reported subsequently will attest to these points.

4.3.3 Analysis of the method of semi-discretization

We consider three approximation schemes for computing the integral

$$\int_{t_i}^{t} e^{\mathbf{A}_i(t-t_i-\hat{t})}\mathbf{A}_d(\hat{t})\mathbf{x}(\hat{t}-\tau)d\hat{t} \tag{4.40}$$

in Eq. (4.23) when $t = t_{i+1}$. The first scheme is the zeroth order method when $\mathbf{A}_d(t)\mathbf{x}(t-\tau)$ is taken to be $\mathbf{A}_d(t_i)\mathbf{x}(t_i-\tau)$. The second scheme is called the improved zeroth order method when $\mathbf{A}_d(t)\mathbf{x}(t-\tau)$ is taken to be $(\mathbf{A}_d(t_i)\mathbf{x}(t_i-\tau)+\mathbf{A}_d(t_{i+1})\mathbf{x}(t_{i+1}-\tau))/2$, i.e. the average of the values at lower and upper ends of the time interval $[t_i, t_{i+1}]$. The third scheme is the first order method when $\mathbf{A}_d(t)\mathbf{x}(t-\tau)$ is assumed to be a line function of time for $t_i \le t \le t_{i+1}$.

To compare the accuracy of the three approximation schemes, we need a measure. Since the exact solutions for periodic systems are not available, we shall consider the following LTI system, for which we can obtain exact stability bounds of the control gains,

$$\ddot{x}(t) + 2\zeta\omega\dot{x}(t) + \omega^2 x(t) = -k_d\dot{x}(t-\tau) - k_p x(t-\tau), \tag{4.41}$$

where ζ is the damping ratio, and ω is the natural frequency. The characteristic equation of the closed loop system is as follows

$$s^2 + 2\zeta\omega s + \omega^2 + k_d s e^{-\tau s} + k_p e^{-\tau s} = 0, \tag{4.42}$$

where s is the Laplace variable. The roots of Eq. (4.42) are the closed loop poles. By studying the stability of the closed loop poles, we can find the exact ranges of the control gains k_d and k_p that stabilize the system. Let k_d^* and k_p^* be a pair of control gains within the stable boundary of the controlled system. Keeping either of the control gains constant, we can determine the upper and lower limits of the other control gain that renders the system marginally stable. We label these exact gains as k_d^u, k_d^l, k_p^u and k_p^l where the superscript u and l respectively stands for upper and lower bounds.

Because the system in Eq. (4.41) is autonomous, we can arbitrarily select a period $T > \tau$ to construct the mapping $\mathbf{\Phi}$. By fixing one of k_d^* and k_p^* in turn and varying the other, we can obtain an approximate value for the upper and lower bound of the control gains crossing the stability boundary defined by $|\lambda|_{\max} = 1$. These approximate gains corresponding to the exact ones k_d^u, k_d^l, k_p^u and k_p^l are denoted as $\tilde{k}_d^u$, $\tilde{k}_d^l$, $\tilde{k}_p^u$ and $\tilde{k}_p^l$.

We introduce the following root mean square error as a measure of accuracy of the semi-discretization method,

$$k_{er} = \frac{1}{2}\sqrt{\left(k_d^u - \tilde{k}_d^u\right)^2 + \left(k_d^l - \tilde{k}_d^l\right)^2 + \left(k_p^u - \tilde{k}_p^u\right)^2 + \left(k_p^l - \tilde{k}_p^l\right)^2}. \tag{4.43}$$

We can also study the effects of the approximation schemes on the optimal gains. The convergence of the control gains and $|\lambda|_{\max}$ as a function of discretization level offers a qualitative measure for comparison, and will be considered hereafter. Finally, we simulate the system response and compare the decay rate of the response to that predicted by the semi-discretization method with different approximation schemes. In the numerical examples, we examine the decay rate of the L_2 norm of the state vector $\mathbf{y}$. This comparison is amenable to both time-invariant and periodic systems.

4.3.3.1 Linear time-invariant second order system

We first consider a second-order autonomous system defined in Eq. (4.41) with $\zeta = 0.05$, $\omega = 2$ and $\tau = \pi/2$. We have selected a period $T = \pi > \tau$ to construct the mapping.

In Table 4.1, we present the solutions for the upper and lower stability bounds of the control gains with different discretization levels. We used $k_d^* = -0.1356$ and $k_p^* = -0.3898$, the optimal control gains by the zeroth order approximation with $N = 20$. These solutions are compared with the exact values. The results in the table are also plotted in Fig. 4.3. The figure shows that the convergence of the first order approximation is far superior to that of the zeroth order approximations. At $N = 10$, its error is comparable to that of the zeroth order approximation at

Table 4.1 Exact and approximate stability bounds of control gains with varying discretization levels.

Discretization	Solution	$k_p = -0.1356$		$k_d = -0.3898$		Error
Level	Method	lower	upper	lower	upper	k_{er}
	Exact	−1.75862692	0.19715358	−1.85247965	1.364948976	*n/a*
10	zeroth	−1.556284	0.1918105	−1.5558121	1.4998466	1.9182036e-1
	improved	−1.792398	0.19921478	−1.84305351	1.41216126	2.9421873e-2
	first order	−1.793502	0.19887852	−1.8705159	1.37106345	1.9887050e-2
20	zeroth	−1.6557696	0.1938158	−1.69536752	1.43163415	9.9651774e-2
	improved	−1.7635742	0.1973303	−1.8401601	1.38495437	1.2005144e-2
	first order	−1.7672229	0.1975833	−1.85698090	1.36647265	4.9157433e-3
40	zeroth	−1.7068267	0.19531915	−1.77153523	1.39809059	5.0835515e-2
	improved	−1.7581437	0.19703111	−1.84436498	1.37407107	6.1096020e-3
	first order	−1.7607684	0.19726091	−1.85360447	1.36532958	1.2255089e-3
60	zeroth	−1.7240170	0.1958938	−1.79796493	1.3869977	3.4122699e-2
	improved	−1.7576467	0.1970252	−1.84662891	1.3708362	4.1793607e-3
	first order	−1.7595781	0.1972013	−1.85297953	1.3651181	5.4439084e-4
80	zeroth	−1.732641	0.1961951	−1.81138360	1.38146835	2.5680546e-2
	improved	−1.7576447	0.1970397	−1.84792522	1.36929181	3.1851518e-3
	first order	−1.7591618	0.1971804	−1.85276083	1.36504411	3.0616550e-4
100	zeroth	−1.7378257	0.196380	−1.81950120	1.37815617	2.0586993e-2
	improved	−1.7577224	0.1970541	−1.84875598	1.36838844	2.5750569e-3
	first order	−1.7589692	0.1971708	−1.85265961	1.36500986	1.9593027e-4

$N = 100$. Since the dimension of the matrix $\mathbf{A}_i$ is $(N+2) \times (N+2)$, and $\mathbf{\Phi}$ is a product of $k > N$ matrices $\mathbf{A}_i$, the computational effort to form $\mathbf{\Phi}$ is proportional to $(N+2) \times (N+2) \times k \sim O\left(N^3\right)$. Thus, the first order approximation provides about 1000 fold computational efficiency increase as compared to the zeroth order scheme. The increase in the computational efficiency significantly speeds up the optimization solution process, which involves repeated calculations of $\mathbf{\Phi}$ and its eigenvalues.

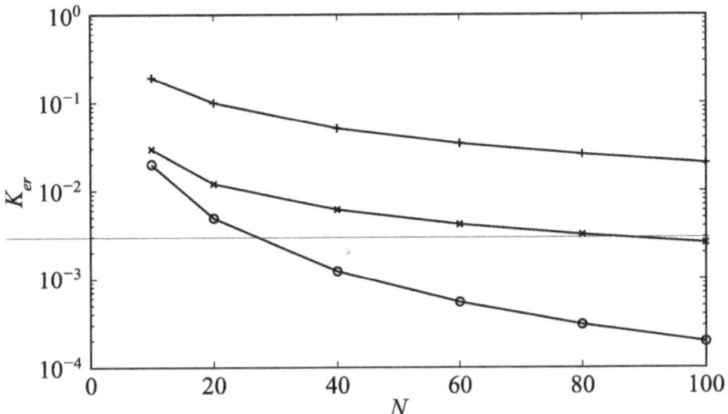

Fig. 4.3 Variation of the control gain error k_{er} with discretization level N. (–+–+–): zeroth order, (–x–x–): improved zeroth order, (–o–o–): first order.

Figure 4.4 shows the effect of the three approximation schemes on the stability boundary in the control gain space. The stable region is inside the closed curve. For $N = 20$, there is a substantial difference between the stability boundary predicted

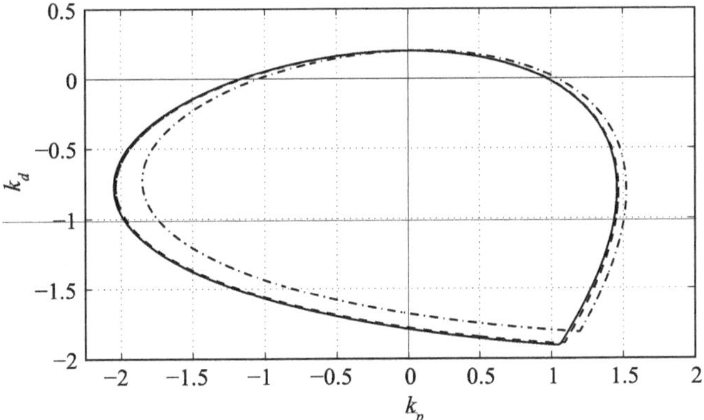

Fig. 4.4 Stability boundaries of the second order linear time-invariant system with time delay. $\tau = \pi/2$ and $N = 20$. (– · – · –):zeroth order approximation, (– – –):improved zeroth order approximation, (------) : first order approximation.

by the first order approach and that by the zeroth order method. It should be noted that there is a slight difference between the improved zeroth order solution and the first order solution. The stability boundary obtained by the zeroth order approximation approaches to that by the first order approximation as N increases beyond 40, and deviates substantially when $N < 40$.

4.3.3.2 Mathieu equation

Next, we consider the Mathieu equation with a delayed feedback control

$$\ddot{x}(t) + (\delta + 2\varepsilon \cos 2t)x(t) = -k_d\dot{x}(t-\tau) - k_p x(t-\tau), \tag{4.44}$$

where $\varepsilon = 1$, $\delta = 4$, and the period of the system is $T = \pi$. We assume a time delay $\tau = \pi/4$. The uncontrolled system is parametrically unstable.

Table 4.2 shows the optimal feedback gains and associated largest absolute value $|\lambda|_{\max}$ of eigenvalues of $\mathbf{\Phi}$. The variation of $|\lambda|_{\max}$ with discretization level is depicted in Fig. 4.5. The solutions obtained by the first order and improved zeroth order approximation converge much faster than that by the zeroth order approximation. Figure 4.6 shows the time history of the norm of the state vector $\mathbf{y}(t)$. The figure validates that the decay rate, characterized by $|\lambda|_{\max}$, obtained by the first order and improved zeroth order approximation converges to the exact ones.

Table 4.2 Optimal control gains and the largest absolute value of eigenvalues of $\mathbf{\Phi}$ with varying discretization levels.

Discretization Level	Approximation	Optimal gains k_p	k_d	$\lvert\lambda\rvert_{\max}$
20	zeroth	−2.016893	−0.3090877	0.00155351
	improved zeroth	−2.035019	−0.2839209	0.00344162
	first order	−2.034412	−0.2832894	0.00335362
40	zeroth	−2.027560	−0.2973511	0.00222317
	improved zeroth	−2.034192	−0.2830767	0.00336122
	first order	−2.034040	−0.2829187	0.00333904
60	zeroth	−2.029776	−0.2925680	0.0026507
	improved zeroth	−2.034037	−0.2829196	0.0033463
	first order	−2.033967	−0.2828478	0.0033365
80	zeroth	−2.030838	−0.2901451	0.0028373
	improved zeroth	−2.033984	−0.2828655	0.0033410
	first order	−2.033947	−0.2828261	0.0033354
100	zeroth	−2.031465	−0.2886839	0.0029433
	improved zeroth	−2.033959	−0.2828401	0.0033386
	first order	−2.033935	−0.2828150	0.0033350
120	zeroth	−2.031879	−0.2877072	0.0030119
	improved zeroth	−2.033946	−0.2828265	0.0033372
	first order	−2.033929	−0.2828085	0.00333479

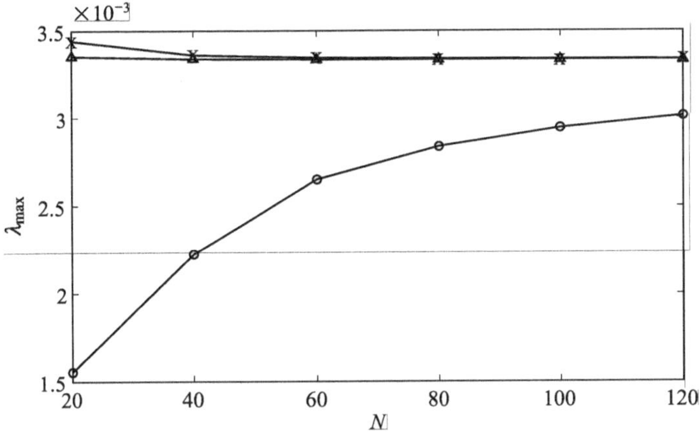

Fig. 4.5 Variation of the largest absolute value of eigenvalues of $\mathbf{\Phi}$ with discretization level N. (–o–o–): zeroth order, (–x–x–): improved zeroth order, (-Δ-Δ-): first order approximation.

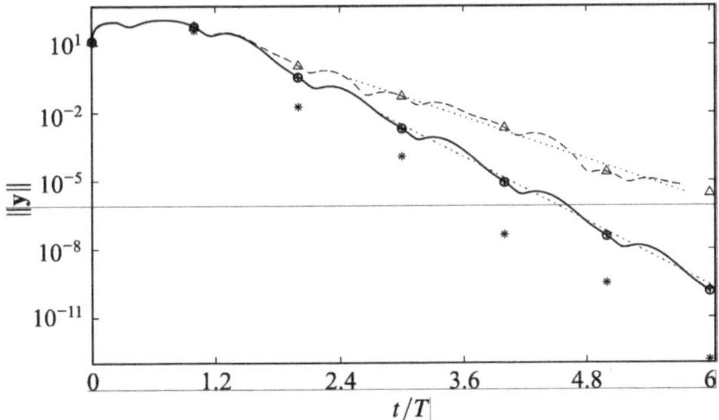

Fig. 4.6 Time history of the norm of the state vector $\mathbf{y}(t)$. (———): Time simulation of the system with optimal gains computed by the first order approximation; $(k_p, k_d) = (-2.03441, -0.28329)$ and $\lambda_{\max} = 3.3536\text{e}{-3}$. Corresponding mapping by the improved zeroth order approximation $(+)$ and by the first order approximation (o). $(- - - -)$: Time simulation of the system with optimal gains computed by the zeroth order approximation, $(k_p, k_d) = (-2.01689, -0.30909)$ and $\lambda_{\max} = 1.5535\text{e}{-3}$. Corresponding mapping by the zeroth order approximation $(*)$, and mapping by the first order approximation $(\triangle)$. $(\cdots\cdots)$: The logarithmic curve fit. In each of the mappings $N = 20$ is used.

Finally, we present the stability boundaries of the control gains with $|\lambda|_{\max} = 1$ using different approximations in Fig. 4.7. The shape of the stability region is more complex than that of the time-invariant system. The irregular geometry is a reflection of the complex behavior of the periodic system. This figure demonstrates again that the proposed first order and improved zeroth approximations substantially improve the accuracy and efficiency of the semi-discretization method even for periodic systems.

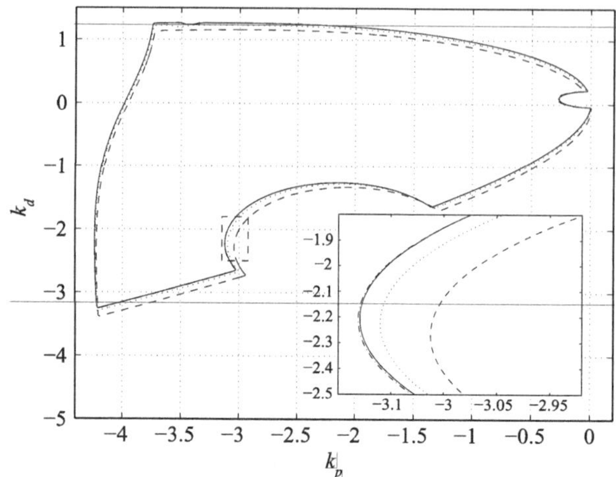

Fig. 4.7 Stability boundary in (k_p, k_d) plane, a section of the graph is enlarged to show the detail. (———): first order approximation with $N = 12$, $(- \cdot - \cdot -)$: improved zeroth order approximation with $N = 12$. (- - - -): zeroth order approximation with $N = 12$, $(\cdot \cdot \cdot \cdot \cdot)$: zeroth order approximation with $N = 40$.

4.3.4 High order control

In the previous discussion of semi-discretization, the delayed response $\mathbf{x}(t - \tau)$ is discretized. In the following, we discretize the delayed control $\mathbf{u}(t - \tau)$ instead. This leads to a control of higher order and with better performance. Consider a linear system with a delayed control,

$$\dot{\mathbf{x}} = \mathbf{A}\mathbf{x}(t) + \mathbf{B}\mathbf{u}(t - \tau), \tag{4.45}$$

where $\mathbf{x} \in \Re^n$ and $\mathbf{u} \in \Re^m$. Let $\Delta\tau = \tau/N$ be the sample time of the digital control system and $t = k\Delta\tau$. We denote $\mathbf{u}(t - \tau) = \mathbf{u}(k - N)$. Following the concept of semi-discretization, we construct a mapping from Eq. (4.45) as

$$\mathbf{x}(k + 1) = \tilde{\mathbf{A}}\mathbf{x}(k) + \tilde{\mathbf{B}}\mathbf{u}(k - N), \tag{4.46}$$

where

$$\tilde{\mathbf{A}}=e^{\mathbf{A}\Delta\tau}, \quad \tilde{\mathbf{B}} = \int_0^{\Delta\tau} e^{\mathbf{A}(\Delta\tau - s)}\mathbf{B}ds. \tag{4.47}$$

Introduce the extended state $(n + Nm) \times 1$ vector

$$\mathbf{y}(k) = [\mathbf{x}(k), \mathbf{u}(k - N), \mathbf{u}(k - N + 1), \ldots, \mathbf{u}(k - 1)]^{\mathrm{T}}. \tag{4.48}$$

Then, Equation (4.46) can be written in terms of the extended vector without time delay as

$$\mathbf{y}(k + 1) = \bar{\mathbf{A}}\mathbf{y}(k) + \bar{\mathbf{B}}\mathbf{u}(k), \tag{4.49}$$

where

$$\bar{\mathbf{A}} = \begin{bmatrix} \tilde{\mathbf{A}} & \tilde{\mathbf{B}} & \mathbf{0} & \cdots & \mathbf{0} \\ \mathbf{0} & \mathbf{0} & \mathbf{I} & \cdots & \mathbf{0} \\ \vdots & \vdots & \vdots & \ddots & \mathbf{0} \\ \mathbf{0} & \mathbf{0} & \mathbf{0} & \cdots & \mathbf{I} \\ \mathbf{0} & \mathbf{0} & \mathbf{0} & \cdots & \mathbf{0} \end{bmatrix}, \quad \bar{\mathbf{B}} = \begin{bmatrix} \mathbf{0} \\ \mathbf{0} \\ \vdots \\ \mathbf{0} \\ \mathbf{I} \end{bmatrix}. \tag{4.50}$$

Consider the full state feedback control (Kwon and Pearson, 1980)

$$\begin{aligned} \mathbf{u}(k) &= -\mathbf{K}\mathbf{y}(k) \\ &= -\mathbf{K}_1\mathbf{x}(k) - \mathbf{K}_2\mathbf{u}(k-N) - \mathbf{K}_2\mathbf{u}(k-N+1) - \cdots - \mathbf{K}_{N+1}\mathbf{u}(k-1). \end{aligned} \tag{4.51}$$

We refer to this control as a higher order control because the gain matrix $\mathbf{K}$ is $m \times (n+Nm)$. The gain matrix for the state feedback control $\mathbf{u}(k) = -\mathbf{K}\mathbf{x}(k)$ is $m \times n$.

The feedback gain can be designed with the digital LQR optimal control to minimize a cost function (Cai et al., 2003)

$$J = \frac{1}{2}\sum_{k=0}^{\infty} \left[\mathbf{y}^{\mathrm{T}}(k)\mathbf{Q}\mathbf{y}(k) + \mathbf{u}^{\mathrm{T}}(k)\mathbf{R}\mathbf{u}(k)\right], \tag{4.52}$$

where $\mathbf{Q}$ is non-negative definite symmetric matrix and $\mathbf{R}$ is a positive definite symmetric matrix. We obtain $\mathbf{K} = [\mathbf{R} + \bar{\mathbf{B}}^{\mathrm{T}}\mathbf{S}_\infty\bar{\mathbf{B}}]^{-1}\bar{\mathbf{B}}^{\mathrm{T}}\mathbf{S}_\infty\bar{\mathbf{A}}$ where $\mathbf{S}_\infty$ satisfies the algebraic Riccati equation (Franklin et al., 1998),

$$\mathbf{S}_\infty = \bar{\mathbf{A}}^{\mathrm{T}}[\mathbf{S}_\infty - \mathbf{S}_\infty\bar{\mathbf{B}}[\mathbf{R} + \bar{\mathbf{B}}^{\mathrm{T}}\mathbf{S}_\infty\bar{\mathbf{B}}]^{-1}\bar{\mathbf{B}}^{\mathrm{T}}\mathbf{S}_\infty]\bar{\mathbf{A}} + \mathbf{Q}. \tag{4.53}$$

4.3.5 Optimal estimation

Assume that the system has an output $\mathbf{z}(k)$ with measurement noise $\mathbf{v}(k)$

$$\mathbf{z}(k) = \mathbf{C}\mathbf{y}(k) + \mathbf{v}(k). \tag{4.54}$$

Furthermore, we assume that the system (4.49) is subject to a white noise disturbance

$$\mathbf{y}(k+1) = \bar{\mathbf{A}}\mathbf{y}(k) + \bar{\mathbf{B}}\mathbf{u}(k) + \mathbf{\Gamma}\mathbf{w}(k), \tag{4.55}$$

where the process noise $\mathbf{w}(k)$ and the measurement noise $\mathbf{v}(k)$ are random processes with zero mean

$$E[\mathbf{w}(k)] = E[\mathbf{v}(k)] = 0, \tag{4.56}$$

and the delta correlation function

$$E[\mathbf{w}(i)\mathbf{w}^{\mathrm{T}}(j)] = E[\mathbf{v}(i)\mathbf{v}^{\mathrm{T}}(j)] = 0, \text{ if } i \neq j, \tag{4.57}$$

$$E[\mathbf{w}(k)\mathbf{w}^{\mathrm{T}}(k)] = \mathbf{R}_{\mathbf{w}}, \ E[\mathbf{v}(k)\mathbf{v}^{\mathrm{T}}(k)] = \mathbf{R}_{\mathbf{v}}. \tag{4.58}$$

Define an estimation system as

$$\bar{\mathbf{y}}(k+1) = \bar{\mathbf{A}}\hat{\mathbf{y}}(k) + \bar{\mathbf{B}}\mathbf{u}(k), \tag{4.59}$$

$$\bar{\mathbf{z}}(k) = \mathbf{C}\bar{\mathbf{y}}(k), \tag{4.60}$$

where $\bar{\mathbf{y}}(k)$ denotes an estimate of $\mathbf{y}(k)$, and $\hat{\mathbf{y}}(k)$ is an update of $\bar{\mathbf{y}}(k)$ given by

$$\begin{aligned}\hat{\mathbf{y}}(k) &= \bar{\mathbf{y}}(k) + \mathbf{L}(k)(\mathbf{z}(k) - \bar{\mathbf{z}}(k)) \\ &= \bar{\mathbf{y}}(k) + \mathbf{L}(k)(\mathbf{z}(k) - \mathbf{C}\bar{\mathbf{y}}(k)),\end{aligned} \tag{4.61}$$

and $\mathbf{L}$ is the estimate gain. $\mathbf{L}$ is determined by considering a recursive least squares problem leading to the following equations (Franklin et al., 1998)

$$\mathbf{L}(k) = \mathbf{M}(k)\mathbf{C}^{\mathrm{T}}(\mathbf{C}\mathbf{M}(k)\mathbf{C}^{\mathrm{T}} + \mathbf{R}_v)^{-1}, \tag{4.62}$$

$$\begin{aligned}\mathbf{M}(k+1) = {} & \bar{\mathbf{A}}[\mathbf{M}(k) - \mathbf{M}(k)\mathbf{C}^{\mathrm{T}}(\mathbf{C}\mathbf{M}(k)\mathbf{C}^{\mathrm{T}} + \mathbf{R}_v)^{-1}\mathbf{C}\mathbf{M}(k)]\bar{\mathbf{A}}^{\mathrm{T}} \\ & + \mathbf{\Gamma}\mathbf{R}_{\mathbf{w}}^{\mathrm{T}}\mathbf{\Gamma}.\end{aligned} \tag{4.63}$$

In the steady state as $k \to \infty$, we have $\mathbf{L} = \mathbf{L}_\infty$ and

$$\mathbf{L}_\infty = \mathbf{M}_\infty^{\mathrm{T}}\mathbf{C}(\mathbf{C}\mathbf{M}_\infty\mathbf{C}^{\mathrm{T}} + \mathbf{R}_v)^{-1}, \tag{4.64}$$

$$\mathbf{M}_\infty = \bar{\mathbf{A}}[\mathbf{M}_\infty - \mathbf{M}_\infty\mathbf{C}^{\mathrm{T}}(\mathbf{C}\mathbf{M}_\infty\mathbf{C}^{\mathrm{T}} + \mathbf{R}_v)^{-1}\mathbf{C}\mathbf{M}_\infty]\bar{\mathbf{A}}^{\mathrm{T}} + \mathbf{\Gamma}\mathbf{R}_{\mathbf{w}}^{\mathrm{T}}\mathbf{\Gamma}. \tag{4.65}$$

4.3.6 Comparison of semi-discretization and higher order control

Reconsider the system (4.41) in terms of the delayed control

$$\ddot{x}(t) + 2\zeta\omega\dot{x}(t) + \omega^2 x(t) = u(t-\tau). \tag{4.66}$$

We compare the performance of the higher order control with that of the PD feedback control with the optimal gains designed by semi-discretization. Let $\tau = \pi/2$, $\omega = 2$, $\zeta = 0.05$ and the discretization level $N = 2^5$. The optimal feedback gains are $k_p = -0.1356$ and $k_d = -0.3898$. Note that the system has one control input $m = 1$ and the weighting matrix $\mathbf{R}$ becomes a scalar denoted as R. We take $\mathbf{Q}$ such that $Q_{ij} = 0$ except $Q_{11} = 4$. The weighting factor R changes in the examples. Figure 4.8 shows the comparison of the impulse response of the system and the controls implemented. In this case, the PD control with optimal gains designed by the semi-discretization method and the higher order control have the same order of magnitude, resulting the similar control performance. Figure 4.9 shows the same comparison where the higher order control is about one order of magnitude larger than the PD control with optimal gains designed by the semi-discretization method resulting much better performance.

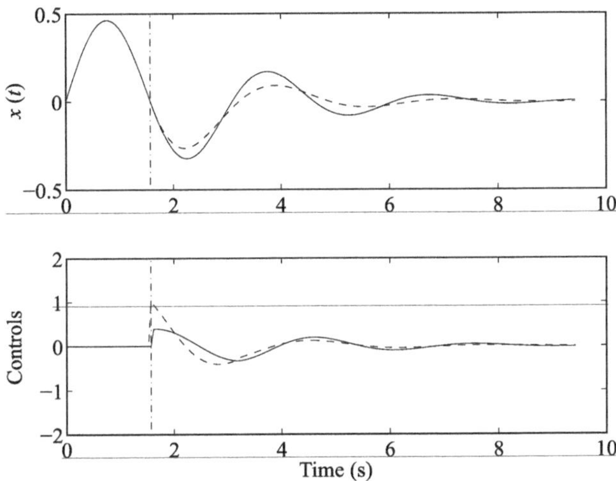

Fig. 4.8 Comparison of impulse responses (top) and controls (bottom) of the LTI system under a PD control (solid line) with optimal gains designed by semi-discretization and a higher order control (dashed line) with $R = 1$.

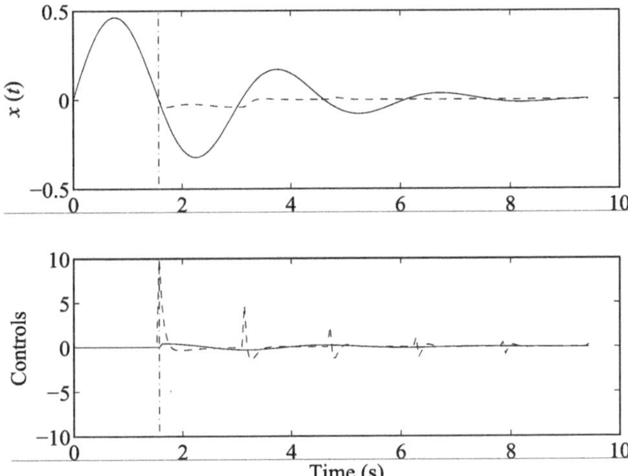

Fig. 4.9 Comparison of impulse responses (top) and controls (bottom) of the LTI system under a PD control (solid line) with optimal gains designed by semi-discretization and a higher order control (dashed line) with $R = 0.005$.

Figure 4.10 shows the comparison of the step response of the system and the controls. As is the case in the previous example, the PD control with optimal gains designed by the semi-discretization method and the higher order control have the same order of magnitude, resulting the similar control performance. Figure 4.11 shows the same comparison where the higher order control is about two orders of

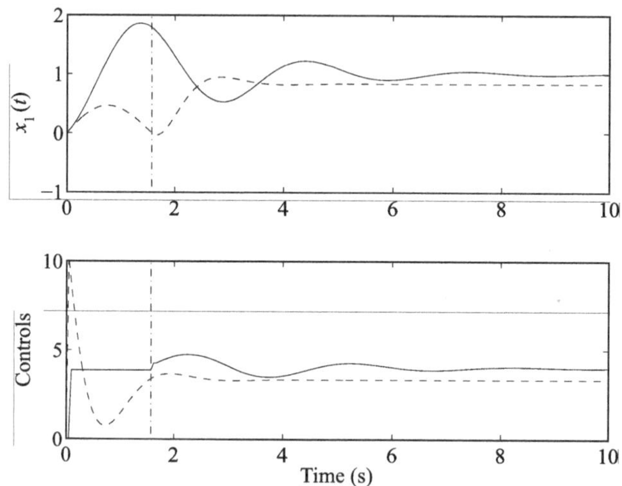

Fig. 4.10 Comparison of step responses (top) and controls (bottom) of the LTI system under a PD control (solid line) with optimal gains designed by semi-discretization and a higher order control (dashed line) with $R = 0.05$.

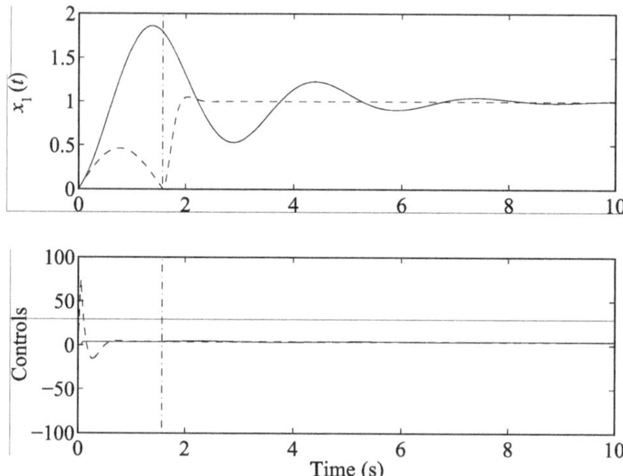

Fig. 4.11 Comparison of step responses (top) and controls (bottom) of the LTI system under a PD control (solid line) with optimal gains designed by semi-discretization and a higher order control (dashed line) with $R = 0.0005$.

magnitude larger than the PD control with optimal gains designed by the semi-discretization method resulting much better tracking control.

The advantage of the higher order control design lies in that it allows the time delay system to have relatively large feedback gains while still keeping the system stable. This is not the case with the delayed PD control as indicated by the stability domains in the gain space presented in this section and Section 4.7.

4.4 Method of continuous time approximation

Consider a dynamical system with one time delay τ given by,

$$\dot{\mathbf{x}} = \mathbf{f}\left(\mathbf{x}\left(t\right), \mathbf{x}\left(t-\tau\right), t\right) + \mathbf{B}\mathbf{u}\left(t\right), \tag{4.67}$$

where $\mathbf{x} \in \Re^n$, $\mathbf{u} \in \Re^m$, $\mathbf{f}$ describes the system dynamics with time delay, and $\mathbf{B} = \{B_{ij}\}$ is the control influence matrix. Following the idea of semi-discretization, we discretize the delayed part of the state vector $(\mathbf{x}\left(t-t_1\right), 0 < t_1 \leq \tau)$. Let N be an integer such that $\Delta\tau = \tau/N$. $\tau_i = i\Delta\tau$ $(i = 1, 2, \ldots, N)$. We introduce a finite forward difference approximation of the derivatives of $(\mathbf{x}\left(t-\tau_i\right), 1 \leq i \leq N)$ as

$$\dot{\mathbf{x}}\left(t-\tau_i\right) = \frac{1}{\Delta\tau}\left[\mathbf{x}\left(t-\tau_{i-1}\right) - \mathbf{x}\left(t-\tau_i\right)\right]. \tag{4.68}$$

Note that other approximation schemes including, for example, the central difference and Gear integration method for ordinary differential equations, can be used, and that the discretization of the time delay interval can be non-uniform. Higher order Runge-Kutta algorithms, Chebyshev nodes to replace the uniform sampled points τ_i and implicit-explicit methods can also be applied leading to better approximation of $\dot{\mathbf{x}}\left(t-\tau_i\right)$ and more accurate solutions overall in frequency and time domain (Bellen and Maset, 2000; Engelborghs and Roose, 2002; Maset, 2003; Koto, 2004; Breda et al., 2004, 2005; Koto, 2009).

Recall the finite dimensional extended state vector $\mathbf{y}\left(t\right)$ defined in Eq. (4.12). We obtain an equation for the vector $\mathbf{y}\left(t\right)$,

$$\begin{aligned}\dot{\mathbf{y}}\left(t\right) &= \begin{bmatrix} \mathbf{f}\left(\mathbf{y}_1\left(t\right), \mathbf{y}_{N+1}\left(t\right), t\right) \\ \frac{1}{\Delta\tau}\left[\mathbf{y}_1\left(t\right) - \mathbf{y}_2\left(t\right)\right] \\ \vdots \\ \frac{1}{\Delta\tau}\left[\mathbf{y}_N\left(t\right) - \mathbf{y}_{N+1}\left(t\right)\right] \end{bmatrix} + \begin{bmatrix} \mathbf{B} \\ \mathbf{0} \\ \vdots \\ \mathbf{0} \end{bmatrix} \mathbf{u}\left(t\right) \\ &\equiv \hat{\mathbf{f}}(\mathbf{y}, t) + \hat{\mathbf{B}}\mathbf{u}\left(t\right).\end{aligned} \tag{4.69}$$

For a linear system

$$\dot{\mathbf{x}} = \mathbf{A}\mathbf{x}(t) + \mathbf{A}_\tau \mathbf{x}(t-\tau) + \mathbf{B}\mathbf{u}(t), \tag{4.70}$$

where $\mathbf{A}$ is the state matrix and $\mathbf{A}_\tau$ is the state matrix related to the delayed response, we have an equation for $\mathbf{y}(t)$ as

$$\begin{aligned} \dot{\mathbf{y}}(t) &= \begin{bmatrix} \mathbf{A} & \mathbf{0} & \cdots & \mathbf{0} & \mathbf{A}_\tau \\ \frac{1}{\Delta\tau}\mathbf{I} & -\frac{1}{\Delta\tau}\mathbf{I} & \mathbf{0} & \cdots & \mathbf{0} \\ \vdots & \vdots & & & \vdots \\ \mathbf{0} & \cdots & \mathbf{0} & \frac{1}{\Delta\tau}\mathbf{I} & -\frac{1}{\Delta\tau}\mathbf{I} \end{bmatrix} \mathbf{y}(t) + \begin{bmatrix} \mathbf{B} \\ \mathbf{0} \\ \vdots \\ \mathbf{0} \end{bmatrix} \mathbf{u}(t) \\ &\equiv \hat{\mathbf{A}}\mathbf{y}(t) + \hat{\mathbf{B}}\mathbf{u}(t). \end{aligned} \tag{4.71}$$

Recall that τ_i need not to be spaced uniformly in the time interval $[0, \tau]$. Non-uniform sampling allows the method to handle more than one independent time delay (Sun, 2009). In the following, we present several control formulations in the extended state space.

4.4.1 *Control problem formulations*

4.4.1.1 Full-state feedback optimal control

Define a performance index as

$$J = \frac{1}{2}\int_0^\infty \left(\mathbf{y}^{\mathrm{T}}\mathbf{Q}\mathbf{y} + \mathbf{u}^{\mathrm{T}}\mathbf{R}\mathbf{u}\right) dt, \tag{4.72}$$

where $\mathbf{Q} = \mathbf{Q}^{\mathrm{T}} \geq 0$ and $\mathbf{R} = \mathbf{R}^{\mathrm{T}} > 0$. When the linear system (4.71) is considered, the full state feedback control $\mathbf{u} = -\mathbf{K}\mathbf{y}$ is the LQR control determined by the matrices $(\hat{\mathbf{A}}, \hat{\mathbf{B}}, \mathbf{Q}, \mathbf{R})$ (Lewis and Syrmos, 1995). When the nonlinear system (4.69) is considered, we have a nonlinear optimal control problem on hand (Slotine and Li, 1991).

Note that the extended state vector $\mathbf{y}$ contains the current and past system response $\mathbf{x}(t)$. The history of the response $\mathbf{x}(t)$ can be stored in the memory to construct $\mathbf{y}$ in real-time implementation of the control. The full state feedback control does not consider possible transport delays since the current state $\mathbf{x}(t)$ is included in the control.

A theoretical issue to investigate is the controllability of the system $(\hat{\mathbf{A}}, \hat{\mathbf{B}})$ in relation to the controllability of the corresponding linear system $(\mathbf{A}, \mathbf{B})$. A rigorous proof of this relationship is elusive at this time. Many numerical examples suggest that $(\hat{\mathbf{A}}, \hat{\mathbf{B}})$ is uncontrollable with a high deficiency of the controllability matrix. This is because the original system lies in an infinite dimensional state space.

4.4.1.2 Output feedback optimal control

Assume that there is a transport delay τ_p. We consider a control of the form $\mathbf{u} = -\mathbf{K}\mathbf{x}(t-\tau_p)$ for the linear system. First, we select a discretization scheme such that τ_p is one of the points τ_i of the time discretization. Assume that $\tau_p = \tau_k$. Define an output equation as

$$\mathbf{v} = \mathbf{C}\mathbf{y} = \mathbf{y}_{k+1} = \mathbf{x}(t-\tau_p), \tag{4.73}$$

where $\mathbf{y}_{k+1}$ is the $(k+1)^{\text{th}}$ elemental vector defined in Eq. (4.12).

According to (Lewis and Syrmos, 1995), if a control gain $\mathbf{K}$ for the linear system in Eq. (4.71) can be found such that the closed-loop system characterized by the matrix $\hat{\mathbf{A}} - \hat{\mathbf{B}}\mathbf{K}\mathbf{C}$ is stable, the system is output stabilizable. Assume that the system is output stabilizable, an optimal control gain can be found in the following optimization problem: Find a control gain $\mathbf{K}$ such that the performance index

$$J = \frac{1}{2}\mathbf{y}_0^{\mathrm{T}}\mathbf{P}\mathbf{y}_0 = \frac{1}{2}\mathrm{tr}\left[\mathbf{P}\mathbf{y}_0\mathbf{y}_0^{\mathrm{T}}\right], \tag{4.74}$$

is minimized where $\mathbf{y}_0$ is an initial condition of the extended state vector $\mathbf{y}(t)$, subject to the constraint of the Lyapunov equation

$$(\hat{\mathbf{A}} - \hat{\mathbf{B}}\mathbf{K}\mathbf{C})^{\mathrm{T}}\mathbf{P} + \mathbf{P}(\hat{\mathbf{A}} - \hat{\mathbf{B}}\mathbf{K}\mathbf{C}) + \mathbf{C}^{\mathrm{T}}\mathbf{K}^{\mathrm{T}}\mathbf{R}\mathbf{K}\mathbf{C} + \mathbf{Q} = 0. \tag{4.75}$$

This is a nonlinear matrix algebraic optimization problem. The Matlab function `fminsearch` can be used to find the optimal control. The optimal gain is in general a function of the initial condition $\mathbf{y}_0$. This is not a desirable feature of the output feedback control. A common approach to select initial conditions is to replace the term $\mathbf{y}_0\mathbf{y}_0^{\mathrm{T}}$ by its statistical average $E[\mathbf{y}_0\mathbf{y}_0^{\mathrm{T}}]$, i.e., the autocorrelation function of $\mathbf{y}_0$. For more discussions, the reader is referred to (Lewis and Syrmos, 1995).

It should be noted that for a given initial value of the control gain to start searching for the optimal one, even the best searching algorithm only gives a local minimum of the performance index J. There are many research issues with the output feedback design that need further studies. For example, how to help the searching algorithm land on a much deeper local minimum? How to select the design matrices $\mathbf{Q}$ and $\mathbf{R}$ to improve the control performance under certain constraints? In the current formulation, when is the system output stabilizable? These turn out to be tough technical questions to answer.

4.4.1.3 Optimal feedback gains via mapping

Another way to obtain optimal gains for output feedback controls is via mapping. This approach has been studied extensively in (Sheng et al., 2004; Sheng and Sun, 2005) with semi-discretization, and is presented in Section 4.3.

4.5 Spectral properties of the CTA method

As is the case for the method of semi-discretization, the CTA method focuses on approximation of temporal responses of the system over a short time interval. As $N \to \infty$, the approximate solution approaches the exact one in time domain at the rate depending on the order of the local approximation. This has been verified by means of extensive numerical simulations (Insperger and Stepan, 2001; Sheng et al., 2004; Sheng and Sun, 2005; Sun, 2009).

The methods of semi-discretization and CTA are not specifically developed to meet frequency dominant requirements such as accurate representation of the open-loop or closed-loop poles and zeros of the original system. Although the methods have been mostly validated with time domain numerical solutions, their properties in frequency dominant have not been studied carefully. In the following, we shall use numerical examples of a linear time-delayed system to examine this issue. Since the CTA formulation can be made completely equivalent to semi-discretization for linear time-invariant systems with a single time delay, the spectral properties of the CTA method are the same as that of semi-discretization.

Consider a linear spring-mass-dashpot oscillator subject to a delayed PD control. The closed-loop characteristic equation is given by Eq. (4.42). The state matrix with the CTA method reads

$$\hat{\mathbf{A}} = \left[\begin{array}{cccccc} \begin{bmatrix} 0 & 1 \\ -\omega^2 & -2\zeta\omega \end{bmatrix} & \mathbf{0} & \cdots & \mathbf{0} & & \begin{bmatrix} 0 & 0 \\ -k_p & -k_d \end{bmatrix} \\ \frac{1}{\Delta\tau}\mathbf{I} & -\frac{1}{\Delta\tau}\mathbf{I} & \mathbf{0} & \cdots & & \mathbf{0} \\ \hline \mathbf{0} & & \ddots & & & \vdots \\ \vdots & & & & & \mathbf{0} \\ \mathbf{0} & \cdots & & \mathbf{0} & \frac{1}{\Delta\tau}\mathbf{I} & -\frac{1}{\Delta\tau}\mathbf{I} \end{array}\right]. \tag{4.76}$$

From the stability chart in (Sheng and Sun, 2005), we know that the system is stable when $(k_p, k_d) = (-0.5, -0.5)$. There are two pairs of the dominant poles with real parts approximately equal to -1 and -0.4. Figures 4.12 and 4.13 show the roots of the characteristic equation (4.42) and the eigenvalues of the state matrix (4.76) constructed with the forward central difference and fourth order Gear's integration algorithm (Carnahan et al., 1969). Extensive simulations show that the CTA method is able to capture the dominant poles of the system only, and completely misses the fast and high frequency poles.

The central finite difference for CTA yields more accurate solutions in time domain. But it introduces a set of lightly damped poles as shown in Fig. 4.14. This causes difficulties when the method is used in control design. On the other hand, the backward finite difference for CTA is unstable. A recent study uses the Chebyshev polynomials to approximate the dynamics of the delayed response (Butcher and Bobrenkov, 2009). The spectral property of the CTA method with the Chebyshev polynomials is much improved as compared to the results based on finite difference approximations.

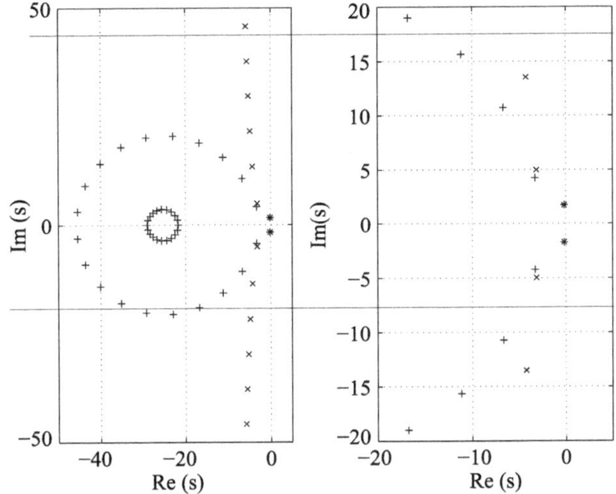

Fig. 4.12 Closed-loop poles of the linear oscillator under a delayed PD control. "x" denotes the roots of the characteristic equation (4.42). "+" denotes the eigenvalues of the state matrix (4.76) constructed with forward finite difference approximation. $(k_p, k_d) = (-0.5, -0.5)$. $\zeta = 0.05$. $\omega = 2$. $\tau = \pi/4$. $N = 20$. (b) is the zoomed view of (a) in the indicated range.

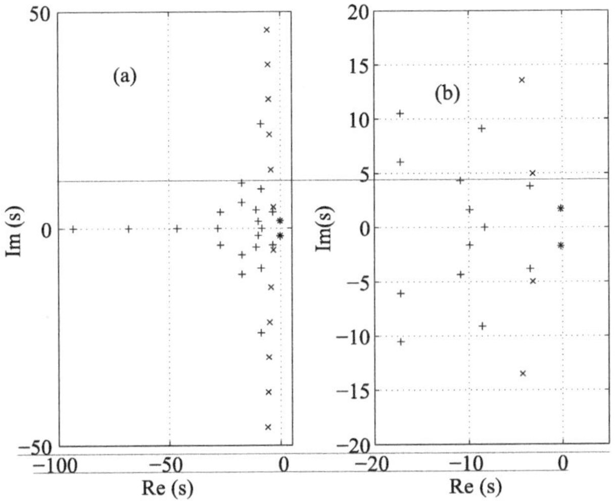

Fig. 4.13 Closed-loop poles of the linear oscillator under a delayed PD control. "x" denotes the roots of the characteristic equation (4.42). "+" denotes the eigenvalues of the state matrix (4.76) constructed with the fourth order Gear's integration scheme. $(k_p, k_d) = (-0.5, -0.5)$. $\zeta = 0.05$. $\omega = 2$. $\tau = \pi/4$. $N = 20$. (b) is the zoomed view of (a) in the indicated range.

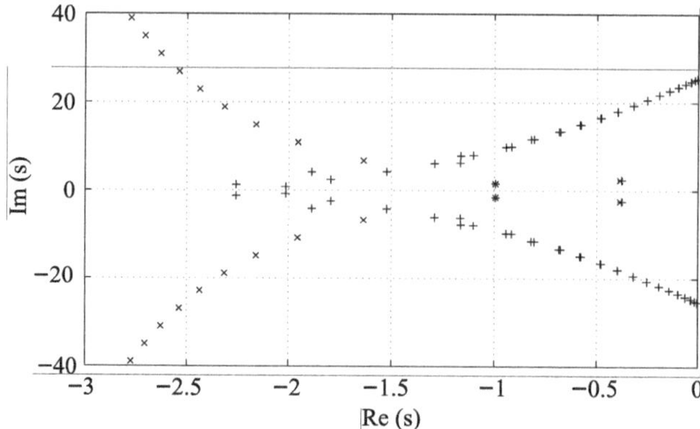

Fig. 4.14 Closed-loop poles of the linear system with central finite difference approximation of the delayed portion of the response under a delayed PD control. "x" denotes the roots of the characteristic equation (4.42). "+" denotes the eigenvalues of the state matrix (4.76). $(k_p, k_d) = (-0.5, -0.5)$. $\zeta = 0.05$. $\omega = 2$. $\tau = \pi/2$. $N = 40$.

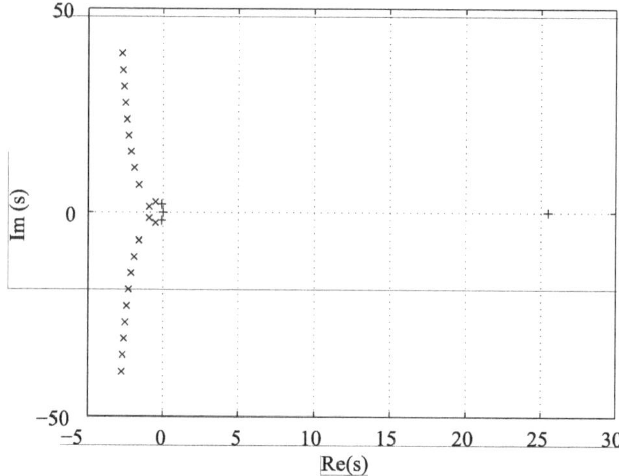

Fig. 4.15 Closed-loop poles of the linear system with backward finite difference approximation of the delayed portion of the response under a delayed PD control. "x" denotes the roots of the characteristic equation (4.42). "+" denotes the eigenvalues of the state matrix (4.76). $(k_p, k_d) = (-0.5, -0.5)$. $\zeta = 0.05$. $\omega = 2$. $\tau = \pi/2$. $N = 40$.

Why can the CTA method accurately predict temporal response $\mathbf{x}(t)$ of the time-delayed system even when it misses all the fast and high frequency poles?

Recall that the response of time-delayed systems lives in an infinite dimensional state space. $\mathbf{x}(t)$ is a projection of the infinite dimensional response $(\mathbf{x}(t), \mathbf{x}(t-t_1), 0 < t_1 \leq \tau)$ on to the finite dimensional space $\Re^n$. *However, different objects in a higher dimensional space can have the same projection in a lower dimensional space.* The CTA method aims at accurate time domain solutions of $\mathbf{x}(t)$ for all $t > 0$, much like the numerical algorithms for integrating ordinary differential equations, such as Runge-Kutta methods whose frequency domain properties are also rarely discussed. These methods in time domain provide one projection of the infinite dimensional response, while the methods in frequency domain such as Padé approximations of the transfer function provide a different projection (Franklin et al., 1986; Vijta, 2000). In principle, the solutions $\mathbf{x}(t)$ obtained by both the time and frequency domain methods with "equivalent accuracy" should be very close to each other, while the solutions obtained by the frequency domain methods may have an advantage of containing more accurate information about the poles and zeros.

The question is then, can we construct a time domain method that accurately predicts both the temporal responses and the poles of the system? In the following, we present a method that is promising.

4.5.1 A low-pass filter based CTA method

Let T be the sample time of a digitized system, and $p > 0$ be a parameter defining an anti-aliasing low pass filter. The derivative of a measured signal can be computed with the following transfer function.

$$H(s) = \frac{p}{s+p} s. \tag{4.77}$$

Hence, $\dot{x}(s) \approx H(s)x(s)$. The digital version of the transfer function with Tustin's approximation reads,

$$H(z) = \frac{2\lambda(z-1)}{(2+\lambda T)z + (\lambda T - 2)}. \tag{4.78}$$

Define a parameter $r = 1/(pT)$. $H(z)$ can be rewritten as

$$H(z) = \frac{1}{T} \frac{z-1}{\left(\frac{1}{2}+r\right) z + \left(\frac{1}{2}-r\right)}. \tag{4.79}$$

In the z-domain, $\dot{x}(z) = H(z)x(z)$, or in the digital time domain,

$$\left(\frac{1}{2}+r\right) \dot{x}(n) + \left(\frac{1}{2}-r\right) \dot{x}(n-1) = \frac{1}{T} \left(x(n) - x(n-1)\right). \tag{4.80}$$

This relationship can be adopted for the CTA method. Equation (4.81) now reads

$$\left(\frac{1}{2}+r\right) \dot{\mathbf{x}}(t-\tau_i) + \left(\frac{1}{2}-r\right) \dot{\mathbf{x}}(t-\tau_{i+1}) = \frac{1}{\Delta\tau} \left[\mathbf{x}\left(t-\tau_i\right) - \mathbf{x}\left(t-\tau_{i+1}\right)\right]. \tag{4.81}$$

where $\Delta\tau = \tau/N$. $\tau_i = i\Delta\tau$ $(i = 0, 1, 2, \ldots, N)$. We have selected the sample time $T = \Delta\tau$. Equation (4.69) becomes

$$\dot{\mathbf{y}}(t) = \mathbf{F}_N(\mathbf{y}(t)) + \mathbf{G}_N\mathbf{u}(t), \tag{4.82}$$

where

$$\mathbf{H}_N = \begin{bmatrix} \mathbf{I} & \mathbf{0} & \cdots & & \mathbf{0} \\ \hline \left(\frac{1}{2}+r\right)\mathbf{I} & \left(\frac{1}{2}-r\right)\mathbf{I} & \mathbf{0} & \cdots & \mathbf{0} \\ \mathbf{0} & \cdots & & & \vdots \\ & & \ddots & & \\ \vdots & & & & \mathbf{0} \\ \mathbf{0} & \cdots & \mathbf{0} & \left(\frac{1}{2}+r\right)\mathbf{I} & \left(\frac{1}{2}-r\right)\mathbf{I} \end{bmatrix}, \tag{4.83}$$

$$\mathbf{F}_N = \mathbf{H}_N^{-1}\begin{bmatrix} \mathbf{f}(\mathbf{y}(t)) \\ \frac{1}{\Delta\tau}\left[\mathbf{y}_1(t) - \mathbf{y}_2(t)\right] \\ \vdots \\ \frac{1}{\Delta\tau}\left[\mathbf{y}_N(t) - \mathbf{y}_{N+1}(t)\right] \end{bmatrix}, \quad \mathbf{G}_N = \mathbf{H}_N^{-1}\begin{bmatrix} \mathbf{B} \\ \mathbf{0} \\ \vdots \\ \mathbf{0} \end{bmatrix}, \tag{4.84}$$

and $\mathbf{y}(t)$ is defined in Eq. (4.12). Note that $\mathbf{H}_N$ is a lower triangular matrix and is non-singular as long as $r \neq 1/2$. When $r = 1/2$, the LPCTA is reduced to be the backward finite difference method, which is unstable for the time-delayed system (Sun, 2009).

Recall that $r = 1/(p\Delta\tau)$. Note that $1/\Delta\tau$ defines the bandwidth of the discretization and p denotes the bandwidth of the lowpass filter. The parameter r is therefore a bandwidth ratio. The bandwidth $1/\Delta\tau$ determines how high the frequency of the time-delayed dynamical system is captured by the numerical solution. Note that the response of time-delayed dynamical systems contains infinitely high frequency components. Since the lowpass filter should preserve the fidelity of high frequency components in the numerical solution, the bandwidth p must be much larger than $1/\Delta\tau$.

Equation (4.82) is based on a first order lowpass filter. Higher order lowpass filters of various type can also be applied.

4.5.2 Example of a first order linear system

Consider a first-order system,

$$\dot{x}(t) = -0.5x(t) + 0.5x(t-\tau), \tag{4.85}$$

where $\tau = \pi/2$ is a constant time delay. We select $r = 0.01$ and $N = 2^{10}$. Figure 4.16 shows the poles of the system by the low pass filter based CTA (LPCTA) method and the fourth order Padé approximation. It is clear that the low pass filter

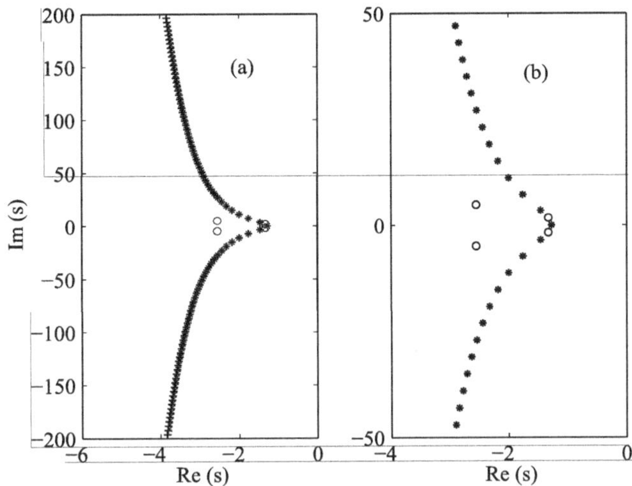

Fig. 4.16 Comparison of poles of the first order time-delayed system. "x" denotes the exact poles. "+" denotes the poles obtained by the low pass filtered CTA method. "o" denotes the poles by the Padé approximation with R_{44}. (b) is the zoomed-in view of (a).

helps the CTA method to substantially improve the spectral prediction capability by accurately computing many higher order poles of the system.

Let us now consider a nonlinear case with two time delays,

$$\dot{x}(t) = -x(t) + x(t-1) - 0.5x(t-\pi/2) - x^3(t) + \sin\omega t. \tag{4.86}$$

To measure the accuracy of the LPCTA solution in time domain, we define a normalized RMS error over a duration of time T_1

$$e_{RMS} = \sqrt{\frac{1}{T_1}\int_0^{T_1} (x(t) - x_{AP}(t))^2\, dt} \div \sqrt{\frac{1}{T_1}\int_0^{T_1} x(t)^2 dt}, \tag{4.87}$$

where $x(t)$ is obtained from the direct numerical integration and $x_{AP}(t)$ denotes the solution obtained with an approximation method.

Figure 4.17 shows all the poles of the linear part of Eq. (4.86) obtained by the LPCTA method. Figure 4.18 shows the responses by the direct integration and by the LPCTA method with $N = 2^6$ and $\omega = 1$. The initial condition is $x(0) = 0$. The agreement between the responses is excellent. Figure 4.19 shows the normalized RMS errors for the LPCTA method and the method with Chebyshev nodes discussed in Section 4.2 as a function of the driving frequency ω. From the figure, we can see that the RMS errors of both the methods have the same trend, and peak around $\omega = 1$. At $\omega = 10^5$, the RMS errors have a sharp jump. We note that $\omega = 10^5$ is larger than the highest frequency of the poles obtained by the approximate methods. In other words, this frequency is out of the bandwidth of the discrete solution. In order to obtain accurate solutions for even higher frequencies, we must increase N or the bandwidth of the method.

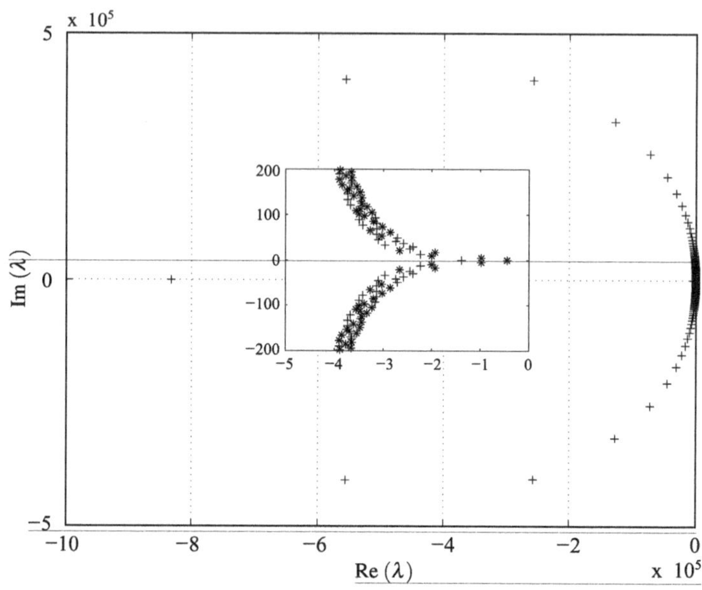

Fig. 4.17 The exact and approximate poles predicted by the LPCTA method with $N = 2^9$ for the linear part of Equation (4.86). 'x': the exact poles; '+': the predicted poles.

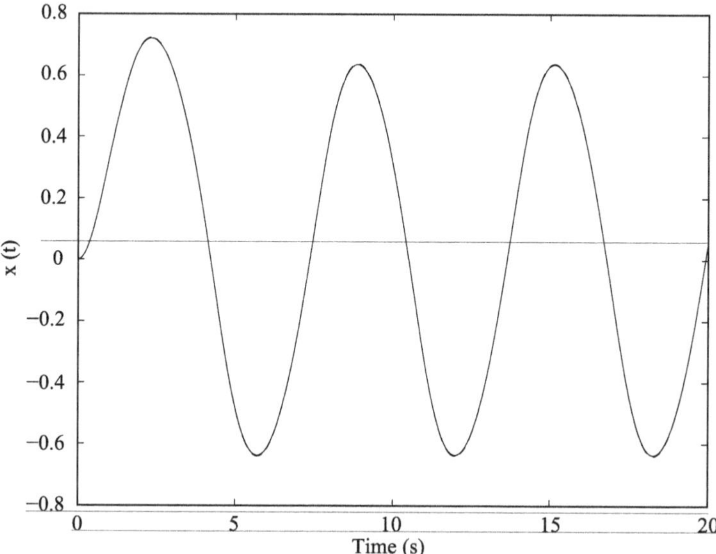

Fig. 4.18 The temporal response of Eq. (4.86). '- -': the direct integration; '-' the LPCTA method with $N = 2^6$. The agreement between the results is excellent.

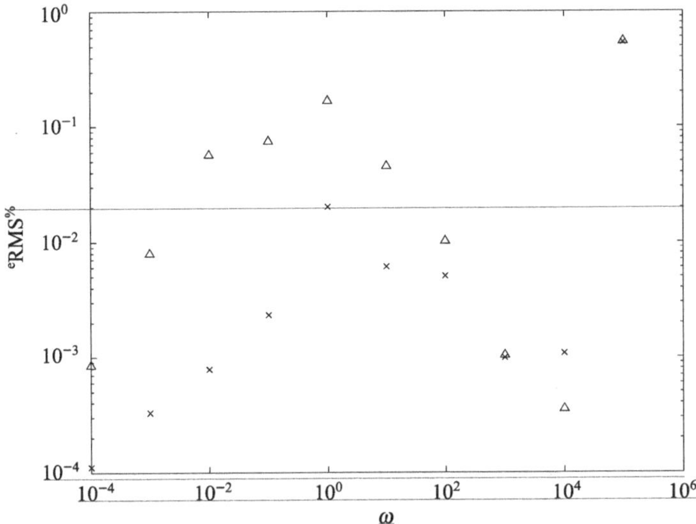

Fig. 4.19 Variation of the normalized RMS errors with ω of Eq. (4.86). 'Δ': the LPCTA method; 'x': the method with Chebyshev nodes. $N = 2^6$ is for both the methods.

4.6 Stability studies of time delay systems

This section presents a few numerical examples of stability boundaries of linear time-invariant systems in the feedback gain space by various methods including Lyapunov, semi-discretization and Padé approximation.

4.6.1 Stability with Lyapunov-Krasovskii functional

Consider a time-delayed linear time-invariant system

$$\dot{\mathbf{x}}(t) = \mathbf{A}\mathbf{x}(t) + \mathbf{A}_d\mathbf{x}(t - \tau), \tag{4.88}$$

subject to the initial condition

$$\mathbf{x}(\sigma) = \phi(\sigma), \ \forall \sigma \in [-\tau, 0], \tag{4.89}$$

where $\phi(\sigma)$ is a given function of time.

4.6.1.1 Delay independent stability conditions

Define a Lyapunov-Krasovskii functional (Ivanescu et al., 2000)

$$V = \mathbf{x}^{\mathrm{T}}(t)\mathbf{P}\mathbf{x}(t) + \int_{t-\tau}^{t} \mathbf{x}^{\mathrm{T}}(s)\mathbf{S}\mathbf{x}(s)ds, \tag{4.90}$$

where $\mathbf{P} = \mathbf{P}^{\mathrm{T}} > 0$ and $\mathbf{S} = \mathbf{S}^{\mathrm{T}} > \mathbf{0}$.

$$\dot{V} = \dot{\mathbf{x}}^{\mathrm{T}}(t)\mathbf{P}\mathbf{x}(t) + \mathbf{x}^{\mathrm{T}}(t)\mathbf{P}\dot{\mathbf{x}}(t) + \mathbf{x}^{\mathrm{T}}(t)\mathbf{S}\mathbf{x}(t) - \mathbf{x}^{\mathrm{T}}(t-\tau)\mathbf{S}\mathbf{x}(t-\tau). \tag{4.91}$$

We write Eq. (4.91) in the matrix form as,

$$\dot{V} = \begin{bmatrix} \mathbf{x}^{\mathrm{T}}(t) \ \mathbf{x}^{\mathrm{T}}(t-\tau) \end{bmatrix} \begin{bmatrix} \mathbf{A}^{\mathrm{T}}\mathbf{P} + \mathbf{P}\mathbf{A} + \mathbf{S} & \mathbf{P}\mathbf{A}_d \\ \mathbf{A}_d^{\mathrm{T}}\mathbf{P} & -\mathbf{S} \end{bmatrix} \begin{bmatrix} \mathbf{x}(t) \\ \mathbf{x}(t-\tau) \end{bmatrix}. \tag{4.92}$$

The stability condition is

$$\begin{bmatrix} \mathbf{A}^{\mathrm{T}}\mathbf{P} + \mathbf{P}\mathbf{A} + \mathbf{S} & \mathbf{P}\mathbf{A}_d \\ \mathbf{A}_d^{\mathrm{T}}\mathbf{P} & -\mathbf{S} \end{bmatrix} < 0. \tag{4.93}$$

A necessary condition is $\mathbf{A}^{\mathrm{T}}\mathbf{P} + \mathbf{P}\mathbf{A} + \mathbf{S} < \mathbf{0}$. Let $\left[\mathbf{A}^{\mathrm{T}}\mathbf{P} + \mathbf{P}\mathbf{A} + \mathbf{S}\right] = -\mathbf{Q}$ where $\mathbf{Q} = \mathbf{Q}^{\mathrm{T}} > \mathbf{0}$. Hence, we have

$$\mathbf{A}^{\mathrm{T}}\mathbf{P} + \mathbf{P}\mathbf{A} = -\mathbf{Q} - \mathbf{S} < \mathbf{0}. \tag{4.94}$$

This is the Lyapunov equation for the linear system without time delay $\dot{\mathbf{x}}(t) = \mathbf{A}\mathbf{x}(t)$, which implies that $\mathbf{A}$ must be Hurwitz stable. This is obviously too strict. Furthermore, the stability condition stated above is independent of the time delay τ. Hence, it must hold for any time delay, including arbitrarily large delay. In the numerical results presented later, we shall show that this stability condition is too conservative.

4.6.1.2 Delay dependent stability conditions

To overcome the limitation of the stability condition independent of time delay, we can modify the Lyapunov-Krasovskii functional to introduce the time delay dependence in the stability conditions (Ivanescu et al., 2000). The system (4.88) can be rewritten in the following form,

$$\frac{d\mathbf{L}}{dt} = \hat{\mathbf{A}}\mathbf{x}(t), \quad t \geqslant 0, \tag{4.95}$$

where $\mathbf{L} = \mathbf{x}(t) + \mathbf{A}_d \int_{t-\tau}^{t} \mathbf{x}(s)ds$ and $\hat{\mathbf{A}} = \mathbf{A} + \mathbf{A}_d$. Consider a Lyapunov-Krasovskii functional as $V = V_1 + V_2$ where

$$V_1 = \mathbf{L}^{\mathrm{T}}\mathbf{P}\mathbf{L}, \tag{4.96}$$

$$V_2 = \int_{t-\tau}^{t} (s - t + \tau)\mathbf{x}^{\mathrm{T}}(s)\mathbf{P}_1\mathbf{x}(s)ds. \tag{4.97}$$

The derivatives of the Lyapunov-Krasovskii functional read,

$$\dot{V}_1 = \dot{\mathbf{L}}^{\mathrm{T}}\mathbf{P}\mathbf{L} + \mathbf{L}^{\mathrm{T}}\mathbf{P}\dot{\mathbf{L}} = [\hat{\mathbf{A}}\mathbf{x}(t)]^{\mathrm{T}}\mathbf{P}\mathbf{L} + \mathbf{L}^{\mathrm{T}}\mathbf{P}[\hat{\mathbf{A}}\mathbf{x}(t)] \tag{4.98}$$

$$\leqslant \mathbf{x}^{\mathrm{T}}(\hat{\mathbf{A}}^{\mathrm{T}}\mathbf{P} + \mathbf{P}\hat{\mathbf{A}} + \tau\hat{\mathbf{A}}^{\mathrm{T}}\mathbf{P}\mathbf{A}_d\mathbf{P}_1^{-1}\mathbf{A}_d^{\mathrm{T}}\mathbf{P}\hat{\mathbf{A}})\mathbf{x} + \int_{t-\tau}^{t} \mathbf{x}^{\mathrm{T}}(s)\mathbf{P}_1\mathbf{x}(s)ds,$$

$$\dot{V}_2 = \tau\mathbf{x}^{\mathrm{T}}\mathbf{P}_1\mathbf{x} - \int_{t-\tau}^{t} \mathbf{x}^{\mathrm{T}}(s)\mathbf{P}_1\mathbf{x}(s)ds, \tag{4.99}$$

$$\dot{V} = \dot{V}_1 + \dot{V}_2 \leqslant \mathbf{x}^{\mathrm{T}}(\hat{\mathbf{A}}^{\mathrm{T}}\mathbf{P} + \mathbf{P}\hat{\mathbf{A}} + \tau\mathbf{P}_1 + \tau\hat{\mathbf{A}}^{\mathrm{T}}\mathbf{P}\mathbf{A}_d\mathbf{P}_1^{-1}\mathbf{A}_d^{\mathrm{T}}\mathbf{P}\hat{\mathbf{A}})\mathbf{x}. \tag{4.100}$$

The stability condition is

$$\dot{V} < \mathbf{0}, \text{ or } \hat{\mathbf{A}}^{\mathrm{T}}\mathbf{P} + \mathbf{P}\hat{\mathbf{A}} + \tau\mathbf{P}_1 + \tau\hat{\mathbf{A}}^{\mathrm{T}}\mathbf{P}\mathbf{A}_d\mathbf{P}_1^{-1}\mathbf{A}_d^{\mathrm{T}}\mathbf{P}\hat{\mathbf{A}} < \mathbf{0}. \tag{4.101}$$

The Schur complement arguments lead to the matrix inequality,

$$\begin{bmatrix} (\mathbf{A}+\mathbf{A}_d)^{\mathrm{T}}\mathbf{P} + \mathbf{P}(\mathbf{A}+\mathbf{A}_d) + \tau\mathbf{P}_1 & \tau(\mathbf{A}+\mathbf{A}_d)^{\mathrm{T}}\mathbf{P}\mathbf{A}_d \\ \tau\mathbf{A}_d^{\mathrm{T}}\mathbf{P}(\mathbf{A}+\mathbf{A}_d) & -\tau\mathbf{P}_1 \end{bmatrix} < \mathbf{0}, \tag{4.102}$$

where a necessary condition is $\left[(\mathbf{A}+\mathbf{A}_d)^{\mathrm{T}}\mathbf{P} + \mathbf{P}(\mathbf{A}+\mathbf{A}_d) + \tau\mathbf{P}_1\right] < \mathbf{0}$.

Let $\left[(\mathbf{A}+\mathbf{A}_d)^{\mathrm{T}}\mathbf{P} + \mathbf{P}(\mathbf{A}+\mathbf{A}_d) + \tau\mathbf{P}_1\right] = -\mathbf{Q}$ where $\mathbf{Q} = \mathbf{Q}^{\mathrm{T}} > \mathbf{0}$. We have

$$\begin{bmatrix} -\mathbf{Q} & \tau(\mathbf{A}+\mathbf{A}_d)^{\mathrm{T}}\mathbf{P}\mathbf{A}_d \\ \tau\mathbf{A}_d^{\mathrm{T}}\mathbf{P}(\mathbf{A}+\mathbf{A}_d) & -\tau\mathbf{P}_1 \end{bmatrix} < \mathbf{0}. \tag{4.103}$$

Another condition stated in (Fan et al., 2002) reads,

$$T(\tau) = \tau \left\|\mathbf{P}_1\right\| < 1. \tag{4.104}$$

4.6.2 Stability with Padé approximation

Consider the same time-invariant linear system with an output $y(t)$ and control $u(t)$. The transfer function $G(s)$ relates the input and output as

$$Y(s) = G(s)U(s) \equiv \frac{N(s)}{D(s)}U(s). \tag{4.105}$$

Consider a feedback control with time delay as $U(s) = e^{-\tau s}C(s)[R(s) - Y(s)]$. The time delay term $e^{-\tau s}$ is approximated with the Padé expansion as

$$e^{-\tau s} = \frac{N_p(s)}{D_p(s)} \equiv R_{n,d}(s), \tag{4.106}$$

where the subscripts of $R_{n,d}(s)$ indicate the orders of the polynomials $N_p(s)$ and $D_p(s)$. Several different order Padé approximations are listed in Table (4.3). The closed-loop system response is given by

$$Y(s) = \frac{N(s)N_p(s)C(s)}{N(s)N_p(s)C(s) + D(s)D_p(s)} R(s). \tag{4.107}$$

The closed-loop characteristic equation reads

$$N(s)N_p(s)C(s) + D(s)D_p(s) = 0. \tag{4.108}$$

In the stability analysis, we can apply the Routh-Hurwitz criterion (Franklin et al., 1986).

Table 4.3 Various order of Padé approximations of the time delay term $e^{-\tau s}$ (Vijta, 2000).

Order	(0,1)	(1,1)	(2,2)	(1,2)	(2,3)
$N_p(s)$	1	$2 - s\tau$	$12 - 6s\tau + (s\tau)^2$	$6 - 2\tau$	$60 - 24s\tau + 3(s\tau)^2$
$D_p(s)$	$1 + s\tau$	$2 + s\tau$	$12 + 6s\tau + (s\tau)^2$	$6 + 4s\tau + (s\tau)^2$	$60 + 36s\tau + 9(s\tau)^2 + (s\tau)^3$

Order	(3,3)	(4,4)
$N_p(s)$	$120 - 60s\tau + 12(s\tau)^2 - (s\tau)^3$	$1680 - 840s\tau + 180(s\tau)^2 - 20(s\tau)^3 + (s\tau)^4$
$D_p(s)$	$120 + 60s\tau + 12(s\tau)^2 + (s\tau)^3$	$1680 + 840s\tau + 180(s\tau)^2 + 20(s\tau)^3 + (s\tau)^4$

4.6.3 Stability with semi-discretization

As discussed in Section 4.3, the method of semi-discretization leads to a mapping of the state vector over the interval $[t_i, t_{i+1}]$ as

$$\mathbf{y}_{i+1} = \mathbf{\Phi}\mathbf{y}_i. \tag{4.109}$$

The stability boundary is given by

$$|\lambda|_{\max} = 1. \tag{4.110}$$

We should comment that the method of semi-discretization has been well studied and proven to be effective and reliable (Elbeyli and Sun, 2004). We shall use this method as a yardstick to compare with other methods in the stability analysis.

4.6.4 Stability of a second order LTI system

Next, we use the second order LTI system in Eq. (4.41) subject to a delayed PD control as an example to compare the stability results in the PD control gain space by various methods discussed above.

Delay independent Lyapunov stability

We first compare the stability condition (4.93) with the method of semi-discretization. In the numerical studies, we have chosen different values of time delays τ. The $\mathbf{Q}$ and $\mathbf{S}$ matrices for the stability condition (4.93) are taken to be the identity matrix. We have found that the stability domains determined by the Lyapunov method are insensitive to the choices of these matrices. This is because we are only interested in the stability domain determined by $\dot{V} = 0$.

Figure 4.20 shows the stability domains by both the methods for $\tau = \pi/30$. Since the Lyapunov stability condition is independent of time delay, it must hold for all time delays including arbitrarily large ones. The figure suggests that as the time delay $\tau \to \infty$, the regions of stability domains for different time delays as obtained by the method of semi-discretization must converge to the domain by the time-independent Lyapunov approach.

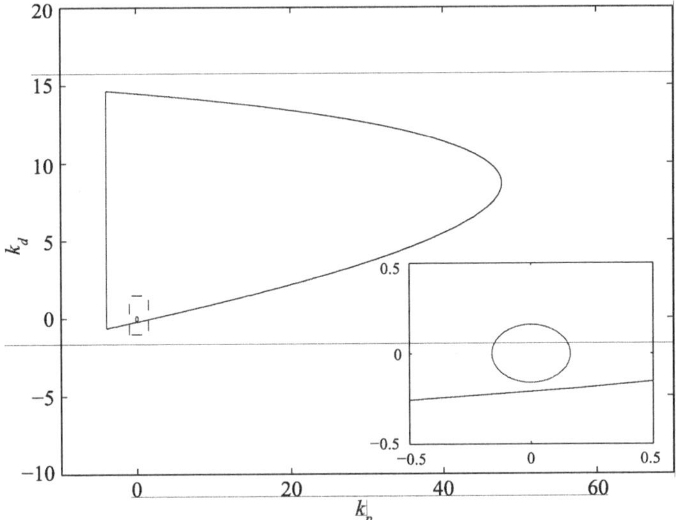

Fig. 4.20 Stability domains in the $k_p - k_d$ gain space ($\tau = \pi/30$). The large domain is obtained with the semi-discretization mothed. The small domain in the sub-figure is determined by Eq. (4.93) of the Lyapunov method independent of time delay.

Delay dependent Lyapunov stability

The delay dependent stability condition (4.103) is considered next. Figures 4.21 to 4.23 show the comparison of the stability domains by the delay dependent Lyapunov method and the semi-discretization with $\tau = \pi/30$, $\pi/4$ and $\pi/2$. We have

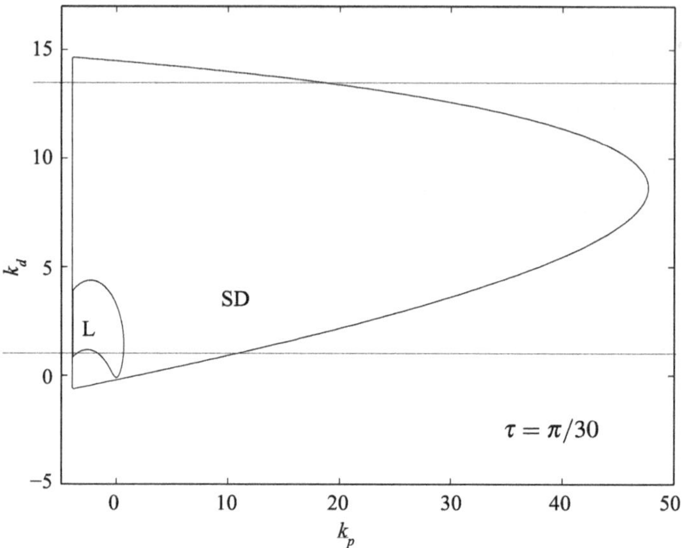

Fig. 4.21 Stability domains in the PD control gain space for the LTI system with time delay. The closed region marked with SD is by the semi-discretization. The closed region marked with L is determined by the delay dependent Lyapunov method. $\tau = \pi/30$.

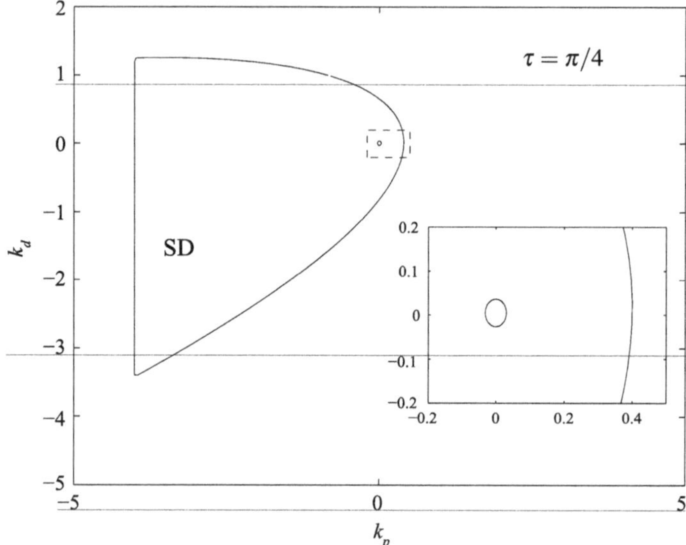

Fig. 4.22 Stability domains in the PD control gain space for the LTI system with time delay. The closed region marked with SD is by the semi-discretization. The small circle is the stability domain determined by the delay dependent Lyapunov method. $\tau = \pi/4$.

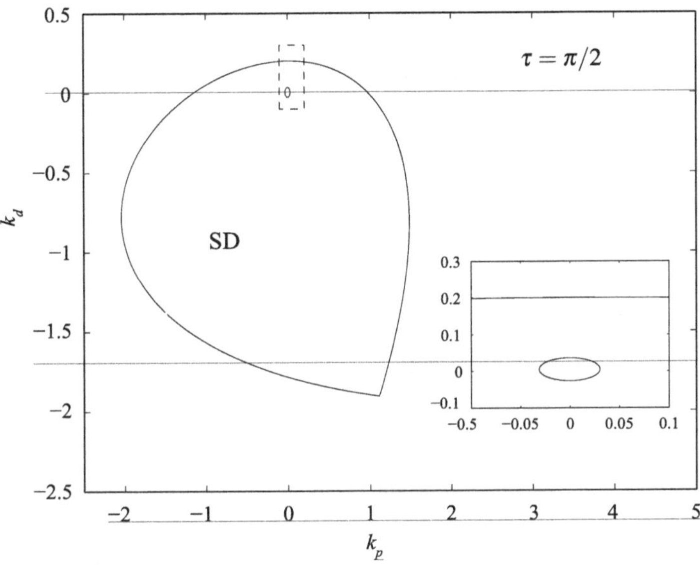

Fig. 4.23 Stability domains in the PD control gain space for the LTI system with time delay. The closed region marked with SD is by the semi-discretization. The small circle is the stability domain determined by the delay dependent Lyapunov method. $\tau = \pi/2$.

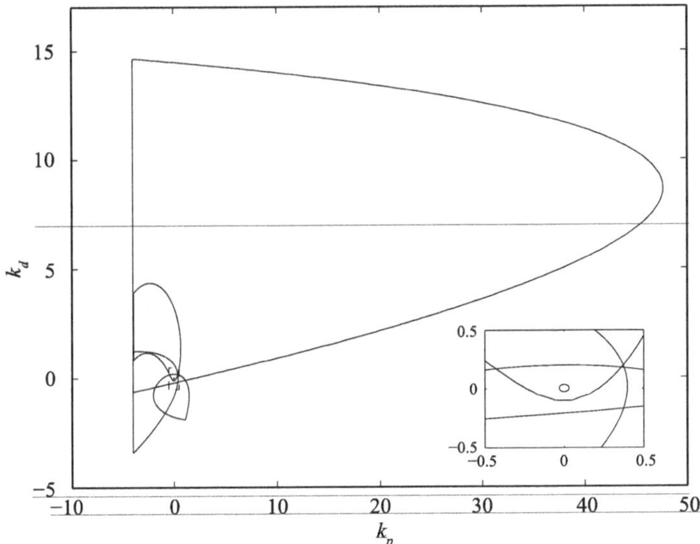

Fig. 4.24 Superposition of all the stability domains in Figs. 4.21 to 4.23. The small circle represents the stability domain by the delay dependent Lyapunov method for all large time delays.

computed the stability domains for a number of time delays. We have found that when the time delay is small, the stability domain by the delay dependent Lyapunov method is indeed enlarged as compared to that by the delay independent Lyapunov method, and is still contained within the domain obtained by the semi-discretization.

Figure 4.24 superimposes all the stability domains in Figs. 4.21 to 4.23. It appears that the stability domains by the delay dependent Lyapunov method converge to that by the delay independent Lyapunov method as the time delay increases. The stability results are still very conservative.

Stability by Padé approximation

Figures 4.25 to 4.29 show the stability domains for $\tau = \pi/30$ and $\pi/2$ obtained by the semi-discretization and the Padé approximation of orders (1,2), (2,3) and (4,4). It is interesting to note that the stability domains by the Padé approximation agree with that by the semi-discretization only when the delay is small. As the time delay increases, the order of the Padé approximation has to be increased also. The Padé approximation gives slightly less conservative stability results because its stability domains don't completely overlap with that by the semi-discretization. When the Padé approximation is used to design feedback controls for systems with time delay, we must pick the gains as much to the centroid of the domian as possible to avoid instability.

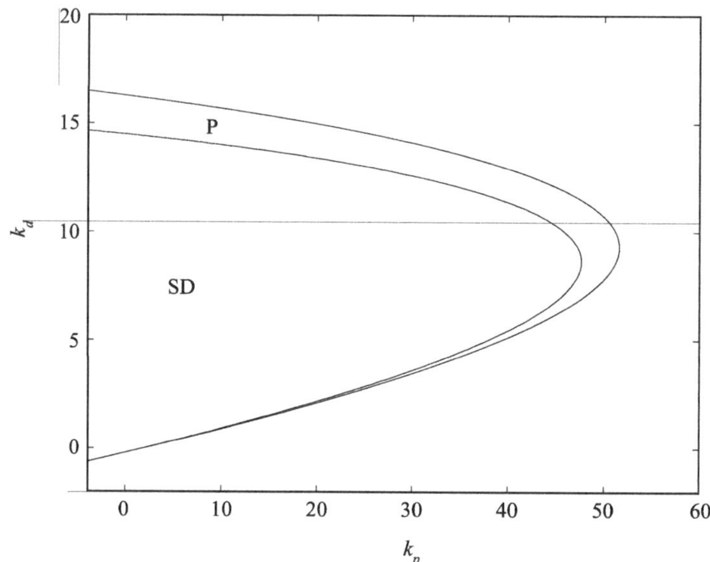

Fig. 4.25 Stability domains in the PD control gain space of the LTI system. The closed region marked as SD is obtained by the semi-discretization and the one marked as P is by the Padé approximation of order (1,2). $\tau = \pi/30$.

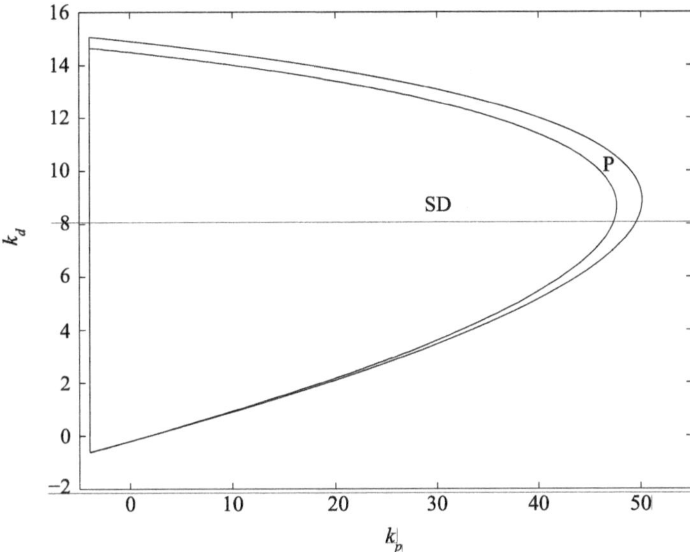

Fig. 4.26 Stability domains in the PD control gain space of the LTI system. The closed region marked as SD is obtained by the semi-discretization and the one marked as P is by the Padé approximation of order (2,3). $\tau = \pi/30$.

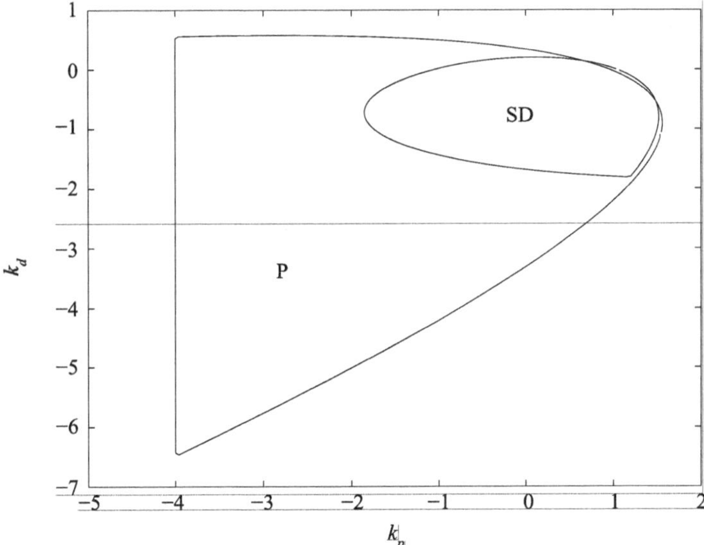

Fig. 4.27 Stability domains in the PD control gain space of the LTI system. The closed region marked as SD is obtained by the semi-discretization and the one marked as P is by the Padé approximation of order (1,2). $\tau = \pi/2$.

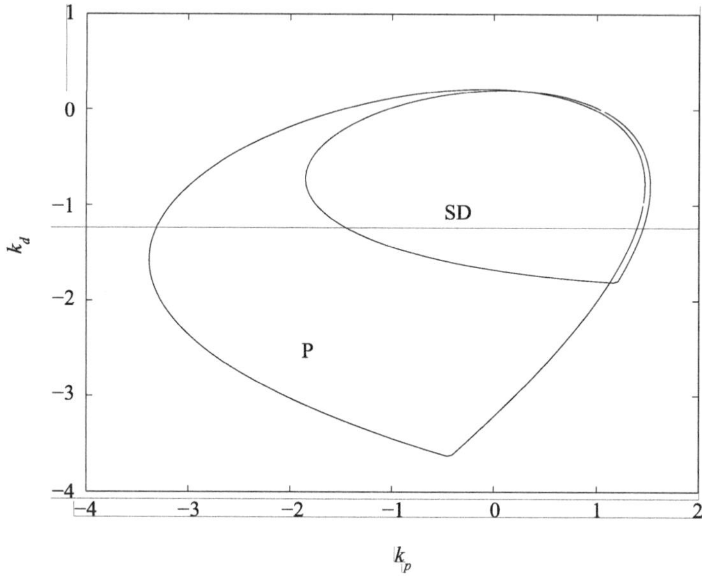

Fig. 4.28 Stability domains in the PD control gain space of the LTI system. The closed region marked as SD is obtained by the semi-discretization and the one marked as P is by the Padé approximation of order (2,3). $\tau = \pi/2$.

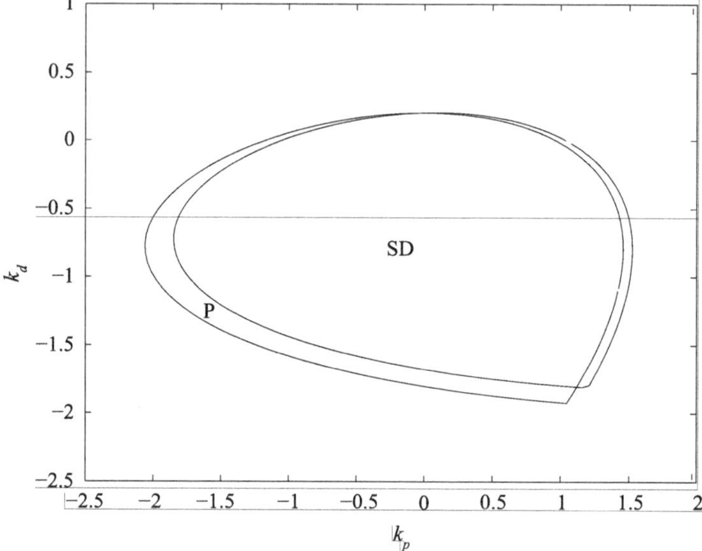

Fig. 4.29 Stability domains in the PD control gain space of the LTI system. The closed region marked as SD is obtained by the semi-discretization and the one marked as P is by the Padé approximation of order (4,4). $\tau = \pi/2$.

In summary, we have compared the stability results obtained by the Lyapunov method, semi-discretization, and the Padé approximation. From extensive numerical simulations, we have found that the Lyapunov method, whether it is delay dependent or independent, leads to conservative results as compared with those obtained by the semi-discretization. As the time delay increases, the Lyapunov results for a LTI system converge to a small region which is the stability domain for arbitrary time delay. The Lyapunov method appears to be quite conservative for LTI systems with time delay and is not effective for feedback control design. On the other hand, the Padé approximation gives less conservative and reasonably accutrate stability results when the order of the method is high enough in comparison to the time delay.

4.7 Control of LTI systems

Consider a delayed PID control of the linear time-invariant (LTI) system in Eq. (4.35). We first examine PD controls. The system parameters are chosen as $\zeta = 0.05$, $\omega = 2$ and $N = 50$. The undamped period of the system is $T = \pi$.

Figure 4.30 shows the upper and lower bounds of the control gain k_p as a function of the time delay τ when $k_d = 0$. This result is fully agreeable to that obtained by the Nyquist criterion. Note a periodic change of the bounds related to T in the figure. The bounds become narrower as the time delay increases. In (Filipovic and Olgac, 1998), a similar result is achieved. This trivial example helps to study the issues such as the accuracy of semi-discretization as a function of the time step Δt.

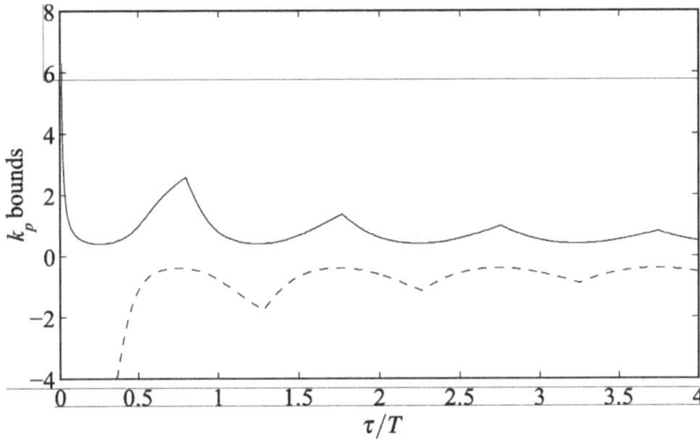

Fig. 4.30 Upper and lower bounds of control gain k_p vs. time delays when $k_d = 0$ for the time-invariant second order system. Solid line is the upper bound. Dashed line is the lower bound.

In Fig. 4.31, the stability boundaries on the $k_p - k_d$ plane are plotted for the system with different time delays. It is interesting to point out that the size of the stability region decreases when the time delay increases. Figure 4.32 shows the performance contours of the feedback control as measured by the maximum absolute value of eigenvalues of the mapping matrix $\mathbf{\Phi}$ for a time delay $\tau = \pi/2$. This chart clearly indicates that for a given delay, there is a finite optimal pair of control gains k_p and k_d that will lead to the best control performance with the smallest $|\lambda|_{\max}$. Solving for the optimization problem in Eq. (4.34), we have found the optimal gains to be $(k_p, k_d) = (-0.1356, -0.3898)$ with $|\lambda|_{\max} = 0.1103$.

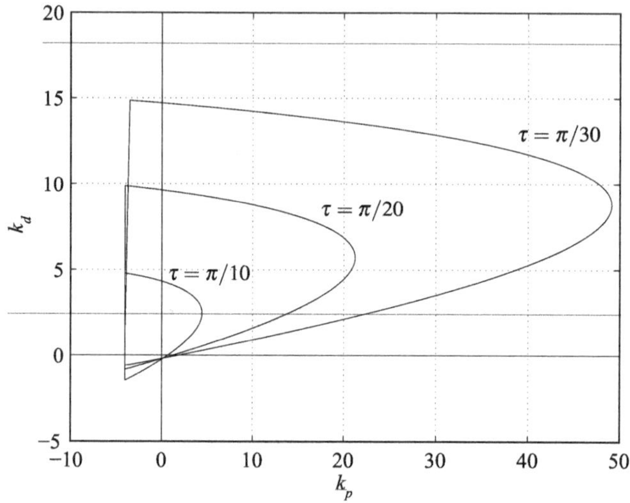

Fig. 4.31 Stability boundaries on the $k_p - k_d$ plane of the time-invariant system with different time delays. The inner part of the closed contour is the stable region.

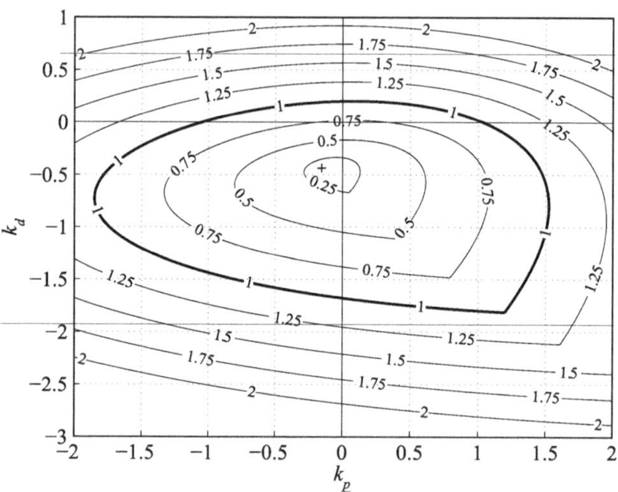

Fig. 4.32 Performance contours of the second order linear time-invariant system with a time delay $\tau = \pi/2$. The labels of the contours are the maximum absolute value $\lambda_{\max}$ of eingenvalues of the mapping $\mathbf{\Phi}$. "+" indicates the optimal control gains $(k_p, k_d) = (-0.1356, -0.3898)$ with the smallest $\lambda_{\max} = 0.1103$.

Figure 4.33 shows a comparison of time histories of the amplitude of the response vector $(x, \dot{x})$ of the system under the optimal and non-optimal feedback controls. The optimal control clearly provides superior performance to that of the non-optimal ones.

Next, we consider PI control.

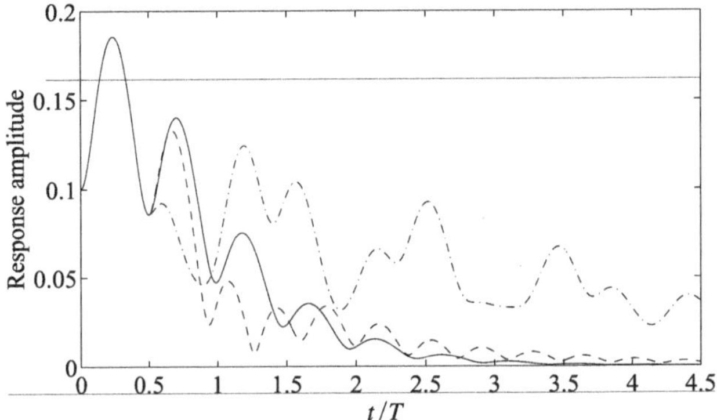

Fig. 4.33 Time history of the response amplitude of the autonomus system. Solid line: the optimal gains $(k_p, k_d) = (-0.1356, -0.3898)$ and $\lambda_{\max} = 0.1103$. Dashed line: $(k_p, k_d) = (-0.5, -0.8)$ and $\lambda_{\max} \approx 0.5$. Dashed-dotted line: $(k_p, k_d) = (0.8, -1.45)$ and $\lambda_{\max} \approx 0.75$.

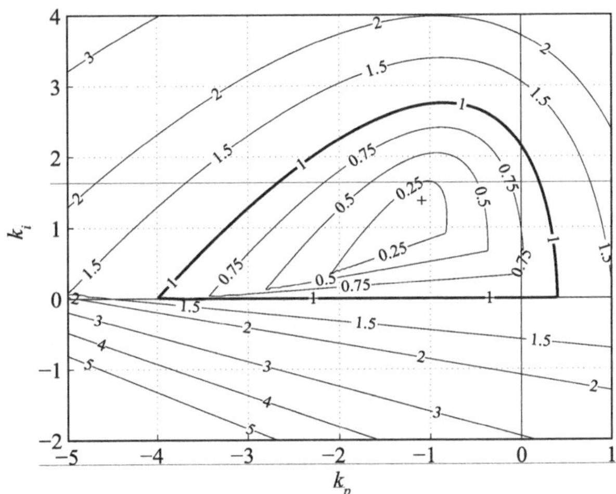

Fig. 4.34 Performance contours of the second order linear time-invariant system with a time delay $\tau = \pi/4$. The labels of the contours are the maximum absolute value $\lambda_{\max}$ of eigenvalues of the mapping $\mathbf{\Phi}$. "+" indicates the optimal control gains $(k_p, k_i) = (-1.1217, 1.3682)$ with the smallest $\lambda_{\max} = 0.0537$.

Figure 4.34 shows the performance contours of the feedback control as measured by the largest absolute value of eigenvalues of the mapping matrix $\mathbf{\Phi}$ for a time delay $\tau = \pi/4$. This chart clearly indicates that for a given delay, there is a finite optimal pair of control gains k_p and k_i that will lead to the best control performance with the smallest $|\lambda|_{\max}$. We have indeed found the optimal gains to be $(k_p, k_i) = (-1.1217, 1.3682)$ with $|\lambda|_{\max} = 0.0537$.

Figure 4.35 shows a comparison of time histories of the amplitude of the response vector $(x, \dot{x}, \ddot{x})$ of the system under the optimal and non-optimal feedback controls. The optimal control clearly provides superior performance to that of the non-optimal ones.

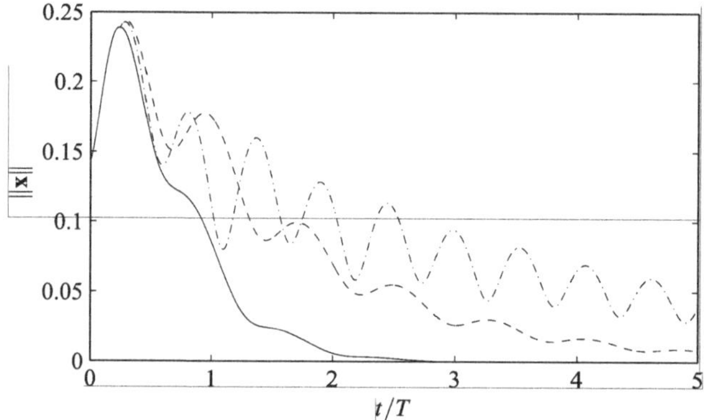

Fig. 4.35 Time history of the response amplitude $||\mathbf{x}||$ of the autonomous system. Solid line: the optimal gains $(k_p, k_i) = (-1.1217, 1.3682)$ and $\lambda_{\max} = 0.0537$. Dashed line: $(k_p, k_i) = (-1.0, 2.0)$ and $\lambda_{\max} \approx 0.5$. Dashed-dotted line: $(k_p, k_i) = (0.0, 1.0)$ and $\lambda_{\max} \approx 0.75$.

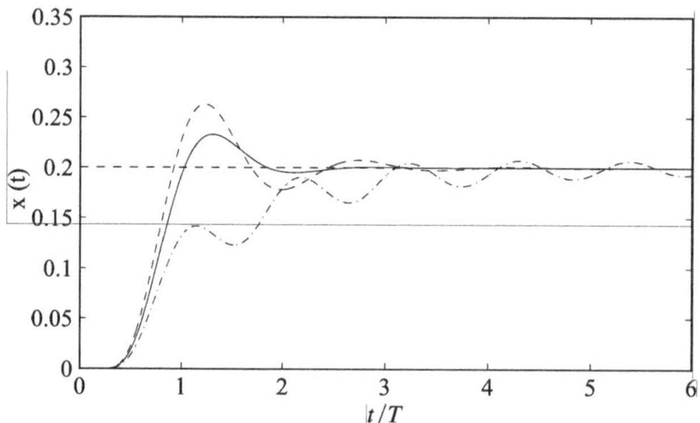

Fig. 4.36 Tracking performance of the response $x(t)$ of the autonomous system. The reference is a step input with $r = 0.2$. Solid line: the optimal gains $(k_p, k_i) = (-1.1217, 1.3682)$ and $\lambda_{\max} = 0.0537$. Dashed line: $(k_p, k_i) = (-1.0, 1.64)$ and $\lambda_{\max} \approx 0.25$. Dashed-dotted line: $(k_p, k_i) = (0.0, 1.0)$ and $\lambda_{\max} \approx 0.75$.

In the classic PID control theory, the steady state tracking error can be eliminated by the integral control. This is still true for time delayed linear systems. Figure 4.36 shows the response under the optimal and non-optimal feedback controls tracking a step reference. Notice that the steady state tracking error is zero.

4.8 Control of the Mathieu system

Consider the Mathieu equation with a delayed PID feedback control

$$\dot{\mathbf{x}}(t) = \begin{bmatrix} 0 & 1 & 0 \\ 0 & 0 & 1 \\ 4\varepsilon \sin 2t & -(\delta + 2\varepsilon \cos 2t) & 0 \end{bmatrix} \mathbf{x}(t) - \begin{bmatrix} 0 & 0 & 0 \\ 0 & 0 & 0 \\ k_i & k_p & k_d \end{bmatrix} \mathbf{x}(t-\tau), \quad (4.111)$$

where $\mathbf{x} = (x, \dot{x}, \ddot{x})^{\mathrm{T}}$. The period of the system is $T = \pi$. We select the parameters to be $\varepsilon = 1$, $\delta = 1, 3$ and 4. When $\delta = 1$ and 4, the uncontrolled system is parametrically unstable. Consider a delay $\tau = \pi/4$ and choose $N = 20$.

Figures 4.37 and 4.38 present the stability boundaries on the $k_p - k_d$ plane with different parameters. As is in the case of the time-invariant system, the size of the stability region decreases with the increase of time delay. The shape of the stability region is more complex than that of the time-invariant system. The irregular geometry of the contours in the figures is a reflection of the complex behavior of the time-varying system.

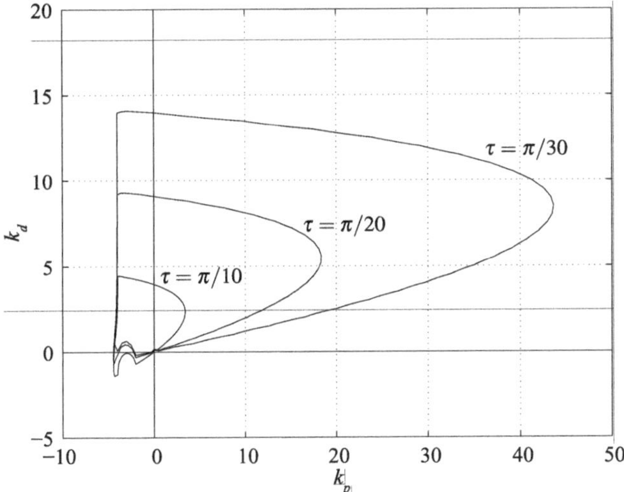

Fig. 4.37 Stability boundaries on the $k_p - k_d$ plane for the periodic system with different time delays. The inner part of the closed contour is the stable region. $\delta = 4$ and $\varepsilon = 1$.

In the following discussions, the parameters are fixed to be $\delta = 4$, $\varepsilon = 1$ and $N = 20$. Figure 4.39 shows the performance contours of the control system with a time delay $\tau = \pi/4$. For the periodic system, there is also a finite optimal pair of the feedback gains that will lead to the best control performance as measured by $|\lambda|_{\max}$. We have found the optimal gains to be $(k_p, k_d) = (-2.0169, -0.3091)$ with $|\lambda|_{\max} = 1.5535\text{e}{-3}$.

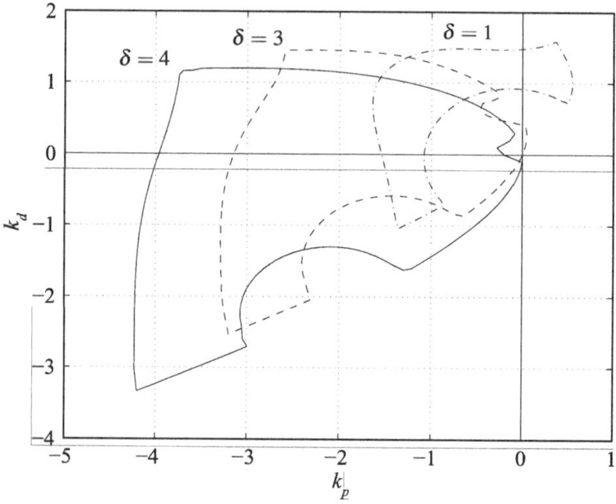

Fig. 4.38 Stability boundaries of the periodic system with a time delay $\tau = \pi/4$. Solid line: $\delta = 4$ and $\varepsilon = 1$. Dashed line: $\delta = 3$ and $\varepsilon = 1$. Dashed-dotted line: $\delta = 1$ and $\varepsilon = 1$. The inner part of the closed contour is the stable region.

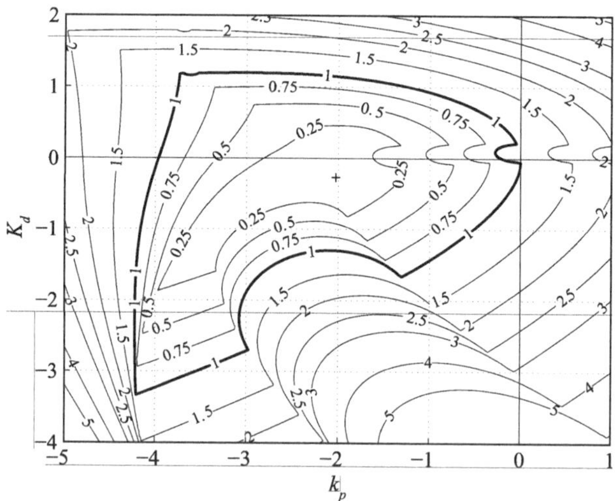

Fig. 4.39 Performance contours of the closed loop periodic system with a time delay $\tau = \pi/4$. The labels of the contours are the maximum absolute value $\lambda_{\max}$ of eingenvalues of the mapping **Φ**. "+" indicates the optimal control gains $(k_p, k_d) = (-2.0169, -0.3091)$ with the smallest $\lambda_{\max} = 1.5535\text{e-}3$.

Figure 4.40 shows a comparison of time histories of the amplitude of the response vector $(x, \dot{x})$ of the system under the optimal and non-optimal feedback controls. The optimal control again provides superior performance to that of the non-optimal ones.

Next, we consider PI control. Figure 4.41 presents the stability boundaries on the $k_p - k_i$ plane with $k_d = 0$. The shape of the stability region is more complex than that of the time-invariant system. When $\tau = \pi/20$, the region breaks into two

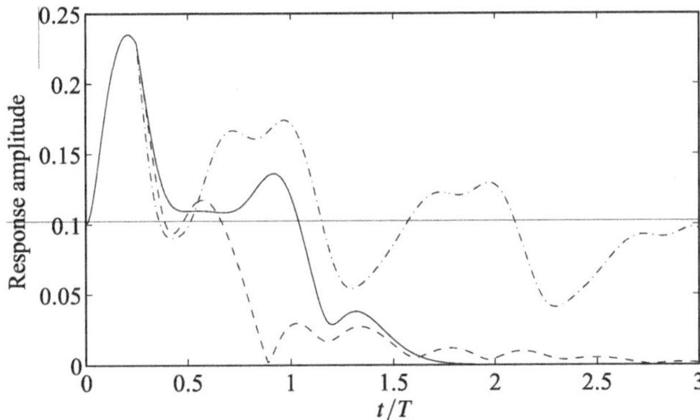

Fig. 4.40 Time history of the response amplitude of the periodic system. Solid line: the optimal gains $(k_p, k_d) = (-2.0169, -0.3091)$ and $\lambda_{\max} = 1.5535e\text{-}3$. Dashed line: $(k_p, k_d) = (-2.0, 0.4)$ and $\lambda_{\max} \approx 0.25$. Dashed-dotted line: $(k_p, k_d) = (-4.0, -1.0)$ and $\lambda_{\max} \approx 0.75$.

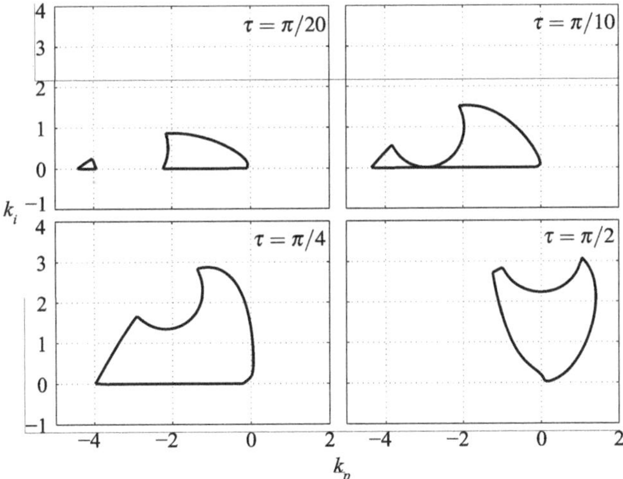

Fig. 4.41 Stability boundaries on the $k_p - k_i$ plane for the periodic system with different time delays when $k_d = 0$. The inner part of the closed contour is the stable region. $\delta = 4$ and $\varepsilon = 1$.

disjoint parts. Figure 4.42 presents the stability boundaries on the $k_p - k_d$ plane with $k_i = 0$. In this case, the size of the stability region decreases with the increase of time delay. Figure 4.43 shows minimal $|\lambda|_{\max}$ as a function of time delay for both the PI and PD controls considered herein.

The irregular geometry of the stability boundary in Figs. 4.41 and 4.42 is a reflection of the complex behavior of the time-varying system. Furthermore, significant variations of minimal $|\lambda|_{\max}$ with time delay suggests the difficulty in designing controls to reach certain performance targets.

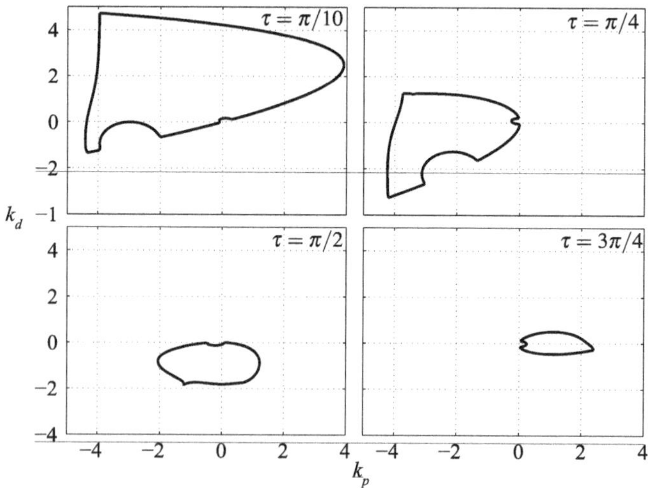

Fig. 4.42 Stability boundaries on the $k_p - k_d$ plane for the periodic system with different time delays when $k_i = 0$. The inner part of the closed contour is the stable region. $\delta = 4$ and $\varepsilon = 1$.

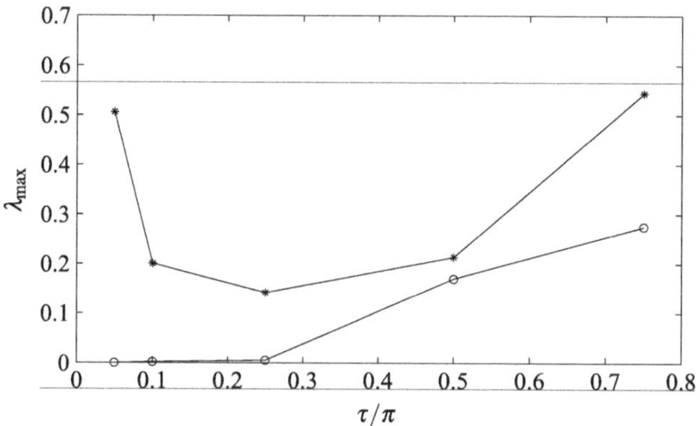

Fig. 4.43 Minimal $\lambda_{\max}$ for the periodic system with different time delays. (-*-): PI control. (-o-): PD control.

Next, we consider the time domain performance of a PI control with time delay $\tau = \pi/4$. Figure 4.44 shows the performance contours of the system. In this case, there exists one finite optimal pair of the feedback gains that will lead to a minimal $|\lambda|_{\max}$. We have found the optimal gains to be $(k_p, k_i) = (-2.2628, 0.9483)$ with $|\lambda|_{\max} = 0.1605$. Figure 4.45 shows a comparison of time histories of the amplitude of the response vector $\mathbf{x}(t)$ of the system under the optimal and non-optimal

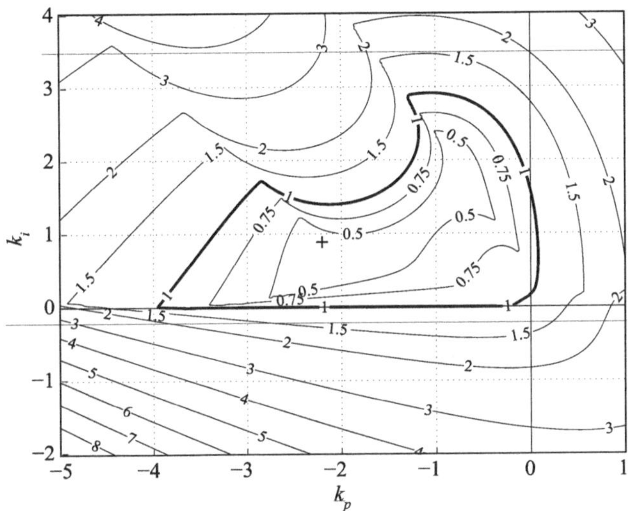

Fig. 4.44 Performance contours of the closed loop periodic system with a time delay $\tau = \pi/4$. The labels of the contours are the maximum absolute value $\lambda_{\max}$ of eigenvalues of the mapping $\mathbf{\Phi}$. "+" indicates the optimal control gains $(k_p, k_i) = (-2.2628, 0.9483)$ with the smallest $\lambda_{\max} = 0.1605$.

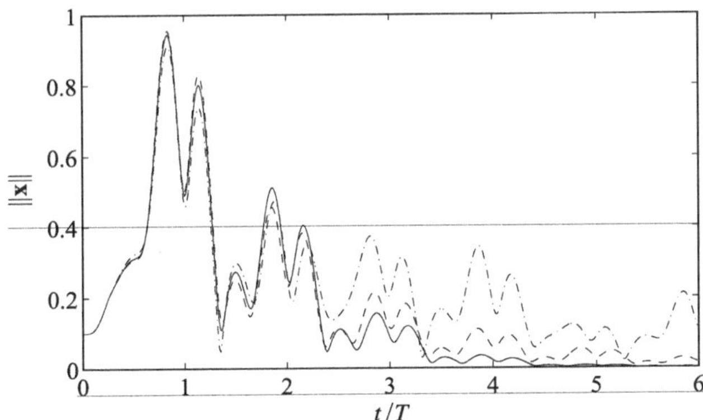

Fig. 4.45 Time history of the response amplitude $||\mathbf{x}||$ of the periodic system under a delayed PI control. Solid line: the optimal gains $(k_p, k_i) = (-2.2628, 0.9483)$ and $\lambda_{\max} = 0.1605$. Dashed line: $(k_p, k_i) = (-2.0, 1.0)$ and $\lambda_{\max} \approx 0.5$. Dashed-dotted line: $(k_p, k_i) = (-3.0, 0.8)$ and $\lambda_{\max} \approx 0.75$.

feedback controls. As pointed earlier, once the system is in steady state, the magnitude of $\mathbf{x}(t)$ decays at the faster rate due to the optimal control gains than that due to the non-optimal ones.

4.9 An experimental validation

We apply the semi-discretization method to the design of the feedback control of a rotary flexible joint experiment made by Quanser. The experimental apparatus is shown in Fig.4.46, which consists of a rotary flexible joint mounted on top of a rigid rotary platform. Two encoders are used in the system. One measures the angular position of the platform, the other measures the angular displacement of the flexible joint relative to the platform. The state equation of the system is of fourth order given by

$$\dot{\mathbf{x}} = \mathbf{A}\mathbf{x} + \mathbf{b}u, \tag{4.112}$$

Fig. 4.46 The rotary flexible joint experiment.

where

$$\mathbf{x} = \left[\theta \; \alpha \; \dot{\theta} \; \dot{\alpha}\right]^{\mathrm{T}}, \tag{4.113}$$

$$\mathbf{A} = \begin{bmatrix} 0 & 0 & 1 & 0 \\ 0 & 0 & 0 & 1 \\ 0 & 689.86 & -57.658 & 0 \\ 0 & -1359.2 & 57.658 & 0 \end{bmatrix}, \tag{4.114}$$

$$\mathbf{b} = \left[0 \; 0 \; 107.39 \; -107.39\right]^{\mathrm{T}}. \tag{4.115}$$

θ is the angular position of the platform, and α is the angular position of the flexible joint relative to the platform. A state feedback control $u = -\mathbf{k}^{\mathrm{T}}\mathbf{x}$ is designed by using the LQR method with the state weighting matrix $\mathbf{Q}$ and the control weighting factor R given by

$$\mathbf{Q} = \mathrm{diag}(\left[3000\ 14500\ 10\ 0\right]),\ R = 2. \tag{4.116}$$

The resulting optimal feedback gains are

$$\mathbf{k} = \left[k_1\ k_2\ k_3\ k_4\right]^{\mathrm{T}} = \left[38.7298\ -73.2505\ 3.3657\ 1.1871\right]^{T}. \tag{4.117}$$

By examining the measured step response of the open-loop system, we have found that the system has a time delay of 0.002 second. An additional transport delay 0.008 second between the input and the output of the system is digitally introduced, leading to a total time delay $\tau = 0.01$ second.

The top plot of Fig. 4.47 shows the step response of the experimental system with time delay under the LQR control. Recall that the LQR control is designed without consideration of time delay. The time delay has already destabilized the control system in simulations and introduces oscillations in the experimental response. The experimental system has a hardware saturation limit of the control output, which keeps the system stable in this case.

The closed-loop system with time delay reads

$$\dot{\mathbf{x}} = \mathbf{A}\mathbf{x} - \mathbf{b}\mathbf{k}^{\mathrm{T}}\mathbf{x}(t - \tau). \tag{4.118}$$

We shall look for the optimal feedback gains $\mathbf{k}$.

Here, the gain space is four dimensional. In this example, we restrict our interest in the domain defined by

$$0 < k_1 < 100,\quad -50 < k_2 < 50,\quad 0 < k_3 < 5,\quad 0 < k_4 < 5. \tag{4.119}$$

We discretize the domain by dividing each gain range into ten partitions. We then choose the grid points of the partition as initial guess of the optimal gains, and a nonlinear search algorithm is run to find the optimal gains that deliver the minimal $|\lambda|_{\mathrm{max}}$. It turns out that there are many local minimums of $|\lambda|_{\mathrm{max}}$. The smallest one among all the local minimums is taken to be the global minimum of $|\lambda|_{\mathrm{max}}$ in the domain (4.119) of the gain space.

We have found the optimal feedback gains to be

$$\mathbf{k}_{\mathrm{min}} = \left[12.32067\ -15.43203\ 0.86435\ 0.45176\right]^{\mathrm{T}}, \tag{4.120}$$

with $|\lambda|_{\mathrm{max}} = 0.6511$. The bottom plot of Fig. 4.47 shows the step response of the closed-loop system with the feedback gains $\mathbf{k}_{\mathrm{min}}$. The current control design explicitly takes into account of the effect of time delay, and improves the tracking performance. Because the minimum of $|\lambda|_{\mathrm{max}}$ is used in the control, the system has plenty damping as shown in the response.

The present control design procedure does not require a significant computational power. After the formulation of the mapping matrix, searching for one optimal gain in the four dimensional space takes less than 10 seconds for a discretization level $N \leq 40$, which gives accurate solutions as demonstrated in (Elbeyli and Sun, 2004).

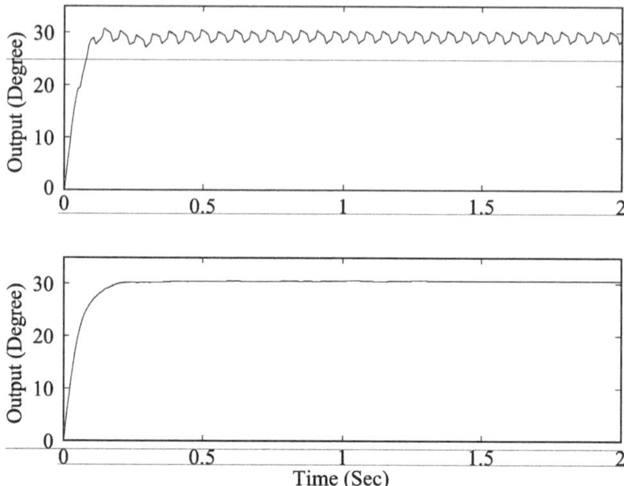

Fig. 4.47 Comparison of the experimental step response of the LQR control (top) and the feedback control designed with the semi-discretization method (bottom) of the rotary flexible joint system with a time delay $\tau = 0.01$ second.

Figure 4.48 shows the computational time for finding one local minimum as a function of the discretization level N. The computation is performed on a 2GHz PC.

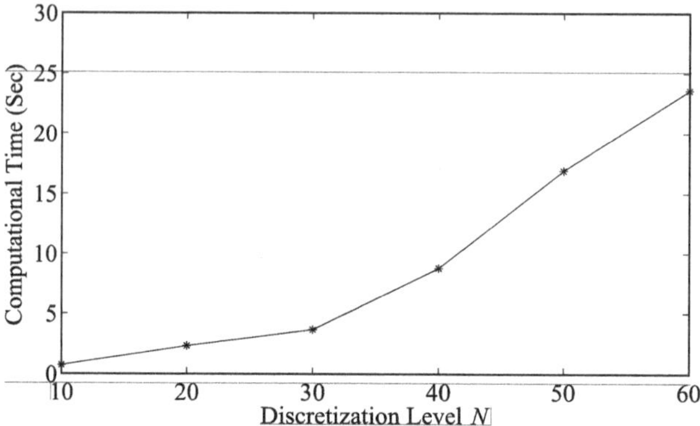

Fig. 4.48 Computational time for one nonlinear search for the optimal gains starting from $\mathbf{k} = [80\ {-20}\ 0.5\ 0.5]$ as a function of the discretization level N.

4.10 Supervisory control

Recall the system in Eq. (4.67). The time delay τ is assumed to be slowly time-varying, and lies in an interval $[\tau_{\min}, \tau_{\max}]$ where the minimum and maximum time delays are assumed to be known. Assume that we have obtained a set of optimal feedback gains for the set of time delays sampled in the interval $[\tau_{\min}, \tau_{\max}]$. We present the switching algorithm for selecting a gain to implement in real time.

The actual time delay τ is such that $\tau_{\min} \leq \tau \leq \tau_{\max}$. We descritize $[\tau_{\min}, \tau_{\max}]$ into $M_\tau - 1$ intervals so that $\tau_{\min} = \tau_1 < \tau_2 < \cdots < \tau_{M_\tau} = \tau_{\max}$. Consider M_τ models of the time-delayed system as

$$\dot{\mathbf{x}}_i = \mathbf{f}\left(\mathbf{x}_i\left(t\right), \mathbf{x}_i\left(t - \tau_i\right), t\right) + \mathbf{B}\mathbf{u}_i\left(t\right), \quad 1 \leq i \leq M_\tau. \tag{4.121}$$

Consider the feedback control $\mathbf{u}_i = -\mathbf{K}_i\mathbf{x}_i\left(t - \tau_i\right)$ where $\mathbf{K}_i \in \mathbf{\Omega}$. Each $\mathbf{K}_i$ is found by imposing Eq. (4.34) subject to an additional constraint: $\mathbf{K}_i$ must be stable for all τ_j $(1 \leq j \leq M_\tau)$. Let $\mathbf{K}_{iOpt} \in \mathbf{\Omega}$ be the optimal gain for τ_i and the associate eigenvalue with the smallest magnitude $|\lambda_i(\mathbf{\Phi})|_{\min} < 1$. Check if $\mathbf{K}_{iOpt}$ stablizes the system in Eq. (4.121) for all other time delays τ_j $(1 \leq j \leq M_\tau)$.

Following the concept of the supervisory control (Morse, 1996, 1997; Hespanha et al., 1999, 2003), we define an estimation error as

$$\mathbf{e}_i = \mathbf{x}_i\left(t\right) - \mathbf{x}\left(t\right), \quad 1 \leq i \leq M_\tau, \tag{4.122}$$

where $\mathbf{x}\left(t\right)$ is the output of the system with unknown time delay. In the experiment, $\mathbf{x}\left(t\right)$ would be obtained from measurements. Consider a positive function of the estimation error $F_i(\mathbf{e}_i) > 0$. An example is $F_i(\mathbf{e}_i) = ||\mathbf{e}_i||^2$. Define a switching index $\pi_i(t)$ such that

$$\begin{aligned} \dot{\pi}_i(t) + \lambda_i\pi_i(t) &= F_i(\mathbf{e}_i), \quad (\lambda_i > 0) \\ \pi_i(0) &= 0, \end{aligned} \tag{4.123}$$

where the parameter λ_i defines the bandwidth of the low pass filter. The general solution for $\pi_i(t)$ can be obtained as

$$\pi_i(t) = e^{-\lambda_i t}\pi_i(0) + \int_0^t e^{-\lambda_i(t-\tau)}F_i(\mathbf{e}_i(\tau))d\tau. \tag{4.124}$$

The hysteretic switching algorithm in (Hespanha et al., 1999, 2003) is stated as follows. Assume that the system is sampled at time interval Δt. At the kth time step, the system is under control with the gain $\mathbf{K}_j$ and the associated switching signal is $\pi_j(k)$. At the $(k+1)^{\text{th}}$ step, if there is an index i such that $\pi_i(k) < (1-\eta)\pi_j(k)$ where $\eta > 0$ is a small number, we switch to the gain $\mathbf{K}_i$. Otherwise, we continue with the gain $\mathbf{K}_j$. η is known as the hysteretic parameter and prevents the system from switching too frequently.

4.10.1 Supervisory control of the LTI system

Consider a second-order autonomous system (4.35) under a delayed PI control. The feedback control reads $u = -[k_i, k_p, 0]\mathbf{x}(t-\tau)$ where $\mathbf{x} = (x, \dot{x}, \ddot{x})^{\mathrm{T}}$. Take $\omega = 2$ and $\zeta = 0.05$. The discretization number of the time delayed response is set to be $N = 20$ for all sampled time delays (Sheng et al., 2004; Sheng and Sun, 2005;

Sun, 2009). $\tau_{min} = 0.0419$ and $\tau_{max} = 0.2094$. We pick five different time delays to design optimal feedback gains according to the method outlined in the previous section.

Table 4.4 Optimal PI feedback gains and $|\lambda|_{max}$ for the four sampled time delays of the linear time invariant system.

Time Delay	k_i	k_p	$\lvert\lambda\rvert_{max}$
τ_1 =0.0419	0.2000	−1.8400	0.9992
τ_2 =0.0733	0.2000	−2.3500	0.9982
τ_3 =0.1047	0.2000	−2.8600	0.9966
τ_4 =0.1571	0.2000	−2.8600	0.9938
τ_5 =0.2094	0.2000	−3.3700	0.9907

The optimal gains associated with the five time delays are listed in Table 4.4. The associated stability domains in the gain space are shown in Fig. 4.49. It should be pointed out that when the optimal gains of all the controls with different time delays fall in the intersection of the stability domains, it is possible to use the hysteretic algorithm to switch among the pre-designed controls and to keep the system stable all the time. When an optimal gain is out of the intersection, the control with that gain can destabilize the system with some time delay in the range $[\tau_{min}, \tau_{max}]$. This property limits the size of the unknown time delay range $[\tau_{min}, \tau_{max}]$ because the stability domains change significantly with the time delay, particularly for periodic systems (Sheng and Sun, 2005).

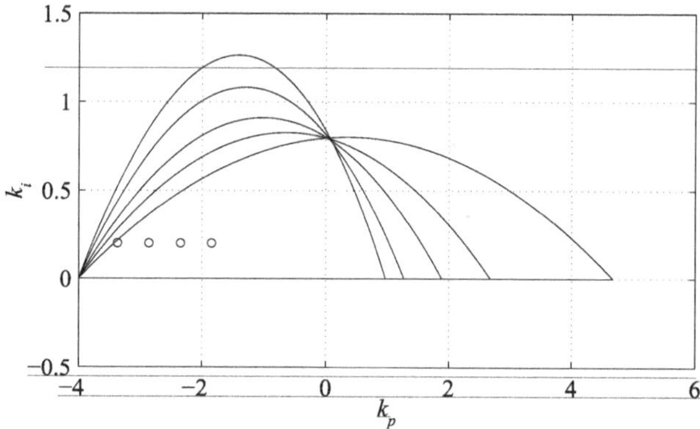

Fig. 4.49 Stability domains (lines) in the gain space and the optimal feedback gains (o) for the autonomous systems with five different time delays τ_i $(i = 1, 2, 3, 4, 5)$. The stability boundaries become taller and narrower, and move upward along k_i axis as time delay increases.

Figure 4.50 shows the closed loop response of the system under the feedback control with all five different time delays when the system true time delay is taken to be τ_1 and is assumed to be unknown. As it can be seen from the figure, when the control designed for the time delay that is close to τ_4 is implemented, the performance is acceptable. Otherwise, the performance can deteriorate as seen in the left-upper sub-figure.

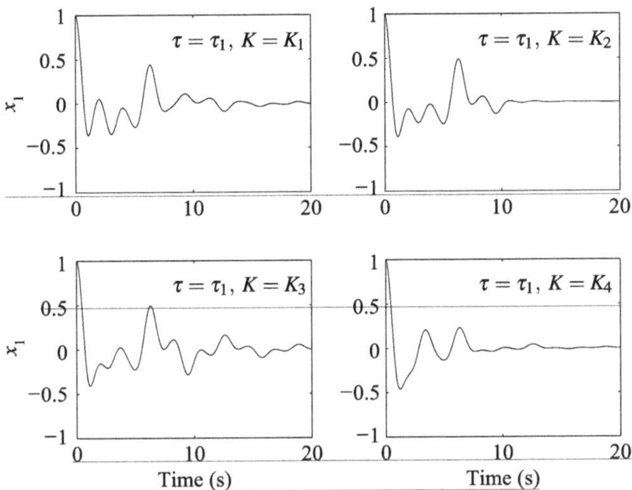

Fig. 4.50 Response of the autonomous system under feedback controls designed for a specific gain when the system true time delay is τ_1 and is assumed to be unknown. When the feedback gains (K_2, K_3, K_4) are designed for the time delay close to the actual one, the control performance is quite good. K_4 and K_5 are the same. When the mismatch gap is large, i.e. when K_4 designed for τ_4 is implemented for the system with time delay τ_1, the performance deteriorates.

Next, we examine how well the hysteretic switch algorithm works. Assume that we start with a control gain K_5 designed for τ_5 while the system delay is τ_1. Figure 4.51 shows that the hysteretic algorithm is able to switch the control to K_1. Figure 4.52 shows the switch signal $\pi(t)$ and the control index.

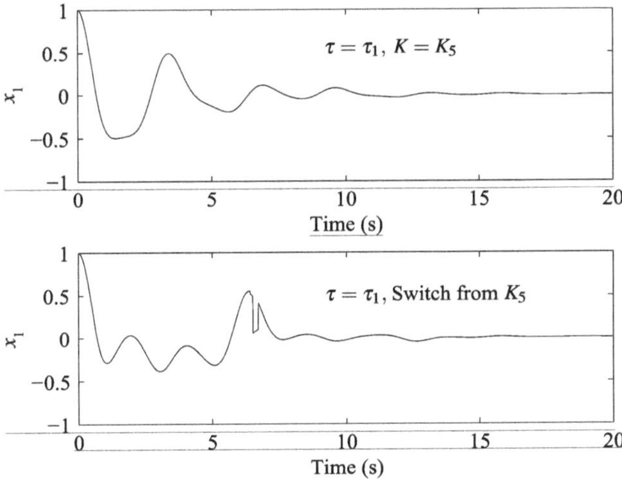

Fig. 4.51 The closed loop response of the system under the switched control when the initial gain of the control is K_5 designed for τ_5 while the system true time delay is τ_1 (bottom), as compared to the case when the gain is fiexed at K_5 (top).

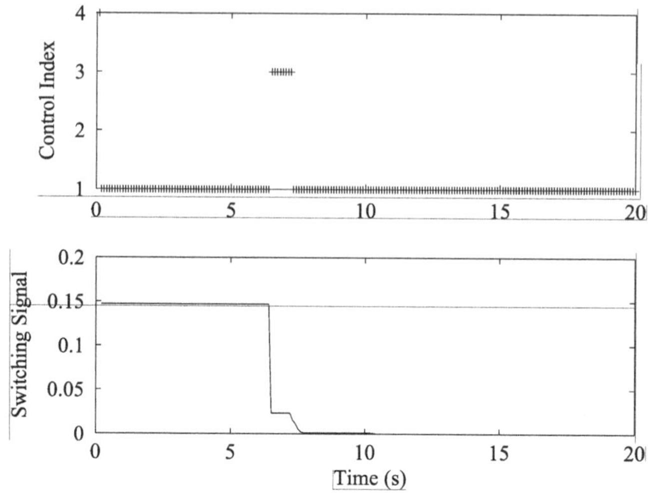

Fig. 4.52 Switch signal (lower figure) and the control index (upper figure) of the hysteretic switching algorithm for the closed loop response in Fig. 4.51.

4.10.2 Supervisory control of the periodic system

Consider the Mathieu equation (4.111) with a delayed PID feedback control. The period of the system is $T = \pi$. We select the parameters to be $\varepsilon = 1$, $\delta = 4$ and $N = 20$ for all sampled time delays. The uncontrolled system is parametrically unstable. Next, we show the closed-loop response of the system under a switching PD control with time delay in the range [0.5498, 1.0210]. Five time delays are sampled from the interval and their optimal gains are listed in Table 4.5. The stability domains in the gain space are shown in Fig. 4.53. Note that the optimal gains of all the controls with different time delays fall in the intersection of the stability domains. Hence, it is possible to use the hysteretic algorithm to switch among the pre-designed controls and to keep the system stable all the time. Another interesting phenomenon as shown in Fig. 4.53 and also in (Sheng and Sun, 2005) is that the stability domain in $k_p - k_d$ gain space grows along the k_d direction as time delay increases.

Table 4.5 Optimal PD feedback gains and $|\lambda|_{\max}$ for the five sampled time delays of the periodic system. Note that the mapping step for the periodic system is one period, while the mapping step for the LTI system is only one time delay τ.

Time Delay	k_p	k_d	$\|\lambda\|_{\max}$
τ_1 =0.5498	−3.6634	−0.0990	0.0130
τ_2 =0.6676	−4.0000	−0.8000	0.0083
τ_3 =0.7854	−2.8000	−0.6000	0.0141
τ_4 =0.9032	−2.6000	−0.6000	0.0213
τ_5 =1.0210	−1.8000	−0.6000	0.0347

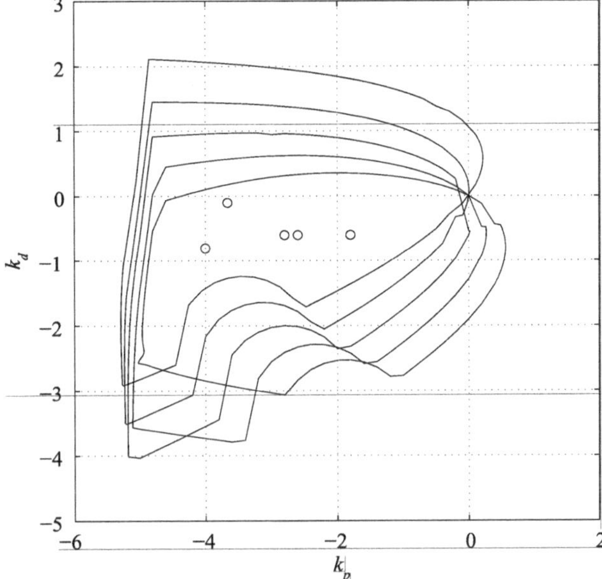

Fig. 4.53 Stability domains (lines) in the gain space and the optimal feedback gains (o) for the periodic system with five different time delays τ_i $(i = 1, 2, 3, 4, 5)$. The stability boundaries move down along k_d axis as time delay increases.

Figure 4.54 shows the closed loop responses of the system under the feedback control with the first four different time delays when the system true time delay is taken to be τ_1 and is assumed to be unknown. As it can be seen from the figure, when the control designed for the time delay that is close to τ_1 is implemented, the performance is better.

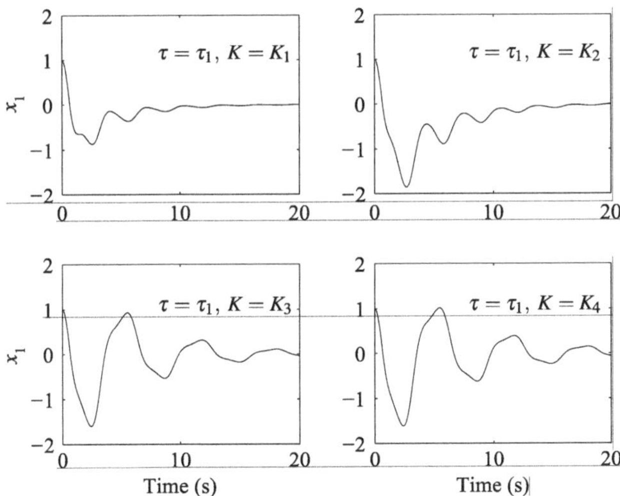

Fig. 4.54 Response of the periodic system under PD feedback controls designed for a specific gain when the system true time delay is τ_1 and is assumed to be unknown.

Next, we start with a control gain K_4 designed for τ_4. Figure 4.55 shows the closed loop response. The hysteretic algorithm switches the gain to reduce the switch signal $\pi(t)$ as shown in Fig. 4.56.

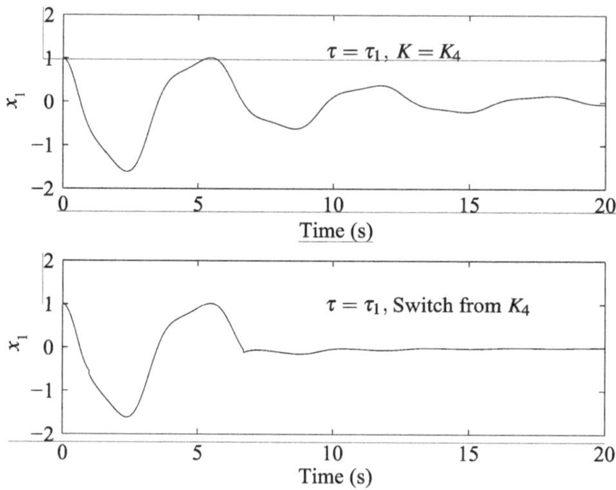

Fig. 4.55 The closed loop response of the periodic system under the switched PD control when the initial gain of the control is K_4 designed for τ_4 while the system true time delay is τ_1 (bottom), as compared to the case when the gain is fiexed at K_4 (top).

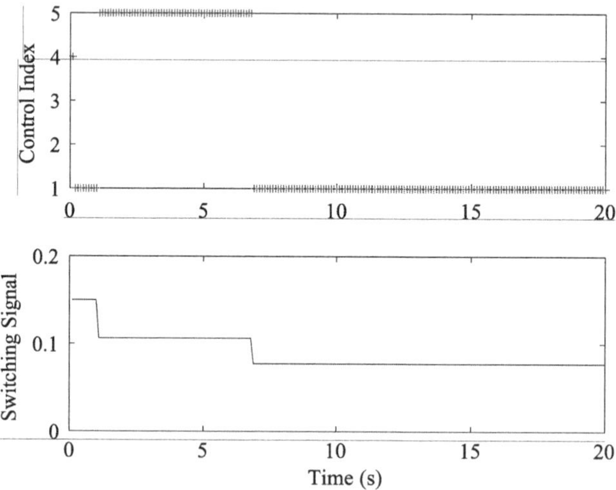

Fig. 4.56 Switch signal (lower figure) and the control index (upper figure) of the hysteretic switching algorithm for the closed loop response in Fig. 4.55.

4.11 Method of semi-discretization for stochastic systems

This section studies systems with time delay subject to both additive and multiplicative stochastic disturbances. Specifically, we present a study of stability analysis with the method of semi-discretization.

4.11.1 Mathematical background

We consider a system

$$\dot{\mathbf{x}} = \mathbf{f}\left(\mathbf{x}\left(t\right), \mathbf{x}\left(t-\tau\right), t\right) + \mathbf{G}\left(\mathbf{x}, t\right) \mathbf{W}\left(t\right), \tag{4.125}$$

where $\mathbf{x} \in \Re^n$, $\mathbf{W} \in \Re^m$, the system dynamics and delayed effects are accounted by the vector $\mathbf{f}$ and $\mathbf{G} = [g_{ij}]$ is the matrix determining the parametric and external random excitations. $W_i\left(t\right)$ are delta correlated Gaussian white noises with $E\left[W_i\left(t\right) W_j\left(t+T\right)\right] = 2\pi K_{ij} \delta\left(T\right)$. Here, we will extend the the semi-discretization method where the general procedure is studied in (Insperger and Stepan, 2002) .

We make use of Itô stochastic calculus to obtain an accurate stochastic discrete map. Equation (4.125) can be converted to Itô differential equation in the following general form,

$$d\mathbf{X} = \mathbf{m}\left(\mathbf{X}\left(t\right), \mathbf{X}\left(t-\tau\right), t\right) dt + \sigma\left(\mathbf{X}, t\right) d\mathbf{B}\left(t\right), \tag{4.126}$$

where $\mathbf{m}$ is the drift including the Wong-Zakai correction term and $\sigma\left(\mathbf{X}, t\right)$ is the diffusion term defined as

$$\sigma_{jl}\left(\mathbf{X}, t\right) \sigma_{kl}\left(\mathbf{X}, t\right) = 2\pi K_{rs} g_{jr} g_{ks}. \tag{4.127}$$

The Brownian motion $d\mathbf{B}\left(t\right)$ has the following properties that we use in the coming sections

$$E\left[\int_{t_i}^{t_i+\Delta t} d\mathbf{B}\left(t\right)\right] = 0, \quad E\left[dB_i\left(t\right) dB_j\left(t\right)\right] = \delta_{ij} dt. \tag{4.128}$$

We restrict ourselves to linear stochastic differential equations with $\mathbf{m} = \mathbf{A}\mathbf{X}\left(t\right) + \mathbf{A}_\tau \mathbf{X}\left(t-\tau\right)$, and $\mathbf{G}\left(\mathbf{X}, t\right)$ is a linear function of $\mathbf{X}$,

$$d\mathbf{X} = \mathbf{A}\mathbf{X}\left(t\right) dt + \mathbf{A}_\tau \mathbf{X}\left(t-\tau\right) dt + \sigma\left(t\right) d\mathbf{B}\left(t\right), \tag{4.129}$$

where $\mathbf{A}$ is the state matrix and $\mathbf{A}_\tau$ is the state matrix related to the time delayed response, σ is the diffusion term due to parametric and external stochastic excitations. The formal solution to this equation in integral form is written as follows

$$\mathbf{X}(t_i+\Delta t)=\mathbf{X}(t_i)+\int_{t_i}^{t_i+\Delta t}\mathbf{A}(t)\,\mathbf{X}(t)\,dt \\ +\int_{t_i}^{t_i+\Delta t}\mathbf{A}_\tau(t)\,\mathbf{X}(t-\tau)\,dt+\int_{t_i}^{t_i+\Delta t}\sigma(t)\,d\mathbf{B}(t)\,, \tag{4.130}$$

where $\mathbf{X}(t_i)$ is the initial value. Note that the fourth term on the RHS is a multi-dimensional stochastic integral that must be interpreted in Itô sense. Although Eq. (4.130) is exact it does not provide any simplifications for the desired mapping. To circumvent this problem we introduce the following notations which will be practical in the formulation,

$$\begin{aligned} &\mathbf{X}(t_i-\tau)=\mathbf{X}((i-N)\Delta t)=\mathbf{X}[i-N]\,,\\ &\mathbf{X}(t_i)=\mathbf{X}[i]\,,\quad \sigma(t_i)=\sigma[i]\,,\\ &\mathbf{A}(t_i)=\mathbf{A}[i]\,,\quad \mathbf{A}_\tau(t_i)=\mathbf{A}_\tau[i]\,, \end{aligned} \tag{4.131}$$

where the time lag τ is divided into an integer N intervals of length Δt such that $\tau = N\Delta t$. Details of this procedure and applications in complex systems can be found in (Insperger and Stepan, 2001, 2002)

Integration of Eq. (4.129) over a short time interval Δt gives:

$$\mathbf{X}[i+1]=\mathbf{X}[i]+\mathbf{A}[i]\,\mathbf{X}[i]\,\Delta t \\ +\mathbf{A}_\tau(t_i)\mathbf{X}[i-N]\,\Delta t+\sigma[i]\int_{t_i}^{t_i+\Delta t}d\mathbf{B}(t)\,. \tag{4.132}$$

The last term on the RHS is discretized in such a way that the diffusion term is kept constant during the short interval. This is the essence of the semi-discretization method and allows us to generate a discrete mapping of diffusion terms. We would like to draw the readers attention to two points at this step. Firstly note that as Δt gets smaller the accuracy of short time integrals improves. Therefore, the discrete map (4.132) approaches to continuous process. Secondly the diffusion term is not linear for the first moments but produces linear relationship for the second order moments. This is due to the linear nature of the $\mathbf{G}$ matrix in Equation (4.125) and we demonstrate it in the following sections. Define a $(N+1)n\times 1$ dimensional state vector as

$$\mathbf{Y}[i]=\left\{\mathbf{X}^{\mathrm{T}}[i]\,,\mathbf{X}^{\mathrm{T}}[i-1]\,,\ldots,\mathbf{X}^{\mathrm{T}}[i-N]\right\}^{\mathrm{T}}. \tag{4.133}$$

A mapping of $\mathbf{Y}[i]$ over the interval $[t_i, t_{i+1}]$ becomes

$$\mathbf{Y}[i+1]=\mathbf{\Phi}[i]\,\mathbf{Y}[i]+\mathbf{R}[i]\,, \tag{4.134}$$

where $\mathbf{\Phi}[i]$ is the mapping that accounts for the system dynamics, delayed effects and relationship of the delayed states and $\mathbf{R}[i]$ is the combined stochastic influence vector defined as

$$\mathbf{R}[i] = \left\{ \left[\sigma[i] \int_{t_i}^{t_i+\Delta t} d\mathbf{B}(t) \right]^{\mathrm{T}}, \overbrace{0,\ldots,0}^{N\cdot n} \right\}^{\mathrm{T}}. \tag{4.135}$$

4.11.2 Stability analysis

To formulate the second order moment stability we define following matrix $\mathbf{Z}[i]$ that represents the second order moments as

$$[\mathbf{Z}[i]]_{M\times M} = E\left(\mathbf{Y}[i]\,\mathbf{Y}^{\mathrm{T}}[i]\right), \tag{4.136}$$

where $M = (n+1)N \times 1$. Then the mapping of second order moments for one step becomes

$$\begin{aligned} \mathbf{Z}[i+1]_{kp} &= E\left(\boldsymbol{\Phi}[i]_{kl}\,\boldsymbol{\Phi}[i]_{qp}\right)\mathbf{Z}[i]_{lq} \\ &+ E\left(\boldsymbol{\Phi}[i]_{kl}\,\mathbf{Y}[i]_l\,\mathbf{R}[i]_p + \boldsymbol{\Phi}[i]_{qp}\,\mathbf{Y}[i]_q\,\mathbf{R}[i]_k\right) + E\left(\mathbf{R}[i]_k\,\mathbf{R}[i]_p\right). \end{aligned} \tag{4.137}$$

Note that due to the linear nature of $\mathbf{G}$ the last term on the RHS can be at most quadratic

$$E\left(\mathbf{R}[i]_k\,\mathbf{R}[i]_p\right) = \mathbf{H}[i]_{klpq}\,\mathbf{Z}[i]_{lq} + \Theta_{kpl} E\left(\mathbf{Y}[i]_l\right) + \boldsymbol{\Gamma}[i]_{kp}, \tag{4.138}$$

where Θ_{kpl} is the term for the coupling of first order moments to the second order moments. We group Eq. (4.137) as follows,

$$\mathbf{Z}[i+1]_{kp} = \boldsymbol{\Psi}[i]_{klpq}\,\mathbf{Z}[i]_{lq} + \Theta_{kpl} E\left(\mathbf{Y}[i]_l\right) + \boldsymbol{\Gamma}[i], \tag{4.139}$$

where $\boldsymbol{\Psi}[i]_{klpq} = \mathbf{H}[i]_{klpq} + E\left(\boldsymbol{\Phi}[i]_{kl}\,\boldsymbol{\Phi}[i]_{qp}\right)$. This step allows us to write a simple stability condition because the effects of stochastic excitations linear in second order moments become explicit. The stability condition for the system in (4.139) is

$$\lim_{i\to\infty} \|\mathbf{Z}[i+1]\| \le |\lambda|_{\max}(i)\,\|\mathbf{Z}[i]\|,$$

where $\|.\|$ is the norm. This condition is guaranteed if

$$|\lambda|_{\max}(i) < 1, \quad \text{for all } i, \tag{4.140}$$

where $|\lambda|_{\max}(i)$ is the maximum absolute value of the linear transformation $\boldsymbol{\Psi}[i]$. Moreover, when limits

$$\lim_{i\to\infty} \mathbf{Z}(i) = \mathbf{Z}, \quad \lim_{i\to\infty} \boldsymbol{\Psi}[i] = \boldsymbol{\Psi}, \quad \lim_{i\to\infty} \boldsymbol{\Gamma}[i] = \boldsymbol{\Gamma}, \quad \lim_{i\to\infty} E[\mathbf{Y}[i]] = 0,$$

exist Eq. (4.139) makes it possible to calculate the stochastically steady state response of second order moments using

$$\mathbf{Z} = [\mathbf{I} - \mathbf{\Psi}]^{-1} \mathbf{\Gamma},$$

where $\mathbf{I}$ is the identity matrix with the same dimension as $\mathbf{\Psi}$.

4.12 Method of finite-dimensional markov process (FDMP)

Consider Itô equation (4.126). Recall that the system lives in an infinite dimensional state space. In general, $\mathbf{X}(t)$ is no longer a Markov process because it depends on its history. Following the idea of the CTA method, we discretize the delayed part of the state vector $(\mathbf{X}(t-t_1), 0 < t_1 \leq \tau)$. Let N be integer such that $\Delta\tau = \tau/N$. $\tau_i = i\Delta\tau$ $(i = 1, 2, \ldots, N)$. Then, we introduce a finite difference approximation of the derivatives of $(\dot{\mathbf{X}}(t-\tau_i), 1 \leq i \leq N)$ as in Eq. (4.81). Define a discrete vector as

$$\begin{aligned} \mathbf{Y}(t) &= \left[\mathbf{X}^{\mathrm{T}}(t), \mathbf{X}^{\mathrm{T}}(t-\Delta\tau), \mathbf{X}^{\mathrm{T}}(t-2\Delta\tau), \ldots, \mathbf{X}^{T}(t-N\Delta\tau)\right]^{\mathrm{T}} \\ &\equiv \left[\mathbf{Y}_1^{\mathrm{T}}(t), \mathbf{Y}_2^{\mathrm{T}}(t), \mathbf{Y}_3^{\mathrm{T}}(t), \ldots, \mathbf{Y}_{N+1}^{\mathrm{T}}(t)\right]^{\mathrm{T}}. \end{aligned} \tag{4.141}$$

We obtain an Itô stochastic equation for the vector $\mathbf{Y}(t)$,

$$\begin{aligned} d\mathbf{Y}(t) &= \begin{bmatrix} \mathbf{m}(\mathbf{Y}_1(t), \mathbf{Y}_{N+1}(t), t) \\ \frac{1}{2\Delta\tau}[\mathbf{Y}_1(t) - \mathbf{Y}_3(t)] \\ \vdots \\ \frac{1}{\Delta\tau}[\mathbf{Y}_N(t) - \mathbf{Y}_{N+1}(t)] \end{bmatrix} dt + \begin{bmatrix} \sigma(\mathbf{Y}_1(t), \mathbf{Y}_{N+1}(t), t) \\ \mathbf{0} \\ \vdots \\ \mathbf{0} \end{bmatrix} d\mathbf{B}(t) \\ &\equiv \hat{\mathbf{m}}(\mathbf{Y}, t)dt + \hat{\sigma}(\mathbf{Y}, t)\, d\mathbf{B}(t). \end{aligned} \tag{4.142}$$

Note that the low-pass filtered CTA approximation in Section 4.5.1 can be applied here also. Consider a linear stochastic differential equation as an example,

$$d\mathbf{X} = \mathbf{A}\mathbf{X}(t)\, dt + \mathbf{A}_\tau \mathbf{X}(t-\tau)\, dt + \sigma(t)\, d\mathbf{B}(t), \tag{4.143}$$

where $\mathbf{A}$ is the state matrix and $\mathbf{A}_\tau$ is the state matrix related to the delayed response. We have an equation for $\mathbf{Y}(t)$ as

$$\begin{aligned} d\mathbf{Y}(t) &= \begin{bmatrix} \mathbf{A} & \mathbf{0} & \cdots & \mathbf{0} & \mathbf{A}_\tau \\ \frac{1}{2\Delta\tau}\mathbf{I} & \mathbf{0} & -\frac{1}{2\Delta\tau}\mathbf{I} & \cdots & \mathbf{0} \\ & & \ddots & & \\ \mathbf{0} & \cdots & \mathbf{0} & \frac{1}{\Delta\tau}\mathbf{I} & -\frac{1}{\Delta\tau}\mathbf{I} \end{bmatrix} \mathbf{Y}(t)\, dt + \begin{bmatrix} \sigma(t) \\ \mathbf{0} \\ \vdots \\ \mathbf{0} \end{bmatrix} d\mathbf{B}(t) \\ &\equiv \hat{\mathbf{A}}\mathbf{Y}(t)\, dt + \hat{\sigma}(t)\, d\mathbf{B}(t), \end{aligned} \tag{4.144}$$

Equation (4.142) indicates that $\mathbf{Y}(t)$ is a Markov process when $d\mathbf{B}(t)$ is a Brownian motion (Lin and Cai, 1995). The conditional probability density function of $\mathbf{Y}(t)$ satisfies a FPK equation as well as backward Kolmogorov equation.

4.12.1 Fokker-Planck-Kolmogorov (FPK) equation

The conditional probability density function $p_{\mathbf{Y}}(\mathbf{y}, t|\mathbf{y}_0, t_0)$ for the stochastic system (4.142) satisfies the FPK equation given by

$$\frac{\partial}{\partial t} p_{\mathbf{Y}}(\mathbf{y}, t|\mathbf{y}_0, t_0) = -\frac{\partial}{\partial y_j}[\hat{m}_j(\mathbf{y}, t) p_{\mathbf{Y}}] + \frac{\partial^2}{\partial y_j \partial y_k}\left[\frac{\hat{b}_{jk}(\mathbf{y}, t)}{2!} p_{\mathbf{Y}}\right], \quad (4.145)$$

where the index j runs from 1 to $M = n(N+1)$ and $\hat{b}_{jk} = \hat{\sigma}_{jl}\hat{\sigma}_{kl}$ or $\hat{\mathbf{b}} = \hat{\sigma}\hat{\sigma}^{\mathrm{T}}$. The index l runs from 1 to p, the dimension of $d\mathbf{B}(t)$. The FPK equation is subject to an initial condition, for example,

$$p_{\mathbf{Y}}(\mathbf{y}, t|\mathbf{y}_0, t_0) = \delta(\mathbf{y} - \mathbf{y}_0). \quad (4.146)$$

Note that since the stochastic excitations only act on the vector $\mathbf{X}(t) = \mathbf{Y}_1(t)$, the second order derivatives of the FPK equation only involve the components of $\mathbf{Y}_1(t)$. Hence, there are only $n \times n$ diffusion terms, instead of $M \times M$. In other words, the time-delay within the FDMP method only affects the drift term of the FPK equation. This is also true with the backward Kolmogorov equation and its derivatives in the study of reliability and first-passage time probability.

Example of the linear system

Recall the linear system in Eq. (4.144). The FPK equation reads

$$\frac{\partial}{\partial t} p_{\mathbf{Y}}(\mathbf{y}, t|\mathbf{y}_0, t_0) = -\frac{\partial}{\partial y_j}[\hat{a}_{jk} y_k p_{\mathbf{Y}}] + \frac{\hat{b}_{jk}}{2!}\frac{\partial^2 p_{\mathbf{Y}}}{\partial y_j \partial y_k}. \quad (4.147)$$

Assume that the matrix $\hat{\mathbf{A}}$ is nondefective. Then, there exist m eigenvalues λ_k and eigenvectors $\mathbf{c}_k$ such that

$$\hat{\mathbf{A}}\mathbf{c}_k = \lambda_k \mathbf{c}_k \text{ (no sum)}, \quad (4.148)$$

and

$$\hat{\mathbf{A}}\mathbf{C} = \mathbf{C}\boldsymbol{\Lambda}, \quad (4.149)$$

where $\boldsymbol{\Lambda} = \mathrm{diag}\{\lambda_k\}$ and $\mathbf{C} = \{\mathbf{c}_1, \mathbf{c}_1, \ldots, \mathbf{c}_m\}$ (Golub and Loan, 1983). Note that $\mathbf{C}$ is nonsingular. Introduce a transformation such that

$$\mathbf{y} = \mathbf{C}\mathbf{z}. \tag{4.150}$$

Let $\mathbf{H} = \mathbf{C}^{-1}\hat{\mathbf{b}}(\mathbf{C}^{-1})^{\mathrm{T}} \equiv \{h_{jk}\}$. We obtain a Gaussian probability density function (Sun, 2006)

$$p_{\mathbf{Z}}(\mathbf{z}, t|\mathbf{z}_0, t_0) = \frac{1}{(2\pi)^{m/2}(\det(\mathbf{C}_{\mathbf{ZZ}}))^{1/2}} \cdot \exp\left(-\frac{1}{2}(\mathbf{z} - \mu_{\mathbf{Z}})^T \mathbf{C}_{\mathbf{ZZ}}^{-1}(\mathbf{z} - \mu_{\mathbf{Z}})\right), \tag{4.151}$$

where

$$(\mu_{\mathbf{Z}})_k = z_{0k}e^{\lambda_k t}, \quad (\mathbf{C}_{\mathbf{ZZ}})_{jk} = -\frac{h_{jk}}{\lambda_j + \lambda_k}\left(1 - e^{(\lambda_j+\lambda_k)t}\right) \quad \text{(no sum)}. \tag{4.152}$$

Equation (4.152) indicates that the first and second order moments of the system are stable if the real part of each eigenvalue λ_k of the matrix $\hat{\mathbf{A}}$ is negative or zero.

Hence, $p_{\mathbf{Y}}(\mathbf{y}, t|\mathbf{y}_0, t_0)$ is also Gaussian with the mean and variance given by

$$\mu_{\mathbf{Y}} = \mathbf{C}\mu_{\mathbf{Z}}, \quad \mathbf{C}_{\mathbf{YY}} = \mathbf{C}\mathbf{C}_{\mathbf{ZZ}}\mathbf{C}^{\mathrm{T}}. \tag{4.153}$$

Recall that $\mathbf{y} \equiv [\mathbf{y}_1, \mathbf{y}_2, \ldots, \mathbf{y}_{N+1}]^{\mathrm{T}}$ for the case of one time delay and $\mathbf{x} = \mathbf{y}_1$. The marginal probability density function of the responses $\mathbf{x}$ of the original time-delayed system is given by

$$p_{\mathbf{X}}(\mathbf{x}, t|\mathbf{y}_0, t_0) = \int_{\Re^n}\int_{\Re^n}\cdots\int_{\Re^n} p_{\mathbf{Y}}(\mathbf{y}, t|\mathbf{y}_0, t_0)d\mathbf{y}_2\cdots d\mathbf{y}_{N+1}. \tag{4.154}$$

It is not hard to show recursively that $p_{\mathbf{X}}(\mathbf{x}, t|\mathbf{y}_0, t_0)$ for the linear time-delayed system is still Gaussian in $\mathbf{x}$ although it is not Markovian. This is an interesting result.

4.12.2 Moment equations

Recall that in the Itô sense, $dB_k(t)$ is defined as the forward difference and $\hat{\sigma}_{jk}(\mathbf{Y}, t)$ is independent of $dB_k(t)$. Also, $E[dB_k(t)] = 0$. Taking the mathematical expectation on both sides of Equation (4.142), we have

$$\frac{dE[Y_j(t)]}{dt} = E[\hat{m}_j(\mathbf{Y}, t)]. \tag{4.155}$$

Consider a function $F(\mathbf{Y}, t) = Y_j Y_k$. According to Itô's lemma, we have

$$d(Y_jY_k) = \left(\hat{m}_jY_k + \hat{m}_kY_j + \frac{1}{2}\hat{b}_{jk}\right)dt + (\hat{\sigma}_{jl}Y_k + \hat{\sigma}_{kl}Y_j)\,dB_l(t). \tag{4.156}$$

Taking the expectation of the equation, and recalling that

$$E[(\hat{\sigma}_{jl}Y_k + \hat{\sigma}_{kl}Y_j)\,dB_l(t)] = 0, \tag{4.157}$$

because of the independence of the terms $\hat{\sigma}_{jl}Y_k + \hat{\sigma}_{kl}Y_j$ and $dB_l(t)$ in the Itô sense, we have the equation for the correlation function

$$\frac{dE[Y_jY_k]}{dt} = E\left[\hat{m}_jY_k + \hat{m}_kY_j + \frac{1}{2}\hat{b}_{jk}\right]. \tag{4.158}$$

By following the same steps, we can construct differential equations governing the evolution of the moments of any order. Consider again the linear example. The moment equations of the first and second orders are readily obtained.

$$\frac{dE[Y_j(t)]}{dt} = \hat{A}_{jk}E[Y_k(t)], \tag{4.159}$$

$$\frac{dE[Y_jY_k]}{dt} = \hat{A}_{jl}E[Y_lY_k] + \hat{A}_{kl}E[Y_lY_j] + \frac{1}{2}\hat{b}_{jk}. \tag{4.160}$$

In the matrix form, they read

$$\frac{d\mu_{\mathbf{Y}}}{dt} = \hat{\mathbf{A}}\mu_{\mathbf{Y}}, \tag{4.161}$$

$$\frac{d\mathbf{R}_{\mathbf{YY}}}{dt} = \hat{\mathbf{A}}\mathbf{R}_{\mathbf{YY}} + \mathbf{R}_{\mathbf{YY}}\hat{\mathbf{A}}^{\mathrm{T}} + \frac{1}{2}\hat{\mathbf{b}}, \tag{4.162}$$

where $\mu_{\mathbf{Y}} = E[\mathbf{Y}]$ and $\mathbf{R}_{\mathbf{YY}} = E[\mathbf{YY}^{\mathrm{T}}]$.

4.12.3 Reliability

The backward Kolmogorov equation for the process $\mathbf{Y}(t)$ can be derived as (Sun, 2006)

$$-\frac{\partial}{\partial t_0}p_{\mathbf{Y}}(\mathbf{y},t|\mathbf{y}_0,t_0) = \hat{m}_j(\mathbf{y}_0,t_0)\frac{\partial p_{\mathbf{Y}}}{\partial y_{0j}} + \frac{\hat{b}_{jk}(\mathbf{y}_0,t_0)}{2!}\frac{\partial^2 p_{\mathbf{Y}}}{\partial y_{0j}\partial y_{0k}}. \tag{4.163}$$

Note that $t_0 \leq t < \infty$. Integrating Equation (4.163) with respect to the delayed components $(\mathbf{y}_2, \mathbf{y}_3, \ldots, \mathbf{y}_{N+1})$ of the state vector leads to the backward equation for the marginal probability density function,

$$-\frac{\partial}{\partial t_0}p_{\mathbf{X}}(\mathbf{x},t|\mathbf{y}_0,t_0) = \hat{m}_j(\mathbf{y}_0,t_0)\frac{\partial p_{\mathbf{X}}}{\partial y_{0j}} + \frac{\hat{b}_{jk}(\mathbf{y}_0,t_0)}{2!}\frac{\partial^2 p_{\mathbf{X}}}{\partial y_{0j}\partial y_{0k}}. \tag{4.164}$$

An important application of the backward Kolmogorov equation is the reliability study. Consider the state vector $\mathbf{X}(t)$ of the original system. Let $\mathcal{S} \subseteq \Re^n$ be a domain in which the system is considered to be safe. Γ is the boundary of $\mathcal{S}$. Assume that all the components of $\mathbf{Y}(t_0) = \mathbf{y}_0$ lie inside $\mathcal{S}$ at time t_0. The probability that the system is still in the *safe domain* $\mathcal{S}$ at time t is given by

$$\begin{aligned}\mathcal{R}_{\mathcal{S}}(t, t_0, \mathbf{y}_0) &= P(t < T \cap \mathbf{X}(t) \in \mathcal{S} | \mathbf{Y}(t_0) = \mathbf{y}_0) \\ &= \int_{\mathcal{S}} p_{\mathbf{X}}(\mathbf{x}, t|\mathbf{y}_0, t_0) d\mathbf{x},\end{aligned} \tag{4.165}$$

where T is the first time when $\mathbf{X}(t)$ crosses the boundary Γ. $\mathcal{R}_{\mathcal{S}}(t, t_0, \mathbf{y}_0)$ is also known as the reliability against the *first-passage failure* with respect to the safe domain $\mathcal{S}$.

Integrating Eq. (4.163) over $\mathcal{S}$ with respect to $\mathbf{x}$, we obtain a partial differential equation of the reliability function $\mathcal{R}_{\mathcal{S}}(t, t_0, \mathbf{y}_0)$,

$$\begin{aligned}-\frac{\partial \mathcal{R}_{\mathcal{S}}(t, t_0, \mathbf{y}_0)}{\partial t_0} &= \hat{m}_j(\mathbf{y}_0, t_0)\frac{\partial \mathcal{R}_{\mathcal{S}}(t, t_0, \mathbf{y}_0)}{\partial y_{0j}} \\ &+ \frac{\hat{b}_{jk}(\mathbf{y}_0, t_0)}{2!}\frac{\partial^2 \mathcal{R}_{\mathcal{S}}(t, t_0, \mathbf{y}_0)}{\partial y_{0j}\partial y_{0k}},\end{aligned} \tag{4.166}$$

subject to the following initial and boundary conditions

$$\mathcal{R}_{\mathcal{S}}(t_0, t_0, \mathbf{y}_0) = 1, \ \mathbf{y}_{0i} \in \mathcal{S}, \ (1 \le i \le N+1), \tag{4.167}$$

$$\mathcal{R}_{\mathcal{S}}(t, t_0, \mathbf{y}_0) = 0, \ \mathbf{y}_{0i} \in \Gamma \text{ (for at least one } i). \tag{4.168}$$

4.12.4 First-passage time probability

Denote the complement of $\mathcal{R}_{\mathcal{S}}(t, t_0, \mathbf{y}_0)$ as $F_{\mathcal{S}}(t, t_0, \mathbf{y}_0)$, which is the probability distribution function of the first-passage time. We have

$$F_{\mathcal{S}}(t, t_0, \mathbf{y}_0) = P(t \ge T | \mathbf{Y}(t_0) = \mathbf{y}_0) = 1 - \mathcal{R}_{\mathcal{S}}(t, t_0, \mathbf{y}_0). \tag{4.169}$$

Substituting this relationship to Equation (4.166), we obtain

$$\begin{aligned}-\frac{\partial F_{\mathcal{S}}(t, t_0, \mathbf{y}_0)}{\partial t_0} &= \hat{m}_j(\mathbf{y}_0, t_0)\frac{\partial F_{\mathcal{S}}(t, t_0, \mathbf{y}_0)}{\partial y_{0j}} \\ &+ \frac{\hat{b}_{jk}(\mathbf{y}_0, t_0)}{2!}\frac{\partial^2 F_{\mathcal{S}}(t, t_0, \mathbf{y}_0)}{\partial y_{0j}\partial y_{0k}}.\end{aligned} \tag{4.170}$$

The probability density function of the first-passage time denoted by $p_T(t|\mathbf{y}_0, t_0)$ is given by

$$p_T(t|\mathbf{y}_0, t_0) = \frac{\partial F_{\mathcal{S}}(t, t_0, \mathbf{y}_0)}{\partial t} = -\frac{\partial \mathcal{R}_{\mathcal{S}}(t, t_0, \mathbf{y}_0)}{\partial t}. \tag{4.171}$$

Differentiating Eq. (4.170) with respect to t, we yield the governing equation for $p_T(t|\mathbf{y}_0, t_0)$

$$-\frac{\partial p_T(t|\mathbf{y}_0,t_0)}{\partial t_0} = \hat{m}_j(\mathbf{y}_0,t_0)\frac{\partial p_T(t|\mathbf{y}_0,t_0)}{\partial y_{0j}} + \frac{\hat{b}_{jk}(\mathbf{y}_0,t_0)}{2!}\frac{\partial^2 p_T(t|\mathbf{y}_0,t_0)}{\partial y_{0j}\partial y_{0k}}. \tag{4.172}$$

Since, at a given time $t > t_0$ and when $\mathbf{y}_{0i} \in \Gamma$ (for at least one i), the reliability of the system vanishes $\mathcal{R}_S(t,t_0,\mathbf{y}_0) = 0$, this suggests the boundary condition

$$p_T(t|\mathbf{y}_0,t_0) = 0,\ t > t_0,\ \ \mathbf{y}_{0i} \in \Gamma \text{ (for at least one } i). \tag{4.173}$$

Assume that initially, the system starts from a point in the safe domain with probability one, we have an initial condition

$$p_T(t_0|\mathbf{y}_0,t_0) = \delta(\mathbf{y}_0),\ \mathbf{y}_{0i} \in \mathcal{S},\ (1 \le i \le N+1). \tag{4.174}$$

4.12.5 Pontryagin-Vitt equations

The first-passage time is a random variable and its rth order moment can be defined as

$$M_r(\mathbf{y}_0,t_0) = E[(T-t_0)^r|\mathbf{y}_0,t_0] = \int_{t_0}^{\infty}(t-t_0)^r p_T(t|\mathbf{y}_0,t_0)\,dt. \tag{4.175}$$

From Eq. (4.172), we obtain a set of integral-partial differential equations for the *moments of the first-passage time* as

$$-\int_{t_0}^{\infty}(t-t_0)^r\frac{\partial p_T(t|\mathbf{y}_0,t_0)}{\partial t_0}dt = \hat{m}_j(\mathbf{y}_0,t_0)\frac{\partial M_r(\mathbf{y}_0,t_0)}{\partial y_{0j}} + \frac{\hat{b}_{jk}(\mathbf{y}_0,t_0)}{2!}\frac{\partial^2 M_r(\mathbf{y}_0,t_0)}{\partial y_{0j}\partial y_{0k}}. \tag{4.176}$$

This equation is in general difficult to solve. Assume that $\mathbf{X}(t)$ is a stationary process such that $p_T(t|\mathbf{y}_0,t_0) = p_T(\tau|\mathbf{y}_0) = -\frac{\partial \mathcal{R}_S(\tau,\mathbf{y}_0)}{\partial \tau}$, $\hat{m}_j(\mathbf{y}_0,t_0) = \hat{m}_j(\mathbf{y}_0)$ and $\hat{b}_{jk}(\mathbf{y}_0,t_0) = \hat{b}_{jk}(\mathbf{y}_0)$ where $\tau = t - t_0$. Let

$$M_r(\mathbf{y}_0) = \int_0^{\infty}\tau^r p_T(\tau|\mathbf{y}_0)\,d\tau. \tag{4.177}$$

Assume that $\lim_{\tau\to\infty}\tau^r p_T(\tau|\mathbf{y}_0) = 0$. We can derive a set of the *generalized Pontryagin-Vitt equations* for the moments of the first-passage time as

$$-rM_{r-1}(\mathbf{y}_0) = \hat{m}_j(\mathbf{y}_0)\frac{\partial M_r(\mathbf{y}_0)}{\partial y_{0j}} + \frac{\hat{b}_{jk}(\mathbf{y}_0)}{2!}\frac{\partial^2 M_r(\mathbf{y}_0)}{\partial y_{0j}\partial y_{0k}}. \tag{4.178}$$

All the moments satisfy the same boundary condition

$$M_r(\mathbf{y}_0) = 0, \ \ \mathbf{y}_{0i} \in \Gamma \text{ (for at least one } i), \ r = 1, 2, 3, ... \tag{4.179}$$

Note that $M_0(\mathbf{y}_0) = 1$ because $p_T(\tau|\mathbf{y}_0)$ is a probability density function of τ. Hence, the mean of the first-passage time satisfies the *Pontryagin-Vitt* equation with $r = 1$,

$$-1 = \hat{m}_j(\mathbf{y}_0)\frac{\partial M_1(\mathbf{y}_0)}{\partial y_{0j}} + \frac{\hat{b}_{jk}(\mathbf{y}_0)}{2!}\frac{\partial^2 M_1(\mathbf{y}_0)}{\partial y_{0j}\partial y_{0k}}. \tag{4.180}$$

4.13 Analysis of stochastic systems with time delay

4.13.1 Stability of second order stochastic systems

Consider a second order linear system subject to both additive and multiplicative stochastic excitations.

$$\begin{aligned}\ddot{x}(t) + a_1\dot{x}(t) + a_2x(t) = {} & W_1(t)\,x(t) + W_2(t)\,\dot{x}(t) \\ & + W_3(t) - k_px(t-\tau) - k_d\dot{x}(t-\tau),\end{aligned} \tag{4.181}$$

where a_1 and a_2 are constant system parameters, k_p and k_d are the delayed feedback control parameters. We first compare the stability results of semi-discretization method with the known analytical solution in the literature. To do so we restrict ourselves to a non-delayed case without feedback and show the essential steps of semi-discretization method. The Itô equations representing this system can be characterized with

$$\mathbf{m} = \begin{bmatrix} X_2 \\ -a_1X_2 - a_2X_1 + \pi\left(K_{22}X_2 + K_{12}X_1 + K_{23}\right)\end{bmatrix} \tag{4.182}$$

$$\mathbf{G} = \begin{bmatrix} 0 & 0 & 0 \\ X_1 & X_2 & 1\end{bmatrix}, \quad \sigma_{1j} = 0 \ \text{ for } \ j = 1, 2, 3 \tag{4.183}$$

$$\begin{aligned}\sigma_{21}^2 + \sigma_{22}^2 + \sigma_{23}^2 = {} & 2\pi K_{11}X_1^2 + K_{22}X_2^2 + K_{33} \\ & + 2\left(K_{12}X_1X_2 + K_{13}X_1 + K_{23}X_2\right).\end{aligned} \tag{4.184}$$

The derivation of these quantities are well known in the literature, see for example (Lin and Cai, 1995) and not presented here. K_{13} and K_{23} are the coupling terms for second order moments with first order moments and would produce terms similar to Θ_{kpl} in Equation (4.139). To keep the example simple we restrict ourselves to the cases where K_{13} and K_{23} are zero. The integration over a short time interval Δt

yields following map

$$\begin{bmatrix} X_1(i+1) \\ X_2(i+1) \end{bmatrix} = \left[\begin{bmatrix} 1 & 0 \\ 0 & 1 \end{bmatrix} + \Delta t \begin{bmatrix} 0 & 1 \\ -a_2 + \pi K_{12} & -a_1 + \pi K_{22} \end{bmatrix} \right] \begin{bmatrix} X_1(i) \\ X_2(i) \end{bmatrix} + \begin{bmatrix} 0 & 0 & 0 \\ \sigma[i]_{21} & \sigma[i]_{22} & \sigma[i]_{23} \end{bmatrix}_i \int_{t_i}^{t_i+\Delta t} d\mathbf{B}(t). \tag{4.185}$$

In the coming sections, we refer to this formulation as direct Itô integration. Comparing Eq. (4.185) to Eq. (4.134) we see that first two matrices on the RHS constitute $\mathbf{\Phi}[i]_{2\times 1}$, the third one constitutes $\mathbf{R}[i]_{2\times 1}$. Naturally due to the absence of the delayed terms the vector sizes are much smaller. Notice that the terms like $\sigma[i]_{21}$ are the discretized constant quantities $\sigma[i]_{21} = \sigma(t_i)_{21}$ for $t \in (t_i, t_i + \Delta t)$. Explicitly, Equation (4.138) reads

$$E\left(\mathbf{R}[i]\mathbf{R}^T[i]\right) = \tag{4.186}$$
$$\begin{bmatrix} \sigma[i]_{11}^2 + \sigma[i]_{12}^2 + \sigma[i]_{13}^2 & \begin{array}{c}\sigma[i]_{11}\sigma[i]_{21} + \sigma[i]_{12}\sigma[i]_{22} \\ +\sigma[i]_{13}\sigma[i]_{23}\end{array} \\ symmetric & \sigma[i]_{21}^2 + \sigma[i]_{22}^2 + \sigma[i]_{23}^2 \end{bmatrix} \Delta t.$$

Here we take advantage of the symmetric nature of the second order moments and use equivalent vector form for $E\left(\mathbf{R}[i]\mathbf{R}[i]^{\mathrm{T}}\right)$ as follows

$$\frac{d}{dt}\begin{bmatrix} E[X_1^2] \\ E[X_1X_2] \\ E[X_2^2] \end{bmatrix} = 2\pi\Delta t \begin{bmatrix} 0 & 0 & 0 \\ 0 & 0 & 0 \\ K_{11} & 2K_{12} & K_{22} \end{bmatrix} E\begin{bmatrix} X_1^2[i] \\ X_1[i]X_2[i] \\ X_2^2[i] \end{bmatrix} + 2\pi\Delta t \begin{bmatrix} 0 \\ 0 \\ K_{33} \end{bmatrix}. \tag{4.187}$$

Note that, the first matrix is the linear contribution $\mathbf{H}$ in Eq. (4.138) whereas the second term is $\mathbf{\Gamma}$.

We take K_{11} and $K_{12} = 0$, and construct the map of second order moments following the procedure described earlier.

$$E\begin{bmatrix} X_1^2 \\ X_1X_2 \\ X_2^2 \end{bmatrix}_{i+1} =$$
$$\left\{ \begin{bmatrix} 1 & 2\Delta t & \Delta t^2 \\ -a_2\Delta t & 1 + (\pi K_{22} - a_1)\Delta t - a_2\Delta t^2 & \Delta t + \Delta t^2(\pi K_{22} - a_1) \\ a_2^2\Delta t^2 & -2\left(a_2\Delta t + a_2\Delta t^2(a_1 - \pi K_{22})\right) & (1 + \Delta t(\pi K_{22} - a_1))^2 \end{bmatrix} + 2\pi\Delta t \begin{bmatrix} 0 & 0 & 0 \\ 0 & 0 & 0 \\ 0 & 0 & K_{22} \end{bmatrix} \right\} E\begin{bmatrix} X_1^2 \\ X_1X_2 \\ X_2^2 \end{bmatrix}_i + 2\pi\Delta t \begin{bmatrix} 0 \\ 0 \\ K_{33} \end{bmatrix}. \tag{4.188}$$

Using the Itô differential rule (Sun, 2006), we can obtain the equations for the second order moments in the continuous time domain.

$$\frac{d}{dt}\begin{bmatrix} E\left[X_1^2\right] \\ E\left[X_1X_2\right] \\ E\left[X_2^2\right] \end{bmatrix} = \begin{bmatrix} 0 & 2 & 0 \\ -a_2 & -a_1+\pi K_{22} & 1 \\ 0 & -2a_2 & -2a_1+4\pi K_{22} \end{bmatrix} \begin{bmatrix} E\left[X_1^2\right] \\ E\left[X_1X_2\right] \\ E\left[X_2^2\right] \end{bmatrix}. \tag{4.189}$$

The stability of the second order moments is guaranteed if all the eigenvalues of the matrix have negative real parts.

The integration of Eq. (4.189) over a short time Δt leads to exactly the same discrete map as in Eq. (4.188) when the higher order terms involving Δt^2 are neglected.

We take this results as our benchmark and investigate the stability boundary of the semi discretized system. Here we briefly compare the stability boundaries obtained by three approaches. To do so for the system in Eq. (4.181) we take $a_2 = 1$ and vary the damping a_1. Figure 4.57 shows the upper bound for K_{22}. The numerical result obtained by Equation (4.13.1) is very close to that of the exact analytical results. Note that the numerical results render discrepancies when a_1 is very small that is when the system parameters are of the same order as Δt.

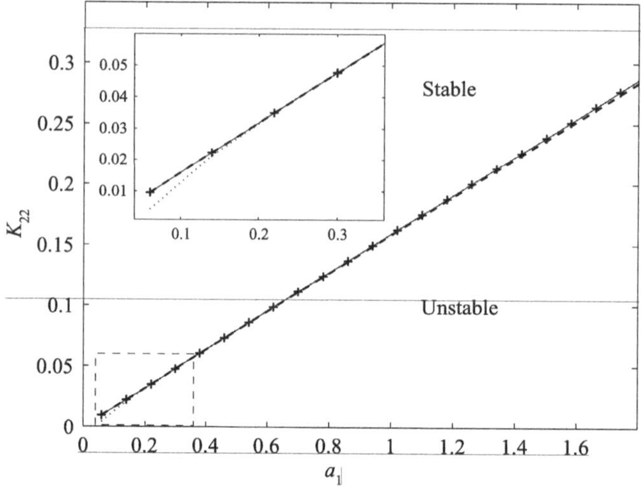

Fig. 4.57 Stability boundary for the strength of parametric excitation K_{22} with varying system damping a_1, for $a_2 = 1$. (+): Exact solution, (- - - -): exact drift mapping with $\Delta t = 0.02$, (· · · ·): direct Itô integration with $\Delta t = 0.02$, (———): direct Itô integration with $\Delta t = 0.0001$

We found that accuracy of the drift term in Eq. (4.130) can be further improved if the drift term is treated separately. The reason behind that is that those terms are of order Δt. We further take advantage of the semi-discretization method and construct a more accurate map of the drift terms using an equivalent continuous model

$$\ddot{x} + (a_1 - \pi K_{22})\,\dot{x} + (a_2 - \pi K_{12})\,x = f\,[i-N] \text{ for } t_i - \Delta t \le t \le t_i, \quad (4.190)$$

where $f\,[i-N] = k_p x\,[i-N] - k_d \dot{x}\,[i-N]$. Such a formulation is significantly efficient and accurate if first order approximation is used for the feedback terms. We call this approach as exact drift mapping and obtain the following 2D map

$$\begin{Bmatrix} x\,[i+1] \\ \dot{x}\,[i+1] \end{Bmatrix} = \begin{bmatrix} \alpha_{1i} & \alpha_{2i} \\ \beta_{1i} & \beta_{2i} \end{bmatrix} \begin{Bmatrix} x\,[i] \\ \dot{x}\,[i] \end{Bmatrix} + \begin{Bmatrix} \alpha_{3i} & \alpha_{4i} \\ \beta_{3i} & \beta_{4i} \end{Bmatrix} \begin{Bmatrix} f\,[i-N] \\ f\,[i-N+1] \end{Bmatrix} + \begin{Bmatrix} R_1\,[i] \\ R_2\,[i] \end{Bmatrix}. \quad (4.191)$$

Note that in order to achieve an efficient numeric formulation we collapsed down the delayed state effects into a scalar quantity $f\,[i-N]$. The explicit expressions of the coefficients α_{ji} and β_{ji} for a deterministic system is studied in the literature (Insperger and Stepan, 2001, 2002). When $W_i\,(t)$ are constants the problem simplifies to deterministic delayed feedback control and R_{1i}, R_{2i} are due to the external stochastic excitations. The transition matrix in Eq. (4.134) $\mathbf{\Phi}\,[i]$ becomes

$$\mathbf{\Phi}\,[i] = \begin{bmatrix} \beta_{2i} & \beta_{1i} & 0 & 0 & \cdots & \beta_{4i} & \beta_{3i} \\ \alpha_{2i} & \alpha_{1i} & 0 & 0 & \cdots & \alpha_{4i} & \alpha_{3i} \\ -k_d & -k_p & 0 & 0 & \cdots & 0 & 0 \\ 0 & 0 & 1 & 0 & \cdots & 0 & 0 \\ \vdots & \vdots & \vdots & \vdots & \ddots & \vdots & \vdots \\ 0 & 0 & 0 & 0 & \cdots & 0 & 0 \\ 0 & 0 & 0 & 0 & \cdots & 1 & 0 \end{bmatrix}. \quad (4.192)$$

Now, we take $W_1 = W_2 = 0$ in Eq. (4.181), and set the feedback gains k_p, k_d to zero. In this case, the steady-state second moments of the system are known as

$$E\left[x_1^2\right] = \frac{\pi K_{33}}{2a_1 a_2}, \quad E\left[x_1 x_2\right] = 0, \quad E\left[x_2^2\right] = \frac{\pi K_{33}}{2a_1}. \quad (4.193)$$

In Fig. 4.58 we vary the discretization time step Δt and obtain the second moments using the proposed method. We employ two formulations as described before: i) direct Itô integration and ii) exact drift mapping. Clearly, the latter yields superior results where the results of $E\left[x_1^2\right]$ and $E\left[x_1 x_2\right] = 0$ are virtually the same as the exact values. For $E\left[x_2^2\right]$, the convergence rate of the exact drift mapping is significantly higher than that of the direct Itô integration.

We first investigate the system in Eq. (4.181) using the system parameters $\tau = 0.16$, $a_1 = 0.4$, $a_2 = 1$ and the strength of stochastic excitations $K_{22} = 0.5/\sqrt{2\pi}$, $K_{11} = 1/\sqrt{2\pi}$, $K_{33} = 0$. Figure 4.59 shows the stability boundary as well as the equal performance curves for the delayed feedback gains. We used a discretization level of $N = 8$. The region outside the solid line is unstable. The equal performance

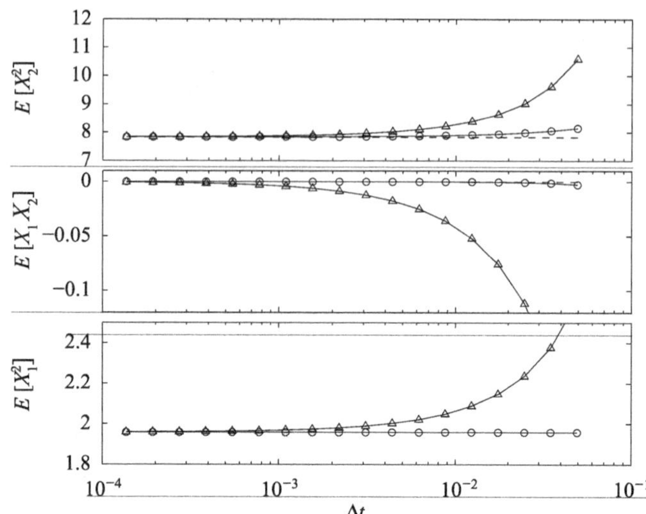

Fig. 4.58 Second order moments versus the discrete time step Δt, $K_{33} = \frac{5}{\sqrt{2\pi}}$, system damping $a_1 = 0.8$ and $a_2 = 4$, all other parameters are set to zero. (- - - -): exact values, (o-o-o): exact drift mapping, (Δ-Δ-Δ): direct Itô integration.

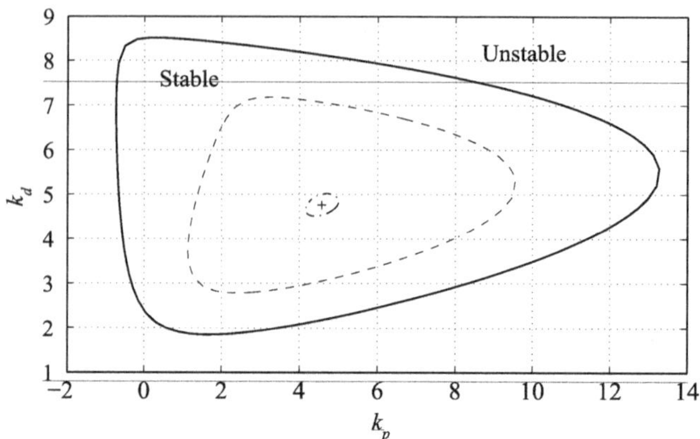

Fig. 4.59 Stability and performance boundaries of the delayed system in Eq. (4.181) in mean square sense with $a_1 = 0.4$, $a_2 = 1$, $\tau = 0.16$, $N = 8$ and stochastic excitations $K_{22} = 0.5/\sqrt{2\pi}$, $K_{11} = 1/\sqrt{2\pi}$, $K_{33} = 0$. (———): Stability boundary where $\lambda_{\max} = 1$, (- - - -):$\lambda_{\max} = 0.875$, (·— ·—): $\lambda_{\max} = 0.75$.

lines correspond to the decay rate of the second moments of the system response. The fastest decay rate can be achieved for $k_p = 4.566586$ and $k_d = 4.763169$ with a corresponding $|\lambda|_{\max} = 0.748013$. Note that, the origin in the unstable region corresponds to the uncontrolled case.

The semi-discretization method applied to stochastic systems is very versatile that one can obtain auto-correlation and cross-correlation functions of the system response in steady state without difficulty. In Fig. 4.60, we demonstrate this phenomenon where we used the same parameters as in Fig. 4.58. This time we stacked up the state vector and discretized the period of the system into a number of parts. We turn off the feedback gains which effectively sets α_{3j}, α_{4j}, β_{3j}, and β_{4i} to zero. We present the auto-correlation of the displacement, $E\left[x_1(t)\,x_1(t-T)\right]$. It should be noted that the results when $T = 0$ correspond to $E\left[x_1^2\right]$. In other words, there is one to one relation between Figs. 4.60 and 4.58.

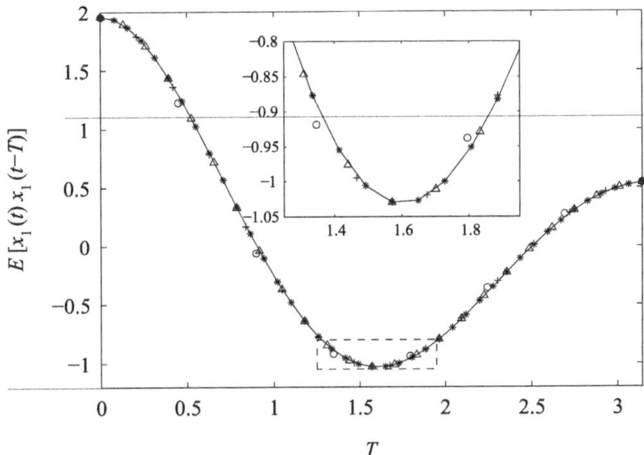

Fig. 4.60 Variation of auto-correlation of the second moment of displacement with varying discretezation, N. $a_1 = 0.8$, $a_2 = 4$, $k_p = 0$, $k_d = 0$, $K_{11} = 0$, $K_{22} = 0$ $K_{33} = \frac{5}{\sqrt{2\pi}}$.(o o): $N = 7$, (+ +): $N = 15$, (Δ Δ): $N = 24$, ($*$ $-$ $*$): $N = 40$.

Figure 4.61 depicts the most general case where we have additive and multiplicative stochastic disturbances. This time we use the same parameter set as in

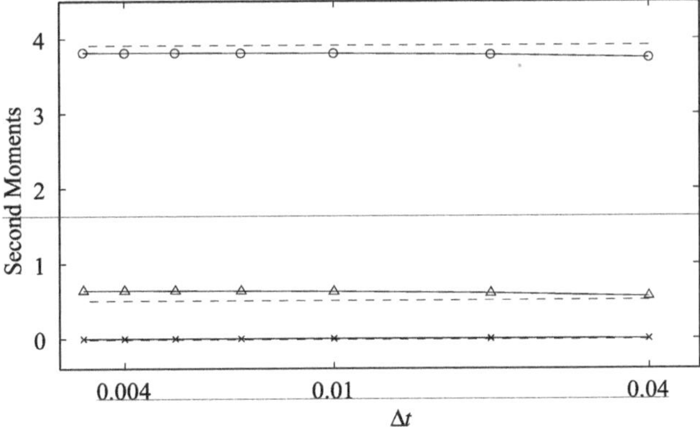

Fig. 4.61 Delayed feedback control response with $k_p = 2$, $k_d = 4$ $a_1 = 0.4$, $a_2 = 1$, $\tau = 0.16$, $N = 8$ and stochastic excitations $K_{22} = 0.5/\sqrt{2\pi}$, $K_{11} = 1/\sqrt{2\pi}$. Second order moments versus the discrete time step Δt; (Δ-Δ-Δ): $E\left[x_1^2\right]$, (o-o-o): $E\left[x_2^2\right]$, (x-x-x): $E\left[x_1 x_2\right]$. Dashed lines indicate the outcome of the Monte Carlo simulations.

Fig. 4.59. The system is unstable when the controller is turned of. We use a feedback gain pair $(k_p, k_d) = (4, 2)$ to stabilize the system. The dashed lines indicate the outcome of the Monte Carlo simulations where the initial conditions for the delayed vector is assumed to be zero. The Monte Carlo simulations yield $0.511044, -0.0103565$ and 3.91148 for $E\left[x_1^2\right]$, $E\left[x_1 x_2\right]$ and $E\left[x_2^2\right]$, respectively.

4.13.2 One Dimensional Nonlinear System

Consider a system defined by

$$dX(t) = [-aX(t) - \varepsilon X^3(t) + bX(t-\tau)]dt + \sigma dB(t), \tag{4.194}$$

where $dB(t)$ is a unit Brownian motion such that

$$E\left[dB\left(t\right)\right] = 0,\ E\left[dB\left(t_1\right) dB\left(t_2\right)\right] = \begin{cases} 0, & t_1 \neq t_2, \\ dt, & t_1 = t_2 = t. \end{cases} \tag{4.195}$$

Equation (4.144) for this example reads

$$\begin{aligned} d\mathbf{Y}\left(t\right) &= \left\{ \begin{bmatrix} -a & 0 & \cdots & 0 & b \\ \frac{1}{2\Delta\tau} & 0 & -\frac{1}{2\Delta\tau} & \cdots & 0 \\ & & \ddots & & \\ 0 & \cdots & 0 & \frac{1}{\Delta\tau} & -\frac{1}{\Delta\tau} \end{bmatrix} \mathbf{Y}\left(t\right) - \begin{bmatrix} \varepsilon Y_1^3(t) \\ 0 \\ \vdots \\ 0 \end{bmatrix} \right\} dt + \begin{bmatrix} \sigma \\ 0 \\ \vdots \\ 0 \end{bmatrix} dB\left(t\right) \\ &\equiv \hat{\mathbf{m}}(\mathbf{Y}\left(t\right), t)dt + \hat{\sigma} dB\left(t\right). \end{aligned} \tag{4.196}$$

The vector $\mathbf{Y}\left(t\right)$ reads

$$\begin{aligned} \mathbf{Y}\left(t\right) &= \left[X_1\left(t\right), X_1\left(t-\Delta\tau\right), X_1\left(t-2\Delta\tau\right), \ldots, X_1\left(t-N\Delta\tau\right)\right]^{\mathrm{T}} \\ &\equiv \left[Y_1\left(t\right), Y_2\left(t\right), \ldots, Y_{N+1}\left(t\right)\right]^{\mathrm{T}}. \end{aligned} \tag{4.197}$$

The FPK equation of the system reads

$$\begin{aligned} \frac{\partial}{\partial t} p_{\mathbf{Y}}(\mathbf{y}, t|\mathbf{y}_0, t_0) &= -\frac{\partial}{\partial y_j}[\hat{m}_j(\mathbf{y}, t) p_{\mathbf{Y}}] + \frac{\partial^2}{\partial y_j \partial y_k}\left[\frac{\hat{b}_{jk}(\mathbf{y}, t)}{2} p_{\mathbf{Y}}\right] \\ &= -\frac{\partial}{\partial y_1}[(-a y_1 + b y_m - \varepsilon y_1^3) p_{\mathbf{Y}}] + \frac{\sigma^2}{2}\frac{\partial^2 p_{\mathbf{Y}}}{\partial y_1^2} \\ &\quad - \frac{1}{2\Delta\tau}\frac{\partial}{\partial y_2}[(a_1 y_1 - a_2 y_3) p_{\mathbf{Y}}] \cdots \\ &\quad - \frac{1}{\Delta\tau}\frac{\partial}{\partial y_m}[(a_{m-1} y_{m-1} - a_m y_m) p_{\mathbf{Y}}]. \end{aligned} \tag{4.198}$$

When $\varepsilon = 0$, the FPK admits an exact solution as outlined in Section 4.12.1. When the state matrix $\hat{\mathbf{A}}$ is stable, the steady state solutions of the system exist. An example of the steady state mean square response of Y_i is shown in Fig. 4.62. The mean square responses of all the components Y_i fluctuate $\pm 1.7\%$ about the average value over the time delay interval. When $b = 0$, $E[Y_1^2]$ matches the exact solution.

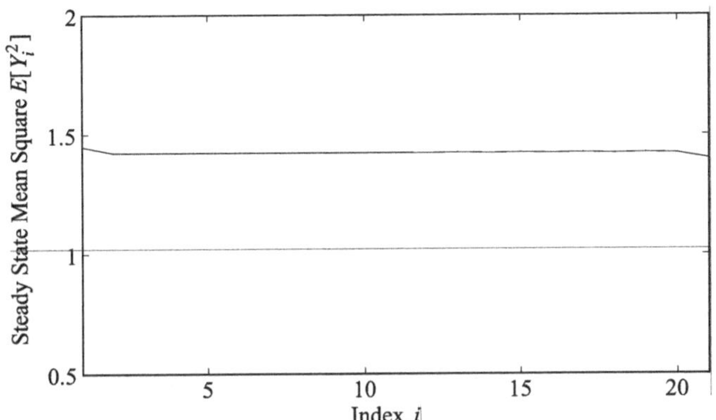

Fig. 4.62 Steady state mean square responses of Y_i of the first order linear system with time delay. $a = -1$. $b = 0.5$. $\epsilon = 0$. $N = 20$.

The FDMP method can also be applied to the stochastic system even when there is no time delay $b = 0$. Note that $E[Y_1 Y_k] = R_{XX}(\tau_{k-1})$ is the correlation function of $X(t)$ where $k = 1, 2, \ldots, N+1$, and $\tau_{k-1} = (k-1)\Delta\tau/N$. Figure 4.63 shows the correlation function $R_{XX}(\tau)$ when $b = 0$. The solution obtained by the FDMP approximation matches perfectly with the exact solution.

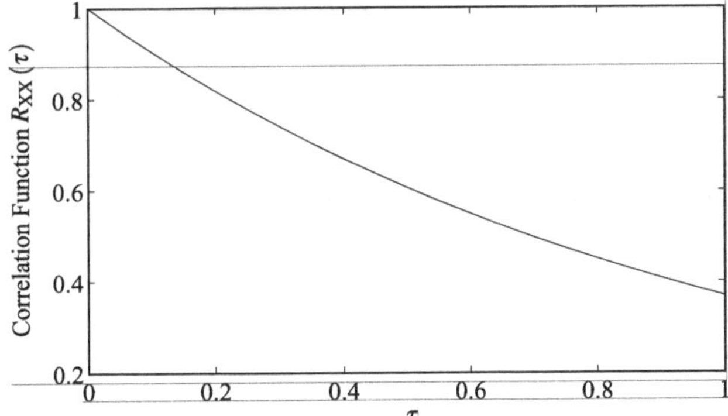

Fig. 4.63 Comparison of the exact correlation function of the first order linear system with the solution by the FDMP method. The agreement is perfect. $a = 1$. $\sigma = \sqrt{2}$. $\epsilon = 0$. $b = 0$.

It is noted that the exact solution of the steady state probability density function governed by the FPK equation (4.198) is illusive at this time, although its non-delayed version is well known.

Acknowledgements This chapter contains the doctoral research by JQS's former graduate students at the University of Delaware. They are Dr. Ozer Elbeyli and Dr. Jie Sheng. Their contributions are deeply appreciated.

References

Ali M.S., Hou Z.K. and Noori M.N., 1998, Stability and performance of feedback control systems with time delays, *Computers and Structures*, **66**(2-3), 241-248.

Atay F.M., 2002, Delayed-feedback control of oscillations in non-linear planar systems, *International Journal of Control*, **75**, 297-304.

Bellen A. and Maset S., 2000, Numerical solution of constant coefficient linear delay differential equations as abstract Cauchy problems, *Numerische Mathematik*, **84**, 351-374.

Bellen A. and Zennaro M., 2003, *Numerical Methods for Delay Differential Equations*, Oxford University Press, New York.

Breda D., Maset S. and Vermiglio R., 2004, Computing the characteristic roots for delay differential equations, *IMA Journal of Numerical Analysis*, **24**, 1-19.

Breda D., Maset S. and Vermiglio R., 2005, Pseudospectral differencing methods for characteristic roots of delay differential equations, *SIAM Journal Science Computing*, **27**, 482-495.

Buckwar E., 2000, Introduction to the numerical analysis of stochastic delay differential equations, *Journal of Computational and Applied Mathematics*, **125**, 297-307.

Butcher E. and Bobrenkov O., 2009, The chebyshev spectral continuous time approximation for periodic delay differential equations, In: Proceedings of ASME 2009 International Design Engineering Technical Conferences (IDETC) and Computers and Information in Engineering Conference (CIE), San Diego, California.

Cai G. and Huang J., 2002,Optimal control method with time delay in control, *Journal of Sound and Vibration*, **251**, 383-394.

Cai G.P., Huang J.Z. and Yang S.X., 2003, An optimal control method for linear systems with time delay, *Computers & structures*, **81**, 1539-1546.

Camacho E.F., Bordons C., 1999, *Model Predictive Control*, Springer, New York.

Cao Y.Y., Lin Z.L. and Hu T.S., 2002, Stability analysis of linear time-delay systems subject to input saturation, *IEEE Transactions on Circuits and Systems I-Fundamental Theory and Applications*, **49**, 233-240.

Carnahan B., Luther H.A. and Wilkes J.O., 1969, *Applied Numerical Methods*, John Wiley and Sons, New York.

Chen W.H., Guan Z.H. and Lu X.M., 2004, Delay-dependent output feedback guaranteed cost control for uncertain time-delay systems, *Automatica*, **40**, 1263-1268.

Deshmukh V., Ma H. and Butchern E.A., 2006, Optimal control of parametrically excited linear delay differential systems via Chebyshev polynomials, *Optimal Control Applications and Methods*, **27**, 123-136.

Deshmukh V., Butcher E.A. and Bueler E., 2008, Dimensional reduction of nonlinear delay differential equations with periodic coefficients using Chebyshev spectral collocation,*Nonlinear Dynamics*, **52**.

Dumont G.A., Elnaggar A. and Elshafei A., 1993, Adaptive predictive control of systems with time-varying time delay, *International Journal of Adaptive Control and Signal Processing*, **7**, 91-101.

Elbeyli O. and Sun J.Q., 2004, On the semi-discretization method for feedback control design of linear systems with time delay, *Journal of Sound and Vibration*, **273**, 429-440.

Engelborghs K. and Roose D., 2002, On stability of LMS methods and characteristic roots of delay differential equations, *IMA Journal of Numerical Analysis*, **40**, 629-650.

Fan K.K., Chen J.D., Lien C.H. and Hsieh J.G., 2002, Delay-dependent stability criterion for neutral time-delay systems via linear matrix inequality approach, *Journal of Mathematical Analysis and Applications*, **273**, 580-589.

Filipovic D., Olgac N., ,1998, Torsional delayed resonator with velocity feedback, *IEEE/ASME Transactions on Mechatronics*, **3**, 67-72.

Frank T.D. and Beek P.J., 2001, Stationary solutions of linear stochastic delay differential equations: Applications to biological systems, *Physical Review E Statistical Physics, Plasmas, Fluids, and Related Interdisciplinary Topics*, **64**, 219,171/1–12.

Franklin G.F., Powell J.D. and Emami-Naeini A., 1986, *Feedback Control of Dynamic Systems*, Addison-Wesley, Reading, Massachusetts.

Franklin G.F., Powell J.D. and Workman M.L., 1998, *Digital Control of Dynamic Systems*, Addison Wesley Longman, Inc., Berkeley, California.

Fridman E. and Orlov Y., 2009, Exponential stability of linear distributed parameter systems with time-varying delays, *Automatica*, **45**, 194-201.

Fu Y., Tian Z. and Shi S., 2003, State feedback stabilization for a class of stochastic time-delay nonlinear systems, *IEEE Transactions on Automatic Control*, **48**, 282-286.

Fujii H.A., Ichiki W., Suda S.I. and Watanabe T.R., 2000, Chaos analysis on librational control of gravity-gradient satellite in elliptic orbit, *Journal of Guidance, Control, and Dynamics*, **23**, 145-146.

Gao H.J., Chen T.W. and Lam J., 2008,A new delay system approach to network-based control, *Automatica*, **44**, 39-52.

Garg N.K., Mann B.P., Kim N.H. and Kurdi M.H., 2007, Stability of a time-delayed system with parametric excitation, *Journal of Dynamic Systems, Measurement, and Control*, **129**, 125-135.

Golub G.H. and Loan C.F.V., 1983, *Matrix Computations*,The Johns Hopkins University Press, Baltimore, Maryland.

Gu K. and Niculescu S.I., 2003, Survey on recent results in the stability and control of time-delay systems,*Journal of Dynamic Systems, Measurement, and Control*, **125**, 158-165.

Guillouzic S., L'Heureux I. and Longtin A., 1999, Small delay approximation of stochastic delay differential equations, *Physical Review E Statistical Physics, Plasmas, Fluids, and Related Interdisciplinary Topics*, **59**, 3970-3982.

Ha C. and Ly U.L., 1996, Sampled-data system with computation time delay: optimal w-synthesis method, *Journal of Guidance, Control, and Dynamics*, **19**, 584-591.

Han Q.L., 2009, A discrete delay decomposition approach to stability of linear retarded and neutral systems, *Automatica*, **45**, 517-524.

He Y., Wang Q.G., Lin C. and Wu M., 2007, Delay-range-dependent stability for systems with time-varying delay, *Automatica*, **43**, 371-376.

Hespanha J.P., Liberzon D. and Morse A.S., 1999, Logic-based switching control of a nonholomic system with parametric modeling uncertainty, *Systems & Control Letters*, **38**, 167-177.

Hespanha J.P., Liberzon D. and Morse A.S, 2003, Hysteresis-based switching algorithms for supervisory control of uncertain systems, *Automatica*, **39**, 263-272.

Hu X.B. and Chen W.H., 2004, Model predictive control for constrained systems with uncertain state-delays,*International Journal of Robust and Nonlinear Control*, **14**, 1421-1432.

Insperger T. and Stepan G., 2001, Semi-discretization of delayed dynamical systems, In: Proceedings of ASME 2001 Design Engineering Technical Conferences and Computers and Information in Engineering Conference, Pittsburgh, Pennsylvania.

Insperger T. and Stepan G., 2002, Semi-discretization method for delayed systems, *International Journal for Numerical Methods in Engineering*, **55**, 503-518.

Ivanescu D., Dion J.M., Dugard L. and Niculescu S.I., 2000, Dynamical compensation for time-delay systems: An LMI approach, *International Journal of Robust and Nonlinear Control*, **10**, 611-628.

Ji G. and Luo Q., 2006, Iterative learning control for uncertain time-delay systems. *Dynamics of Continuous Discrete and Impulsive Systems – Series A Mathematical Analysis*, **13**, 1300-1306.

Jiang X.F. and Han Q.L., 2008, New stability criteria for linear systems with interval time-varying delay, *Automatica*, **44**, 2680-2685.

Kalmar-Nagy T., 2005, A novel method for efficient numerical stability analysis of delay-differential equations, In: Proceedings of American Control Conference, Portland, Oregon, pp 2823–2826.

Kao C.Y. and Rantzer A, 2007, Stability analysis of systems with uncertain time-varying delays, *Automatica*, **43**, 959–970

Kapila V. and Haddad W.M., 1999, Robust stabilization for systems with parametric uncertainty and time delay, *Journal of the Franklin Institute*, **336**, 473-480.

Kim J.H., 2008, Improved ellipsoidal bound of reachable sets for time-delayed linear systems with disturbances,*Automatica*, **44**, 2940-2943.

Klein E.J. and Ramirez W.F., 2001, State controllability and optimal regulator control of time-delayed systems, *International Journal of Control*, **74**, 281-289.

Kolmanovskii V.B. and Richard J.P.,1999, Stability of some linear systems with delays, *IEEE Transactions on Automatic Control*, **44**, 984-989.

Koto T.,2004, Method of lines approximations of delay differential equations, *Computers and Mathematics with Applications*, **48**, 45-59.

Koto T., 2009, Stability of implicit-explicit linear multistep methods for ordinary and delay differential equations, *Frontiers of Mathematics in China*, **4**, 113-129.

Kuchler U. and Platen E., 2002, Weak discrete time approximation of stochastic differential equations with time delay, *Mathematics and Computers in Simulation*, **59**, 497-507.

Kwon O.M., Park J.H. and Lee S.M., 2008, On delay-dependent robust stability of uncertain neutral systems with interval time-varying delays,*Applied Mathematics and Computation*, **203**, 843-853.

Kwon W. and Pearson A., ,1980,Feedback stabilization of linear systems with delayed control. *IEEE Transactions on Automatic Control*, **25**, 266–269

Kwon W.H., Lee G.W. and Kim S.W., 1990, Performance improvement using time delays in multivariable controller design, *International Journal of Control*, **52**, 1455-1473.

Leugering G., 2000,On the semi-discretization of optimal control problems for networks of elastic strings: global optimality systems and domain decomposition, *Journal of Computational and Applied Mathematics*, **120**, 133-157.

Lewis F.L. and Syrmos V.L., 1995, *Optimal Control*, John Wiley and Sons, New York.

Li Z., Ye L. and Liu Y., 1989, Unconditional stability of discrete systems with any time delay,*Advances in Modelling and Simulation*, **17**, 11-18.

Lin C.L., Chen C.H. and Huang H.C., 2008, Stabilizing control of networks with uncertain time varying communication delays, *Control Engineering Practice*, **16**, 56-66

Lin Y.K. and Cai G.Q., 1995, *Probabilistic Structural Dynamics — Advanced Theory and Applications*, McGraw-Hill, New York.

Ma H., Butcher E.A. and Bueler E., 2003, Chebyshev expansion of linear and piecewise linear dynamic systems with time delay and periodic coefficients under control excitations, *Journal of Dynamic Systems, Measurement, and Control*, **125**, 236-243.

Ma H., Deshmukh V., Butcher E.A. and Averina V., 2005, Delayed state feedback and chaos control for time periodic systems via a symbolic approach, *Communications in Nonlinear Science and Numerical Simulation*, **10**, 479-497.

Maset S, 2003, Numerical solution of retarded functional differential equations as abstract Cauchy problems, *Journal of Computational and Applied Mathematics*, **161**, 259-282.

Miller D.E. and Davison D.E., 2005, Stabilization in the presence of an uncertain arbitrarily large delay, *IEEE Transactions on Automatic Control*, **50**, 1074-1089.

Morse A.S., 1996, Supervisory control of families of linear set-point controllers – Part 1: Exact matching, *IEEE Transactions on Automatic Control*, **41**, 1413-1431.

Morse A.S., 1997, Supervisory control of families of linear set-point controllers - Part 2: Robustness, *IEEE Transactions on Automatic Control*, **42**, 1500-1515.

Niculescu S.I., Verriest E.I., Dugard L. and Dion J.M., 1998, Stability of linear systems with delayed state: A guided tour. In: Proceedings of the IFAC Workshop: Linear Time Delay Systems, Grenoble, France, pp 31-38.

Nohmi M. and Matsumoto K., 2002, Teleoperation of a truss structure by force command in ets-vii robotics mission, *AIAA Journal*, **40**, 334-339.

Normey-Rico O.E. and Camacho E.F., 1999, Robustness effects of a prefilter in a smith predictor-based generalized predictive controller, *IEE Proceedings: Control Theory and Applications*, **146**, 179-185.

de Oliveira MC and Geromel JC, 2004, Synthesis of non-rational controllers for linear delay systems, *Automatica*, **40**, 171-188.

Pfeiffer B.M. and Marquardt W., 1996, Symbolic semi-discretization of partial differential equation systems, *Mathematics and Computers in Simulation*, **42**, 617-628.

Pinto O.C. and Goncalves P.B., 2002, Control of structures with cubic and quadratic non-linearities with time delay consideration, *Journal of the Brazilian Society of Mechanical Sciences*, **24**, 99-104.

Rawlings J.B., 2000, Tutorial overview of model predictive control, *IEEE Control Systems Magazine*, **20**, 38–52

Shanmugathasan N. and Johnston R.D., 1988, Exploitation of time delays for improved process control, *International Journal of Control*, **48**, 1137-1152.

Shao H.Y., 2008, Improved delay-dependent stability criteria for systems with a delay varying in a range, *Automatica*, **44**, 3215-3218.

Sheng J. and Sun J.Q., 2005, Feedback controls and optimal gain design of delayed periodic linear systems, *Journal of Vibration and Control*, **11**, 277-294.

Sheng J., Elbeyli O. and Sun J.Q., 2004, Stability and optimal feedback controls for time-delayed linear periodic systems, *AIAA Journal*, **42**, 908-911.

Singh T., 1995, Fuel/time optimal control of the benchmark problem, *Journal of Guidance, Control, and Dynamics*, **18**, 1225-1231.

Slotine J.J.E. and Li W., 1991, *Applied Nonlinear Control*, Prentice Hall, New Jersey.

Smith O.J.M., 1957, Closer control of loops with dead time, *Chemical Engineering Progress*, **53**, 217-219.

Stepan G., 1998, Delay-differential equation models for machine tool chatter, In: Moon FC (ed.) Dynamics and Chaos in Manufacturing Processes, Wiley, New York, pp 165–192.

Suh I.H. and Bien Z., 1979, Proportional minus delay controller, *IEEE Transactions on Automatic Control*, **24**, 370-372.

Suh I.H. and Bien Z., 1980, Use of time-delay actions in the controller design, *IEEE Transactions on Automatic Control*, **25**, 600-603.

Sun J.Q., 2006, *Stochastic Dynamics and Control*, Elsevier Science, Ltd., Oxford, U.K.

Sun J.Q., 2009, A method of continuous time approximation of delayed dynamical systems, *Communications in Nonlinear Science and Numerical Simulation*, **14**, 998-1007.

Tallman G.H. and Smith O.J.M.,1958, Analog study of dead-beat posicast control, *IEEE Transactions on Automatic Control*, **3**, 14-21.

Vijta M., 2000, Some remarks on the Padé-approximations, In: *Proceedings of the 3rd TEMPUS-INTCOM Symposium*, Veszpré,. Hungary, pp 1-6

Wu H. and Mizukami K., 1995, Robust stability criteria for dynamical systems including delayed perturbations, *IEEE Transactions on Automatic Control*, **40**, 487-490.

Xia Y.Q. and Jia Y.M., 2003, Robust control of state delayed systems with polytopic type uncertainties via parameter-dependent lyapunov functionals, *Systems & Control Letters*, **50**, 183-193.

Xiao H. and Liu Y, 1994, The stability of linear time-varying discrete systems with time-delay, *Journal of Mathematical Analysis and Applications*, **188**, 66-77.

Yang B. and Wu X., 1998, Modal expansion of structural systems with time delays, *AIAA Journal*, **36**, 2218-2224.

Yue D., Tian E., Zhang Y. and Peng C., 2009, Delay-distribution-dependent robust stability of uncertain systems with time-varying delay, *International Journal of Robust and Nonlinear Control*, **19**, 377-393.

Zhang X.P., Tsiotras P. and Knospe C., 2002, Stability analysis of lpv time-delayed systems, *International Journal of Control*, **75**, 538-558.

Chapter 5
Synchronization of Dynamical Systems in Sense of Metric Functionals of Specific Constraints

Albert C.J. Luo

Abstract In this chapter, a theory for synchronization of multiple dynamical systems under specific constraints is developed from a theory of discontinuous dynamical systems. The metric functionals based on specific constraints are proposed to describe the synchronicity of the two or more dynamical systems to such specific constraints. The synchronization, desynchronization and penetration of multiple dynamical systems to multiple specified constraints are discussed through such metric functionals, and the necessary and sufficient conditions for such synchronicity are developed. The synchronicity of two dynamical systems to a single specific constraint is presented, and the synchronicity of the two systems to multiple specific constraints is investigated as well. The chapter provides a theoretic frame work in order to control slave systems which can be synchronized with master systems though specific constraints in a general sense.

5.1 Introduction

The study on synchronization should go back to the 17th century. Huygens (1673) gave a detailed description of the synchronization of two pendulum clocks with a weak interaction. Once the coupled pendulums possess small oscillations with the same initial conditions or the initial phase difference is zero, the two pendulums will be synchronized. However, when the initial phase difference is 180°, the anti-synchronization of two pendulums can be observed. For a general case, the motion of the two pendulums will be combined by the synchronization and anti-synchronization modes of vibration. Four types of synchronizations of two or more dynamical systems have been considered: (i) identical or complete synchronization,

Albert C.J. Luo
Department of Mechanical and Industrial Engineering, Southern Illinois University Edwardsville, IL 62026-1805, USA.
Email: aluo@siue.edu

(ii) generalized synchronization, (iii) phase synchronization, (iv) anticipated and lag synchronization and amplitude envelope synchronization. All the synchronizations of two or more systems at least possess one constraint for synchronicity, and the aforementioned synchronizations possess the characteristics of asymptotic stability. Once the two or more dynamical systems generate a state of synchronization under a specific constraint, such a synchronized state should be stable, as referred to Pikovsky et al. (2001) and Boccaletti (2008). After the Huygens's studies, Rayleigh described the synchronization phenomena in acoustic systems (Rayleigh J., 1945). Due to electrical and radio wave propagations, the wave synchronization was of great interest in 1920s. For early studies on synchronizations, one focused on the limit cycles in self-excited dynamical systems, resonance phenomena in multiple-degrees of freedom systems and, steady-state motion in forced vibration. The limit cycle in self-excited dynamical systems was discussed (e.g., van der Pol B., 1927), which is a kind of synchronization and such synchronization can be stabilized. The other discussions on steady-state motion and resonance in nonlinear oscillations can be referred in many books (e.g., Stocker J.J., 1950; Hayashi C., 1964). Recently, one tried to control a flow of dynamical systems with attractors. Such an investigation is actually to look into a dynamical system synchronizing with a goal dynamics, as discussed in (Jackson, 1991).

Pecora and Carroll (1990) investigated the synchronization of two systems connected with common signals and gave a criterion of the Lyapunov exponents. The common signals are as constraints between the two systems. Based on this idea, the synchronized circuits for chaos were developed by Carroll and Pecaora (1991). Since then, one focused on developing the corresponding control methods and schemes to achieve the synchronization of two dynamical systems with constraints. Pyragas (1992) presented two methods to obtain the synchronization of two chaotic dynamical systems with a small time continuous perturbation. Kapitaniak (1994) used a continuous control to present the synchronization of two chaotic systems. Ding and Ott (1994) stated that a slave system (receiver system) is not necessary to be a replica of part of master systems. Such a synchronization of two systems is called an identical (or complete) synchronization. However, Rulkov et al. (1995) discussed a generalized synchronization of chaos in directionally coupled chaotic systems. Kocarev and Parlitz (1995) developed a general method to construct chaotic synchronized systems, which decomposes the given systems into the active and passive systems. Peng et al. (1996) presented the chaotic synchronization of n-dimensional systems, and Pyragas (1996) discussed the weak and strong synchronizations of chaos by the coupling strength of two dynamical systems. Ding et al. (1997) reviewed the control and synchronization of chaos in high-dimensional dynamical systems. Boccaletti et al. (1997) presented an adaptive synchronization of chaos for secure communication. Abarbanel et al. (1997) used a small force to control a dynamical system to specific orbits. Pyragas (1998) systematically introduced the basic ideas of the generalized synchronization of chaos. Yang and Chua (1999) used linear transformations to study generalized synchronization. Zhan et al. (2003) investigated the complete and generalized synchronizations of coupled time-delay systems. Campos and Urias (2004) presented a mathematical description of multi-modal synchronization with chaos. The definition of master-slave synchronization was presented and a multivalued, synchronized function was introduced. Koronovskii et al. (2004) discussed the duration of a process of complete synchronization of two coupled, identical chaotic systems. Mosekilde et al. (2001) discussed chaotic synchronization and applied such concepts to living systems, and recent contribution on synchronization in biosystems can also be found (e.g., Wang et al., 2008; Enjieu Kadji et al., 2008; Peng et al., 2009). Newell et al. (1994) investigated synchronization in chaotic diode resonator. Kocarev and Parlitz (1996) investigated synchronizing spatiotemporal chaos in coupled nonlinear oscillators. Teufel et al. (1996) presented the synchronization of two flow-excited pendula as similar to Huygens' work (1986) . Yamapi and Woafo (2006) investigated synchronizations in a ring of four mutually coupled self-sustained electromechanical devices. Boccaletti et al. (2002) gave a systematical review about the synchronization of chaotic systems. The definitions and concepts are further clarified. Chen et al. (2006) discussed on stability of synchronized dynamics and pattern formation in coupled systems. As aforementioned, the phase synchronization exists in self-excited vibration systems, forced nonlinear vibrating systems and coupled nonlinear systems. For such an synchronization one employed the perturbation techniques (e.g., Stocker, 1950; Hayashi, 1964). Kuramoto (1984) investigated the waves and turbulence in chemical oscillations by use of the phase synchronization (or entertainments). Zaks et al. (1999) studied the imperfect phase synchronization through the alternative locking ratios. Feng and Shen (2005) investigated phase synchronization and anti-phase synchronization of chaos in degenerate optical parametric oscillator.

On the other hand, one has been interested in the synchronization of discrete systems with mappings. Pecora et al. (1997) discussed the volume-preserving and volume-expanding synchronized chaotic systems through discrete maps. Stocjanovski et al. (1979) used the symbolic dynamics to study chaos synchronization, and the information entropy was used to the synchronization of chaotic systems through discrete maps. Rulkov (2001) discussed a regularization of synchronized chaotic bursts. Afraimovich et al. (2002) studied the generalized synchronization of chaos of noninvertible maps in mathematics. Barreto et al. (2003) discussed the geometrical behavior of chaos synchronization through discrete maps. Hu et al. (2008) investigated the hybrid projective synchronization of a general class of chaotic maps. Pareek et al. (2005) used multiple one-dimensional chaotic maps to investigated cryptography, and the extension of such a research can be found in (Xiang et al., 2008). Bowong et al. (2006) adopted the parameter modulation of a chaotic system for secure communications. Fallahi et al. (2008) adopted the extended Kalman filter and multi-shift cipher algorithm for secure chaotic communication, and Kiani-B et al. (2009) used fractional chaotic systems to secure communication through an extended fractional Kalman filter. Wang and Yu (2009) used multiple-chaotic systems to develop a block encryption algorithm with a dynamical sequence. Soto-Crespo and Akhmediev (2005) showed nonlinear synchronization and chaos through solitons as strange attractors. Hung et al. (2006) discussed chaos synchronization of two stochastically coupled random Boolean networks. The more discussion about phase synchronization in oscillatory networks was presented in Osipov et al. (2007). The investigations on synchronization on the dynamical systems with time-delay were very active and the recent results can be found in (e.g., Zhan et al., 2003; Bowong et al., 2006; Ghosh et al., 2007; Wang et al., 2008, Cruz-Hernández and Romero-Haros, 2008; Lu, 2008).

From the above discussions, the synchronization of two or more dynamical systems is that the corresponding flows of the two or more dynamical systems are constrained under specific constraint conditions for a time interval. If the constraint conditions are considered as constraint boundaries, the synchronization of the two or more dynamical systems can be investigated by the theory of discontinuous dynamical systems. In Luo (2009), a theory for synchronization of dynamical systems with specific constraints was presented by the theory for discontinuous dynamical systems in (Luo, 2005, 2006, 2008). The general concept of synchronization was presented. The necessary and sufficient conditions for the synchronization, desynchronization and penetration were developed, and the synchronization complexity for multiple slave systems with multiple master systems will be discussed under specific constraints. In this chapter, the metric functional will be introduced for the synchronicity of two dynamical systems with single and multiple constraints. With such metric functional, the necessary and sufficient conditions for the synchronization, desynchronization and penetration will be presented.

5.2 System synchronization

As in Luo (2009), the basic concepts and definitions about the synchronization of dynamical systems will be presented. To solve the over constraints of slave systems, a generalized synchronization will be introduced. For slave and master systems with full constraints, the static synchronization will be introduced. To describe the synchronization of two systems (e.g., slave and master systems) under specific constraints, the corresponding domains and boundaries relative to the constraints will be discussed through discontinuous dynamical systems. To discuss the synchronicity of the two systems, the metric functional will be introduced.

5.2.1 Synchronization of slave and master systems

Consider two dynamic systems as

$$\dot{\mathbf{x}} = \mathbf{F}(\mathbf{x}, t, \mathbf{p}) \in \mathfrak{R}^n \tag{5.1}$$

and

$$\dot{\tilde{\mathbf{x}}} = \tilde{\mathbf{F}}(\tilde{\mathbf{x}}, t, \tilde{\mathbf{p}}) \in \mathfrak{R}^{\tilde{n}} \tag{5.2}$$

where $\mathbf{x} = (x_1, x_2, \ldots, x_n)^{\mathrm{T}}$ and $\mathbf{p} = (p_1, p_2, \ldots, p_k)^{\mathrm{T}}$; $\tilde{\mathbf{x}} = (\tilde{x}_1, \tilde{x}_2, \ldots, \tilde{x}_{\tilde{n}})^{\mathrm{T}}$ and $\tilde{\mathbf{p}} = (\tilde{p}_1, \tilde{p}_2, \ldots, \tilde{p}_{\tilde{k}})^{\mathrm{T}}$. The two vector functions $\mathbf{F} = (F_1, F_2, \ldots, F_n)^{\mathrm{T}}$ and $\tilde{\mathbf{F}} = (\tilde{F}_1, \tilde{F}_2, \ldots,$

$\tilde{F}_{\tilde{n}})^{\mathrm{T}}$ can be either time-dependent or time-independent. Consider a time interval $I_{12} \equiv (t_1, t_2) \subset \Re$ and domains $U_{\mathbf{x}} \subseteq \Re^n$ and $\tilde{U}_{\tilde{\mathbf{x}}} \subseteq \Re^{\tilde{n}}$. For initial conditions $(t_0, \mathbf{x}_0) \in I_{12} \times U_{\mathbf{x}}$ and $(t_0, \tilde{\mathbf{x}}_0) \in I_{12} \times \tilde{U}_{\tilde{\mathbf{x}}}$, the corresponding flows of the two systems are $\mathbf{x}(t) = \boldsymbol{\Phi}(t, \mathbf{x}_0, t_0, \mathbf{p})$ and $\tilde{\mathbf{x}}(t) = \tilde{\boldsymbol{\Phi}}(t, \tilde{\mathbf{x}}_0, t_0, \tilde{\mathbf{p}})$ for $(t, \mathbf{x}) \in I_{12} \times U_{\mathbf{x}}$ and $(t, \tilde{\mathbf{x}}) \in I_{12} \times \tilde{U}_{\tilde{\mathbf{x}}}$ with $\mathbf{p} \in U_{\mathbf{p}} \subseteq \Re^k$ and $\tilde{\mathbf{p}} \in U_{\tilde{\mathbf{p}}} \subseteq \Re^{\tilde{k}}$. The semi-group properties for two flows hold (i.e., $\boldsymbol{\Phi}(t+s, \mathbf{x}_0, t_0, \mathbf{p}) = \boldsymbol{\Phi}(t, \boldsymbol{\Phi}(s, \mathbf{x}_0, t_0, \mathbf{p}), s, \mathbf{p})$ with $\mathbf{x}(t_0) = \boldsymbol{\Phi}(t_0, \mathbf{x}_0, t_0, \mathbf{p})$, and $\tilde{\boldsymbol{\Phi}}(t+s, \tilde{\mathbf{x}}_0, t_0, \tilde{\mathbf{p}}) = \tilde{\boldsymbol{\Phi}}(t, \tilde{\boldsymbol{\Phi}}(s, \tilde{\mathbf{x}}_0, t_0, \tilde{\mathbf{p}}), s, \tilde{\mathbf{p}})$ with $\tilde{\mathbf{x}}(t_0) = \tilde{\boldsymbol{\Phi}}(t_0, \tilde{\mathbf{x}}_0, t_0, \tilde{\mathbf{p}})$).

Consider the synchronization of the two systems in Eqs.(5.1) and (5.2), the slave and master systems are defined as follows.

Definition 5.1. A system in Eq.(5.2) is called a *master* system if its flow $\tilde{\mathbf{x}}(t)$ is independent. A system in Eq.(5.1) is called a *slave* system of the master system if its flow $\mathbf{x}(t)$ is constrained by a flow $\tilde{\mathbf{x}}(t)$ of the master system.

From the foregoing definition, a *slave* system is constrained by a *master* system through a specific condition. In other words, a slave system will be controlled by a master system under a specific constraint. Such a phenomenon is called the synchronization of the slave and master systems under such a specific condition. The corresponding definition is given as follows.

Definition 5.2. If a flow $\mathbf{x}(t)$ of a slave system in Eq.(5.1) is constrained by a flow $\tilde{\mathbf{x}}(t)$ of a *master* system in Eq.(5.2) through

$$\varphi(\mathbf{x}(t), \tilde{\mathbf{x}}(t), t, \boldsymbol{\lambda}) = 0, \quad \boldsymbol{\lambda} \in \Re^{n_0}, \tag{5.3}$$

for time $t \in [t_{m_1}, t_{m_2}]$, then the slave system is said to be *synchronized* with the master system in the sense of Eq.(5.3) for time $t \in [t_{m_1}, t_{m_2}]$, which is also called an $(n : \tilde{n})$-dimensional synchronization of the slave and master systems in the sense of Eq.(5.3). There are four special cases:

(i) If $t_{m_2} \to \infty$, the slave system is said to be absolutely synchronized with the master system in the sense of Eq.(5.3) for time $t \in [t_{m_1}, \infty)$.

(ii) If $t_{m_1} \to \infty$, the slave system is said to *be asymptotically synchronized* with the master system in the sense of Eq.(5.3).

(iii) For $n = \tilde{n}$, such a synchronization of the slave and master systems is called an *equi-dimensional* system synchronization in the sense of Eq.(5.3) for time $t \in [t_{m_1}, t_{m_2}]$.

(iv) For $n = \tilde{n}$, such a synchronization of the slave and master systems is called an *absolute*, equi-dimensional system synchronization in the sense of Eq.(5.3) for time $t \in [t_{m_1}, \infty)$. If $n \neq \tilde{n}$, the $(n : \tilde{n})$-dimensional synchronization is called a *non-equi-dimensional system synchronization*.

From the previous definition, the state variables in a slave system can be less or more than those in the master system. Therefore, it is not necessary to require the slave and master systems have the same dimensions in state space for synchronization. Under a certain rule in Eq.(5.3), it is interesting that a slave system can follow another completely different master system to synchronize. From the proceeding definition, a slave system can be synchronized with a master system under a constraint condition. In fact, constraints for such a synchronization phenomenon can be more than one. In other words, a slave system is synchronized with a master system under multiple constraints. Thus, the synchronization of a slave system with a master system under multiple constraints is described as follows:

Definition 5.3. An n-dimensional slave system in Eq.(5.1) is called to be synchronized with an $\tilde{n}$-dimensional master system in Eq.(5.2) of the $(n:\tilde{n};l)$-*type* (or an $(n:\tilde{n};l)$-*synchronization*) if there are l-linearly independent functions $\varphi_j(\mathbf{x}(t),\tilde{\mathbf{x}}(t),t,\boldsymbol{\lambda}_j)$ ($j\in\mathbb{L}$ and $\mathbb{L}=\{1,2,\ldots,l\}$) to make two flows $\tilde{\mathbf{x}}(t)$ and $\mathbf{x}(t)$ of the master and slave systems satisfy

$$\varphi_j(\mathbf{x}(t),\tilde{\mathbf{x}}(t),t,\boldsymbol{\lambda}_j)=0 \text{ for } \boldsymbol{\lambda}_j\in\mathfrak{R}^{n_j} \text{ and } j\in\mathbb{L} \tag{5.4}$$

for time $t\in[t_{m_1},t_{m_2}]$. There are eight special cases:

(i) If $t_{m_2}\to\infty$, the slave system is said to be *absolutely* synchronized of the $(n:\tilde{n};l)$-*type* with the master system (or an $(n:\tilde{n};l)$-absolute synchronization) in the sense of Eq.(5.4) for time $t\in[t_{m_1},\infty)$.

(ii) If $t_{m_1}\to\infty$, the slave system is said to be *asymptotically* synchronized of the $(n:\tilde{n};l)$-*type* with the master system (*or* an $(n:\tilde{n};l)$-*asymptotic synchronization*) in the sense of Eq.(5.4).

(iii) For $l=n$, the slave system is said to be *completely* synchronized of the $(n:\tilde{n};n)$-*type* with the master system (*or an* $(n:\tilde{n};n)$-*complete synchronization*) in the sense of Eq.(5.4) for time $t\in[t_{m_1},t_{m_2}]$.

(iv) For $l=n$ and $t_{m_2}\to\infty$, the synchronization of the slave and master systems is called an $(n:\tilde{n};n)$-*absolute, complete synchronization* in the sense of Eq. (5.4) for time $t\in[t_{m_1},\infty)$.

(v) If $n=\tilde{n}>l$, the synchronization of the slave and master systems is called an equi-dimensional system synchronization (or an $(n:n;l)$-synchronization) in the sense of Eq.(5.4) for time $t\in[t_{m_1},t_{m_2}]$.

(vi) If $n=\tilde{n}>l$ and $t_{m_1}\to\infty$, the synchronization of the slave and master systems is called an equi-dimensional, $(n:n;l)$-absolute synchronization in the sense of Eq.(5.4) for time $t\in[t_{m_1},\infty)$.

(vii) If $n=\tilde{n}=l$, the synchronization of the slave and master systems is called an equi-dimensional, complete synchronization (simply said *a synchronization*) in the sense of Eq.(5.4) for time $t\in[t_{m_1},t_{m_2}]$.

(viii) If $n=\tilde{n}=l$ and $t_{m_2}\to\infty$, the synchronization of the slave and master systems is called an equi-dimensional, absolute, complete synchronization (simply said an *absolute* synchronization) in the sense of Eq.(5.4) for time $t\in[t_{m_1},\infty)$.

In the foregoing definition, if the l-nonlinear equations are linearly independent, then there is a set of constants k_j and only $k_j=0$ for all $j\in\mathbb{L}$ exists to make the following equation hold for all the domains and time,

$$\sum\nolimits_{j=1}^{l}k_j\varphi_j(\mathbf{x}(t),\tilde{\mathbf{x}}(t),t,\boldsymbol{\lambda}_j)=0. \tag{5.5}$$

In addition, the independence of functions $\varphi_j(\mathbf{x}(t),\tilde{\mathbf{x}}(t),t,\boldsymbol{\lambda}_j)$ (for all $j \in \mathbb{L}$) is checked through the corresponding normal vectors. The normal vector of $\varphi_j(\mathbf{x}(t), \tilde{\mathbf{x}}(t),t,\boldsymbol{\lambda}_j)$ is computed by

$$\mathbf{n}_{\varphi_j} = \nabla\varphi_j(\mathbf{x},\tilde{\mathbf{x}},t,\boldsymbol{\lambda}_j) = \left(\frac{\partial\varphi_j}{\partial x_1},\frac{\partial\varphi_j}{\partial x_2},\ldots,\frac{\partial\varphi_j}{\partial x_m};\frac{\partial\varphi_j}{\partial \tilde{x}_1},\frac{\partial\varphi_j}{\partial \tilde{x}_2},\ldots,\frac{\partial\varphi_j}{\partial \tilde{x}_n}\right)^{\mathrm{T}}_{(t,\mathbf{x},\tilde{\mathbf{x}})}. \quad (5.6)$$

For all domains and time, if all the normal vectors $\mathbf{n}_{\varphi_j}(j \in \mathbb{L})$ are linearly-independent, i.e.,

$$\sum\nolimits_{j=1}^{l} k_j\mathbf{n}_{\varphi_j} = 0 \text{ only if } k_j = 0 \text{ for all } j \in \mathbb{L}, \quad (5.7)$$

then the functions $\varphi_j(\mathbf{x}(t),\tilde{\mathbf{x}}(t),t,\boldsymbol{\lambda}_j)$ are linearly-independent.

The foregoing definition tells that the slave and master systems are synchronized under l-constraints whatever the state-space dimension of the slave system is higher or lower than the master system. For $l < n$, the l-variables of the n-state variables of the slave system can be expressed by the $\tilde{n}$-state variables of the master system via the l-constraints. Select any l-variables $x_{[j]}$ and the rest $(n-l)$ variables $x_{(k)}$ of the n-state variables, i.e.,

$$\begin{aligned} &x_{[j]} \in \{x_i, i=1,2,\ldots,n\} \text{ for } j=1,2,\ldots,l,\\ &x_{(k)} \in \{x_i, i=1,2,\ldots,n\} \text{ for } k=l+1,l+2,\ldots,n. \end{aligned} \quad (5.8)$$

From Eq.(5.4), because of the linear-independence of functions $\varphi_j(\mathbf{x}(t),\tilde{\mathbf{x}}(t),\boldsymbol{\lambda}_j)$ $(j=1,2,\ldots,l)$, the constraint conditions gives

$$x_{[j]} = f_{[j]}(x_{(l+1)},x_{(l+2)},\ldots,x_{(n)},\tilde{\mathbf{x}},\boldsymbol{\lambda}) \text{ for } j \in \mathbb{L}. \quad (5.9)$$

Thus, the state variables $x_{[j]}$ of the slave system for $j \in \mathbb{L}$ can be said to be synchronized with the master system in the conditions of Eq.(5.4). The subscripts $[\cdot]$ and $(\cdot)$ of the state variables of the slave systems represent the *synchronizable* and *non-synchronizable* variables of the slave system to the master systems, respectively. If $l=1$, this definition reduces to Definition 5.2 and $(n:\tilde{n};1) \equiv (n:\tilde{n})$, the $(n:\tilde{n};l)$-synchronization reduce to the $(n:\tilde{n})$-synchronization. However, for $l=n$, the n-linearly independent conditions constrain the responses of the master and slave flows in the $\tilde{n}$-dimensional systems. Thus, the n-components of the slave flow can be completely determined by the $\tilde{n}$-components of a flow in the master system. Therefore, for the complete synchronization of the slave and master systems, a flow of the slave system is completely controlled by the master system through the constraint conditions in Eq.(5.4). For $l > n$, the slave system is overconstrained by the master system. Such a case will be discussed later. For $n = \tilde{n} = l$, an equi-dimensional, complete synchronization of the slave and master systems is obtained. For this case, n-components of a flow in the slave system are controlled by the n-components of a flow in the master system through the n-constraint equations in Eq.(5.4). Because the n-constraint equations in Eq.(5.4) are linearly-independent, the determinant of the Jocabian matrix of functions in Eq.(5.4) in neighborhood of the master flow $\tilde{\mathbf{x}}$ is non-zero. Therefore, there is a one-to-one relation between the slave and master flows $\mathbf{x}$ and $\tilde{\mathbf{x}}$. It implies that the slave flow is completely controlled by the master flow. From the above discussion, one obtains

$$\begin{aligned} &\mathbf{x}(t) = \mathbf{h}(\tilde{\mathbf{x}}(t), \boldsymbol{\lambda}) \text{ or} \\ &x_i(t) = h_i(\tilde{\mathbf{x}}(t), \boldsymbol{\lambda}) \text{ for } i = 1, 2, \ldots, n. \end{aligned} \tag{5.10}$$

Introduce a set of new variables with n-linear, independent relations between the slave and master systems. So one obtains

$$\begin{aligned} &\mathbf{z}(t) = \mathbf{x}(t) - \mathbf{B}\tilde{\mathbf{x}}(t) = \mathbf{h}(\tilde{\mathbf{x}}(t)) - \mathbf{B}\tilde{\mathbf{x}}(t) \text{ or} \\ &z_i(t) = x_i(t) - b_i\tilde{x}_i(t) = h_i(\tilde{\mathbf{x}}(t)) - b_i\tilde{x}_i(t) \text{ for } i = 1, 2, \ldots, n. \end{aligned} \tag{5.11}$$

where a constant diagonal matrix $\mathbf{B} = \mathrm{diag}(b_1, b_2, \ldots, b_n)$. One likes to consider the synchronization of two systems to be $z_i(t) \to 0$ for $t \to t_{m_1}$ and $z_i(t) = 0$ for $t \in [t_{m_1}, t_{m_2}]$, from which the slave and master system are synchronized. The n-equations (i.e., $b_i\tilde{x}_i(t) = h_i(\tilde{\mathbf{x}}(t)) \equiv h_i(\tilde{x}_1, \tilde{x}_2, \ldots, \tilde{x}_n)$ for $i = 1, 2, \ldots, n$) give the synchronization state independent of time. Such a concept can be extended to the affine synchronization, i.e., for $z_i(t) \to c_i$ (constant) for $t \to t_m$ and $z_i(t) = c_i$ for $t \in [t_{m_1}, t_{m_2}]$. The definition is given as follows:

Definition 5.4. For the slave and master in Eqs.(5.1) and (5.2) with $n = \tilde{n}$, if the slave and master flows satisfy

$$\mathbf{x}(t) - \mathbf{B}\tilde{\mathbf{x}}(t) = \mathbf{c} \text{ (constant)} \tag{5.12}$$

with a constant diagonal matrix $\mathbf{B} = \mathrm{diag}(b_1, b_2, \ldots, b_n)$ and a constant vector $\mathbf{c} = (c_1, c_2, \ldots, c_n)^{\mathrm{T}}$ for $t \in [t_{m_1}, t_{m_2}]$, then the slave and master systems are equi-dimensionally synchronized in such a affine transformation sense. If $t_{m+1} \to \infty$, the synchronization of the slave and master systems is absolutely and equi-dimensionally synchronized in the linear sense for time $t \in [t_{m_1}, \infty)$. Three important synchronizations are also given as follows.

(i) If $\mathbf{c} = \mathbf{0}$ and $b_i = 1$ $(i = 1, 2, \ldots, n)$, the synchronization of the slave and master systems is called an identical synchronization.
(ii) If $\mathbf{c} = \mathbf{0}$ and $b_i = -1$ $(i = 1, 2, \ldots, n)$, the synchronization of the slave and master systems is called an anti-symmetric synchronization.
(iii) If $\mathbf{c} = \mathbf{0}$, and $b_i \in \{1, -1\}$ $(i = 1, 2, \ldots, n)$, the synchronization of the slave and master systems is called a mixed, identical and anti-symmetric synchronization.

Notice that the matrix $\mathbf{B}$ can be a full matrix. To extend the above idea, new variables are introduced as

$$\begin{aligned} &z_j = \varphi_j(\mathbf{x}(t), \tilde{\mathbf{x}}(t), t, \boldsymbol{\lambda}_j) \text{ for } j \in \mathbb{L}, \\ &\text{or } \mathbf{z} = \varphi(\mathbf{x}(t), \tilde{\mathbf{x}}(t), t, \boldsymbol{\lambda}). \end{aligned} \tag{5.13}$$

If $z_j = c_j$ (const) or $z_j = 0$, Equation (5.13) can be used as the constraint condition in Eq.(5.4). If the slave and master systems are not synchronized, the new variables $(z_j \neq c_j, j = 1, 2, \ldots, l)$ will change with time t. The corresponding time-change rate is given by

$$\begin{aligned}
\dot{z}_j &= \frac{d}{dt}\varphi_j(\mathbf{x}(t), \tilde{\mathbf{x}}(t), \boldsymbol{\lambda}_j) = \sum\nolimits_{i=1}^{n} \frac{\partial \varphi_j}{\partial x_i}\dot{x}_i + \sum\nolimits_{k=1}^{\tilde{n}} \frac{\partial \varphi_j}{\partial \tilde{x}_k}\dot{\tilde{x}}_k + \frac{\partial \varphi_j}{\partial t} \\
&= \sum\nolimits_{i=1}^{n} \frac{\partial \varphi_j}{\partial x_i} F_i + \sum\nolimits_{k=1}^{\tilde{n}} \frac{\partial \varphi_j}{\partial \tilde{x}_k}\tilde{F}_k + \frac{\partial \varphi_j}{\partial t}, (j = 1, 2, \ldots, l); \text{ or} \\
\dot{\mathbf{z}} &= \frac{d}{dt}\boldsymbol{\varphi}(\mathbf{x}(t), \tilde{\mathbf{x}}(t), \boldsymbol{\lambda}) = \frac{\partial \boldsymbol{\varphi}}{\partial \mathbf{x}} \cdot \dot{\mathbf{x}}(t) + \frac{\partial \boldsymbol{\varphi}}{\partial \tilde{\mathbf{x}}} \cdot \dot{\tilde{\mathbf{x}}}(t) + \frac{\partial \boldsymbol{\varphi}}{\partial t} \\
&= \frac{\partial \boldsymbol{\varphi}}{\partial \mathbf{x}} \cdot \mathbf{F} + \frac{\partial \boldsymbol{\varphi}}{\partial \tilde{\mathbf{x}}} \cdot \tilde{\mathbf{F}} + \frac{\partial \boldsymbol{\varphi}}{\partial t}.
\end{aligned} \tag{5.14}$$

If the slave and master systems are continuous, the time-change rate of the new variables for the constraint conditions in Eq.(5.4) should be zero, i.e., $\dot{z}_j = 0\ (j \in \mathbb{L})$ or $\dot{\mathbf{z}} = \mathbf{0} \in \mathfrak{R}^l$. However, if the slave and master systems are discontinuous to the constraint conditions, the time-change rate of the new variables for the constraint conditions in Eq.(5.4) may not be zero. To investigate the synchronization, the constraint conditions are considered as boundaries for discontinuous dynamical systems.

The slave and master flows $\mathbf{x}(t)$ and $\tilde{\mathbf{x}}(t)$ are determined by differential equations in Eqs.(5.1) and (5.2). Suppose at least there is a point $\mathbf{x}_m$ at time t_m to satisfy the constraint condition in Eq.(5.3), i.e.,

$$z_m = \varphi(\mathbf{x}(t_m), \tilde{\mathbf{x}}(t_m), t_m, \boldsymbol{\lambda}) = 0. \tag{5.15}$$

For $t > t_m$, the synchronization between the slave and master systems requires the slave and master flows to satisfy the constraint condition in Eq.(5.3). Because the master flow is independent, only the slave flow can be changed for the condition in Eq.(5.3). If the constraint condition in Eq.(5.3) is treated as a super-surface, the slave system should be switched to the super-surface. If the slave and master systems are C^r-continuous and differentiable $(r \geq 1)$ to the super-surface, the slave and master flows will pass through the super-surface instead of staying on the super-surface because of the continuity and differentiation of the slave and master flows. Otherwise, on the super-surface, one obtains $\dot{z} = d\varphi/dt = 0$ for all time $t > t_m$ and $d^k\varphi/dt^k = 0$ $(k = 1, 2, \ldots)$. From a theory of discontinuous dynamical system in Luo (2006, 2008), at least the slave system possesses discontinuous vector fields to make the slave and master flows stay on the super-surface, which means that the slave and master systems to the constraint can keep the synchronization on the super-surface. Therefore, the constraints can be used as super-surfaces to investigate the synchronization of slave and master systems.

5.2.2 Generalized synchronization

As discussed in the previous section, if the number of constraints for slave and master systems is over the number of state variables of the slave system (i.e., $l > n$), the slave system is overcontrained under the constraint conditions by the master system. In other words, if all the constraint conditions are satisfied, the master system should be partially constrained also for $n < l \leq n + \tilde{n}$. Otherwise, the constraint conditions cannot be satisfied for the synchronization of the slave and master systems. The overconstrained synchronization for slave and master systems can be defined from Definition 5.3, i.e.,

Definition 5.5. If $l > n$, an $(n : \tilde{n}; l)$-synchronization of the slave and master systems in Eqs.(5.1) and (5.2) in sense of Eq.(5.4) for time $t \in [t_{m_1}, t_{m_2}]$ is said an $(n : \tilde{n}; l)$-overconstrained synchronization.

To make an overconstrained slave system be synchronized with a master system, the flow of the master system should be controlled by the constraints. Generally speaking, the slave system can be partially controlled by some constraints in Eq.(5.4), and the master system can be partially controlled by the rest constraints in Eq.(5.4) as well. For some time intervals, the slave system can be controlled by the master system under the constraints. With time varying, for some time intervals, the master system can also be controlled by the slave system. For this case, it is very difficult to know which one of two systems is a slave or master system. In fact, it is not necessary to distinguish slave and master systems from two dynamical systems. To investigate the synchronization of two or more systems, Definition 5.2 can be generalized as follows.

Definition 5.6. If a flow $\mathbf{x}(t)$of a system in Eq.(5.1) with a flow $\tilde{\mathbf{x}}(t)$ of a system in Eq.(5.2) is constrained by a single constraint in Eq.(5.3) for time $t \in [t_{m_1}, t_{m_2}]$, then the two systems are said to be *synchronized* in the sense of Eq.(5.3) for time $t \in [t_{m_1}, t_{m_2}]$. There are five special cases:

(i) If $t_{m_2} \to \infty$, the two systems are said to be *absolutely* synchronized in the sense of Eq.(5.3) for time$t \in [t_{m_1}, \infty)$.

(ii) If $t_{m_1} \to \infty$, the two systems are said to be *asymptotically* synchronized in the sense of Eq.(5.3).

(iii) For $n = \tilde{n}$, the two *equi-dimensional* systems are said to be synchronized in the sense of Eq.(5.3) for time $t \in [t_{m_1}, t_{m_2}]$.

(iv) For $n = \tilde{n}$ and $t_{m_2} \to \infty$, the two *equi-dimensional* systems are said to be *absolutely* synchronized in the sense of Eq.(5.3) for time $t \in [t_{m_1}, \infty)$.

(v) For $n = \tilde{n}$ and $t_{m_1} \to \infty$, the two *equi-dimensional* systems are said to be *asymptotically* synchronized in the sense of Eq.(5.3).

In an alike fashion, the synchronization of slave and master systems in Definition 5.3 should be generalized for the synchronization of slave and master systems with or without overconstraints.

Definition 5.7. An n-dimensional system in Eq.(5.1) with an $\tilde{n}$-dimensional system in Eq.(5.2) is said to be synchronized with l-constraints (or an l-constraint *synchronization*) for time $t \in [t_{m_1}, t_{m_2}]$ if there are l-linearly independent functions $\varphi_j(\mathbf{x}(t), \tilde{\mathbf{x}}(t), t, \boldsymbol{\lambda}_j)$ ($j \in \mathbb{L}$ and $\mathbb{L} = \{1, 2, \ldots, l\}$ with $l < n + \tilde{n}$) to make two flows $\tilde{\mathbf{x}}(t)$ and $\mathbf{x}(t)$ of the two systems satisfy the constraints in Eq.(5.4) for time $t \in [t_{m_1}, t_{m_2}]$. There are five special cases:

(i) If $t_{m_2} \to \infty$, the two systems are said to be *absolutely* synchronized with l-constraints (or an absolute, l-constraint synchronization) in the sense of Eq.(5.4) for time $t \in [t_{m_1}, \infty)$.

(ii) If $t_{m_1} \to \infty$, the two systems are said to be *asymptotically* synchronized with l-constraints (or an *asymptotic* l-constraint synchronization) in the sense of Eq.(5.4).

(iii) If $n = \tilde{n}$, the two equi-dimensional systems are said to be synchronized with l-constraints in the sense of Eq.(5.4) for time $t \in [t_{m_1}, t_{m_2}]$.

(iv) If $n = \tilde{n}$ and $t_{m_2} \to \infty$, the two equi-dimensional systems are said to be *absolutely* synchronized with l-constraints in the sense of Eq.(5.4) for time $t \in [t_{m_1}, \infty)$.

(v) If $n = \tilde{n}$ and $t_{m_1} \to \infty$, the two equi-dimensional systems are said to be *asymptotically* synchronized with l-constraints in the sense of Eq.(5.4) for time $t \in [t_{m_1}, \infty)$.

From the above definition, the number of constraints in Eq.(5.4) can be greater than the dimension number of state space for one of the two systems in Eqs.(5.1) and (5.2) (i.e., $l > n$ or $l > \tilde{n}$). For such a case, one cannot control only one of the two systems to make them be synchronized through the constraints. In other words, one must control both of two systems to make the corresponding synchronization occur. Of course, if $l \leq n$ or $l \leq \tilde{n}$, one can control only one of two systems to make them be synchronized through the constraints in Eq.(5.4). If the constraint functions $\varphi_j(\mathbf{x}(t), \tilde{\mathbf{x}}(t), t, \boldsymbol{\lambda}_j)$ (for all $j \in \mathbb{L}$) are time-independent for $l = n + \tilde{n}$, Equation (5.4) will give a set of fixed values of $\mathbf{x}^*$ and $\tilde{\mathbf{x}}^*$, which are independent of time. The constraints yield the values-fixed, static points in the resultant sate space. To make the two systems in Eqs.(5.1) and (5.2) be synchronized at the static points in phase space, such a synchronization can be called *a static synchronization* of two systems in Eqs.(5.1) and (5.2). For $l > n + \tilde{n}$, the time-independent constraints in Eq.(5.4) will give the statically overconstrained synchronization, which may not be meaningful for practical problems. Such a case will not be discussed any more. If the constraint functions of $\varphi_j(\mathbf{x}(t), \tilde{\mathbf{x}}(t), t, \boldsymbol{\lambda}_j)$ (for all $j \in \mathbb{L}$) are time-dependent for $l = n + \tilde{n}$, Equation (5.4) will give a flow of $\mathbf{x}^*$ and $\tilde{\mathbf{x}}^*$ relative to time. To eliminate time, the constraints in Eq.(5.4) give a one-dimensional flow in the resultant phase space. If the time-dependent constraint functions of $\varphi_j(\mathbf{x}(t), \tilde{\mathbf{x}}(t), t, \boldsymbol{\lambda}_j)$ (for all $j \in \mathbb{L}$) are of l-dimensions with $l = n + \tilde{n} + 1$, Equation (5.4) will give a set of fixed values of $\mathbf{x}^*$ and $\tilde{\mathbf{x}}^*$ at a specific time t^* in the resultant phase space, which is an instantaneous fixed point only at time t^*. For this case, it is very difficult for the two systems to be synchronized for such an instantaneous point. Such a case may not be too meaningful, which will not be discussed. Therefore, the following two definitions are given to describe the afore-discussed cases.

Definition 5.8. An n-dimensional system in Eq.(5.1) with an $\tilde{n}$-dimensional system in Eq.(5.2) is said to be statically synchronized with l-constraints (or *a static synchronization*) for time $t \in [t_{m_1}, t_{m_2}]$ if there are l-linearly independent and *time-independent* functions $\varphi_j(\mathbf{x}(t), \tilde{\mathbf{x}}(t), \boldsymbol{\lambda}_j)$ ($j \in \mathbb{L}$ and $\mathbb{L} = \{1, 2, \ldots, l\}$ with $l = n + \tilde{n}$) to make two flows $\tilde{\mathbf{x}}(t)$ and $\mathbf{x}(t)$ of the two systems satisfy the constraints in Eq.(5.4) for time $t \in [t_{m_1}, t_{m_2}]$. There are two special cases:

(i) If $t_{m_2} \to \infty$, the two systems are said to be *absolutely* and *statically* synchronized with l-constraints (or an *absolute* and *static* synchronization) in the sense of Eq.(5.4) for time $t \in [t_{m_1}, \infty)$.

(ii) If $t_{m_1} \to \infty$, the two systems are said to be *asymptotically* and *statically* synchronized with l-constraints (or an *asymptotic and static* synchronization) in the sense of Eq.(5.4).

Definition 5.9. An n-dimensional system in Eq.(5.1) with an $\tilde{n}$-dimensional system in Eq.(5.2) is said to be synchronized with a one-dimensional constraint-flow (or a *$1-D$ constraint-flow synchronization*) for time $t \in [t_{m_1}, t_{m_2}]$ if there are l-linearly independent and *time-dependent* functions $\varphi_j(\mathbf{x}(t), \tilde{\mathbf{x}}(t), t, \boldsymbol{\lambda}_j)$ ($j \in \mathbb{L}$ and $\mathbb{L} = \{1, 2, \ldots, l\}$ with $l = n + \tilde{n}$) to make two flows $\tilde{\mathbf{x}}(t)$ and $\mathbf{x}(t)$ of the two systems satisfy constraints in Eq.(5.4) for time $t \in [t_{m_1}, t_{m_2}]$. Two special cases are given as follows:

(i) If $t_{m_2} \to \infty$, the two systems are said to be *absolutely* synchronized with a one-dimensional constraint-flow (or an *absolute, $1-D$ constraint-flow synchronization*) in the sense of Eq.(5.4) for time $t \in [t_{m_1}, \infty)$.

(ii) If $t_{m_1} \to \infty$, the two systems are said to be *asymptotically* synchronized with a one-dimensional constraint-flow (an *asymptotic, $1-D$ constraint-flow synchronization*) in the sense of Eq.(5.4).

5.2.3 Resultant dynamical systems

From the theory of discontinuous dynamical systems in Luo (2006, 2008), the synchronization of two or more dynamical systems with specific constraints can be investigated through a resultant dynamical system. The constraint conditions can be considered as a set of super-surfaces. If the resultant system to the constraints is discontinuous, the resultant discontinuous dynamical system can be adjusted on both sides of each super-surface for such synchronization. For doing so, a set of new state variables for the resultant discontinuous system will be introduced, and the sub-domains and boundaries relative to the constraints will be presented. For the synchronization of slave and master systems on the constraint surfaces, only the slave system can be adjusted, and the master system cannot be adjusted. In other words, the slave system can be controlled in order to make it be synchronized with the master system through the constraints. That is, the slave system can be expressed by discontinuous vector fields to all the constraint surfaces for such synchronization, but the master system should keep a continuous vector field to such constraint surfaces. However, for a resultant system formed by two systems with constraints, one can adjust two dynamical systems to make them be synchronized on the constraint conditions in general.

A new vector of state variables of two dynamical systems in Eqs.(5.1) and (5.2) is introduced as

$$\mathbf{X} = (\mathbf{x};\tilde{\mathbf{x}})^{\mathrm{T}} = (x_1, x_2, \ldots, x_n; \tilde{x}_1, \tilde{x}_2, \ldots, \tilde{x}_{\tilde{n}})^{\mathrm{T}} \in \mathfrak{R}^{n+\tilde{n}} \tag{5.16}$$

The notation $(\cdot;\cdot) \equiv (\cdot,\cdot)$ is just for a combined vector of state vectors of two dynamical systems. From the constraint condition in Eq.(5.3), a constraint boundary for the discontinuous description of the synchronization of two dynamical systems in Eqs.(5.1) and (5.2) can be defined, and the corresponding domains separated by such a constraint boundary can be obtained.

Definition 5.10. A constraint boundary in an $(n+\tilde{n})$-dimensional phase space for the synchronization of two dynamical systems in Eqs.(5.1) and (5.2) to constraint condition in Eq.(5.3) is defined as

$$\begin{aligned} \partial\Omega_{12} &= \bar{\Omega}_1 \cap \bar{\Omega}_2 \\ &= \left\{ \mathbf{X}^{(0)} \left| \begin{array}{l} \varphi(\mathbf{X}^{(0)}, t, \boldsymbol{\lambda}) \equiv \varphi(\mathbf{x}^{(0)}(t), \tilde{\mathbf{x}}^{(0)}(t), t, \boldsymbol{\lambda}) = 0, \\ \varphi \text{ is } C^r\text{-continuous } (r \geq 1) \end{array} \right. \right\} \\ &\subset \mathfrak{R}^{n+\tilde{n}-1}; \end{aligned} \tag{5.17}$$

and two corresponding domains for a resultant system of two dynamical systems in Eqs.(5.1) and (5.2) are defined as

$$\begin{aligned} \Omega_1 &= \left\{ \mathbf{X}^{(1)} \left| \begin{array}{l} \varphi(\mathbf{X}^{(1)}, t, \boldsymbol{\lambda}) \equiv \varphi(\mathbf{x}^{(1)}(t), \tilde{\mathbf{x}}^{(1)}(t), t, \boldsymbol{\lambda}) > 0, \\ \varphi \text{ is } C^r\text{-continuous}(r \geq 1) \end{array} \right. \right\} \subset \mathfrak{R}^{n+\tilde{n}}; \\ \Omega_2 &= \left\{ \mathbf{X}^{(2)} \left| \begin{array}{l} \varphi(\mathbf{X}^{(2)}, t, \boldsymbol{\lambda}) \equiv \varphi(\mathbf{x}^{(2)}(t), \tilde{\mathbf{x}}^{(2)}(t), t, \boldsymbol{\lambda}) < 0, \\ \varphi \text{ is } C^r\text{-continuous}(r \geq 1) \end{array} \right. \right\} \subset \mathfrak{R}^{n+\tilde{n}}. \end{aligned} \tag{5.18}$$

On the two domains, the resultant system of two dynamical systems is discontinuous to the constraint boundary, defined by

$$\dot{\mathbf{X}}^{(\alpha)} = \mathbb{F}^{(\alpha)}(\mathbf{X}^{(\alpha)}, t, \boldsymbol{\pi}^{(\alpha)}) \text{ in } \Omega_\alpha \ (\alpha = 1, 2), \tag{5.19}$$

where $\mathbb{F}^{(\alpha)} = (\mathbf{F}^{(\alpha)}; \tilde{\mathbf{F}})^{\mathrm{T}} = (F_1^{(\alpha)}, F_2^{(\alpha)}, \ldots, F_n^{(\alpha)}; \tilde{F}_1, \tilde{F}_2, \ldots, \tilde{F}_{\tilde{n}})^{\mathrm{T}}$ and $\boldsymbol{\pi}^{(\alpha)} = (\mathbf{p}_\alpha, \tilde{\mathbf{p}})^{\mathrm{T}}$. Suppose there is a vector field $\mathbb{F}^{(0)}(\mathbf{X}^{(0)}, t, \boldsymbol{\lambda})$ on the constraint boundary with $\varphi(\mathbf{X}^{(0)}, t, \boldsymbol{\lambda}) = 0$, and the corresponding dynamical system on such a boundary is expressed by

$$\dot{\mathbf{X}}^{(0)} = \mathbb{F}^{(0)}(\mathbf{X}^{(0)}, t, \boldsymbol{\lambda}) \text{ on } \partial\Omega_{12}. \tag{5.20}$$

The domains Ω_α ($\alpha = 1,2$) are separated by the constraint boundary $\partial\Omega_{12}$, as shown in Fig.5.1. For a point $(\mathbf{x}^{(1)},\tilde{\mathbf{x}}^{(1)}) \in \Omega_1$ at time t, one obtains $\varphi(\mathbf{x}^{(1)},\tilde{\mathbf{x}}^{(1)},t,\boldsymbol{\lambda}) > 0$. For a point $(\mathbf{x}^{(2)},\tilde{\mathbf{x}}^{(2)}) \in \Omega_2$ at time t, one obtains $\varphi(\mathbf{x}^{(2)},\tilde{\mathbf{x}}^{(2)},t,\boldsymbol{\lambda}) < 0$. However, on the boundary $(\mathbf{x}^{(0)},\tilde{\mathbf{x}}^{(0)}) \in \partial\Omega_{12}$ at time t, the constraint condition for synchronization should be satisfied (i.e., $\varphi(\mathbf{x}^{(0)},\tilde{\mathbf{x}}^{(0)},t,\boldsymbol{\lambda}) = 0$). If the constraint condition is time-independent, the constraint boundary determined by the constraint condition is invariant. If there are many constraint conditions for the synchronization of two dynamical systems, the above definition can be extended.

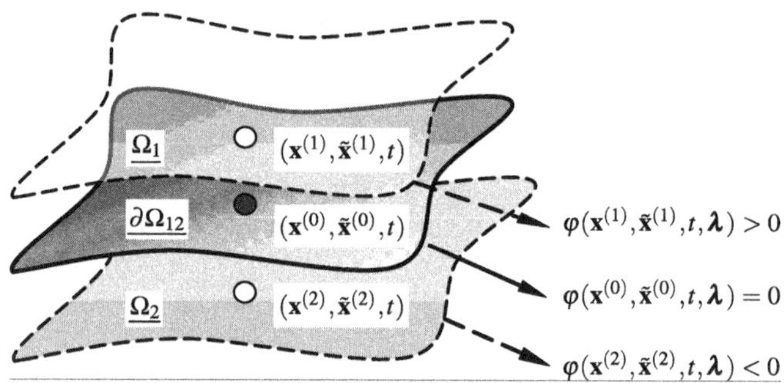

Fig. 5.1 Constraint boundary and domains in $(n+\tilde{n})$-dimensional state space.

Definition 5.11. The jth-constraint boundary in an $(n+\tilde{n})$-dimensional phase space for the synchronization of two dynamical systems in Eqs.(5.1) and (5.2), relative to the jth-constraint of the constraint conditions in Eq.(5.4), is defined as

$$\begin{aligned}\partial\Omega_{12(j)} &= \bar{\Omega}_{(1,j)} \cap \bar{\Omega}_{(2,j)} \\ &= \left\{ \mathbf{X}^{(0,j)} \left| \begin{array}{l} \varphi_j(\mathbf{X}^{(0,j)},t,\boldsymbol{\lambda}_j) \equiv \varphi_j(\mathbf{x}^{(0,j)}(t),\tilde{\mathbf{x}}^{(0,j)}(t),t,\boldsymbol{\lambda}_j) = 0, \\ \varphi_j \text{ is } C^{r_j}\text{-continuous}(r_j \geq 1) \end{array} \right. \right\} \\ &\subset \mathfrak{R}^{n+\tilde{n}-1};\end{aligned} \tag{5.21}$$

and two domains pertaining to the jth-boundary for a resultant system of two dynamical systems in Eqs.(5.1) and (5.2) are defined as

$$\begin{aligned}\Omega_{(1,j)} &= \left\{ \mathbf{X}^{(1,j)} \left| \begin{array}{l} \varphi_j(\mathbf{X}^{(1,j)},t,\boldsymbol{\lambda}_j) \equiv \varphi_j(\mathbf{x}^{(1,j)}(t),\tilde{\mathbf{x}}^{(1,j)}(t),t,\boldsymbol{\lambda}_j) > 0, \\ \varphi_j \text{ is } C^{r_j}\text{-continuous}(r_j \geq 1) \end{array} \right. \right\} \\ &\subset \mathfrak{R}^{n+\tilde{n}}; \\ \Omega_{(2,j)} &= \left\{ \mathbf{X}^{(2,j)} \left| \begin{array}{l} \varphi_j(\mathbf{X}^{(2,j)},t,\boldsymbol{\lambda}_j) \equiv \varphi_j(\mathbf{x}^{(2,j)}(t),\tilde{\mathbf{x}}^{(2,j)}(t),t,\boldsymbol{\lambda}_j) < 0, \\ \varphi_j \text{ is } C^{r_j}\text{-continuous}(r_j \geq 1) \end{array} \right. \right\} \\ &\subset \mathfrak{R}^{n+\tilde{n}}.\end{aligned} \tag{5.22}$$

On the two domains relative to the jth-constraint boundary, a discontinuous resultant system of two dynamical systems in Eqs.(5.1) and (5.2) with the jth-constraint in Eq.(5.4) is defined by

$$\dot{\mathbf{X}}^{(\alpha_j,j)} = \mathbb{F}^{(\alpha_j,j)}(\mathbf{X}^{(\alpha_j,j)},t,\boldsymbol{\pi}_j^{(\alpha_j)}) \text{ in } \Omega_{(\alpha_j,j)}, \tag{5.23}$$

where $\mathbb{F}^{(\alpha_j,j)} = (\mathbf{F}^{(\alpha_j,j)};\tilde{\mathbf{F}}^{(\alpha_j,j)})^{\mathrm{T}} = (F_1^{(\alpha_j,j)},\ldots,F_n^{(\alpha_j,j)};\tilde{F}_1^{(\alpha_j,j)},\ldots,\tilde{F}_{\tilde{n}}^{(\alpha_j,j)})^{\mathrm{T}}$ and $\boldsymbol{\pi}_j^{(\alpha_j)} = (\mathbf{p}_j^{(\alpha_j)},\tilde{\mathbf{p}}_j^{(\alpha_j)})^{\mathrm{T}}$. Suppose there is a vector field of $\mathbb{F}^{(0,j)}(\mathbf{X}^{(0,j)},t,\boldsymbol{\lambda}_j)$ on the jth-constraint boundary with $\varphi_j(\mathbf{X}^{(0,j)},t,\boldsymbol{\lambda}_j) = 0$, and the corresponding dynamical system on the jth-boundary is expressed by

$$\dot{\mathbf{X}}^{(0,j)} = \mathbb{F}^{(0,j)}(\mathbf{X}^{(0,j)},t,\boldsymbol{\lambda}_j) \text{ on } \partial\Omega_{12(j)}. \tag{5.24}$$

Since l-constraint conditions are linearly independent, any two boundaries are intersected each other. Consider two constraint boundaries of $\partial\Omega_{12(j)}$ and $\partial\Omega_{12(k)}$ for synchronization. The intersection of the two constraint boundaries is given by

$$\partial\Omega_{12(jk)} = \partial\Omega_{12(j)} \cap \partial\Omega_{12(k)} \subset \Re^{n+\tilde{n}-2} \tag{5.25}$$

and the corresponding domain in phase space is separated into four sub-domains

$$\Omega_{(\alpha_j\alpha_k,jk)} = \Omega_{(\alpha_j,j)} \cap \Omega_{(\alpha_k,k)} \subset \Re^{n+\tilde{n}} \text{ for } j,k = 1,2,\ldots \text{ and } \alpha_j,\alpha_k = 1,2. \tag{5.26}$$

Such a partition of the domain in state space for a resultant system of two dynamical systems is sketched in Fig. 5.2. The intersection of the two constraint boundaries in state space for a resultant system of two dynamical systems is depicted by an $(n+\tilde{n}-2)$-manifold, depicted by a dark curve. For the l-linearly independent constraints, the state space partition can be completed via such l-linearly independent constraint boundaries. Based on the l-constraint conditions, the corresponding intersection of boundaries is

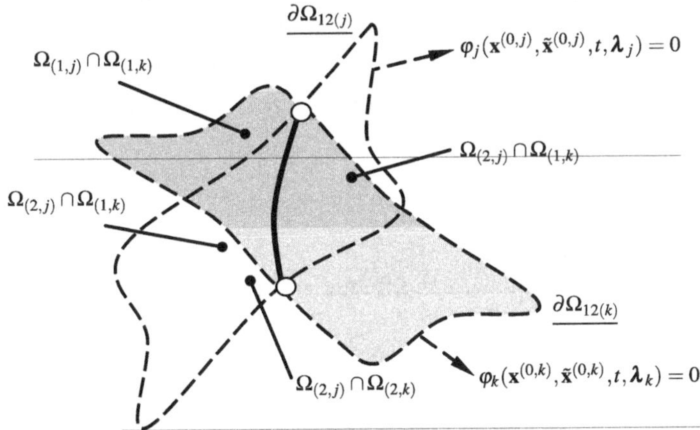

Fig. 5.2 An intersection of two boundaries with $\varphi_j = 0$ and $\varphi_k = 0$ for $j,k \in \mathbb{L}$ and $j \neq k$.

$$\partial\Omega_{12(\mathbf{J})} = \cap_{j=1}^{l} \partial\Omega_{12(j)} \subset \mathfrak{R}^{n+\tilde{n}-l} \tag{5.27}$$

which gives an $(n+\tilde{n}-l)$-dimensional manifold. Consider the synchronization of the slave and master systems for discussion. If $n = l$, the intersection manifold of the constraints is an $\tilde{n}$-dimensional state space. In other words, the slave system can be completely controlled through the n-constraints to be synchronized with the master system. From the l-constraint conditions in Eq.(5.4), the domain in $(n+\tilde{n})$-dimensional state space is partitioned into many sub-domains for the resultant system of two dynamical systems, i.e.,

$$\Omega_{\alpha} = \Omega_{(\alpha_1\alpha_2\cdots\alpha_l)} = \cap_{j=1}^{l} \Omega_{(\alpha_j,j)} \subset \mathfrak{R}^{n+\tilde{n}} \text{ for } \alpha_j = 1,2 \text{ and } j \in \mathbb{L}. \tag{5.28}$$

The total domain $\mho = \cup_{j=1}^{l} \cup_{\alpha_j=1}^{2} \left(\cap_{j=1}^{l} \Omega_{(\alpha_j,j)}\right) \subset \mathfrak{R}^{n+\tilde{n}}$ is a union of all the sub domains.

From the foregoing description of a resultant dynamical system, the synchronization of two systems under constraints can be investigated through such a resultant dynamical system with the constraint boundaries as in Luo (2006, 2008). The constraint boundaries can be either of one-side or of two sides. If the resultant system for the synchronization of two systems can be defined (or can exist) in one of the two sub-domains only, such a constraint boundary is called one-side boundary. Otherwise, the constraint boundary is called two-side constraint boundary. If a flow of the resultant system can approach to a constraint flow on the constraint boundaries as $t \to \infty$, for such a case, the synchronization of two systems to the constraint boundaries is asymptotic.

5.2.4 Metric functionals

For a better description of the synchronicity of two dynamical systems, a metric functional based on the constraint boundaries can be introduced. The metric functionals are a set of non-negative functions of constraint functions. For a constraint function in Eq.(5.3), the definition of a metric function is given as follows:

Definition 5.12. A metric function for two dynamical systems in Eqs.(5.1) and (5.2) is defined by a non-negative functional

$$V(\mathbf{X},t,\boldsymbol{\lambda}) = f(\varphi(\mathbf{X},t,\boldsymbol{\lambda})) \tag{5.29}$$

with the following conditions

(i) $V(\mathbf{X},t,\boldsymbol{\lambda})$ is continuous and differentiable for time t and $\mathbf{X}$,

(ii) for $\mathbf{X} = \mathbf{X}^{(\alpha)} \in \Omega_{\alpha}$, $V(\mathbf{X}^{(\alpha)},t,\boldsymbol{\lambda}) > 0$,

(iii) only for $\mathbf{X} = \mathbf{X}^{(0)} \in \partial\Omega_{12}$, $V(\mathbf{X}^{(0)},t,\boldsymbol{\lambda}) = 0$;

and the time-change rates of the metric functional are for $k = 1,2,\ldots$

$$V^{(k)}(\mathbf{X},t,\boldsymbol{\lambda}) = \frac{d^k}{dt^k}V(\mathbf{X},t,\boldsymbol{\lambda}) = \frac{d}{dt}\left[V^{(k-1)}(\mathbf{X},t,\boldsymbol{\lambda})\right] \tag{5.30}$$

where

$$V^{(0)}(\mathbf{X},t,\boldsymbol{\lambda}) = V(\mathbf{X},t,\boldsymbol{\lambda}). \tag{5.31}$$

From the foregoing definition, consider a metric functional as

$$V(\mathbf{X},t,\boldsymbol{\lambda}) = \frac{1}{2}\left[\varphi(\mathbf{X},t,\boldsymbol{\lambda})\right]^2. \tag{5.32}$$

For two dynamical systems in Eqs.(5.1) and (5.2), from the foregoing definition, one obtains

$$V(\mathbf{X}^{(\alpha)},t,\boldsymbol{\lambda}) = \frac{1}{2}\left[\varphi(\mathbf{X}^{(\alpha)},t,\boldsymbol{\lambda})\right]^2 > 0 \tag{5.33}$$

for $\mathbf{X}^{(\alpha)} \in \Omega_\alpha$ and

$$V(\mathbf{X}^{(0)},t,\boldsymbol{\lambda}) = \frac{1}{2}\left[\varphi(\mathbf{X}^{(0)},t,\boldsymbol{\lambda})\right]^2 = 0 \tag{5.34}$$

for $\mathbf{X}^{(0)} \in \partial\Omega_{12}$. From Eq.(5.34), one obtains $\varphi(\mathbf{X}^{(0)},t,\boldsymbol{\lambda}) = 0$. So the conditions in Eq.(5.34) is equivalent to the constraint condition in Eq.(5.3). The time-change rate of the metric function is

$$\begin{aligned} V^{(1)}(\mathbf{X},t,\boldsymbol{\lambda}) &= \frac{d\varphi}{dt}\varphi = \varphi\left[\nabla_{\mathbf{X}}\varphi\cdot\dot{\mathbf{X}} + \frac{\partial\varphi}{\partial t}\right] = \varphi\left[\nabla_{\mathbf{X}}\varphi\cdot\mathbb{F}(\mathbf{X},t,\boldsymbol{\lambda}) + \frac{\partial\varphi}{\partial t}\right] \\ &= \varphi\left[\nabla_{\mathbf{x}}\varphi\cdot\dot{\mathbf{x}} + \nabla_{\tilde{\mathbf{x}}}\varphi\cdot\dot{\tilde{\mathbf{x}}} + \frac{\partial\varphi}{\partial t}\right] = \varphi\left[\nabla_{\mathbf{x}}\varphi\cdot\mathbf{F} + \nabla_{\tilde{\mathbf{x}}}\varphi\cdot\tilde{\mathbf{F}} + \frac{\partial\varphi}{\partial t}\right]. \end{aligned} \tag{5.35}$$

In fact, the metric functional in Eq.(5.29) can be also defined by the other non-negative functions. For instance, using the absolute function, we have

$$V(\mathbf{X},t,\boldsymbol{\lambda}) = |\varphi(\mathbf{X},t,\boldsymbol{\lambda})|. \tag{5.36}$$

In Definition 5.12, the metric functional for the single constraint is presented. For multiple constraints in Eq.(5.4), the corresponding metric functionals can be defined in order to describe the synchronicity of two dynamical systems under such constraints. From Eq.(5.4), a set of metric functionals for two dynamical systems in Eq.(5.1) and (5.2) are introduced.

Definition 5.13. The jth- individual metric functional and a resultant metric functional for two dynamical systems in Eq.(5.1) and (5.2) are

$$\begin{aligned} {}^{(n:\tilde{n})}V_j(\mathbf{X}_j,t,\boldsymbol{\lambda}_j) &= f_j(\varphi_j(\mathbf{X},t,\boldsymbol{\lambda}_j)) \text{ for } j \in \mathbb{L} = \{1,2,\ldots,l\} \\ {}^{(n:\tilde{n}:l)}V(\mathbf{X},t,\boldsymbol{\lambda}) &= \textstyle\sum_{j=1}^{l} {}^{(n:\tilde{n})}V_j(\mathbf{X},t,\boldsymbol{\lambda}_j) \\ &= \textstyle\sum_{j=1}^{l} f_j(\varphi_j(\mathbf{X},t,\boldsymbol{\lambda}_j)) \text{ for } l \in \{1,2,\ldots,n\} \end{aligned} \tag{5.37}$$

with the following conditions for $(j = 1,2,\ldots,l)$

(i) ${}^{(n:\tilde{n})}V_j(\mathbf{X},t,\boldsymbol{\lambda}_j)$ is continuous and differentiable for time t and $\mathbf{X}$,

(ii) for $\mathbf{X} = \mathbf{X}^{(\alpha_j:j)} \in \Omega_{(\alpha_j,j)}$, ${}^{(n:\tilde{n})}V_j(\mathbf{X}^{(\alpha_j:j)},t,\boldsymbol{\lambda}_j) > 0$,

(iii) for $\mathbf{X} = \mathbf{X}^{(0,j)} \in \partial\Omega_{12(j)}$, ${}^{(n:\tilde{n})}V_j(\mathbf{X}^{(0,j)},t,\boldsymbol{\lambda}_j) = 0$;

and the time-change rates are for $k = 1,2,\ldots$

$$ {}^{(n:\tilde{n})}V_j^{(k)}(\mathbf{X},t,\boldsymbol{\lambda}) = \frac{d^k}{dt^k}{}^{(n:\tilde{n})}V_j(\mathbf{X},t,\boldsymbol{\lambda}) = \frac{d}{dt}\left[{}^{(n:\tilde{n})}V_j^{(k-1)}(\mathbf{X},t,\boldsymbol{\lambda})\right]. \tag{5.38} $$

For the jth-metric functional relative to the jth-constraint, it is similar to the metric functional pertaining to the single constraint, as discussed before. Herein, consider a resultant metric function as

$$ {}^{(n:\tilde{n}:l)}V(\mathbf{X},t,\boldsymbol{\lambda}) = \frac{1}{2}\sum\nolimits_{j=1}^{l}\left[{}^{(n:\tilde{n})}V_j(\mathbf{X},t,\boldsymbol{\lambda}_j)\right]^2 \text{ for } l = 1,2,\ldots,n \tag{5.39} $$

For two dynamical systems in Eqs.(5.1) and (5.2), from the foregoing definition, one obtains

$$ {}^{(n:\tilde{n}:l)}V(\mathbf{X}^{(\alpha)},t,\boldsymbol{\lambda}) = \frac{1}{2}\sum\nolimits_{j=1}^{l}\left[\varphi_j(\mathbf{X}^{(\alpha_j:j)},t,\boldsymbol{\lambda}_j)\right]^2 > 0 \text{ for } l = 1,2,\ldots,n \tag{5.40} $$

for $\mathbf{X}^{(\alpha_j,j)} \in \Omega_{(\alpha_j,j)}$ $(j = 1,2,\ldots,l;\alpha_j = 1,2), \mathbf{X}^{(\alpha)} \in \Omega_{(\boldsymbol{\alpha})}$ and

$$ {}^{(n:\tilde{n}:l)}V(\mathbf{X}^{(0)},t,\boldsymbol{\lambda}) = \frac{1}{2}\sum\nolimits_{j=1}^{l}\left[\varphi_j(\mathbf{X}^{(0,j)},t,\boldsymbol{\lambda}_j)\right]^2 = 0 \text{ for } l = 1,2,\cdots,n \tag{5.41} $$

for $\mathbf{X}^{(0,j)} \in \partial\Omega_{12(j)}$ $(j = 1,2,\ldots,l;\alpha_j = 1,2)$ and $\mathbf{X}^{(0)} \in \partial\Omega_{12(\mathbf{J})}$. From Eq.(5.41), one obtains

$$ \varphi_j(\mathbf{X}^{(0,j)},t,\boldsymbol{\lambda}_j) = 0 \text{ for } j = 1,2,\ldots,l \text{ and } l = 1,2,\ldots,n. \tag{5.42} $$

So the conditions in Eq.(5.42) is equivalent to that in Eq.(5.4).

Similarly, we discuss another metric functional as

$$ {}^{(n:\tilde{n}:l)}V(\mathbf{X},t,\boldsymbol{\lambda}) = \sum\nolimits_{j=1}^{l}|\varphi_j(\mathbf{X}^{(\alpha_j,j)},t,\boldsymbol{\lambda}_j)| \text{ for } l = 1,2,\ldots,n. \tag{5.43} $$

For the slave and master systems in Eqs.(5.1) and (5.2), from the foregoing definitions, one obtains

$$ {}^{(n:\tilde{n}:l)}V(\mathbf{X}^{(\alpha)},t,\boldsymbol{\lambda}) = \sum\nolimits_{j=1}^{l}|\varphi_j(\mathbf{X}^{(\alpha_j:j)},t,\boldsymbol{\lambda}_j)| > 0 \text{ for } l = 1,2,\ldots,n \tag{5.44} $$

for $\mathbf{X}^{(\alpha_j,j)} \in \Omega_{(\alpha_j,j)}(j = 1,2,\ldots,l), \mathbf{X}^{(\alpha)} \in \Omega_{(\alpha)}$ and

$$ {}^{(n:l)}V(\mathbf{X}^{(0)},t,\boldsymbol{\lambda}) = \sum\nolimits_{j=1}^{l}|\varphi_j(\mathbf{X}^{(0:j)},t,\boldsymbol{\lambda}_j)| = 0 \text{ for } l = 1,2,\ldots,n \tag{5.45} $$

for $\mathbf{X}^{(0,j)} \in \partial\Omega_{12(j)}$ $(j = 1,2,\ldots,l)$ and $\mathbf{X}^{(0)} \in \partial\Omega_{12(\mathbf{J})}$. From Eq.(5.45), one obtains

$$\varphi_j(\mathbf{X}^{(0:j)}, t, \boldsymbol{\lambda}_j) = 0 \text{ for } j = 1, 2, \ldots, l \text{ and } l = 1, 2, \ldots, n. \tag{5.46}$$

So the conditions in Eq.(5.46) are equivalent to those constraint conditions in Eq.(5.4). The resultant metric functional is a kind of a generalized Lyanunov function, which cannot be use to determine the synchronization in general. However, the resultant metric functional can be only used to two special cases: (i) full synchronization, and (ii) full desynchronization. The detailed discussion about the synchronization and desynchronization will be presented in the following section. For any general case of synchronization, all the individual metric functions should be adopted. As afore-discussed, each individual metric functional in Eq.(5.37) can be the same as the resultant metric functional. However, those individual metric functionals will provide more possibility for one to discuss the synchronicity of two dynamical systems.

5.3 Single-constraint synchronization

In this section, under a single constraint, the synchronicity of two dynamical systems will be discussed. Based on the metric functional in Eq.(5.29), the proper definitions relative to the synchronicity of two dynamical systems in Eqs.(5.1) and (5.2) to a constraint in Eq.(5.3) will be presented. The necessary and sufficient conditions for the synchronicity of two dynamical systems to the constraint are developed.

5.3.1 *Synchronicity*

Before discussing the synchronicity of two dynamical systems to the constraint boundary, the neighborhood of the constraint boundary should be introduced through a typical point on such a constraint boundary for time t_m. For any small $\varepsilon > 0$, the neighborhood of a constraint boundary is defined as follows.

Definition 5.14. For $\mathbf{X}_m^{(\alpha)} \in \Omega_\alpha$ $(\alpha \in \{1,2\})$ and $\mathbf{X}_m^{(0)} \in \partial\Omega_{12}$ at time $t_m, \mathbf{X}_m^{(\alpha)} = \mathbf{X}_m^{(0)}$. For any small $\varepsilon > 0$, there is a time interval $[t_{m-\varepsilon}, t_m)$ or $(t_m, t_{m+\varepsilon}]$. The ε-neighborhood of the constraint boundary $\partial\Omega_{12}$ is defined as

$$\begin{aligned}\Omega_\alpha^{-\varepsilon} &= \left\{\mathbf{X}^{(\alpha)} \middle| \|\mathbf{X}^{(\alpha)}(t) - \mathbf{X}_m^{(0)}\| \leq \delta, \delta > 0, t \in [t_{m-\varepsilon}, t_m)\right\}, \\ \Omega_\alpha^{+\varepsilon} &= \left\{\mathbf{X}^{(\alpha)} \middle| \|\mathbf{X}^{(\alpha)}(t) - \mathbf{X}_m^{(0)}\| \leq \delta, \delta > 0, t \in (t_m, t_{m+\varepsilon}]\right\}.\end{aligned} \tag{5.47}$$

For a point $\mathbf{X}_m^{(0)} = (\mathbf{x}_m^{(0)}, \tilde{\mathbf{x}}_m^{(0)})^{\mathrm{T}} \in \partial\Omega_{12}$ at time t_m, a surface of the constraint boundary $\partial\Omega_{12}$ at the instantaneous time t_m is governed by $\varphi(\mathbf{x}^{(0)}, \tilde{\mathbf{x}}^{(0)}, t_m, \boldsymbol{\lambda}) = \varphi(\mathbf{x}_m^{(0)}, \tilde{\mathbf{x}}_m^{(0)}, t_m, \boldsymbol{\lambda}) = 0$. If the constraint function φ is time-independent, such a constraint surface for the synchronization of two dynamical systems is invariant with respect to time. Otherwise, this constraint surface changes with the instantaneous time t_m. In addition to the constraint surface, two boundaries of domain $\Omega_{\alpha}^{-\varepsilon}(\alpha = 1,2)$ are determined by $\varphi(\mathbf{x}^{(\alpha)}, \tilde{\mathbf{x}}^{(\alpha)}, t_{m-\varepsilon}, \boldsymbol{\lambda}) = \varphi(\mathbf{x}_{m-\varepsilon}^{(\alpha)}, \tilde{\mathbf{x}}_{m-\varepsilon}^{(\alpha)}, t_{m-\varepsilon}, \boldsymbol{\lambda}) = C$, as shown in Fig.5.3. In the ε-neighborhood of a constraint boundary, if the resultant system of two dynamical systems is attractive to such a constraint boundary, any a flow in the two ε-domains will approach to the constraint boundary. Further, the synchronicity of two dynamical systems to the constraint boundary can be investigated. In other words, the attractivity of the resultant system to the constraint boundary requires that any flow in the two ε-domains of $\Omega_{\alpha}(\alpha = 1,2)$ approach the constraint boundary $\partial\Omega_{12}$ as $t \to t_m$. From Luo (2006, 2008), the synchronization of two dynamical systems to the constraint needs that any flows of the resultant system in the two ε-domains of $\Omega_{\alpha}(\alpha = 1,2)$ are attractive to the constraint boundary.

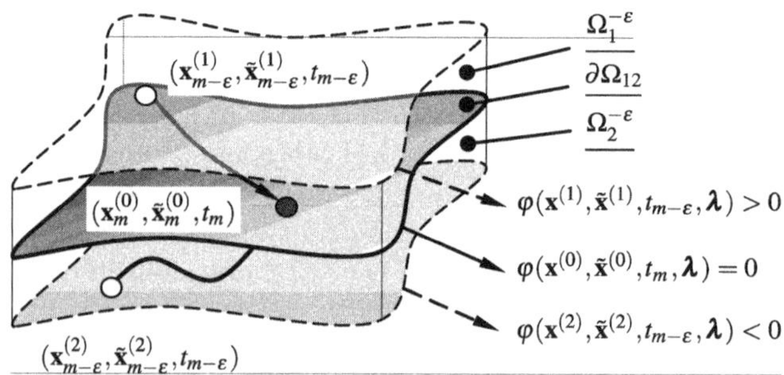

Fig. 5.3 A neighborhood of the constraint boundary and the attractivity of a resultant flow to the constraint boundary in $(n+\tilde{n})$-dimensional state space.

Definition 5.15. For two dynamical systems in Eqs.(5.1) and (5.2) with a constraint in Eq.(5.3), there is a metric functional of $V(\mathbf{X}, t, \boldsymbol{\lambda}) = f(\varphi(\mathbf{X}, t, \boldsymbol{\lambda}))$ in Eq.(5.29). For $\mathbf{X}_m^{(\alpha)} \in \Omega_{\alpha}$ $(\alpha \in \{1,2\})$ and $\mathbf{X}_m^{(0)} \in \partial\Omega_{12}$ at time t_m, $\mathbf{X}_m^{(\alpha)} = \mathbf{X}_m^{(0)}$. For any small $\varepsilon > 0$, there is a time interval $[t_{m-\varepsilon}, t_m)$. The two systems in Eqs.(5.1) and (5.2) to constraint in Eq.(5.3) is called to be *synchronized* in sense of the metric functional for time $t_m \in [t_{m_1}, t_{m_2}]$ if

$$
\begin{aligned}
& V(\mathbf{X}_{m-}^{(\alpha)}, t_{m-}, \boldsymbol{\lambda}) = V(\mathbf{X}_{m}^{(0)}, t_m, \boldsymbol{\lambda}) = 0; \\
& V(\mathbf{X}_{m-}^{(\alpha)}, t_{m-}, \boldsymbol{\lambda}) - V(\mathbf{X}_{m-\varepsilon}^{(\alpha)}, t_{m-\varepsilon}, \boldsymbol{\lambda}) < 0 \text{ for } \alpha = 1,2.
\end{aligned}
\tag{5.48}
$$

In addition to the attractivity of a flow of the resultant system to the constraint boundary, the repulsion of a flow of the resultant system to the constraint boundary can be defined. Because such a repulsion, any flows of the resultant system in the two ε-domains of $\Omega_{\alpha}(\alpha = 1,2)$ can never approach to the constraint boundary. In other words, two dynamical systems in Eqs.(5.1) and (5.2) cannot make the constraint condition in Eq.(5.3) be satisfied. Thus the repulsion of a flow of the resultant system to the constraint boundary should be introduced. Such a repulsion phenomenon is sketched in Fig.5.4. The constraint boundary $\partial\Omega_{12}$ is governed by $\varphi(\mathbf{x}^{(0)}, \tilde{\mathbf{x}}^{(0)}, t_m, \boldsymbol{\lambda}) = 0$. The boundary of the ε-neighborhood of the constraint is obtained by $\varphi(\mathbf{x}^{(\alpha)}, \tilde{\mathbf{x}}^{(\alpha)}, t_{m+\varepsilon}, \boldsymbol{\lambda}) = \varphi(\mathbf{x}^{(\alpha)}_{m+\varepsilon}, \tilde{\mathbf{x}}^{(\alpha)}_{m+\varepsilon}, t_{m+\varepsilon}, \boldsymbol{\lambda}) = C$. Two flows of the resultant system on both sides of the constraint boundary $\partial\Omega_{12}$ move away in two domains Ω_α $(\alpha = 1, 2)$, which means that no any flows of the resultant system can arrive to the constraint boundary. So the synchronization of two dynamical systems in Eqs.(5.1) and (5.2) to the constraint in Eq.(5.3) cannot be achieved. Such a repulsion of a resultant system to the constraint boundary implies that the two dynamical systems is desynchronized to the constraint in Eq.(5.3). To descript the desynchronization of two systems to a constraint, a mathematical description can be given as follows.

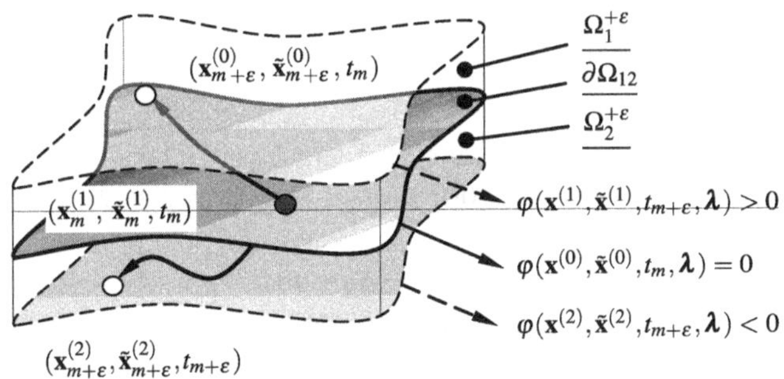

Fig. 5.4 The repulsion of a resultant flow to the constraint boundary in$(n+\tilde{n})$-dimensional state space.

Definition 5.16. For two systems in Eqs.(5.1) and (5.2) with constraint in Eq.(5.3), there is a metric functional of $V(\mathbf{X}, t, \boldsymbol{\lambda}) = f(\varphi(\mathbf{X}, t, \boldsymbol{\lambda}))$ in Eq.(5.29). For $\mathbf{X}^{(\alpha)}_m \in \Omega_\alpha$ $(\alpha \in \{1, 2\})$ and $\mathbf{X}^{(0)}_m \in \partial\Omega_{12}$ at time t_m, $\mathbf{X}^{(\alpha)}_m = \mathbf{X}^{(0)}_m$. For any small $\varepsilon > 0$, there is a time interval $[t_m, t_{m+\varepsilon}]$. The two dynamical system systems in Eqs.(5.1) and (5.2) to constraint in Eq.(5.3) are said to *be repelled* (*or desynchronized*) in sense of the metric fucntional for $t_m \in [t_{m_1}, t_{m_2}]$ if

$$\begin{aligned} &V(\mathbf{X}^{(\alpha)}_{m+}, t_{m+}, \boldsymbol{\lambda}) = V(\mathbf{X}^{(0)}_m, t_m, \boldsymbol{\lambda}) = 0;\\ &V(\mathbf{X}^{(\alpha)}_{m+\varepsilon}, t_{m+\varepsilon}, \boldsymbol{\lambda}) - V(\mathbf{X}^{(\alpha)}_{m+}, t_{m+}, \boldsymbol{\lambda}) > 0 \text{ for } \alpha = 1, 2. \end{aligned} \tag{5.49}$$

From the theory of discontinuous dynamical systems in Luo (2006, 2008), a resultant system of two dynamical systems in Eqs.(5.1) and (5.2) may pass through the constraint boundary from a domain to another. For this case, the penetration synchronicity of two dynamical systems can occur, as sketched in Fig. 5.5. Such synchronization can be called *an instantaneous synchronization*. A flow of a resultant system to the constraint boundary for time $t < t_m$ and $t > t_m$ lies in the two domains

Ω_1 and Ω_2. In sense of Eq.(5.3), a definition of such penetration synchronicity is given as follows.

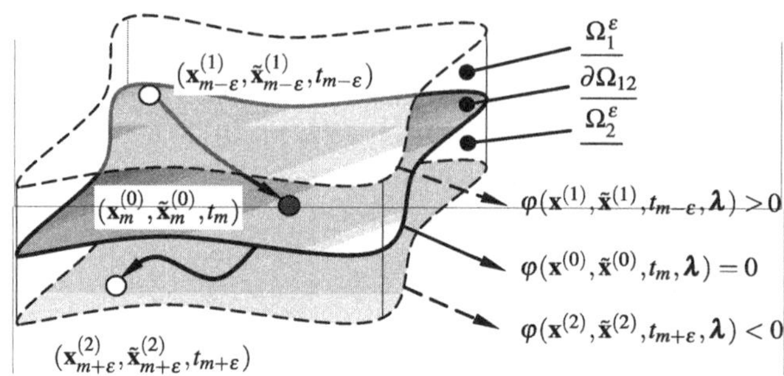

Fig. 5.5 A penetration of a resultant flow to the constraint boundary in $(n+\tilde{n})$-dimensional state space.

Definition 5.17. For two dynamical systems in Eqs.(5.1) and (5.2) with constraint in Eq.(5.3), there is a metric functional of $V(\mathbf{X},t,\boldsymbol{\lambda}) = f(\varphi(\mathbf{X},t,\boldsymbol{\lambda}))$ in Eq.(5.29). For $\mathbf{X}_m^{(\alpha)} \in \Omega_\alpha$ $(\alpha \in \{1,2\})$ and $\mathbf{X}_m^{(0)} \in \partial\Omega_{12}$ at time t_m, $\mathbf{X}_m^{(\alpha)} = \mathbf{X}_m^{(0)}$. For any small $\varepsilon > 0$, there is a time interval $[t_{m-\varepsilon}, t_{m+\varepsilon}]$. A resultant flow of two dynamical systems in Eqs.(5.1) and (5.2) is said to be *penetrated* to the constraint boundary $\partial\Omega_{\alpha\beta}$ from Ω_α to Ω_β at time t_m in the sense of the metric fucntional if for $\alpha,\beta \in \{1,2\}$ and $\alpha \neq \beta$

$$
\begin{aligned}
&V(\mathbf{X}_{m-}^{(\alpha)}, t_{m-}, \boldsymbol{\lambda}) = V(\mathbf{X}_{m+}^{(\beta)}, t_{m+}, \boldsymbol{\lambda}) = V(\mathbf{X}_m^{(0)}, t_m, \boldsymbol{\lambda}) = 0;\\
&V(\mathbf{X}_{m-\varepsilon}^{(\alpha)}, t_{m-\varepsilon}, \boldsymbol{\lambda}) - V(\mathbf{X}_{m-}^{(\alpha)}, t_{m-}, \boldsymbol{\lambda}) > 0;\\
&V(\mathbf{X}_{m+\varepsilon}^{(\beta)}, t_{m+\varepsilon}, \boldsymbol{\lambda}) - V(\mathbf{X}_{m+}^{(\beta)}, t_{m+}, \boldsymbol{\lambda}) > 0.
\end{aligned}
\tag{5.50}
$$

In Definition 5.17, the incoming flow with "−" and outgoing flow with "+" to the boundary are prescribed. From the above definition, a penetration flow of the resultant system of two dynamical systems to the constraint boundary can be considered to be formed by the semi-synchronization and semi-desynchronization. Such a penetration flow of the resultant system to the constraint boundary can also be called *an instantaneous synchronization* of two dynamical systems in Eqs (5.1) and (5.2) to constraint in Eq.(5.3). Such an instantaneous synchronization will disappear because the semi-desynchronization exists. From the definition of a penetration flow, a flow of the resultant system in domain Ω_α approaches to the constraint boundary. However, in domain Ω_β, such a flow will leave from the constraint boundary.

To investigate the relations among three types of synchronicity of two dynamical systems to the constraint in Eq.(5.3), the switchability of the synchronization, desynchronization and penetration is of great interest, which can be discussed through the singularity of the resultant system to the constraint boundary.

5.3.2 Singularity to constraint

From a theory of discontinuous dynamical systems in Luo (2006, 2008), a flow of a resultant system of two dynamical systems may be tangential to the constraint boundary governed by the constraint condition in Eq.(5.3). For this case, the synchronicity of two dynamical systems to the constraint occurs only at one point and then returns back to the same domain. Such an *instantaneous* synchronization is different from a penetration flow of the resultant system to the constraint boundary. The tangential synchronization of two dynamical systems to the constraint is sketched in Fig. 5.6. In domain Ω_1, the tangential synchronization of the two systems to the constraint boundary $\partial\Omega_{12}$ is presented. The two boundaries at time $t_{m-\varepsilon}$ and $t_{m+\varepsilon}$ are given by the two different surfaces. For such synchronicity, the following definition is given.

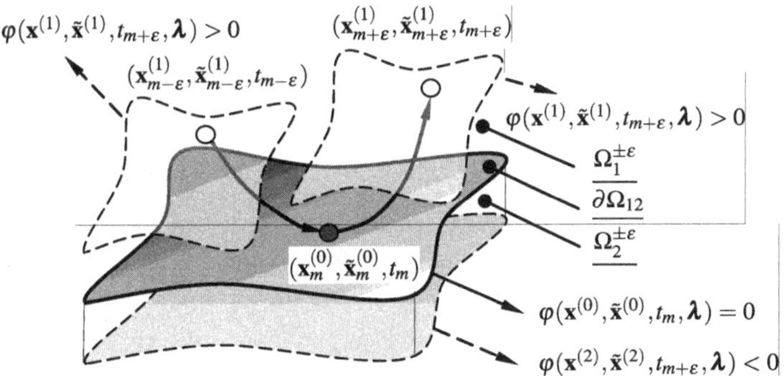

Fig. 5.6 Tangential synchronization to the constraint in an $(n+\tilde{n})$-dimensional state space.

Definition 5.18. For two dynamical systems in Eqs.(5.1) and (5.2) with constraint in Eq.(5.3), there is a metric functional of $V(\mathbf{X},t,\boldsymbol{\lambda}) = f(\varphi(\mathbf{X},t,\boldsymbol{\lambda}))$ in Eq.(5.29). For $\mathbf{X}_m^{(\alpha)} \in \Omega_\alpha$ $(\alpha \in \{1,2\})$ and $\mathbf{X}_m^{(0)} \in \partial\Omega_{12}$ at time t_m, $\mathbf{X}_m^{(\alpha)} = \mathbf{X}_m^{(0)}$. For any small $\varepsilon > 0$, there is a time interval $[t_{m-\varepsilon}, t_{m+\varepsilon}]$. At $\mathbf{X}^{(\alpha)} \in \Omega_\alpha^{\pm\varepsilon}$ for $t \in [t_{m-\varepsilon}, t_{m+\varepsilon}]$, the functional $V(\mathbf{X}^{(\alpha)},t,\boldsymbol{\lambda})$ is C^{r_α}-continuous $(r_\alpha \geq 2)$ and $|V^{(r_\alpha+1)}(\mathbf{X}^{(\alpha)},t,\boldsymbol{\lambda})| < \infty$. A flow of a resultant system of two dynamical systems in Eqs.(5.1) and (5.2) is said to be *tangential* (or grazing) to the constraint boundary at time t_m in the sense of the metric fucntional if for $\alpha \in \{1,2\}$

$$
\begin{aligned}
&V(\mathbf{X}_{m\pm}^{(\alpha)}, t_{m\pm}, \boldsymbol{\lambda}) = V(\mathbf{x}_m^{(0)}, t_m, \boldsymbol{\lambda}) = 0;\\
&V^{(1)}(\mathbf{X}_{m\pm}^{(\alpha)}, t_{m\pm}, \boldsymbol{\lambda}) = 0;\\
&V(\mathbf{X}_{m\pm\varepsilon}^{(\alpha)}, t_{m\pm\varepsilon}, \boldsymbol{\lambda}) - V(\mathbf{X}_{m\pm}^{(\alpha)}, t_{m\pm}, \boldsymbol{\lambda}) > 0.
\end{aligned}
\tag{5.51}
$$

In Definition 5.18, the incoming flow with "−" and outgoing flow with "+" to the boundary are prescribed. Such a tangency of a resultant flow to the constraint boundary will cause the synchronicity to be changed. The onset and vanishing singularity for synchronizations can be discussed, and the corresponding definition is given as follows.

Definition 5.19. For two dynamical systems in Eqs.(5.1) and (5.2) with constraint in Eq.(5.3), there is a metric functional of $V(\mathbf{X},t,\boldsymbol{\lambda}) = f(\varphi(\mathbf{X},t,\boldsymbol{\lambda}))$ in Eq.(5.29). For $\mathbf{X}_m^{(\alpha)} \in \Omega_\alpha$ ($\alpha \in \{1,2\}$) and $\mathbf{X}_m^{(0)} \in \partial\Omega_{12}$ at time $t_m, \mathbf{X}_m^{(\alpha)} = \mathbf{X}_m^{(0)}$. For any small $\varepsilon > 0$, there is a time interval $[t_{m-\varepsilon}, t_{m+\varepsilon}]$. At $\mathbf{X}^{(\alpha)} \in \Omega_\alpha^{\pm\varepsilon}$ for $t \in [t_{m-\varepsilon}, t_{m+\varepsilon}]$, the functional $V(\mathbf{X}^{(\alpha)},t,\boldsymbol{\lambda})$ is C^{r_α}-continuous and $|V^{(r_\alpha+1)}(\mathbf{X}^{(\alpha)},t,\boldsymbol{\lambda})| < \infty$ ($r_\alpha \geq 2$).

(i) The *synchronization* of two dynamical systems in Eqs.(5.1) and (5.2) with constraint in Eq.(5.3) is called to be *vanishing* to form a penetration from domain Ω_α to Ω_β at the constraint boundary at time t_m in the sense of the metric fucntional if for $\alpha, \beta \in \{1,2\}$ and $\alpha \neq \beta$

$$
\left.\begin{array}{l}
V(\mathbf{X}_{m-}^{(\alpha)},t_{m-},\boldsymbol{\lambda}) = V(\mathbf{X}_{m\mp}^{(\beta)},t_{m\mp},\boldsymbol{\lambda}) = V(\mathbf{X}_m^{(0)},t_m,\boldsymbol{\lambda}) = 0;\\
V^{(1)}(\mathbf{X}_{m-}^{(\alpha)},t_{m-},\boldsymbol{\lambda}) \neq 0,\ V^{(1)}(\mathbf{X}_{m\mp}^{(\beta)},t_{m\mp},\boldsymbol{\lambda}) = 0;\\
V(\mathbf{X}_{m-\varepsilon}^{(\alpha)},t_{m-\varepsilon},\boldsymbol{\lambda}) - V(\mathbf{X}_{m-}^{(\alpha)},t_{m-},\boldsymbol{\lambda}) > 0;\\
V(\mathbf{X}_{m\mp\varepsilon}^{(\beta)},t_{m\mp\varepsilon},\boldsymbol{\lambda}) - V(\mathbf{X}_{m\mp}^{(\beta)},t_{m\mp},\boldsymbol{\lambda}) > 0.
\end{array}\right\} \tag{5.52}
$$

(ii) The synchronization of two dynamical systems in Eqs.(5.1) and (5.2) with constraint in Eq.(5.3) is called to be *onset* from a penetration from domain Ω_α to Ω_β at the constraint boundary at time t_m in the sense of the metric fucntional if for $\alpha, \beta \in \{1,2\}$ and $\alpha \neq \beta$

$$
\left.\begin{array}{l}
V(\mathbf{X}_{m-}^{(\alpha)},t_{m-},\boldsymbol{\lambda}) = V(\mathbf{X}_{m\pm}^{(\beta)},t_{m\pm},\boldsymbol{\lambda}) = V(\mathbf{X}_m^{(0)},t_m,\boldsymbol{\lambda}) = 0;\\
V^{(1)}(\mathbf{X}_{m-}^{(\alpha)},t_{m-},\boldsymbol{\lambda}) \neq 0,\ V^{(1)}(\mathbf{X}_{m\pm}^{(\beta)},t_{m\pm},\boldsymbol{\lambda}) = 0;\\
V(\mathbf{X}_{m-\varepsilon}^{(\alpha)},t_{m-\varepsilon},\boldsymbol{\lambda}) - V(\mathbf{X}_{m-}^{(\alpha)},t_{m-},\boldsymbol{\lambda}) > 0;\\
V(\mathbf{X}_{m\pm\varepsilon}^{(\beta)},t_{m\pm\varepsilon},\boldsymbol{\lambda} - V(\mathbf{X}_{m\pm}^{(\beta)},t_{m\pm},\boldsymbol{\lambda}) > 0.
\end{array}\right\} \tag{5.53}
$$

In Eq.(5.52), the notation "$\mp$" represents the synchronization first with "$-$" and the penetration secondly with "$+$". This condition is called either the *vanishing* condition of synchronization to form a new penetration or the *onset* condition of *penetration* from the synchronization at the boundary of constraint in Eq.(5.3). However, in Eq.(5.53), the notation "$\pm$" represents the penetration first with "$+$" and the synchronization secondly with "$-$". This condition is called the *onset* condition of *synchronization* from a state of penetration to the boundary, which can also be called the *vanishing* condition of *penetration* to form a synchronization at the constraint boundary at time t_m. The switching conditions between the synchronization and desynchronization are presented as follows.

Definition 5.20. For two dynamical systems in Eqs.(5.1) and (5.2) with constraint in Eq.(5.3), there is a metric functional of $V(\mathbf{X},t,\boldsymbol{\lambda}) = f(\varphi(\mathbf{X},t,\boldsymbol{\lambda}))$ in Eq.(5.29).

For $\mathbf{X}_m^{(\alpha)} \in \Omega_\alpha$ ($\alpha \in \{1,2\}$) and $\mathbf{X}_m^{(0)} \in \partial\Omega_{12}$ at time t_m, $\mathbf{X}_m^{(\alpha)} = \mathbf{X}_m^{(0)}$. For any small $\varepsilon > 0$, there is a time interval $[t_{m-\varepsilon}, t_{m+\varepsilon}]$. At $\mathbf{x}^{(\alpha)} \in \Omega_\alpha^{\pm\varepsilon}$ for $t \in [t_{m-\varepsilon}, t_{m+\varepsilon}]$, the functional $V(\mathbf{X}^{(\alpha)}, t, \boldsymbol{\lambda})$ is C^{r_α}-continuous ($r_\alpha \geq 2$) and $|V^{(r_\alpha+1)}(\mathbf{X}^{(\alpha)}, t, \boldsymbol{\lambda})| < \infty$.

(i) The *synchronization of* two dynamical systems in Eqs.(5.1) and (5.2) to constraint in Eq.(5.3) are called to be *onset* from a desynchronization at the constraint boundary at time t_m in the sense of the metric fucntional if for $\alpha = 1,2$

$$\begin{aligned}
&V(\mathbf{X}_{m\pm}^{(\alpha)}, t_{m\pm}, \boldsymbol{\lambda}) = V(\mathbf{X}_m^{(0)}, t_m, \boldsymbol{\lambda}) = 0;\\
&V^{(1)}(\mathbf{X}_{m\pm}^{(\alpha)}, t_{m\pm}, \boldsymbol{\lambda}) = 0;\\
&V(\mathbf{X}_{m\pm\varepsilon}^{(\alpha)}, t_{m\pm\varepsilon}, \boldsymbol{\lambda}) - V(\mathbf{X}_{m\pm}^{(\alpha)}, t_{m\pm}, \boldsymbol{\lambda}) > 0.
\end{aligned} \tag{5.54}$$

(ii) The synchronization of two dynamical systems in Eqs.(5.1) and (5.2) to constraint in Eq.(5.3) is called to be *vanished* to form a desynchronization at the constraint boundary at time t_m in the sense of the metric fucntional if for $\alpha = 1,2$

$$\begin{aligned}
&V(\mathbf{X}_{m\mp}^{(\alpha)}, t_{m\mp}, \boldsymbol{\lambda}) = V(\mathbf{X}_m^{(0)}, t_m, \boldsymbol{\lambda}) = 0;\\
&V^{(1)}(\mathbf{X}_{m\mp}^{(\alpha)}, t_{m\mp}, \boldsymbol{\lambda}) = 0;\\
&V(\mathbf{X}_{m\mp\varepsilon}^{(\alpha)}, t_{m\mp\varepsilon}, \boldsymbol{\lambda}) - V(\mathbf{X}_{m\mp}^{(\alpha)}, t_{m\mp}, \boldsymbol{\lambda}) > 0.
\end{aligned} \tag{5.55}$$

Similarly, in Eq.(5.54), the notation "$\pm$" represents the desynchronization first with "$+$" and the synchronization with "$-$" secondly. This condition is called either *the onset condition of synchronization* from the desynchronization on the boundary or *the vanishing condition of desynchronization* to form a new synchronization on the boundary. In Eq.(5.55), the notation "$\mp$" represents the synchronization first with "$-$" and the desynchronization secondly with "$+$". This condition is called *the vanishing condition of synchronization* to form a new desynchronization, which can also be called *the onset condition of desynchronization* from the synchronization. In other words, the onset and vanishing conditions of the desynchronization from the penetration can be discussed as for the synchronization. The following definition will give the onset and vanishing conditions of desynchronization.

Definition 5.21. For two dynamical systems in Eqs.(5.1) and (5.2) with constraint in Eq.(5.3), there is a metric functional of $V(\mathbf{X}, t, \boldsymbol{\lambda}) = f(\varphi(\mathbf{X}, t, \boldsymbol{\lambda}))$ in Eq.(5.29). For $\mathbf{X}_m^{(\alpha)} \in \Omega_\alpha$ ($\alpha \in \{1,2\}$) and $\mathbf{X}_m^{(0)} \in \partial\Omega_{12}$ at time t_m, $\mathbf{X}_m^{(\alpha)} = \mathbf{X}_m^{(0)}$. For any small $\varepsilon > 0$, there is a time interval $[t_{m-\varepsilon}, t_{m+\varepsilon}]$. At $\mathbf{X}^{(\alpha)} \in \Omega_\alpha^{\pm\varepsilon}$, for $t \in [t_{m-\varepsilon}, t_{m+\varepsilon}]$, the functional $V(\mathbf{X}^{(\alpha)}, t, \boldsymbol{\lambda})$ is C^{r_α}-continuous ($r_\alpha \geq 2$) and $|V^{(r_\alpha+1)}(\mathbf{X}^{(\alpha)}, t, \boldsymbol{\lambda})| < \infty$.

(i) The desynchronization of two dynamical systems in Eqs.(5.1) and (5.2) to constraint in Eq.(5.3) is called to be *vanished* to form a penetration from Ω_α to Ω_β at the constraint boundary at time t_m in the sense of the metric fucntional if for $\alpha, \beta \in \{1,2\}$ and $\alpha \neq \beta$

$$
\begin{aligned}
&V(\mathbf{X}_{m\pm}^{(\alpha)},t_{m\pm},\boldsymbol{\lambda})=V(\mathbf{X}_{m+}^{(\beta)},t_{m+},\boldsymbol{\lambda})=V(\mathbf{X}_{m}^{(0)},t_{m},\boldsymbol{\lambda})=0;\\
&V^{(1)}(\mathbf{X}_{m\pm}^{(\alpha)},t_{m\pm},\boldsymbol{\lambda})=0,\ V^{(1)}(\mathbf{x}_{m+}^{(\beta)},t_{m+},\boldsymbol{\lambda})\neq 0;\\
&V(\mathbf{X}_{m\pm\varepsilon}^{(\alpha)},t_{m\pm\varepsilon},\boldsymbol{\lambda})-V(\mathbf{X}_{m\pm}^{(\alpha)},t_{m\pm},\boldsymbol{\lambda})>0,\\
&V(\mathbf{X}_{m+\varepsilon}^{(\beta)},t_{m+\varepsilon},\boldsymbol{\lambda})-V(\mathbf{X}_{m+}^{(\beta)},t_{m+},\boldsymbol{\lambda})>0.
\end{aligned}
\tag{5.56}
$$

(ii) The desynchronization of two dynamical systems in Eqs.(5.1) and (5.2) to constraint in Eq.(5.3) is called to be *onset* from a penetration from Ω_α to Ω_β at the constraint boundary at time t_m in the sense of the metric fucntional if for $\alpha,\beta \in \{1,2\}$ and $\alpha \neq \beta$

$$
\begin{aligned}
&V(\mathbf{X}_{m\pm}^{(\alpha)},t_{m\pm},\boldsymbol{\lambda})=V(\mathbf{X}_{m+}^{(\beta)},t_{m+},\boldsymbol{\lambda})=V(\mathbf{X}_{m}^{(0)},t_{m},\boldsymbol{\lambda})=0;\\
&V^{(1)}(\mathbf{X}_{m\pm}^{(\alpha)},t_{m\pm},\boldsymbol{\lambda})=0,\ V^{(1)}(\mathbf{X}_{m+}^{(\beta)},t_{m+},\boldsymbol{\lambda})\neq 0;\\
&V(\mathbf{X}_{m\pm\varepsilon}^{(\alpha)},t_{m\pm\varepsilon},\boldsymbol{\lambda})-V(\mathbf{X}_{m\pm}^{(\alpha)},t_{m\pm},\boldsymbol{\lambda})>0;\\
&V(\mathbf{X}_{m+\varepsilon}^{(\beta)},t_{m+\varepsilon},\boldsymbol{\lambda})-V(\mathbf{X}_{m+}^{(\beta)},t_{m+},\boldsymbol{\lambda})>0.
\end{aligned}
\tag{5.57}
$$

Notice that in Eq.(5.56), the notation "$\pm$" represents the desynchronization first with "+" and the penetration secondly with "−". This condition is called the *vanishing* condition of *desynchronization* to form a new penetration on the boundary, and can also be called the *onset* condition of penetration from a synchronization state. However, in Eq.(5.57), the notation "$\pm$" represents the penetration first with "+" and the synchronization secondly with "−". This condition is called *the onset condition of desynchronization* from a penetration and also can be called *the vanishing condition of the penetration* to form a desynchronization state. From the previous three definitions, the switching between synchronization and penetration, between desynchronization and penetration, and between desynchronization and synchronization were presented. However, another switching between two penetrations should be discussed.

Definition 5.22. For two dynamical systems in Eqs.(5.1) and (5.2) with constraint in Eq.(5.3), there is a metric functional of $V(\mathbf{X},t,\boldsymbol{\lambda})=f(\varphi(\mathbf{X},t,\boldsymbol{\lambda}))$ in Eq.(5.29). For $\mathbf{X}_m^{(\alpha)} \in \Omega_\alpha$ ($\alpha \in \{1,2\}$) and $\mathbf{X}_m^{(0)} \in \partial\Omega_{12}$ at t_m, $\mathbf{X}_m^{(\alpha)} = \mathbf{X}_m^{(0)}$. For any small $\varepsilon > 0$, there is a time interval $[t_{m-\varepsilon}, t_{m+\varepsilon}]$. At $\mathbf{X}^{(\alpha)} \in \Omega_\alpha^{\pm\varepsilon}$, for $t \in [t_{m-\varepsilon}, t_{m+\varepsilon}]$, the functional $V(\mathbf{X}^{(\alpha)},t,\boldsymbol{\lambda})$ is C^{r_α}-continuous ($r_\alpha \geq 2$) and $|V^{(r_\alpha+1)}(\mathbf{X}^{(\alpha)},t,\boldsymbol{\lambda})| < \infty$. The penetration of the slave and master systems in Eqs.(5.1) and (5.2) to constraint in Eq.(5.3) are called to be *switched* at the constraint boundary at time t_m in the sense of the metric fucntional if for $\alpha,\beta \in \{1,2\}$

$$\begin{aligned}
&V(\mathbf{X}_{m\mp}^{(\alpha)},t_{m\mp},\boldsymbol{\lambda})=V(\mathbf{X}_{m\pm}^{(\beta)},t_{m\pm},\boldsymbol{\lambda})=V(\mathbf{X}_{m}^{(0)},t_{m},\boldsymbol{\lambda})=0;\\
&V^{(1)}(\mathbf{X}_{m\mp}^{(\alpha)},t_{m\mp},\boldsymbol{\lambda})=V^{(1)}(\mathbf{X}_{m\pm}^{(\beta)},t_{m\pm},\boldsymbol{\lambda})=0;\\
&V(\mathbf{X}_{m\mp\varepsilon}^{(\alpha)},t_{m\mp\varepsilon},\boldsymbol{\lambda})-V(\mathbf{X}_{m\mp}^{(\alpha)},t_{m\mp},\boldsymbol{\lambda})>0,\\
&V(\mathbf{X}_{m\pm\varepsilon}^{(\beta)},t_{m\pm\varepsilon},\boldsymbol{\lambda})-V(\mathbf{X}_{m\pm}^{(\beta)},t_{m\pm},\boldsymbol{\lambda})>0.
\end{aligned} \tag{5.58}$$

Based on the definitions of the tangential (or grazing) and switching singularity, there is a critical parameter $\boldsymbol{\lambda}_{cr}$ from which $\partial V(\mathbf{X}_{m\pm}^{(\alpha)},t_{m\pm},\boldsymbol{\lambda})/\partial\boldsymbol{\lambda}|_{\boldsymbol{\lambda}_{cr}}\neq 0$, such a singularity is called the corresponding bifurcation at $\boldsymbol{\lambda}_{cr}$ for parameter $\boldsymbol{\lambda}$.

5.3.3 Synchronicity with singularity

As similar to discontinuous dynamical systems in Luo (2006, 2008), the above synchronicity of two dynamical systems in Eqs.(5.1) and (5.2) with constraint in Eq.(5.3) can be extended to the case of higher-order singularity. The corresponding definitions can be presented. The definition for the $(2k_\alpha:2k_\beta)$-synchronization of two dynamical systems in Eqs.(5.1) and (5.2) with constraint in Eq.(5.3) at the corresponding constraint boundary for time $t_m\in[t_{m_1},t_{m_2}]$ is presented first.

Definition 5.23. For two dynamical systems in Eqs.(5.1) and (5.2) with constraint in Eq.(5.3), there is a metric functional of $V(\mathbf{X}t,\boldsymbol{\lambda})=f(\varphi(\mathbf{X},t,\boldsymbol{\lambda}))$ in Eq.(5.29). For $\mathbf{X}_m^{(\alpha)}\in\Omega_\alpha$ $(\alpha\in\{1,2\})$ and $\mathbf{X}_m^{(0)}\in\partial\Omega_{12}$ at t_m, $\mathbf{X}_m^{(\alpha)}=\mathbf{X}_m^{(0)}$. For any small $\varepsilon>0$, there is a time interval $[t_{m-\varepsilon},t_m]$. At $\mathbf{X}^{(\alpha)}\in\Omega_\alpha^{-\varepsilon}$, for $t\in[t_{m-\varepsilon},t_m)$, the functional $V(\mathbf{X}^{(\alpha)},t,\boldsymbol{\lambda})$ is C^{r_α}-continuous $(r_\alpha\geq 2k_\alpha+1)$ and $|V^{(r_\alpha+1)}(\mathbf{X}^{(\alpha)},t,\boldsymbol{\lambda})|<\infty$. The two dynamical systems in Eqs.(5.1) and (5.2) with constraint in Eq.(5.3) is called to be synchronized with the $(2k_1:2k_2)$-type to the constraint in Eq.(5.3) for time $t_m\in[t_{m_1},t_{m_2}]$ in the sense of the metric fucntional if for $\alpha=1,2$

$$\begin{aligned}
&V(\mathbf{X}_{m-}^{(\alpha)},t_{m-},\boldsymbol{\lambda})=V(\mathbf{X}_{m}^{(0)},t_{m},\boldsymbol{\lambda})=0;\\
&V^{(s_\alpha)}(\mathbf{X}_{m-}^{(\alpha)},t_{m-},\boldsymbol{\lambda})=0 \text{ for } s_\alpha=1,2,\ldots,2k_\alpha;\\
&V(\mathbf{X}_{m-}^{(\alpha)},t_{m-},\boldsymbol{\lambda})-V(\mathbf{X}_{m-\varepsilon}^{(\alpha)},t_{m-\varepsilon},\boldsymbol{\lambda})<0.
\end{aligned} \tag{5.59}$$

As in the definition for the $(2k_1:2k_2)$-synchronization, the $(2k_1:2k_2)$-desynchronization of two dynamical systems in Eqs.(5.1) and (5.2) with constraint in Eq.(5.3) on the corresponding constraint boundary for time $t_m\in[t_{m_1},t_{m_2}]$ is also presented.

Definition 5.24. For two dynamical systems in Eqs.(5.1) and (5.2) with constraint in Eq.(5.3), there is a metric functional of $V(\mathbf{X},t,\boldsymbol{\lambda})=f(\varphi(\mathbf{X},t,\boldsymbol{\lambda}))$ in Eq.(5.29). For $\mathbf{X}_m^{(\alpha)}\in\Omega_\alpha$ $(\alpha\in\{1,2\})$ and $\mathbf{X}_m^{(0)}\in\partial\Omega_{12}$ at t_m, $\mathbf{X}_m^{(\alpha)}=\mathbf{X}_m^{(0)}$. For any small $\varepsilon>0$, there is a time interval $[t_m,t_{m+\varepsilon}]$. At $\mathbf{X}^{(\alpha)}\in\Omega_\alpha^{+\varepsilon}$ for $t\in(t_m,t_{m+\varepsilon}]$, the functional

$V(\mathbf{X}^{(\alpha)},t,\boldsymbol{\lambda})$ is C^{r_α}-continuous ($r_\alpha \geq 2k_\alpha+1$) and $|V^{(r_\alpha+1)}(\mathbf{X}^{(\alpha)},t,\boldsymbol{\lambda})| < \infty$. The two dynamical systems in Eqs.(5.1) and (5.2) with constraint in Eq.(5.3) is said to be desynchronized (or repelled) with the $(2k_1 : 2k_2)$-type to the constraint in Eq.(5.3) for $t_m \in [t_{m_1}, t_{m_2}]$ in the sense of the metric fucntional if for $\alpha = 1,2$

$$\begin{aligned}
&V(\mathbf{X}_{m+}^{(\alpha)},t_{m+},\boldsymbol{\lambda}) = V(\mathbf{X}_m^{(0)},t_m,\boldsymbol{\lambda}) = 0;\\
&V^{(s_\alpha)}(\mathbf{X}_{m+}^{(\alpha)},t_{m+},\boldsymbol{\lambda}) = 0,\ s_\alpha = 1,2,\ldots,2k_\alpha;\\
&V(\mathbf{X}_{m+\varepsilon}^{(\alpha)},t_{m+\varepsilon},\boldsymbol{\lambda}) - V(\mathbf{x}_{m+}^{(\alpha)},t_{m+},\boldsymbol{\lambda}) > 0.
\end{aligned} \tag{5.60}$$

As discussed before, the penetration on the boundary of constraint is composed of the semi-synchronization and semi-desynchronization. From the foregoing two definitions, the $(2k_\alpha : 2k_\beta)$-penetration of two dynamical systems in Eqs.(5.1) and (5.2) to constraint in Eq.(5.3) at time t_m is described.

Definition 5.25. For two dynamical systems in Eqs.(5.1) and (5.2) with constraint in Eq.(5.3), there is a metric functional of $V(\mathbf{X},t,\boldsymbol{\lambda}) = f(\varphi(\mathbf{X},t,\boldsymbol{\lambda}))$ in Eq.(5.29). For $\mathbf{X}_m^{(\alpha)} \in \Omega_\alpha$ ($\alpha \in \{1,2\}$) and $\mathbf{X}_m^{(0)} \in \partial\Omega_{12}$ at t_m, $\mathbf{X}_m^{(\alpha)} = \mathbf{X}_m^{(0)}$. For any small $\varepsilon > 0$, there is a time interval $[t_{m-\varepsilon}, t_{m+\varepsilon}]$. At $\mathbf{X}^{(\alpha)} \in \Omega_\alpha^{\pm\varepsilon}$, for $t \in [t_{m-\varepsilon}, t_{m+\varepsilon}]$, the functional $V(\mathbf{X}^{(\alpha)},t,\boldsymbol{\lambda})$ is C^{r_α}-continuous and $|V^{(r_\alpha+1)}(\mathbf{X}^{(\alpha)},t,\boldsymbol{\lambda})| < \infty$ ($r_\alpha \geq 2k_\alpha$). A flow of two dynamical systems in Eqs.(5.1) and (5.2) with constraint in Eq.(5.3) is said to be *penetrated with the* $(2k_\alpha : 2k_\beta)$*-type* from domain Ω_α to domain Ω_β at the constraint boundary at time t_m in the sense of the metric fucntional if

$$\begin{aligned}
&V(\mathbf{X}_{m-}^{(\alpha)},t_{m-},\boldsymbol{\lambda}) = V(\mathbf{X}_{m+}^{(\beta)},t_{m+},\boldsymbol{\lambda}) = V(\mathbf{X}_m^{(0)},t_m,\boldsymbol{\lambda}) = 0;\\
&V^{(r_\alpha)}(\mathbf{X}_{m-}^{(\alpha)},t_{m-},\boldsymbol{\lambda}) = 0 \text{ for } s_\alpha = 1,2,\ldots,2k_\alpha;\\
&V^{(s_\beta)}(\mathbf{X}_{m+}^{(\beta)},t_{m+},\boldsymbol{\lambda}) = 0 \text{ for } s_\beta = 1,2,\ldots,2k_\beta;\\
&V(\mathbf{X}_{m-\varepsilon}^{(\alpha)},t_{m-\varepsilon},\boldsymbol{\lambda}) - V(\mathbf{X}_{m-}^{(\alpha)},t_{m-},\boldsymbol{\lambda}) > 0 \text{ for } \alpha \in \{1,2\} \text{ and}\\
&V(\mathbf{X}_{m+\varepsilon}^{(\beta)},t_{m+\varepsilon},\boldsymbol{\lambda}) - V(\mathbf{X}_{m+}^{(\beta)},t_{m+},\boldsymbol{\lambda}) > 0 \text{ for } \alpha \neq \beta \in \{1,2\}.
\end{aligned} \tag{5.61}$$

From the three definitions, the higher-singularity is used for description of the synchronization, desynchronization and penetration at the constraint boundary, and the switching among the three synchronous states can be discussed through the higher-order singularity as well.

5.3.4 Higher-order singularity

From the previous descriptions of the synchronization, desynchronization and penetration with the higher-order singularity for two dynamical systems to the constraint, the higher-order singularity of the two dynamical systems to the constraint boundary should be further discussed as follows.

Definition 5.26. For two dynamical systems in Eqs.(5.1) and (5.2) with constraint in Eq.(5.3), there is a metric functional of $V(\mathbf{X},t,\boldsymbol{\lambda}) = f(\varphi(\mathbf{X},t,\boldsymbol{\lambda}))$ in Eq.(5.29). For $\mathbf{X}_m^{(\alpha)} \in \Omega_\alpha$ $(\alpha \in \{1,2\})$ and $\mathbf{X}_m^{(0)} \in \partial\Omega_{12}$ at t_m, $\mathbf{X}_m^{(\alpha)} = \mathbf{X}_m^{(0)}$. For any small $\varepsilon > 0$, there is a time interval $[t_{m-\varepsilon}, t_{m+\varepsilon}]$. At $\mathbf{X}^{(\alpha)} \in \Omega_\alpha^{\pm\varepsilon}$ for $t \in [t_{m-\varepsilon}, t_{m+\varepsilon}]$, the functional $V(\mathbf{X}^{(\alpha)},t,\boldsymbol{\lambda})$ is C^{r_α}-continuous and $|V^{(r_\alpha+1)}(\mathbf{X}^{(\alpha)},t,\boldsymbol{\lambda})| < \infty$ $(r_\alpha \geq 2k_\alpha)$. A resultant flow of the two dynamical systems in Eqs.(5.1) and (5.2) with constraint in Eq.(5.3) is said to be *tangential* to the constraint boundary with the $(2k_\alpha - 1)^{\text{th}}$-order at time t_m in the sense of the metric fucntional if for $\alpha \in \{1,2\}$

$$
\left.
\begin{aligned}
& V(\mathbf{X}_{m\pm}^{(\alpha)}, t_{m\pm}, \boldsymbol{\lambda}) = V(\mathbf{X}_m^{(0)}, t_m, \boldsymbol{\lambda}) = 0; \\
& V^{(s_\alpha)}(\mathbf{X}_{m\pm}^{(\alpha)}, t_{m\pm}, \boldsymbol{\lambda}) = 0 \; s_\alpha = 1,2,\ldots,2k_\alpha - 1; \\
& V(\mathbf{X}_{m-\varepsilon}^{(\alpha)}, t_{m-\varepsilon}, \boldsymbol{\lambda}) - V(\mathbf{X}_{m-}^{(\alpha)}, t_{m-}, \boldsymbol{\lambda}) > 0 \text{ and} \\
& V(\mathbf{X}_{m+\varepsilon}^{(\alpha)}, t_{m+\varepsilon}, \boldsymbol{\lambda}) - V(\mathbf{X}_{m+}^{(\alpha)}, t_{m+}, \boldsymbol{\lambda}) > 0.
\end{aligned}
\right\} \tag{5.62}
$$

The foregoing definition gives the definition of the $(2k_\alpha - 1)^{\text{th}}$ tangential condition to the constraint boundary. Based on the similar ideas, the switchability of the synchronization, desynchronization and penetration of two dynamical systems to the constraint boundary can be described.

Definition 5.27. For two dynamical systems in Eqs.(5.1) and (5.2) with constraint in Eq.(5.3), there is a metric functional of $V(\mathbf{X},t,\boldsymbol{\lambda}) = f(\varphi(\mathbf{X},t,\boldsymbol{\lambda}))$ in Eq.(5.29). For $\mathbf{X}_m^{(\alpha)} \in \Omega_\alpha$ $(\alpha \in \{1,2\})$ and $\mathbf{X}_m^{(0)} \in \partial\Omega_{12}$ at t_m, $\mathbf{X}_m^{(\alpha)} = \mathbf{X}_m^{(0)}$. For any small $\varepsilon > 0$, there is a time interval $[t_{m-\varepsilon}, t_{m+\varepsilon}]$. At $\mathbf{X}^{(\alpha)} \in \Omega_\alpha^{\pm\varepsilon}$ for $t \in [t_{m-\varepsilon}, t_{m+\varepsilon}]$, the functional $V(\mathbf{X}^{(\alpha)},t,\boldsymbol{\lambda})$ is C^{r_α}-continuous and $|V^{(r_\alpha+1)}(\mathbf{X}^{(\alpha)},t,\boldsymbol{\lambda})| < \infty$ $(r_\alpha \geq 2k_\alpha + 1)$.

(i) The $(2k_\alpha : 2k_\beta)$-synchronization of the two dynamical systems in Eqs.(5.1) and (5.2) with constraint in Eq.(5.3) is said to be *vanishing* to form a $(2k_\alpha : 2k_\beta)$-penetration from domain Ω_α to domain Ω_β at the constraint boundary at time t_m in the sense of the metric fucntional if for $\alpha, \beta \in \{1,2\}$ and $\alpha \neq \beta$

$$
\left.
\begin{aligned}
& V(\mathbf{X}_{m-}^{(\alpha)}, t_{m-}, \boldsymbol{\lambda}) = V(\mathbf{X}_{m\mp}^{(\beta)}, t_{m\mp}, \boldsymbol{\lambda}) = V(\mathbf{X}_m^{(0)}, t_m, \boldsymbol{\lambda}) = 0; \\
& V^{(s_\alpha)}(\mathbf{X}_{m-}^{(\alpha)}, t_{m-}, \boldsymbol{\lambda}) = 0 \text{ for } s_\alpha = 1,2,\ldots,2k_\alpha, \\
& V^{(s_\beta)}(\mathbf{X}_{m\mp}^{(\beta)}, t_{m\mp}, \boldsymbol{\lambda}) = 0 \text{ for } s_\beta = 1,2,\ldots,2k_\beta + 1; \\
& V(\mathbf{X}_{m-\varepsilon}^{(\alpha)}, t_{m-\varepsilon}, \boldsymbol{\lambda}) - V(\mathbf{X}_{m-}^{(\alpha)}, t_{m-}, \boldsymbol{\lambda}) > 0, \\
& V(\mathbf{X}_{m\mp\varepsilon}^{(\beta)}, t_{m\mp\varepsilon}, \boldsymbol{\lambda}) - V(\mathbf{X}_{m\mp}^{(\beta)}, t_{m\mp}, \boldsymbol{\lambda}) > 0.
\end{aligned}
\right\} \tag{5.63}
$$

(ii) The $(2k_\alpha : 2k_\beta)$-synchronization of the two dynamical systems in Eqs.(5.1) and (5.2) with constraint in Eq.(5.3) is said to be *onset* from the $(2k_\alpha : 2k_\beta)$-

penetration from Ω_α to Ω_β at the constraint boundary at time t_m in the sense of the metric fucntional if for $\alpha, \beta \in \{1,2\}$ and $\alpha \neq \beta$

$$\begin{aligned}
&V(\mathbf{X}_{m-}^{(\alpha)}, t_{m-}, \boldsymbol{\lambda}) = V(\mathbf{X}_{m+}^{(\beta)}, t_{m+}, \boldsymbol{\lambda}) = V(\mathbf{X}_m^{(0)}, t_m, \boldsymbol{\lambda}) = 0;\\
&V^{(s_\alpha)}(\mathbf{X}_{m-}^{(\alpha)}, t_{m-}, \boldsymbol{\lambda}) = 0 \text{ for } s_\alpha = 1,2,\ldots,2k_\alpha,\\
&V^{(s_\beta)}(\mathbf{X}_{m\pm}^{(\beta)}, t_{m\pm}, \boldsymbol{\lambda}) = 0 \text{ for } s_\beta = 1,2,\ldots,2k_\beta+1;\\
&V(\mathbf{X}_{m-\varepsilon}^{(\alpha)}, t_{m-\varepsilon}, \boldsymbol{\lambda}) - V(\mathbf{X}_{m-}^{(\alpha)}, t_{m-}, \boldsymbol{\lambda}) > 0,\\
&V(\mathbf{X}_{m\pm\varepsilon}^{(\beta)}, t_{m\pm\varepsilon}, \boldsymbol{\lambda}) - V(\mathbf{X}_{m\pm}^{(\beta)}, t_{m\pm}, \boldsymbol{\lambda}) > 0.
\end{aligned} \tag{5.64}$$

From this definition, this condition in Eq.(5.63) for *the onset of the* $(2k_\alpha : 2k_\beta)$-*synchronization* from the $(2k_\alpha : 2k_\beta)$-penetration on the constraint boundary can be called *the vanishing condition of the* $(2k_\alpha : 2k_\beta)$-*penetration* to form a new $(2k_\alpha : 2k_\beta)$-synchronization on the constraint boundary. In Eq.(5.64), the *vanishing condition of the*$(2k_\alpha : 2k_\beta)$-synchronization to form a new $(2k_\alpha : 2k_\beta)$-penetration can be called *the onset condition of the* $(2k_\alpha : 2k_\beta)$-penetration from the synchronization. The onset and vanishing conditions of the $(2k_\alpha : 2k_\beta)$-desynchronization from the $(2k_\alpha : 2k_\beta)$-penetration can be discussed. The following definition will give the onset and vanishing conditions of the $(2k_\alpha : 2k_\beta)$-desynchronization.

Definition 5.28. For two dynamical systems in Eqs.(5.1) and (5.2) with constraint in Eq.(5.3), there is a metric functional of $V(\mathbf{X}, t, \boldsymbol{\lambda}) = f(\varphi(\mathbf{X}, t, \boldsymbol{\lambda}))$ in Eq.(5.29). For $\mathbf{X}_m^{(\alpha)} \in \Omega_\alpha$ $(\alpha \in \{1,2\})$ and $\mathbf{X}_m^{(0)} \in \partial\Omega_{12}$ at t_m, $\mathbf{X}_m^{(\alpha)} = \mathbf{X}_m^{(0)}$. For any small $\varepsilon > 0$, there is a time interval $[t_{m-\varepsilon}, t_{m+\varepsilon}]$. At $\mathbf{X}^{(\alpha)} \in \Omega_\alpha^{\pm\varepsilon}$ for $t \in [t_{m-\varepsilon}, t_{m+\varepsilon}]$, the functional $V(\mathbf{X}^{(\alpha)}, t, \boldsymbol{\lambda})$ is C^{r_α}-continuous and $|V^{(r_\alpha+1)}(\mathbf{X}^{(\alpha)}, t, \boldsymbol{\lambda})| < \infty$ $(r_\alpha \geq 2k_\alpha + 1)$.

(i) The $(2k_\alpha : 2k_\beta)$-synchronization of the two dynamical systems in Eqs.(5.1) and (5.2) with constraint in Eq.(5.3) is called to be *vanished* to form a $(2k_\alpha : 2k_\beta)$-desynchronization at the constraint boundary at time t_m in the sense of the metric fucntional if for $\alpha, \beta \in \{1,2\}$ and $\alpha \neq \beta$

$$\begin{aligned}
&V(\mathbf{X}_{m\mp}^{(\alpha)}, t_{m\mp}, \boldsymbol{\lambda}) = V(\mathbf{X}_{m\mp}^{(\beta)}, t_{m\mp}, \boldsymbol{\lambda}) = V(\mathbf{X}_m^{(0)}, t_m, \boldsymbol{\lambda}) = 0;\\
&V^{(s_\alpha)}(\mathbf{X}_{m\mp}^{(\alpha)}, t_{m\mp}, \boldsymbol{\lambda}) = 0 \text{ for } s_\alpha = 1,2,\ldots,2k_\alpha+1,\\
&V^{(s_\beta)}(\mathbf{X}_{m\mp}^{(\beta)}, t_{m\mp}, \boldsymbol{\lambda}) = 0 \text{ for } s_\beta = 1,2,\ldots,2k_\beta+1;\\
&V(\mathbf{X}_{m\mp\varepsilon}^{(\alpha)}, t_{m\mp\varepsilon}, \boldsymbol{\lambda}) - V(\mathbf{X}_{m\mp}^{(\alpha)}, t_{m\mp}, \boldsymbol{\lambda}) > 0,\\
&V(\mathbf{X}_{m\mp\varepsilon}^{(\beta)}, t_{m\mp\varepsilon}, \boldsymbol{\lambda}) - V(\mathbf{X}_{m\mp}^{(\beta)}, t_{m\mp}, \boldsymbol{\lambda}) > 0.
\end{aligned} \tag{5.65}$$

(ii) The $(2k_\alpha : 2k_\beta)$-synchronization of the two dynamical systems in Eqs.(5.1) and (5.2) with constraint in Eq.(5.3) is said to be *onset* from the $(2k_\alpha : 2k_\beta)$-desynchronization at the constraint boundary at time t_m in the sense of the metric fucntional if for $\alpha, \beta \in \{1,2\}$ and $\alpha \neq \beta$

$$
\begin{aligned}
&V(\mathbf{X}_{m\pm}^{(\alpha)}, t_{m\pm}, \boldsymbol{\lambda}) = V(\mathbf{X}_{m\pm}^{(\beta)}, t_{m\pm}, \boldsymbol{\lambda}) = V(\mathbf{X}_{m}^{(0)}, t_{m}, \boldsymbol{\lambda}) = 0;\\
&V^{(s_\alpha)}(\mathbf{X}_{m\pm}^{(\alpha)}, t_{m\pm}, \boldsymbol{\lambda}) = 0 \text{ for } s_\alpha = 1, 2, \ldots, 2k_\alpha + 1,\\
&V^{(s_\beta)}(\mathbf{X}_{m\pm}^{(\beta)}, t_{m\pm}, \boldsymbol{\lambda}) = 0 \text{ for } s_\beta = 1, 2, \ldots, 2k_\beta + 1;\\
&V(\mathbf{X}_{m\pm\varepsilon}^{(\alpha)}, t_{m\pm\varepsilon}, \boldsymbol{\lambda}) - V(\mathbf{X}_{m\pm}^{(\alpha)}, t_{m\pm}, \boldsymbol{\lambda}) > 0,\\
&V(\mathbf{X}_{m\pm\varepsilon}^{(\beta)}, t_{m\pm\varepsilon}, \boldsymbol{\lambda}) - V(\mathbf{X}_{m\pm}^{(\beta)}, t_{m\pm}, \boldsymbol{\lambda}) > 0.
\end{aligned}
\tag{5.66}
$$

It is observed that the conditions in Eqs.(5.65) and (5.66) are symmetrically switched. The condition in Eq.(5.65) for *the onset condition of the* $(2k_\alpha : 2k_\beta)$-*synchronization* from the $(2k_\alpha : 2k_\beta)$-desynchronization on the constraint boundary can be called *the vanishing condition of the* $(2k_\alpha : 2k_\beta)$-desynchronization to form a new $(2k_\alpha : 2k_\beta)$-synchronization on such a constraint boundary. However, the condition in Eq.(5.66) for *the vanishing condition of the* $(2k_\alpha : 2k_\beta)$-synchronization to form a new $(2k_\alpha : 2k_\beta)$-penetration can be called *the onset condition of the* $(2k_\alpha : 2k_\beta)$-desynchronization from the synchronization. The switching of desynchronization and penetration on the boundary will be discussed as follows.

Definition 5.29. For two dynamical systems in Eqs.(5.1) and (5.2) with constraint in Eq.(5.3), there is a metric functional of $V(\mathbf{X}, t, \boldsymbol{\lambda}) = f(\varphi(\mathbf{X}, t, \boldsymbol{\lambda}))$ in Eq.(5.29). For $\mathbf{X}_m^{(\alpha)} \in \Omega_\alpha$ $(\alpha \in \{1, 2\})$ and $\mathbf{X}_m^{(0)} \in \partial\Omega_{12}$ at t_m, $\mathbf{X}_m^{(\alpha)} = \mathbf{X}_m^{(0)}$. For any small $\varepsilon > 0$, there is a time interval $[t_{m-\varepsilon}, t_{m+\varepsilon}]$. At $\mathbf{X}^{(\alpha)} \in \Omega_\alpha^{\pm\varepsilon}$ for $t \in [t_{m-\varepsilon}, t_{m+\varepsilon}]$, the functional $V(\mathbf{X}^{(\alpha)}, t, \boldsymbol{\lambda})$ is C^{r_α}-continuous and $|V^{(r_\alpha+1)}(\mathbf{X}^{(\alpha)}, t, \boldsymbol{\lambda})| < \infty (r_\alpha \geq 2k_\alpha + 1)$.

(i) The $(2k_\alpha : 2k_\beta)$-desynchronization of the two dynamical systems in Eqs.(5.1) and (5.2) with constraint in Eq.(5.3) is called to be *vanished* to form a $(2k_\alpha : 2k_\beta)$-penetration from domain Ω_α to domain Ω_β at the constraint boundary at time t_m in the sense of the metric fucntional if for $\alpha, \beta \in \{1, 2\}$ and $\alpha \neq \beta$

$$
\begin{aligned}
&V(\mathbf{X}_{m\pm}^{(\alpha)}, t_{m\pm}, \boldsymbol{\lambda}) = V(\mathbf{X}_{m+}^{(\beta)}, t_{m+}, \boldsymbol{\lambda}) = V(\mathbf{X}_{m}^{(0)}, t_{m}, \boldsymbol{\lambda}) = 0;\\
&V^{(r_\alpha)}(\mathbf{X}_{m\pm}^{(\alpha)}, t_{m\pm}, \boldsymbol{\lambda}) = 0 \text{ for } r_\alpha = 1, 2, \ldots, 2k_\alpha + 1;\\
&V^{(r_\beta)}(\mathbf{X}_{m+}^{(\beta)}, t_{m+}, \boldsymbol{\lambda}) = 0 \text{ for } r_\beta = 1, 2, \ldots, 2k_\beta;\\
&V(\mathbf{X}_{m\pm\varepsilon}^{(\alpha)}, t_{m\pm\varepsilon}, \boldsymbol{\lambda}) - V(\mathbf{X}_{m\pm}^{(\alpha)}, t_{m\pm}, \boldsymbol{\lambda}) > 0,\\
&V(\mathbf{X}_{m+\varepsilon}^{(\beta)}, t_{m+\varepsilon}, \boldsymbol{\lambda}) - V(\mathbf{X}_{m+}^{(\beta)}, t_{m+}, \boldsymbol{\lambda}) > 0.
\end{aligned}
\tag{5.67}
$$

(ii) The $(2k_\alpha : 2k_\beta)$-desynchronization of the two dynamical systems in Eqs.(5.1) and (5.2) with constraint in Eq.(5.3) is said to be *onset* from the $(2k_\alpha : 2k_\beta)$-penetration from domain Ω_α to domain Ω_β at the constraint boundary at time t_m in the sense of the metric fucntional if for $\alpha, \beta \in \{1, 2\}$ and $\alpha \neq \beta$,

$$
\begin{aligned}
&V(\mathbf{X}_{m\mp}^{(\alpha)}, t_{m\mp}, \boldsymbol{\lambda}) = V(\mathbf{X}_{m+}^{(\beta)}, t_{m+}, \boldsymbol{\lambda}) = V(\mathbf{X}_{m}^{(0)}, t_{m}, \boldsymbol{\lambda}) = 0;\\
&V^{(s_\alpha)}(\mathbf{X}_{m\mp}^{(\alpha)}, t_{m\mp}, \boldsymbol{\lambda}) = 0 \text{ for } s_\alpha = 1, 2, \ldots, 2k_\alpha + 1,
\end{aligned}
$$

$$
\begin{aligned}
&V^{(s_\beta)}(\mathbf{X}_{m+}^{(\beta)}, t_{m+}, \boldsymbol{\lambda}) = 0 \text{ for } s_\beta = 1,2,\ldots,2k_\beta;\\
&V(\mathbf{X}_{m\mp\varepsilon}^{(\alpha)}, t_{m\mp\varepsilon}, \boldsymbol{\lambda}) - V(\mathbf{X}_{m\mp}^{(\alpha)}, t_{m\mp}, \boldsymbol{\lambda}) > 0,\\
&V(\mathbf{X}_{m+\varepsilon}^{(\beta)}, t_{m+\varepsilon}, \boldsymbol{\lambda}) - V(\mathbf{X}_{m+}^{(\beta)}, t_{m+}, \boldsymbol{\lambda}) > 0.
\end{aligned}
\tag{5.68}
$$

Similarly, *the onset condition of the* $(2k_\alpha : 2k_\beta)$*-desynchronization* from the $(2k_\alpha : 2k_\beta)$-penetration on the constraint boundary in Eq.(67) can be called *the vanishing condition of the* $(2k_\alpha : 2k_\beta)$-penetration to form a new $(2k_\alpha : 2k_\beta)$-desynchronization on the constraint boundary. However, in Eq.(5.68), *the vanishing condition of the* $(2k_\alpha : 2k_\beta)$-synchronization to form a new $(2k_\alpha : 2k_\beta)$-penetration can be called *the onset condition of the* $(2k_\alpha : 2k_\beta)$-penetration from the $(2k_\alpha : 2k_\beta)$-desynchronization.

Definition 5.30. For two dynamical systems in Eqs.(5.1) and (5.2) with constraint in Eq.(5.3), there is a metric functional of $V(\mathbf{X}, t, \boldsymbol{\lambda}) = f(\varphi(\mathbf{X}, t, \boldsymbol{\lambda}))$ in Eq.(5.29). For $\mathbf{X}_m^{(\alpha)} \in \Omega_\alpha$ $(\alpha \in \{1,2\})$ and $\mathbf{X}_m^{(0)} \in \partial\Omega_{12}$ at t_m, $\mathbf{X}_m^{(\alpha)} = \mathbf{X}_m^{(0)}$. For any small $\varepsilon > 0$, there is a time interval $[t_{m-\varepsilon}, t_{m+\varepsilon}]$. At $\mathbf{X}^{(\alpha)} \in \Omega_\alpha^{\pm\varepsilon}$ for $t \in [t_{m-\varepsilon}, t_{m+\varepsilon}]$, the functional $V(\mathbf{X}^{(\alpha)}, t, \boldsymbol{\lambda})$ is C^{r_α}-continuous and differentiable $(r_\alpha \geq 2k_\alpha + 1)$ and $|V^{(r_\alpha+1)}(\mathbf{X}^{(\alpha)}, t, \boldsymbol{\lambda})| < \infty$. The $(2k_\alpha : 2k_\beta)$-penetration of the two dynamical systems in Eqs.(5.1) and (5.2) with constraint in Eq.(5.3) is called to be *switched* to a new $(2k_\beta : 2k_\alpha)$-penetration at the constraint boundary at time t_m in the sense of the metric fucntional if for $\alpha, \beta \in \{1,2\}$ and $\alpha \neq \beta$

$$
\begin{aligned}
&V(\mathbf{X}_{m\mp}^{(\alpha)}, t_{m\mp}, \boldsymbol{\lambda}) = V(\mathbf{X}_{m\pm}^{(\beta)}, t_{m\pm}, \boldsymbol{\lambda}) = V(\mathbf{X}_m^{(0)}, t_m, \boldsymbol{\lambda}) = 0;\\
&V^{(s_\alpha)}(\mathbf{X}_{m\mp}^{(\alpha)}, t_{m\mp}, \boldsymbol{\lambda}) = 0 \text{ for } s_\alpha = 1,2,\ldots,2k_\alpha + 1;\\
&V^{(s_\beta)}(\mathbf{X}_{m\pm}^{(\beta)}, t_{m\pm}, \boldsymbol{\lambda}) = 0 \text{ for } s_\beta = 1,2,\ldots,2k_\beta + 1;\\
&V(\mathbf{X}_{m\mp\varepsilon}^{(\alpha)}, t_{m\mp\varepsilon}, \boldsymbol{\lambda}) - V(\mathbf{X}_{m\mp}^{(\alpha)}, t_{m\mp}, \boldsymbol{\lambda}) > 0,\\
&V(\mathbf{X}_{m\pm\varepsilon}^{(\beta)}, t_{m\pm\varepsilon}, \boldsymbol{\lambda}) - V(\mathbf{X}_{m\pm}^{(\beta)}, t_{m\pm}, \boldsymbol{\lambda}) > 0.
\end{aligned}
\tag{5.69}
$$

In the foregoing definition, the condition for the $(2k_\alpha : 2k_\beta)$-penetration switching to the $(2k_\beta : 2k_\alpha)$-penetration at the boundary is presented.

5.3.5 Synchronization to constraint

In the previous section, the definitions for the synchronicity and the corresponding singularity of two dynamical systems to a specific constraint were discussed. What conditions can guarantee such synchronicity of the two dynamical systems to the constraint exists? In this section, necessary and sufficient conditions for the synchronization of two dynamical systems to the specific constraint will be presented.

The synchronicity switching is discussed through the singularity of a flow of the resultant system to the constraint boundary.

Theorem 5.1. For two dynamical systems in Eqs.(5.1) and (5.2) with constraint in Eq.(5.3), there is a metric functional of $V(\mathbf{X},t,\boldsymbol{\lambda}) = f(\varphi(\mathbf{X},t,\boldsymbol{\lambda}))$ in Eq.(5.29). For $\mathbf{X}_m^{(\alpha)} \in \Omega_\alpha$ $(\alpha \in \{1,2\})$ and $\mathbf{X}_m^{(0)} \in \partial\Omega_{12}$ at time t_m, $\mathbf{X}_m^{(\alpha)} = \mathbf{X}_m^{(0)}$. For any small$\varepsilon > 0$, there is a time interval $[t_{m-\varepsilon}, t_m)$ or $(t_m, t_{m+\varepsilon}]$. For $\mathbf{X}^{(\alpha)} \in \Omega_\alpha^{\pm\varepsilon}$ at time $t \in [t_{m-\varepsilon}, t_m)$ or $(t_m, t_{m+\varepsilon}]$, the functional $V(\mathbf{X}^{(\alpha)},t,\boldsymbol{\lambda})$ is C^{r_α}-continuous $(r_\alpha \geq k_\alpha + 1)$ and $|V^{(r_\alpha+2)}(\mathbf{X}^{(\alpha)},t,\boldsymbol{\lambda})| < \infty$. For $\mathbf{X}^{(\alpha)} \in \Omega_\alpha$ and $\mathbf{X}^{(0)} \in \partial\Omega_{12}$, suppose $D^{(s_\alpha)}\mathbb{F}^{(\alpha)}(\mathbf{X}^{(\alpha)},t,\boldsymbol{\pi}^{(\alpha)}) = D^{(s_\alpha)}\mathbb{F}^{(0)}(\mathbf{X}^{(0)},t,\boldsymbol{\lambda})$ $(s_\alpha = 0,1,2,\ldots)$ for $\mathbf{X}^{(\alpha)} = \mathbf{X}^{(0)}$ $(\alpha \in \{1,2\})$. The two dynamical systems in Eqs.(5.1) and (5.2) to the constraint in Eq.(5.3) are synchronized for time $t \in [t_{m_1}, t_{m_2}]$ in the sense of the metric fucntional if and only if

(i) for $\mathbf{X}_m^{(\alpha)} \in \Omega_\alpha$ and $\mathbf{X}_m^{(0)} \in \partial\Omega_{12}$ with any time t_m

$$\mathbf{X}_m^{(\alpha)} = \mathbf{X}_m^{(0)},\; V^{(r_\alpha)}(X_m^{(\alpha)}, t_m, \lambda) = 0 \text{ for } \alpha = 1,2 \text{ and } r_\alpha = 0,1,2,\ldots; \tag{5.70}$$

(ii) for $\mathbf{X}_\kappa^{(\alpha)} \in \Omega_\alpha^{-\varepsilon}$ at $t_\kappa^- \in [t_{m-\varepsilon}, t_m)$ and $\mathbf{X}_m^{(0)} \in \partial\Omega_{12}$ with $t_m \in (t_{m_1}, t_{m_2})$

$$\begin{aligned} &\mathbf{X}_\kappa^{(\alpha)} \neq \mathbf{X}_m^{(0)},\; V^{(1)}(\mathbf{X}_\kappa^{(\alpha)}, t_\kappa^-, \boldsymbol{\lambda}) < 0 \text{ and} \\ &\lim_{t_\kappa^- \to t_m} V^{(1)}(\mathbf{X}_\kappa^{(\alpha)}, t_\kappa^-, \boldsymbol{\lambda}) = 0 \text{ for } \alpha = 1,2; \end{aligned} \tag{5.71}$$

(iii) for $\mathbf{X}_\kappa^{(\alpha)} \in \Omega_\alpha^{+\varepsilon}$ at $t_\kappa^+ \in (t_m, t_{m+\varepsilon}]$ and $\mathbf{X}_m^{(0)} \in \partial\Omega_{12}$ with $t_m \notin [t_{m_1}, t_{m_2}]$

$$\begin{aligned} &\mathbf{X}_\kappa^{(\alpha)} \neq \mathbf{X}_m^{(0)},\; V^{(1)}(\mathbf{X}_\kappa^{(\alpha)}, t_\kappa^+, \boldsymbol{\lambda}) > 0 \text{ and} \\ &\lim_{t_\kappa^+ \to t_m} V^{(1)}(\mathbf{X}_\kappa^{(\alpha)}, t_\kappa^+, \boldsymbol{\lambda}) = 0 \text{ for } \alpha = 1,2; \end{aligned} \tag{5.72}$$

(iv) for $\mathbf{X}_\kappa^{(\alpha)} \in \Omega_\alpha^{\pm\varepsilon}$ at $t_\kappa^- \in [t_{m-\varepsilon}, t_{m-})$ or $t_\kappa^+ \in (t_{m+}, t_{m+\varepsilon}]$ and $\mathbf{X}_m^{(0)} \in \partial\Omega_{12}$ with $t_m = t_{m_1}$ and t_{m_2}

$$\begin{aligned} &\mathbf{X}_\kappa^{(\alpha)} \neq \mathbf{X}_m^{(0)},\quad \lim_{t_\kappa^\pm \to t_{m\pm}} V^{(1)}(\mathbf{X}_\kappa^{(\alpha)}, t_\kappa^\pm, \boldsymbol{\lambda}) = 0, \\ &\lim_{t_\kappa^\pm \to t_{m\pm}} V^{(2)}(\mathbf{X}_\kappa^{(\alpha)}, t_\kappa^\pm, \boldsymbol{\lambda}) < 0 \text{ for } \alpha = 1,2. \end{aligned} \tag{5.73}$$

Proof: (i) For two dynamical systems in Eqs.(5.1) and (5.2) with a constraint condition in Eq.(5.3), the boundary $\partial\Omega_{12}$ in Eq.(5.17) and two domains Ω_α $(\alpha = 1,2)$ in Eq.(5.18) are defined. From Definition 5.13, at time t_m, $\mathbf{X}_m^{(\alpha)} = \mathbf{X}_m^{(0)} \in \partial\Omega_{12}$. So one obtains

$$V(\mathbf{X}_m^{(\alpha)}, t_m, \boldsymbol{\lambda}) = 0.$$

From Definition 5.13, for any time t, one gets $\mathbf{X}^{(\alpha)} = \mathbf{X}^{(0)} \in \partial\Omega_{12}$,

$$\varphi(\mathbf{X}^{(\alpha)}(t),t,\boldsymbol{\lambda}) = \varphi(\mathbf{X}^{(0)}(t),t,\boldsymbol{\lambda}) = 0,$$

which implies

$$V(\mathbf{X}^{(\alpha)}(t),t,\boldsymbol{\lambda}) = V(\mathbf{X}^{(0)}(t),t,\boldsymbol{\lambda}) = 0.$$

On the constraint boundary $\partial\Omega_{12}$,

$$D^{(r_\alpha)}\mathbb{F}^{(\alpha)}(\mathbf{X}^{(\alpha)},t,\boldsymbol{\pi}^{(\alpha)}) = D^{(r_\alpha)}\mathbb{F}^{(0)}(\mathbf{X}^{(0)},t,\boldsymbol{\lambda})$$

gives

$$\frac{d^{r_\alpha+1}\mathbf{X}^{(\alpha)}(t)}{dt^{r_\alpha+1}} = \frac{d^{r_\alpha+1}\mathbf{X}^{(0)}(t)}{dt^{r_\alpha+1}}.$$

The foregoing equation gives

$$V^{(r_\alpha)}(\mathbf{X}^{(\alpha)}(t),t,\boldsymbol{\lambda}) = 0.$$

On the other hand, consider $\mathbf{X}^{(\alpha)}(t) = \mathbf{X}^{(0)}(t) \in \partial\Omega_{12}$ at time t. Selecting $t' = t+\varepsilon$ for any small $\varepsilon > 0$, the Taylor series expansion gives

$$V(\mathbf{X}^{(\alpha)},t',\boldsymbol{\lambda}) - V(\mathbf{X}^{(\alpha)},t,\boldsymbol{\lambda}) = \sum\nolimits_{r_\alpha=1} V^{(r_\alpha)}(\mathbf{X}^{(\alpha)},t,\boldsymbol{\lambda})\varepsilon^{r_\alpha}.$$

Using Eq.(5.70), the foregoing equation yields

$$V(\mathbf{X}^{(\alpha)},t',\boldsymbol{\lambda}) - V(\mathbf{X}^{(\alpha)},t,\boldsymbol{\lambda}) = 0.$$

Because of $V(\mathbf{X}^{(\alpha)},t,\boldsymbol{\lambda}) = 0$, for $t' = t+\varepsilon$, the following equation holds

$$V(\mathbf{X}^{(\alpha)},t',\boldsymbol{\lambda}) = V(\mathbf{X}^{(0)},t',\boldsymbol{\lambda}) = 0.$$

Therefore, $\mathbf{X}^{(\alpha)}(t') = \mathbf{X}^{(0)}(t')$, i.e., $\mathbf{X}^{(\alpha)}(t')$ is on the boundary $\partial\Omega_{12}$.

(ii) and (iii) For $\mathbf{x}_\kappa^{(\alpha)} \in \Omega_\alpha^{\pm\varepsilon}$ at $t_\kappa^- \in [t_{m-\varepsilon},t_m)$ or $t_\kappa^+ \in (t_m,t_{m+\varepsilon}]$ and $\mathbf{X}_m^{(0)} \in \mathbf{X}_m^{(0)} \in \partial\Omega_{12}$ with $t_m \in (t_{m_1},t_{m_2})$,

$$V(\mathbf{X}_\kappa^{(\alpha)},t_\kappa^\pm,\boldsymbol{\lambda}) > 0.$$

Introduce $0 < \varepsilon_1 = |t_{m\pm\varepsilon} - t_\kappa^\pm| < |t_{m\pm\varepsilon} - t_m| = \varepsilon$ for $t_m > t_\kappa^-$ and $t_m < t_\kappa^+$. Because of

$$V(\mathbf{X}_{m\pm\varepsilon}^{(\alpha)},t_{m\pm\varepsilon},\boldsymbol{\lambda}) - V(\mathbf{X}_\kappa^{(\alpha)},t_\kappa^\pm,\boldsymbol{\lambda}) = V^{(1)}(\mathbf{X}_\kappa^{(\alpha)},t_\kappa^\pm,\boldsymbol{\lambda})(\pm\varepsilon_1) + o(\varepsilon_1)$$

and once higher-order terms drop, the foregoing equation leads to

$$V(\mathbf{X}_{m\pm\varepsilon}^{(\alpha)},t_{m\pm\varepsilon},\boldsymbol{\lambda}) - V(\mathbf{X}_\kappa^{(\alpha)},t_\kappa^\pm,\boldsymbol{\lambda}) = V^{(1)}(\mathbf{X}_\kappa^{(\alpha)},t_\kappa^\pm,\boldsymbol{\lambda})(\pm\varepsilon_1).$$

From Definition 5.15 for $t_m \in (t_{m_1},t_{m_2})$ with t_κ^-, we have

$$\lim_{t_\kappa^- \to t_{m-}} V^{(1)}(\mathbf{X}_\kappa^{(\alpha)},t_\kappa^-,\boldsymbol{\lambda}) < 0 \text{ and } \lim_{t_\kappa^- \to t_m} V^{(1)}(\mathbf{X}_\kappa^{(\alpha)},t_\kappa^-,\boldsymbol{\lambda}) = V^{(1)}(\mathbf{X}_m^{(\alpha)},t_m,\boldsymbol{\lambda}) = 0$$

However, using Eq.(5.71), the condition in Definition 5.15 is obtained.

From Definition 5.16 for $t_m \notin [t_{m_1}, t_{m_2}]$ with t_{κ}^{+}, we have

$$\lim_{t_{\kappa}^{+}\to t_{m+}} V^{(1)}(\mathbf{X}_{\kappa}^{(\alpha)}, t_{\kappa}^{+}, \boldsymbol{\lambda}) > 0 \text{ and } \lim_{t_{\kappa}^{+}\to t_{m}} V^{(1)}(\mathbf{X}_{\kappa}^{(\alpha)}, t_{\kappa}^{+}, \boldsymbol{\lambda}) = V^{(1)}(\mathbf{X}_{m}^{(\alpha)}, t_m, \boldsymbol{\lambda}) = 0,$$

However, using Eq. (5.72), the condition in Definition 5.16 is obtained.

(iv) For $\mathbf{X}_{\kappa}^{(\alpha)} \in \Omega_{\alpha}^{\pm\varepsilon}$ at time $t_{\kappa}^{-} \in [t_{m-\varepsilon}, t_{m-})$ or $t_{\kappa}^{+} \in (t_{m+}, t_{m+\varepsilon}]$ and $\mathbf{X}_{m}^{(0)} \in \partial\Omega_{12}$ with $t_m = t_{m_1}$ and t_{m_2},

$$\begin{aligned} &\lim_{t_{\kappa}^{\pm}\to t_{m\pm}} \left[V(\mathbf{X}_{m\pm\varepsilon}^{(\alpha)}, t_{m\pm\varepsilon}, \boldsymbol{\lambda}) - V(\mathbf{X}_{\kappa}^{(\alpha)}, t_{\kappa}^{\pm}, \boldsymbol{\lambda}) \right] \\ &= \lim_{t_{\kappa}^{\pm}\to t_{m\pm}} V^{(1)}(\mathbf{X}_{\kappa}^{(\alpha)}, t_{\kappa}^{\pm}, \boldsymbol{\lambda})(\pm\varepsilon_1) + \lim_{t_{\kappa}^{\pm}\to t_{m\pm}} V^{(2)}(\mathbf{X}_{\kappa}^{(\alpha)}, t_{\kappa}^{\pm}, \boldsymbol{\lambda})(\pm\varepsilon_1)^2 + o(\varepsilon_1^2). \end{aligned}$$

Ignoring the third-order term and the higher-order terms of ε_1 yields

$$\begin{aligned} &\lim_{t_{\kappa}^{\pm}\to t_{m\pm}} \left[V(\mathbf{X}_{m\pm\varepsilon}^{(\alpha)}, t_{m\pm\varepsilon}, \boldsymbol{\lambda}) - V(\mathbf{X}_{\kappa}^{(\alpha)}, t_{\kappa}^{\pm}, \boldsymbol{\lambda}) \right] \\ &= \lim_{t_{\kappa}^{\pm}\to t_{m\pm}} V^{(1)}(\mathbf{X}_{\kappa}^{(\alpha)}, t_{\kappa}^{\pm}, \boldsymbol{\lambda})(\pm\varepsilon_1) + \lim_{t_{\kappa}^{\pm}\to t_{m\pm}} V^{(2)}(\mathbf{X}_{\kappa}^{(\alpha)}, t_{\kappa}^{\pm}, \boldsymbol{\lambda})(\pm\varepsilon_1)^2. \end{aligned}$$

Using $\lim_{t_{\kappa_i}^{\pm}\to t_{m_i}\pm} V^{(1)}(\mathbf{X}_{\kappa_i}^{(\alpha)}, t_{\kappa_i}^{\pm}, \boldsymbol{\lambda}) = 0$, the foregoing equation becomes

$$\lim_{t_{\kappa}^{\pm}\to t_{m\pm}} \left[V(\mathbf{X}_{m\pm\varepsilon}^{(\alpha)}, t_{m\pm\varepsilon}, \boldsymbol{\lambda}) - V(\mathbf{X}_{\kappa}^{(\alpha)}, t_{\kappa}^{\pm}, \boldsymbol{\lambda}) \right] = \lim_{t_{\kappa}^{\pm}\to t_{m\pm}} V^{(2)}(\mathbf{X}_{\kappa}^{(\alpha)}, t_{\kappa}^{\pm}, \boldsymbol{\lambda})(\pm\varepsilon_1)^2.$$

If $\lim_{t_{\kappa}^{\pm}\to t_{m\pm}} V^{(2)}(\mathbf{X}_{\kappa}^{(\alpha)}, t_{\kappa}^{\pm}, \boldsymbol{\lambda}) > 0$, we have

$$\lim_{t_{\kappa}^{\pm}\to t_{m\pm}} \left[V(\mathbf{X}_{m\pm\varepsilon}^{(\alpha)}, t_{m\pm\varepsilon}, \boldsymbol{\lambda}) - V(\mathbf{X}_{\kappa}^{(\alpha)}, t_{\kappa}^{\pm}, \boldsymbol{\lambda}) \right] > 0.$$

From Definition 5.18, the point $(\mathbf{X}_{m_i\pm}^{(\alpha)}, t_{m_i\pm})$ $(i = 1, 2)$ is tangential point to the constraint. The synchronization at such a point appears or disappears. However, the conditions in Definition 5.18, Equation (5.73) can be obtained. This theorem is proved. ■

For the point $(\mathbf{X}_{m_1}^{(\alpha)}, t_{m_1})$, the synchronization will be onset. However, for the point $(\mathbf{X}_{m_2}^{(\alpha)}, t_{m_2})$, the synchronization will vanish. For $t_m \in (t_{m_1}, t_{m_2})$, the synchronization at point $(\mathbf{X}_{m}^{(\alpha)}, t_m)$ on the constraint boundary can be formed. For $t_m \notin [t_{m_1}, t_{m_2}]$, the desynchronization at point $(\mathbf{X}_{m}^{(\alpha)}, t_m)$ on the constraint boundary can be formed. If $t_{m_1} \to -\infty$ and $t_{m_2} \to \infty$, the synchronization is absolute. The synchronization of two dynamical systems to the constraint can occur at any time t_m. Once the synchronization is formed on the constraint boundary, such synchronization on the constraint boundary will not disappear. If the higher order singularity on the boundary exists, the corresponding theorem is presented in a similar fashion.

Theorem 5.2. For two dynamical systems in Eqs.(5.1) and (5.2) with constraint in Eq.(5.3), there is a metric functional of $V(\mathbf{X},t,\boldsymbol{\lambda}) = f(\varphi(\mathbf{X},t,\boldsymbol{\lambda}))$ in Eq.(5.29). For $\mathbf{X}_m^{(\alpha)} \in \Omega_\alpha$ $(\alpha \in \{1,2\})$ and $\mathbf{X}_m^{(0)} \in \partial\Omega_{12}$ at t_m, $\mathbf{X}_m^{(\alpha)} = \mathbf{X}_m^{(0)}$. For any small $\varepsilon > 0$, there is a time interval $[t_{m-\varepsilon},t_m)$ or $(t_m,t_{m+\varepsilon}]$. For $\mathbf{X}^{(\alpha)} \in \Omega_\alpha^{\pm\varepsilon}$ at time $t \in [t_{m-\varepsilon},t_m)$ or $(t_m,t_{m+\varepsilon}]$, the functional $V(\mathbf{X}^{(\alpha)},t,\boldsymbol{\lambda})$ is C^{r_α}-continuous and $|V^{(r_\alpha+1)}(\mathbf{X}^{(\alpha)},t,\boldsymbol{\lambda})| < \infty$ $(r_\alpha \geq k_\alpha+1)$. For $\mathbf{X}^{(\alpha)} \in \Omega_\alpha$ and $\mathbf{X}^{(0)} \in \partial\Omega_{12}$, suppose $D^{(s_\alpha)}\mathbb{F}^{(\alpha)}(\mathbf{X}^{(\alpha)},t,\boldsymbol{\pi}^{(\alpha)}) = D^{(s_\alpha)}\mathbb{F}^{(0)}(\mathbf{X}^{(0)},t,\boldsymbol{\lambda})$ $(s_\alpha = 0,1,2,\ldots)$ for $\mathbf{X}^{(\alpha)} = \mathbf{X}^{(0)}$ $(\alpha \in \{1,2\})$. The two dynamical systems in Eqs.(5.1) and (5.2) to the constraint in Eq.(5.3) are synchronized for time $t \in [t_{m_1},t_{m_2}]$ in the sense of the metric fucntional if and only if

(i) for $\mathbf{X}_m^{(\alpha)} \in \Omega_\alpha$ and $\mathbf{X}_m^{(0)} \in \partial\Omega_{12}$ with $t_m \in (t_{m_1},t_{m_2})$

$$\begin{aligned} &\mathbf{X}_m^{(\alpha)} = \mathbf{X}_m^{(0)},\ V^{(r_\alpha)}(\mathbf{X}_m^{(\alpha)},t_m,\boldsymbol{\lambda}) = 0 \\ &\text{for } \alpha = 1,2 \text{ and } r_\alpha = 0,1,2,\ldots \end{aligned} \tag{5.74}$$

(ii) for $\mathbf{X}_\kappa^{(\alpha)} \in \Omega_\alpha^{-\varepsilon}$ at time $t_\kappa^- \in [t_{m-\varepsilon},t_m)$ and $\mathbf{X}_m^{(0)} \in \partial\Omega_{12}$ with $t_m \in (t_{m_1},t_{m_2})$

$$\begin{aligned} &\mathbf{X}_\kappa^{(\alpha)} \neq \mathbf{X}_m^{(0)},\ \lim_{t_\kappa^- \to t_{m-}} V^{(2k_\alpha)}(\mathbf{X}_\kappa^{(\alpha)},t_\kappa^-,\boldsymbol{\lambda}) = 0 \text{ for } s_\alpha = 1,2,\ldots,2k_\alpha; \\ &V^{(2k_\alpha+1)}(\mathbf{X}_\kappa^{(\alpha)},t_\kappa^-,\boldsymbol{\lambda}) < 0 \text{ and } \lim_{t_\kappa^- \to t_m} V^{(2k_\alpha+1)}(\mathbf{X}_\kappa^{(\alpha)},t_\kappa^-,\boldsymbol{\lambda}) = 0 \text{ for } \alpha = 1,2; \end{aligned} \tag{5.75}$$

(iii) for $\mathbf{X}_\kappa^{(\alpha)} \in \Omega_\alpha^{+\varepsilon}$ at time $t_\kappa^+ \in (t_m,t_{m+\varepsilon}]$ and $\mathbf{X}_m^{(0)} \in \partial\Omega_{12}$ with $t_m \notin [t_{m_1},t_{m_2}]$

$$\begin{aligned} &\mathbf{X}_\kappa^{(\alpha)} \neq \mathbf{X}_m^{(0)},\ \lim_{t_\kappa^+ \to t_{m+}} V^{(s_\alpha)}(\mathbf{X}_\kappa^{(\alpha)},t_\kappa^+,\boldsymbol{\lambda}) = 0 \text{ for } s_\alpha = 1,2,\ldots,2k_\alpha; \\ &V^{(2k_\alpha+1)}(\mathbf{X}_\kappa^{(\alpha)},t_\kappa^+,\boldsymbol{\lambda}) > 0 \text{ and } \lim_{t_\kappa^+ \to t_m} V^{(2k_\alpha+1)}(\mathbf{X}_\kappa^{(\alpha)},t_\kappa^+,\boldsymbol{\lambda}) = 0 \text{ for } \alpha = 1,2; \end{aligned} \tag{5.76}$$

(iv) for $\mathbf{X}_\kappa^{(\alpha)} \in \Omega_\alpha^{+\varepsilon}$ at time $t_\kappa^- \in [t_{m-\varepsilon},t_{m-})$ $t_\kappa^+ \in (t_{m+},t_{m+\varepsilon}]$ and $\mathbf{X}_m^{(0)} \in \partial\Omega_{12}$ with $t_m = t_{m_1}$ and t_{m_2}

$$\begin{aligned} &\mathbf{X}_\kappa^{(\alpha)} \neq \mathbf{X}_m^{(0)},\ \lim_{t_\kappa^\pm \to t_{m\pm}} V^{(s_\alpha)}(\mathbf{X}_\kappa^{(\alpha)},t_\kappa^\pm,\boldsymbol{\lambda}) = 0 \text{ for } s_\alpha = 1,2,\ldots,2k_\alpha+1; \\ &\lim_{t_\kappa^\pm \to t_{m\pm}} V^{(2k_\alpha+2)}(\mathbf{X}_\kappa^{(\alpha)},t_\kappa^\pm,\boldsymbol{\lambda}) < 0 \text{ for } \alpha = 1,2. \end{aligned} \tag{5.77}$$

Proof: (i) For two dynamical systems in Eqs.(5.1) and (5.2) with a constraint condition in Eq.(5.3), the boundary $\partial\Omega_{12}$ in Eq.(5.17) and two domains Ω_α $(\alpha = 1,2)$ in Eq.(5.18) are defined. From Definition 5.13, at time t_m, $\mathbf{X}_m^{(\alpha)} = \mathbf{X}_m^{(0)} \in \partial\Omega_{12}$. So the following equation holds,

$$V(\mathbf{X}_m^{(\alpha)},t_m,\boldsymbol{\lambda}) = 0.$$

From Definition 5.13, for any time t, one obtains $\mathbf{X}^{(\alpha)} = \mathbf{X}^{(0)} \in \partial\Omega_{12}$,

$$\varphi(\mathbf{X}^{(\alpha)}(t),t,\boldsymbol{\lambda}) = \varphi(\mathbf{X}^{(0)}(t),t,\boldsymbol{\lambda}) = 0,$$

which is implies

$$V(\mathbf{X}^{(\alpha)}(t),t,\boldsymbol{\lambda}) = V(\mathbf{X}^{(0)}(t),t,\boldsymbol{\lambda}) = 0.$$

On the boundary $\partial\Omega_{12}$,

$$D^{(r_\alpha)}\mathbb{F}^{(\alpha)}(\mathbf{X}^{(\alpha)},t,\boldsymbol{\pi}^{(\alpha)}) = D^{(r_\alpha)}\mathbb{F}^{(0)}(\mathbf{X}^{(0)},t,\boldsymbol{\lambda})$$

i.e.,

$$\frac{d^{r_\alpha}\mathbf{X}^{(\alpha)}(t)}{dt^{r_\alpha}} = \frac{d^{r_\alpha}\mathbf{X}^{(0)}(t)}{dt^{r_\alpha}}.$$

The foregoing equation gives

$$V^{(r_\alpha)}(\mathbf{X}^{(\alpha)}(t),t,\boldsymbol{\lambda}) = 0.$$

On the other hand, consider a point $\mathbf{X}^{(\alpha)}(t) = \mathbf{X}^{(0)}(t) \in \partial\Omega_{12}$ at time t. Selecting $t' = t + \varepsilon$ for any small $\varepsilon > 0$, the Taylor series expansion gives

$$V(\mathbf{X}^{(\alpha)},t',\boldsymbol{\lambda}) - V(\mathbf{X}^{(\alpha)},t,\boldsymbol{\lambda}) = \sum\nolimits_{r_\alpha=1} V^{(r_\alpha)}(\mathbf{X}^{(\alpha)},t,\boldsymbol{\lambda})\varepsilon^{r_\alpha}.$$

Using Eq.(5.74), the foregoing equation gives

$$V(\mathbf{X}^{(\alpha)},t',\boldsymbol{\lambda}) - V(\mathbf{X}^{(\alpha)},t,\boldsymbol{\lambda}) = 0.$$

Because of $V(\mathbf{X}^{(\alpha)},t,\boldsymbol{\lambda}) = 0$, for $t' = t + \varepsilon$, one obtains

$$V(\mathbf{X}^{(\alpha)},t',\boldsymbol{\lambda}) = V(\mathbf{X}^{(0)},t',\boldsymbol{\lambda}) = 0.$$

Therefore, $\mathbf{X}^{(\alpha)}(t') = \mathbf{X}^{(0)}(t')$, i.e., $\mathbf{X}^{(\alpha)}(t')$ is on the boundary $\partial\Omega_{12}$.

(ii) and (iii) For $\mathbf{X}_\kappa^{(\alpha)} \in \Omega_\alpha^{\pm\varepsilon}$ at time $t_\kappa^- \in [t_{m-\varepsilon},t_m)$ or $t_\kappa^+ \in (t_m,t_{m+\varepsilon}]$ and $\mathbf{X}_m^{(0)} \in \partial\Omega_{12}$ with $t_m \in (t_{m_1},t_{m_2})$,

$$V(\mathbf{X}_\kappa^{(\alpha)},t_\kappa^\pm,\boldsymbol{\lambda}) > 0.$$

Introduce $0 < \varepsilon_1 = |t_{m\pm\varepsilon} - t_\kappa^\pm| < |t_{m\pm\varepsilon} - t_m| = \varepsilon$ for $t_m > t_\kappa^-$ and $t_m < t_\kappa^+$. Because of

$$\begin{aligned} &V(\mathbf{X}_{m\pm\varepsilon}^{(\alpha)},t_{m\pm\varepsilon},\boldsymbol{\lambda}) - V(\mathbf{X}_\kappa^{(\alpha)},t_\kappa^\pm,\boldsymbol{\lambda}) \\ &= \sum\nolimits_{s_\alpha}^{2k_\alpha} V^{(s_\alpha)}(\mathbf{X}_\kappa^{(\alpha)},t_\kappa^\pm,\boldsymbol{\lambda})(\pm\varepsilon_1)^{s_\alpha} + V^{(2k_\alpha+1)}(\mathbf{X}_\kappa^{(\alpha)},t_\kappa^\pm,\boldsymbol{\lambda})(\pm\varepsilon_1)^{2k_\alpha+1} + o((\varepsilon_1)^{2k_\alpha+1}) \end{aligned}$$

and once the $(2k_\alpha + 2)$ and higher-order terms drop, one obtains

$$\begin{aligned} &V(\mathbf{X}_{m\pm\varepsilon}^{(\alpha)},t_{m\pm\varepsilon},\boldsymbol{\lambda}) - V(\mathbf{X}_\kappa^{(\alpha)},t_\kappa^\pm,\boldsymbol{\lambda}) \\ &= \sum\nolimits_{s_\alpha}^{2k_\alpha} V^{(s_\alpha)}(\mathbf{X}_\kappa^{(\alpha)},t_\kappa^\pm,\boldsymbol{\lambda})(\pm\varepsilon_1)^{s_\alpha} + V^{(2k_\alpha+1)}(\mathbf{X}_\kappa^{(\alpha)},t_\kappa^\pm,\boldsymbol{\lambda})(\pm\varepsilon_1)^{2k_\alpha+1} \end{aligned}$$

and

$$\lim_{t_\kappa^\pm \to t_{m\pm}} \left[V(\mathbf{X}_{m\pm\varepsilon}^{(\alpha)}, t_{m\pm\varepsilon}, \boldsymbol{\lambda}) - V(\mathbf{X}_\kappa^{(\alpha)}, t_\kappa^\pm, \boldsymbol{\lambda}) \right]$$
$$= \sum_{s_\alpha}^{2k_\alpha} \lim_{t_\kappa^\pm \to t_{m\pm}} V^{(s_\alpha)}(\mathbf{X}_\kappa^{(\alpha)}, t_\kappa^\pm, \boldsymbol{\lambda})(\pm\varepsilon_1)^{s_\alpha} + \lim_{t_\kappa^\pm \to t_{m\pm}} V^{(2k_\alpha+1)}(\mathbf{X}_\kappa^{(\alpha)}, t_\kappa^\pm, \boldsymbol{\lambda})(\pm\varepsilon_1)^{2k_\alpha+1}.$$

Definition 5.23 for $t_m \in (t_{m_1}, t_{m_2})$ with t_κ^- gives

$$\lim_{t_\kappa^- \to t_{m-}} V^{(2k_\alpha+1)}(\mathbf{X}_\kappa^{(\alpha)}, t_\kappa^-, \boldsymbol{\lambda}) < 0 \text{ and}$$

$$\lim_{t_\kappa^- \to t_m} V^{(2k_\alpha+1)}(\mathbf{X}_\kappa^{(\alpha)}, t_\kappa^-, \boldsymbol{\lambda}) = V^{(2k_\alpha+1)}(\mathbf{X}_m^{(\alpha)}, t_m, \boldsymbol{\lambda}) = 0.$$

However, using Eq.(5.75), the condition in Definition 5.15 is obtained. Definition 5.24 for $t_m \notin [t_{m_1}, t_{m_2}]$ with t_κ^+ leads to

$$\lim_{t_\kappa^+ \to t_{m+}} V^{(2k_\alpha+1)}(\mathbf{X}_\kappa^{(\alpha)}, t_\kappa^+, \boldsymbol{\lambda}) > 0$$
$$\text{and } \lim_{t_\kappa^+ \to t_m} V^{(2k_\alpha+1)}(\mathbf{X}_\kappa^{(\alpha)}, t_\kappa^+, \boldsymbol{\lambda}) = V^{(2k_\alpha+1)}(\mathbf{X}_m^{(\alpha)}, t_m, \boldsymbol{\lambda}) = 0.$$

However, using Eq. (5.76), the condition in Definition 5.16 is obtained.

(iv) Similarly, for $\mathbf{X}_\kappa^{(\alpha)} \in \Omega_\alpha^{\pm\varepsilon}$ at time $t_\kappa^- \in [t_{m-\varepsilon}, t_{m-})$ or $t_\kappa^+ \in (t_{m+}, t_{m+\varepsilon}]$ and $\mathbf{X}_m^{(0)} \in \partial\Omega_{12}$ with $t_m = t_{m_1}$ and t_{m_2},

$$V(\mathbf{X}_{m\pm\varepsilon}^{(\alpha)}, t_{m\pm\varepsilon}, \boldsymbol{\lambda}) - V(\mathbf{X}_\kappa^{(\alpha)}, t_\kappa^\pm, \boldsymbol{\lambda})$$
$$= \sum_{s_\alpha}^{2k_\alpha+1} V^{(s_\alpha)}(\mathbf{X}_\kappa^{(\alpha)}, t_\kappa^\pm, \boldsymbol{\lambda})(\pm\varepsilon_1)^{s_\alpha} + V^{(2k_\alpha+2)}(\mathbf{X}_\kappa^{(\alpha)}, t_\kappa^\pm, \boldsymbol{\lambda})(\pm\varepsilon_1)^{2k_\alpha+2} + o(\varepsilon_1^{2k_\alpha+2}),$$

Ignoring the $(2k_\alpha + 3)$term or higher-order terms, we have

$$\lim_{t_\kappa^\pm \to t_{m\pm}} \left[V(\mathbf{X}_{m\pm\varepsilon}^{(\alpha)}, t_{m\pm\varepsilon}, \boldsymbol{\lambda}) - V(\mathbf{X}_\kappa^{(\alpha)}, t_\kappa^\pm, \boldsymbol{\lambda}) \right]$$
$$= \sum_{s_\alpha}^{2k_\alpha+1} \lim_{t_\kappa^\pm \to t_{m\pm}} V^{(s_\alpha)}(\mathbf{X}_\kappa^{(\alpha)}, t_\kappa^\pm, \boldsymbol{\lambda})(\pm\varepsilon_1)^{s_\alpha} + \lim_{t_\kappa^\pm \to t_{m\pm}} V^{(2k_\alpha+2)}(\mathbf{X}_\kappa^{(\alpha)}, t_\kappa^\pm, \boldsymbol{\lambda})(\pm\varepsilon_1)^{2k_\alpha+2}.$$

Using $\lim_{\kappa_i \to m_i\pm} V^{(s_\alpha)}(\mathbf{X}_{\kappa_i}^{(\alpha)}, t_{\kappa_i}^\pm, \boldsymbol{\lambda}) = 0$ $(s_\alpha = 1, 2, \ldots, 2k_\alpha + 1)$, the foregoing equation gives

$$\lim_{t_\kappa^\pm \to t_{m\pm}} \left[V(\mathbf{X}_{m\pm\varepsilon}^{(\alpha)}, t_{m\pm\varepsilon}, \boldsymbol{\lambda}) - V(\mathbf{X}_\kappa^{(\alpha)}, t_\kappa^\pm, \boldsymbol{\lambda}) \right]$$
$$= \lim_{t_\kappa^\pm \to t_{m\pm}} V^{(2k_\alpha+2)}(\mathbf{X}_\kappa^{(\alpha)}, t_\kappa^\pm, \boldsymbol{\lambda})(\pm\varepsilon_1)^{2k_\alpha+2}.$$

If $\lim_{t_\kappa^\pm \to t_{m\pm}} V^{(2k_\alpha+2)}(\mathbf{X}_\kappa^{(\alpha)}, t_\kappa^\pm, \boldsymbol{\lambda}) > 0$, one obtains

$$\lim_{t_\kappa^\pm \to t_{m\pm}} \left[V(\mathbf{X}_{m\pm\varepsilon}^{(\alpha)}, t_{m\pm\varepsilon}, \boldsymbol{\lambda}) - V(\mathbf{X}_\kappa^{(\alpha)}, t_\kappa^\pm, \boldsymbol{\lambda}) \right] > 0.$$

From Definition 5.26, the point $(\mathbf{X}_{m_i\pm}^{(\alpha)}, t_{m_i\pm})$ $(i = 1,2)$ is a tangential point to the constraint. The synchronization at such a point appears or disappears. However, the conditions in Definition 5.26, Eq. (5.73) can be obtained. This theorem is proved. ■

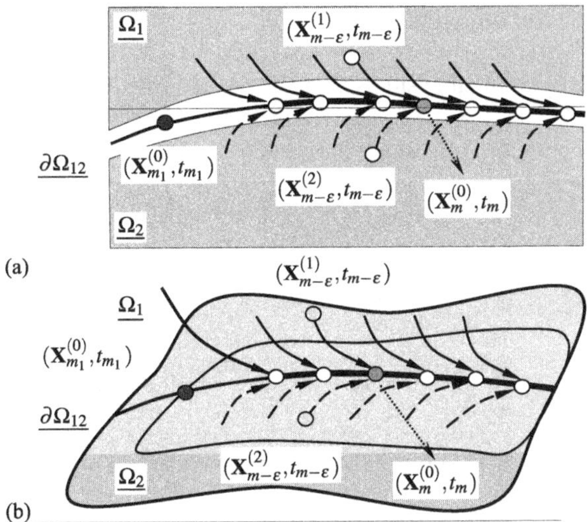

Fig. 5.7 (a) A cross section view and (b) a 3-D view for an absolute synchronization of two dynamical systems to the constraint in vicinity of the constraint boundary $\partial\Omega_{12}$ in $(n+\tilde{n})$-dimensional state space. Any point for synchronization on the constraint boundary, is expressed by $(\mathbf{X}_m^{(0)}, t_m)$. In two domains, the resultant flows in the vicinity of the constraint boundary are expressed by $(\mathbf{X}_{m-\varepsilon}^{(\alpha)}, t_{m-\varepsilon})$ $(\alpha = 1,2)$. The onset point on the constraint boundary is $(\mathbf{X}_{m_1}^{(0)}, t_{m_1})$, depicted by a red circular symbol.

Consider the foregoing two theorems with $t_{m_1} \to -\infty$ and $t_{m_2} \to \infty$. For this case, once the two dynamical systems to the constraint are synchronized, such synchronization can keep forever. To explain the two theorems, the synchronization of the flows of two dynamical systems on the boundary $\partial\Omega_{12}$ are in Fig. 5.7. Any point of a constraint flow on the constraint boundary is expressed by $(\mathbf{X}_m^{(0)}, t_m)$ for synchronization. In the two domains, the reslutant flows in the vicinity of the constraint boundary are expressed by $(\mathbf{X}_{m-\varepsilon}^{(\alpha)}, t_{m-\varepsilon})$ $(\alpha = 1,2)$. The onset point is denoted by $(\mathbf{X}_{m_1}^{(0)}, t_{m_1})$. For $t_m > t_{m_1}$ and $t_{m_2} \to \infty$, all the flows of the resultant system of two dynamical systems will be on the constraint boundary. Thus, the synchronization of the two dynamical systems to the constraint is an absolute synchronization. The starting point of a resultant flow for the synchronization can occur at any time $t_m > t_{m_1}$. However, if t_{m_2} is finite, the two dynamical systems to the constraint can be synchronized only in a finite time interval of $t \in (t_{m_1}, t_{m_2})$. To the point on the boundary at time $t = t_{m_2}$, such synchronization will disappear. Further, the resultant flow on the constraint boundary for synchronization vanishing will enter into the domain, which cannot be synchronized any more in sense of Eq.(5.3). Such synchronization is very easily realized through the discontinuous vector fields to the two dynamical systems to the constraint boundary. For the synchronization of slave and master systems to the constraint, a slave system is controlled by discontinuous, external vector fields in order to make it synchronize with the master system. To answer this question, let us discuss the metric functional first.

For $\mathbb{F}^{(\alpha)}(\mathbf{X}^{(\alpha)},t,\pi^{(\alpha)}) = \mathbb{F}^{(0)}(\mathbf{X}^{(0)},t,\boldsymbol{\lambda})$ at $\mathbf{X}^{(\alpha)} = \mathbf{X}^{(0)}$ $(\alpha \in \{1,2\})$, the synchronization of two dynamical systems to a specific constraint requires $D^{(k)}\varphi(\mathbf{X}^{(\alpha)}(t), t,\boldsymbol{\lambda}) = D^{(k)}\varphi(\mathbf{X}^{(0)}(t),t,\boldsymbol{\lambda}) = 0$. For a metric functional in Eq.(5.32), if $\mathbf{X}^{(\alpha)}(t) = \mathbf{X}^{(0)}(t)$ on the constraint boundary, one gets

$$V^{(1)}(\mathbf{X}^{(\alpha)},t,\boldsymbol{\lambda}) = \frac{d\varphi}{dt} \cdot \varphi \equiv 0, \tag{5.78}$$

because of $\varphi(\mathbf{X}^{(0)},t,\boldsymbol{\lambda}) = 0$. Furthermore, for $\mathbf{X}^{(\alpha)}(t) = \mathbf{X}^{(0)}(t)$, if

$$V^{(2)}\left(\mathbf{X}^{(\alpha)}(t),t,\boldsymbol{\lambda}\right) = \left(\frac{d\varphi}{dt}\right)^2 + \varphi\frac{d^2\varphi}{dt^2} = 0 \tag{5.79}$$

is required, the following condition should be satisfied, i.e.,

$$\frac{d}{dt}\varphi(\mathbf{X}^{(\alpha)}(t),t,\boldsymbol{\lambda}) = 0. \tag{5.80}$$

Continuously, if $\mathbf{X}^{(\alpha)}(t) = \mathbf{X}^{(0)}(t) \in \partial\Omega_{12}$, $V^{(k)}(\mathbf{X}^{(\alpha)}(t),t,\boldsymbol{\lambda}) = 0$ leads to

$$\frac{d^{k-1}}{dt^{k-1}}\varphi(\mathbf{X}^{(\alpha)}(t),t,\boldsymbol{\lambda}) = 0 \text{ for } k = 1,2,\ldots \tag{5.81}$$

Consider another metric functional in Eq.(5.36), and the corresponding time change rate is

$$V^{(1)}(\mathbf{x}^{(\alpha)},t,\boldsymbol{\lambda}) = \mathrm{sgn}(\varphi)\frac{d\varphi}{dt}, \tag{5.82}$$

For $\mathbf{X}^{(\alpha)}(t) = \mathbf{X}^{(0)}(t)$, one requires

$$V^{(1)}(\mathbf{x}^{(\alpha)},t,\boldsymbol{\lambda}) = \mathrm{sgn}(\varphi)\frac{d\varphi}{dt} = 0. \tag{5.83}$$

So one achieves

$$\frac{d}{dt}\varphi(\mathbf{X}^{(\alpha)}(t),t,\boldsymbol{\lambda}) = 0. \tag{5.84}$$

Furthermore, if $\mathbf{X}^{(\alpha)}(t) = \mathbf{X}^{(0)}(t) \in \partial\Omega_{12}$, $V^{(k)}(\mathbf{X}^{(\alpha)}(t),t,\boldsymbol{\lambda}) = 0$ leads to

$$\frac{d^k}{dt^k}\varphi(\mathbf{X}^{(\alpha)}(t),t,\boldsymbol{\lambda}) = 0 \text{ for } k = 1,2,\ldots \tag{5.85}$$

If a resultant system of two different dynamical systems is continuous to the constraint boundary, it is very difficult to make the two different dynamical systems be synchronized with a specific constraint. Most of such synchronization is *asymptotic* as $t \to \infty$. To make the synchronization of two dynamical systems to a specific constraint possible, one often considers control schemes to realize the synchronization through adjusting vector fields. Next, consider the resultant system of two different dynamical systems to be discontinuous to the constraint boundary.

For $\mathbb{F}^{(\alpha)}(\mathbf{X}^{(\alpha)},t,\boldsymbol{\pi}^{(\alpha)}) \neq \mathbb{F}^{(0)}(\mathbf{X}^{(0)},t,\boldsymbol{\lambda})$ for $\mathbf{X}^{(\alpha)} = \mathbf{X}^{(0)}$ $(\alpha \in \{1,2\})$, the synchronization of two dynamical systems with a specific constraint does not require the condition in Eq.(5.85), i.e.,

$$\frac{d^k}{dt^k}\varphi(\mathbf{X}^{(\alpha)}(t),t,\boldsymbol{\lambda}) \neq 0 \text{ for } k = 1,2,\ldots \tag{5.86}$$

To distinguish $\mathbf{X}_{s-}^{(\alpha)}$ from $\mathbf{X}_{s}^{(0)}$ at time $t_s \in [t_m,t_{m+1}]$, a point $\mathbf{X}_{s-}^{(\alpha)} \in \Omega_{\alpha}^{-\varepsilon}$ in the domain infinitesimally approaches a point $\mathbf{X}_{s}^{(0)} \in \partial\Omega_{12}$ on the constraint boundary at time$t$. For $\mathbf{X}_{s-}^{(\alpha)} \in \Omega_{\alpha}^{-\varepsilon}$ (or $\mathbf{X}_{s-}^{(\alpha)} \notin \partial\Omega_{12}$), the corresponding differentiation of vector fields with respect to state variables can be carried out. For $\mathbf{X}_{s}^{(0)} \in \partial\Omega_{12}$ on the constraint boundary, such differentiation cannot be done for $t' \in (t_s - \varepsilon, t_s)$ (any small $\varepsilon > 0$) because the vector fields $(\mathbb{F}^{(\alpha)}(\mathbf{X}^{(\alpha)},t,\boldsymbol{\pi}^{(\alpha)})$, $\alpha \in \{1,2\})$ to the constraint boundary $\partial\Omega_{12}$ are discontinuous (i.e., $\mathbb{F}^{(0)}(\mathbf{X}_{s}^{(0)},t_s,\boldsymbol{\lambda}) \neq \mathbb{F}^{(\alpha)}(\mathbf{X}_{s-}^{(\alpha)},t_{s-},\boldsymbol{\pi}^{(\alpha)})$ for $\mathbf{X}_{s-}^{(\alpha)} = \mathbf{X}_{s}^{(0)}$ at time $t_s = t_{s-}$). Therefore, the time t_s will be replaced by $t_{s-} = t_s - 0$ for a point $\mathbf{X}_{s-}^{(\alpha)} \in \Omega_{\alpha}$. For a metric functional in Eq.(5.32), at $\mathbf{X}^{(\alpha)}(t) = \mathbf{X}^{(0)}(t)$, $V^{(1)}(\mathbf{X}^{(\alpha)}(t),t,\boldsymbol{\lambda}) \equiv 0$ is always observed even if $d\varphi/dt \neq 0$. One also obtains $V^{(2)}(\mathbf{X}^{(\alpha)}(t),t,\boldsymbol{\lambda}) > 0$ if $d\varphi/dt \neq 0$. It implies that $V(\mathbf{X}^{(\alpha)}(t'),t',\boldsymbol{\lambda}) > V(\mathbf{X}^{(0)}(t_m),t_m,\boldsymbol{\lambda})$ for $t' = t_m + \varepsilon$. So one obtains $\mathbf{X}^{(0)}(t) \neq \mathbf{X}^{(\alpha)}(t') \in \Omega_{\alpha}$. It means that the two dynamical systems to a specific constaint cannot be synchronized. For this case, one cannot use such a metric functional in Eq.(5.32) to investigate the synchronization for a discontinuous resultant system. However, for a metric functional in Eq.(5.36), at $\mathbf{X}^{(\alpha)}(t_-) = \mathbf{X}^{(0)}(t)$, one obtains $V^{(1)}(\mathbf{X}^{(\alpha)}(t_{s-}),t_{s-},\boldsymbol{\lambda}) = \text{sgn}(\varphi_{\pm})d\varphi/dt \neq 0$ if $d\varphi/dt \neq 0$. If $d\varphi/dt \leq 0$ and $\varphi_+ = 0_+$ or if $d\varphi/dt \geq 0$ and $\varphi_- = 0_-$, then $V^{(1)}(\mathbf{X}^{(\alpha)}(t),t,\boldsymbol{\lambda}) = \text{sgn}(\varphi_{\pm})d\varphi/dt \leq 0$. From the aforementioned discussion, the metric functional in Eq.(5.36) can be considered as a candidate of metric functional to investigate the synchronization of two dynamical systems with $\mathbb{F}^{(\alpha)}(\mathbf{X}^{(\alpha)},t,\boldsymbol{\pi}^{(\alpha)}) \neq \mathbb{F}^{(0)}(\mathbf{X}^{(0)},t,\boldsymbol{\lambda})$. Under the constraint condition in Eq.(5.3), the corresponding theorem is presented for the synchronization of two dynamical systems in Eqs.(5.1) and (5.2) as follows.

Theorem 5.3. For two dynamical systems in Eqs.(5.1) and (5.2) with constraint in Eq.(5.3), there is a metric functional of $V(\mathbf{X},t,\boldsymbol{\lambda}) = f(\varphi(\mathbf{X},t,\boldsymbol{\lambda}))$ in Eq.(5.29).

For $\mathbf{X}_m^{(\alpha)} \in \Omega_\alpha$ ($\alpha \in \{1,2\}$) and $\mathbf{X}_m^{(0)} \in \partial\Omega_{12}$ at time t_m, $\mathbf{X}_m^{(\alpha)} = \mathbf{X}_m^{(0)}$. For any small $\varepsilon > 0$, there is a time interval $[t_{m-\varepsilon}, t_m)$ or $(t_m, t_{m+\varepsilon}]$. For $\mathbf{X}^{(\alpha)} \in \Omega_\alpha^{\pm\varepsilon}$ at $t \in [t_{m-\varepsilon}, t_m)$ or $(t_m, t_{m+\varepsilon}]$, the functional $V(\mathbf{X}^{(\alpha)}, t, \boldsymbol{\lambda})$ is C^{r_α}-continuous ($r_\alpha \geq 3$) and $|V^{(r_\alpha+1)}(\mathbf{X}^{(\alpha)}, t, \boldsymbol{\lambda})| < \infty$. For $\mathbf{X}^{(\alpha)} \in \Omega_\alpha$ and $\mathbf{X}^{(0)} \in \partial\Omega_{12}$, let $\mathbb{F}^{(\alpha)}(\mathbf{X}^{(\alpha)}, t, \boldsymbol{\pi}^{(\alpha)}) \neq \mathbb{F}^{(0)}(\mathbf{X}^{(0)}, t, \boldsymbol{\lambda})$ for $\mathbf{X}^{(\alpha)} = \mathbf{X}^{(0)}$ and $\alpha \in \{1,2\}$. The two dynamical systems in Eqs.(5.1) and (5.2) to the constraint in Eq.(5.3) are synchronized for time $t \in [t_{m_1}, t_{m_2}]$ in the sense of the metric fucntional if and only if

(i) for $\mathbf{X}_{m-}^{(\alpha)} = \mathbf{X}_m^{(0)}$ and $\mathbf{X}^{(\alpha)}(t) \in \Omega_\alpha$ ($\alpha \in \{1,2\}$) at time $t = t_m \in [t_{m_1}, t_{m_2}]$

$$V(\mathbf{X}_{m-}^{(\alpha)}, t_{m-}, \boldsymbol{\lambda}) = V(\mathbf{X}_m^{(0)}, t_m, \boldsymbol{\lambda}) = 0; \tag{5.87}$$

(ii) for time $t_m \in (t_{m_1}, t_{m_2})$,

$$\mathbf{X}_{m-}^{(\alpha)} = \mathbf{X}_m^{(0)} \text{ and } V^{(1)}(\mathbf{X}_{m-}^{(\alpha)}, t_{m-}, \boldsymbol{\lambda}) < 0 \text{ for } \alpha = 1,2, \tag{5.88}$$

(iii) with penetration at time $t = t_{m_i}$, $\mathbf{X}_{m_i}^{(\alpha)} = \mathbf{X}_{m_i}^{(0)}$ $(i = 1,2)$

$$\left.\begin{array}{l} V^{(1)}(\mathbf{X}_{m_i\pm}^{(\alpha)}, t_{m_i\pm}, \boldsymbol{\lambda}) = 0 \text{ and } V^{(2)}(\mathbf{X}_{m_i\pm}^{(\alpha)}, t_{m_i\pm}, \boldsymbol{\lambda}) > 0 \text{ for } \alpha \in \{1,2\} \\ V^{(1)}(\mathbf{X}_{m_i-}^{(\beta)}, t_{m_i-}, \boldsymbol{\lambda}) < 0 \text{ for } \beta \in \{1,2\} \text{ and } \beta \neq \alpha \end{array}\right\} \tag{5.89}$$

or with desynchronization at time $t = t_{m_i}$, $\mathbf{X}_{m_i}^{(\alpha)} = \mathbf{X}_{m_i}^{(0)}$ $(i = 1,2)$

$$\begin{array}{l} V^{(1)}(\mathbf{X}_{m_i\pm}^{(\alpha)}, t_{m_i\pm}, \boldsymbol{\lambda}) = 0 \text{ and } V^{(2)}(X_{m_i\pm}^{(\alpha)}, t_{m_i\pm}, \boldsymbol{\lambda}) > 0 \text{ for } \alpha \in \{1,2\}, \\ V^{(1)}(\mathbf{X}_{m_i\pm}^{(\beta)}, t_{m_i\pm}, \boldsymbol{\lambda}) = 0 \text{ and } V^{(2)}(X_{m_i\pm}^{(\beta)}, t_{m_i\pm}, \boldsymbol{\lambda}) > 0 \text{ for } \beta \in \{1,2\} \text{ and } \beta \neq \alpha. \end{array} \tag{5.90}$$

Proof: From Definition 5.13, the metric functions for the constraint boundary $\partial\Omega_{12}$ and domains Ω_α $(\alpha = 1,2)$ are given by

$$\begin{array}{ll} V(\mathbf{X}^{(0)}, t, \boldsymbol{\lambda}) = 0 & \text{for } \mathbf{X}^{(0)} \in \partial\Omega_{12}, \\ V(\mathbf{X}^{(\alpha)}, t, \boldsymbol{\lambda}) > 0 & \text{for } \mathbf{X}^{(\alpha)} \in \Omega_\alpha, \alpha = 1,2. \end{array}$$

(i) For $t = t_{m-}$ and $\mathbf{X}^{(\alpha)} = \mathbf{X}_{m-}^{(\alpha)} \in \Omega_\alpha$ ($\alpha \in \{1,2\}$), $\mathbf{X}_{m-}^{(\alpha)} = \mathbf{X}_m^{(0)} \in \partial\Omega_{12}$. Further,

$$V(\mathbf{X}_{m-}^{(\alpha)}, t_{m-}, \boldsymbol{\lambda}) = V(\mathbf{X}_m^{(0)}, t_m, \boldsymbol{\lambda}) = 0,$$

Equation (5.87) is obtained, vice versa.

(ii) For time $t_m \in (t_{m_1}, t_{m_2})$, $\mathbf{X}_{m-}^{(\alpha)} = \mathbf{X}_m^{(0)} \in \partial\Omega_{12}$. Consider a point $\mathbf{X}_{m-\varepsilon}^{(\alpha)} \in \Omega_\alpha^\varepsilon$ for $t_{m-\varepsilon} = t_m - \varepsilon$ in the neighborhood of $\mathbf{X}_m^{(0)} \in \partial\Omega_{12}$ and $\varepsilon > 0$. We have

$$V(\mathbf{X}_{m-\varepsilon}^{(\alpha)}, t_{m-\varepsilon}, \boldsymbol{\lambda}) - V(\mathbf{X}_{m-}^{(\alpha)}, t_{m-}, \boldsymbol{\lambda}) = -V^{(1)}(\mathbf{X}_{m-}^{(\alpha)}, t_{m-}, \boldsymbol{\lambda})\varepsilon + o(\varepsilon).$$

Because of any selection of $\varepsilon > 0$, if

$$V^{(1)}(\mathbf{X}_{m-}^{(\alpha)}, t_{m-}, \boldsymbol{\lambda}) < 0 \text{ for } \alpha = 1, 2,$$

then

$$V(\mathbf{X}_{m-\varepsilon}^{(\alpha)}, t_{m-\varepsilon}, \boldsymbol{\lambda}) - V(\mathbf{X}_{m-}^{(\alpha)}, t_{m-}, \boldsymbol{\lambda}) > 0.$$

From Definition 5.15, the two dynamical systems to a specific constraint are synchronized for time interval of $t_m \in (t_{m_1}, t_{m_2})$. However, Eq. (5.88) is achieved if $V(\mathbf{X}_{m-}^{(\alpha)}, t_{m-}, \boldsymbol{\lambda}) - V(\mathbf{X}_{m-\varepsilon}^{(\alpha)}, t_{m-\varepsilon}, \boldsymbol{\lambda}) < 0$.

(iii) At time $t = t_{m_i}, \mathbf{X}_{m_i\pm}^{(\alpha)} = \mathbf{X}_{m_i}^{(0)} \in \partial\Omega_{12}$. Consider a point $\mathbf{X}_{m_i\pm\varepsilon}^{(\alpha)} \in \Omega_\alpha$ $(\alpha = 1, 2)$ for $t_{m_i\pm\varepsilon} = t_{m_i} \pm \varepsilon$ in the neighborhood of $\mathbf{X}_m^{(0)} \in \partial\Omega_{12}$ and $\varepsilon > 0$. The Taylor series expansion gives

$$\begin{aligned} &V(\mathbf{X}_{m_2\pm\varepsilon}^{(\alpha)}, t_{m_2\pm\varepsilon}, \boldsymbol{\lambda}) - V(\mathbf{X}_{m_i\pm}^{(\alpha)}, t_{m_i\pm}, \boldsymbol{\lambda}) \\ &= \pm V^{(1)}(\mathbf{X}_{m_i\pm}^{(\alpha)}, t_{m_i\pm}, \boldsymbol{\lambda})\varepsilon + V^{(2)}(\mathbf{X}_{m_i\pm}^{(\alpha)}, t_{m_i\pm}, \boldsymbol{\lambda})\varepsilon^2 + o(\varepsilon^2). \end{aligned}$$

If the third and higher order terms are dropped in the foregoing equation in Ω_α $(\alpha = 1, 2)$, with the condition

$$V^{(1)}(\mathbf{X}_{m_i\pm}^{(\alpha)}, t_{m_i\pm}, \boldsymbol{\lambda}) = 0,$$

the following equation is achieved:

$$V(\mathbf{X}_{m_2\pm\varepsilon}^{(\alpha)}, t_{m_2\pm\varepsilon}, \boldsymbol{\lambda}) - V(\mathbf{X}_{m_i\pm}^{(\alpha)}, t_{m_i\pm}, \boldsymbol{\lambda}) = V^{(2)}(\mathbf{X}_{m_i\pm}^{(\alpha)}, t_{m_i\pm}, \boldsymbol{\lambda})\varepsilon^2 + o(\varepsilon^2).$$

If $V^{(1)}(\mathbf{X}_{m_i-}^{(\alpha)}, t_{m_i-}, \boldsymbol{\lambda}) = V^{(1)}(\mathbf{X}_{m_i+}^{(\alpha)}, t_{m_i+}, \boldsymbol{\lambda}) \neq 0$ and only the first order term in the Taylor series expansion is considered, one gets

$$V(\mathbf{X}_{m_2\pm\varepsilon}^{(\alpha)}, t_{m_2\pm\varepsilon}, \boldsymbol{\lambda}) - V(\mathbf{X}_{m_i\pm}^{(\alpha)}, t_{m_i\pm}, \boldsymbol{\lambda}) = \pm V^{(1)}(\mathbf{X}_{m_i-}^{(\alpha)}, t_{m_i-}, \boldsymbol{\lambda})\varepsilon.$$

For $\alpha, \beta \in \{1, 2\}$ and $\alpha \neq \beta$, from Definition 5.19, the disappearance and appearance of synchronization with the penetration require

$$\begin{aligned} &V(\mathbf{X}_{m_i\pm\varepsilon}^{(\alpha)}, t_{m_i\pm\varepsilon}, \boldsymbol{\lambda}) - V(\mathbf{X}_{m_i\pm}^{(\alpha)}, t_{m_i\pm}, \boldsymbol{\lambda}) > 0, \\ &V(\mathbf{X}_{m_i-\varepsilon}^{(\beta)}, t_{m_i-\varepsilon}, \boldsymbol{\lambda}) - V(\mathbf{X}_{m_i-}^{(\beta)}, t_{m_i-}, \boldsymbol{\lambda}) > 0, \end{aligned}$$

from which Eq.(5.89) is obtained, vice versa.

For $\alpha, \beta \in \{1, 2\}$ and $\alpha \neq \beta$, from Definition 5.20, the disappearance and onset of synchronization with the desynchronization require

$$\begin{aligned} &V(\mathbf{X}_{m_i\pm\varepsilon}^{(\alpha)}, t_{m_2\pm\varepsilon}, \boldsymbol{\lambda}) - V(\mathbf{X}_{m_i\pm}^{(\alpha)}, t_{m_i\pm}, \boldsymbol{\lambda}) > 0, \\ &V(\mathbf{X}_{m_i\pm\varepsilon}^{(\beta)}, t_{m_i\pm\varepsilon}, \boldsymbol{\lambda}) - V(\mathbf{X}_{m_i\pm}^{(\beta)}, t_{m_i\pm}, \boldsymbol{\lambda}) > 0, \end{aligned}$$

from which Eq.(5.90) is obtained, vice versa. Therefore, this theorem is proved. ■

From the foregoing theorem, the synchronization of two dynamical systems to a special constraint requires that the first-order derivative of the metric functional be less than zero. The *onset and vanishing* conditions of the synchronization in Eqs.(5.89) and (5.90) are the *vanishing and onset* conditions relative to the penetration and desynchronization, respectively. If the first-order derivative is zero, under what conditions can two dynamical systems to a special constraint be synchronized in sense of Eq.(5.3)? The following theorem will consider the synchronization of two dynamical systems to a special constraint with higher-order singularity.

Theorem 5.4. For two dynamical systems in Eqs.(5.1) and (5.2) with constraint in Eq.(5.3), there is a metric functional of $V(\mathbf{X},t,\boldsymbol{\lambda}) = f(\varphi(\mathbf{X},t,\boldsymbol{\lambda}))$ in Eq.(5.29). For $\mathbf{X}_m^{(\alpha)} \in \Omega_\alpha$ ($\alpha \in \{1,2\}$) and $\mathbf{X}_m^{(0)} \in \partial\Omega_{12}$ at t_m, $\mathbf{X}_m^{(\alpha)} = \mathbf{X}_m^{(0)}$. For any small $\varepsilon > 0$, there is a time interval $[t_{m-\varepsilon}, t_m)$ or $(t_m, t_{m+\varepsilon}]$. For $\mathbf{X}^{(\alpha)} \in \Omega_\alpha^{\pm\varepsilon}$ at $t \in [t_{m-\varepsilon}, t_m)$ or $(t_m, t_{m+\varepsilon}]$, the functional $V(\mathbf{X}^{(\alpha)},t,\boldsymbol{\lambda})$ is C^{r_α}-continuous ($r_\alpha \geq 2k_\alpha + 1$) and $|V^{(r_\alpha+2)}(\mathbf{X},t,\boldsymbol{\lambda})| < \infty$. For $\mathbf{X}^{(\alpha)} \in \Omega_\alpha$ and $\mathbf{X}^{(0)} \in \partial\Omega_{12}$, let $\mathbb{F}^{(\alpha)}(\mathbf{X}^{(\alpha)},t,\boldsymbol{\pi}^{(\alpha)}) \neq \mathbb{F}^{(0)}(\mathbf{X}^{(0)},t,\boldsymbol{\lambda})$ for $\mathbf{X}^{(\alpha)} = \mathbf{X}^{(0)}$ and $\alpha \in \{1,2\}$. The two dynamical systems in Eqs.(5.1) and (5.2) to constraint in Eq.(5.3) are synchronized of the $(2k_\alpha : 2k_\beta)$-type for time $t \in [t_{m_1}, t_{m_2}]$ in the sense of the metric functional if and only if

(i) for $\mathbf{X}_{m-}^{(\alpha)} = \mathbf{X}_m^{(0)}$ and $\mathbf{X}^{(\alpha)}(t) \in \Omega_\alpha$ ($\alpha \in \{1,2\}$) at time $t = t_m \in [t_{m_1}, t_{m_2}]$

$$V(\mathbf{X}_{m-}^{(\alpha)}, t_{m-}, \boldsymbol{\lambda}) = V(\mathbf{X}_m^{(0)}, t_m, \boldsymbol{\lambda}) = 0. \tag{5.91}$$

(ii) for time $t_m \in (t_{m_1}, t_{m_2})$,

$$\left.\begin{aligned} &\mathbf{X}_{m-}^{(\alpha)} = \mathbf{X}_m^{(0)} \text{ and } V^{(s_\alpha)}(\mathbf{X}_{m-}^{(\alpha)}, t_{m-}, \boldsymbol{\lambda}) = 0 \quad (s_\alpha = 1,2,\ldots,2k_\alpha) \\ &V^{(2k_\alpha+1)}(\mathbf{X}_{m-}^{(\alpha)}, t_{m-}, \boldsymbol{\lambda}) < 0 \text{ for } \alpha = 1,2. \end{aligned}\right\} \tag{5.92}$$

(iii) with the $(2k_\alpha : 2k_\beta)$-penetration for time $t = t_{m_i}$, $\mathbf{X}_{m_i}^{(\alpha)} = \mathbf{X}_{m_i}^{(0)}$ $(i = 1,2)$,

$$\left.\begin{aligned} &V^{(s_\alpha)}(\mathbf{X}_{m_i\pm}^{(\alpha)}, t_{m_i\pm}, \boldsymbol{\lambda}) = 0 \ (s_\alpha = 1,2,\ldots,2k_\alpha + 1), \\ &\text{and } V^{(2k_\alpha+2)}(\mathbf{X}_{m_i\pm}^{(\alpha)}, t_{m_i\pm}, \boldsymbol{\lambda}) > 0 \text{ for } \alpha \in \{1,2\}; \\ &V^{(s_\beta)}(\mathbf{X}_{m_i-}^{(\beta)}, t_{m_i-}, \boldsymbol{\lambda}) = 0 \ (s_\beta = 1,2,\ldots,2k_\beta), \\ &V^{(2k_\beta+1)}(\mathbf{X}_{m_i-}^{(\beta)}, t_{m_i-}, \boldsymbol{\lambda}) < 0 \text{ for } \beta \in \{1,2\} \text{ and } \beta \neq \alpha. \end{aligned}\right\} \tag{5.93}$$

or with the $(2k_\alpha : 2k_\beta)$-desynchronization for time $t = t_{m_i}$, $\mathbf{X}_{m_i}^{(\alpha)} = \mathbf{X}_{m_i}^{(0)}$ $(i = 1,2)$,

$$\begin{aligned}
&V^{(s_\alpha)}(\mathbf{X}^{(\alpha)}_{m_i\pm},t_{m_i\pm},\boldsymbol{\lambda})=0\ (s_\alpha=1,2,\ldots,2k_\alpha+1),\\
&\text{and } V^{(2k_\alpha+2)}(\mathbf{X}^{(\alpha)}_{m_i\pm},t_{m_i\pm},\boldsymbol{\lambda})>0 \text{ for } \alpha\in\{1,2\};\\
&V^{(s_\beta)}(\mathbf{X}^{(\beta)}_{m_i\pm},t_{m_i\pm},\boldsymbol{\lambda})=0\ (s_\beta=1,2,\ldots,2k_\beta+1),\\
&V^{(2k_\beta+2)}(\mathbf{X}^{(\beta)}_{m_i\pm},t_{m_i\pm},\boldsymbol{\lambda})>0 \text{ for } \beta\in\{1,2\} \text{ and } \beta\neq\alpha.
\end{aligned} \tag{5.94}$$

Proof: From Definition 5.13, the metric functions for the boundary $\partial\Omega_{12}$ and domains Ω_α $(\alpha=1,2)$ are given by

$$\begin{aligned}
&V(\mathbf{X}^{(0)},t,\boldsymbol{\lambda})=0 \quad \text{for } \mathbf{X}^{(0)}\in\partial\Omega_{12},\\
&V(\mathbf{X}^{(\alpha)},t,\boldsymbol{\lambda})>0 \quad \text{for } \mathbf{X}^{(\alpha)}\in\Omega_\alpha, \alpha=1,2.
\end{aligned}$$

(i) For $t=t_m\in[t_{m_1},t_{m_2}]$ and $\mathbf{X}^{(\alpha)}=\mathbf{X}^{(\alpha)}_m\in\Omega_\alpha$, $\mathbf{X}^{(\alpha)}_{m-}=\mathbf{X}^{(0)}_m\in\partial\Omega_{12}$. Further,

$$V(\mathbf{X}^{(\alpha)}_{m-},t_{m-},\boldsymbol{\lambda})=V(\mathbf{X}^{(0)}_m,t_m,\boldsymbol{\lambda})=0.$$

So Eq.(5.91) is obtained. If Eq.(5.91) exists, from Definition 5.13, $\mathbf{X}^{(\alpha)}_{m-}=\mathbf{X}^{(0)}_m\in\partial\Omega_{12}$.

(ii) For time $t_m\in(t_{m_1},t_{m_2})$, $\mathbf{X}^{(\alpha)}_{m-}=\mathbf{X}^{(0)}_m\in\partial\Omega_{12}$. Consider a point $\mathbf{X}^{(\alpha)}_{m-\varepsilon}\in\Omega^{\varepsilon}_\alpha$ for $t_{m-\varepsilon}=t_m-\varepsilon$ in the neighborhood of $\mathbf{X}^{(0)}_m\in\partial\Omega_{12}$ and $\varepsilon>0$. The following Taylor series expansion is achieved.

$$\begin{aligned}
&V(\mathbf{X}^{(\alpha)}_{m-\varepsilon},t_{m-\varepsilon},\boldsymbol{\lambda})-V(\mathbf{X}^{(\alpha)}_{m-},t_{m-},\boldsymbol{\lambda})\\
&=\sum\nolimits_{s_\alpha=1}^{2k_\alpha}V^{(s_\alpha)}(\mathbf{X}^{(\alpha)}_{m-},t_{m-},\boldsymbol{\lambda})(-\varepsilon)^{s_\alpha}-V^{(2k_\alpha+1)}(\mathbf{X}^{(\alpha)}_{m-},t_{m-},\boldsymbol{\lambda})\varepsilon^{2k_\alpha+1}+o(\varepsilon^{2k_\alpha+1}).
\end{aligned}$$

Due to the higher order singularity, i.e.,

$$V^{(s_\alpha)}(\mathbf{X}^{(\alpha)}_{m-},t_{m-},\boldsymbol{\lambda})=0 \text{ for } s_\alpha=1,2,\ldots,2k_\alpha,$$

and by ignoring of the $(2k_\alpha+2)$-order and higher-order terms, the Taylor series expansion gives

$$V(\mathbf{X}^{(\alpha)}_{m-\varepsilon},t_{m-\varepsilon},\boldsymbol{\lambda})-V(\mathbf{X}^{(\alpha)}_{m-},t_{m-},\boldsymbol{\lambda})=-V^{(2k_\alpha+1)}(\mathbf{X}^{(\alpha)}_{m-},t_{m-},\boldsymbol{\lambda})\varepsilon^{2k_\alpha+1}.$$

From Definition 5.24, the synchronization of two dynamical systems to a specific constraint for time $t_m\in(t_{m_1},t_{m_2})$ requires

$$V(\mathbf{X}^{(\alpha)}_{m-\varepsilon},t_{m-\varepsilon},\boldsymbol{\lambda})-V(\mathbf{X}^{(\alpha)}_{m-},t_{m-},\boldsymbol{\lambda})>0.$$

Thus,

$$V^{(2k_\alpha+1)}(\mathbf{X}^{(\alpha)}_{m-},t_{m-},\boldsymbol{\lambda})<0.$$

However, if

$$V^{(2k_\alpha+1)}(\mathbf{X}_{m-}^{(\alpha)}, t_{m-}, \boldsymbol{\lambda}) < 0,$$

then

$$V\left(\mathbf{X}_{m-\varepsilon}^{(\alpha)}, t_{m-\varepsilon}, \boldsymbol{\lambda}\right) - V(\mathbf{X}_{m-}^{(\alpha)}, t_{m-}, \boldsymbol{\lambda}) > 0$$

is achieved, which implies the two dynamical systems to the specific constraint are synchronized for time $t_m \in (t_{m_1}, t_{m_2})$.

(iii) At time $t = t_{m_i}$, $\mathbf{X}_{m_i\pm}^{(\alpha)} = \mathbf{X}_{m_i}^{(0)} \in \partial\Omega_{12}$. Consider a point $\mathbf{X}_{m_i\pm\varepsilon}^{(\alpha)} \in \Omega_\alpha$ for $t_{m_i\pm\varepsilon} = t_{m_i} \pm \varepsilon$ in the neighborhood of $\mathbf{X}_{m_i}^{(0)} \in \partial\Omega_{12}$ and $\varepsilon > 0$. The Taylor series expansion gives

$$\begin{aligned} &V(\mathbf{X}_{m_i\pm\varepsilon}^{(\alpha)}, t_{m_i\pm\varepsilon}, \boldsymbol{\lambda}) - V(\mathbf{X}_{m_i\pm}^{(\alpha)}, t_{m_i\pm}, \boldsymbol{\lambda}) \\ &= \sum\nolimits_{s_\alpha=1}^{2k_\alpha+1} V^{(r_\alpha)}(\mathbf{X}_{m_i\pm}^{(\alpha)}, t_{m_i\pm}, \boldsymbol{\lambda})(\pm\varepsilon)^{s_\alpha} + V^{(2k_\alpha+2)}(\mathbf{X}_{m_i\pm}^{(\alpha)}, t_{m_i\pm}, \boldsymbol{\lambda})\varepsilon^{2k_\alpha+2} + o(\varepsilon^{2k_\alpha+2}). \end{aligned}$$

Letting $\alpha = \beta$, because of the higher order singularity of the V-function in domain Ω_β, i.e.,

$$V^{(s_\beta)}(\mathbf{X}_{m_i\pm}^{(\beta)}, t_{m_i\pm}, \boldsymbol{\lambda}) = 0 \quad (s_\beta = 1, 2, \ldots, 2k_\beta)$$

and once the higher order terms of $\varepsilon^{2k_\beta+1}$ are dropped, one obtains

$$V(\mathbf{X}_{m_i\pm\varepsilon}^{(\beta)}, t_{m_i\pm\varepsilon}, \boldsymbol{\lambda}) - V(\mathbf{X}_{m_i\pm}^{(\beta)}, t_{m_i\pm}, \boldsymbol{\lambda}) = \pm V^{(2k_\beta+1)}(\mathbf{X}_{m_i\pm}^{(\beta)}, t_{m_i\pm}, \boldsymbol{\lambda})\varepsilon^{2k_\beta+1}.$$

Similarly, if the following equation exists

$$V^{(s_\alpha)}(\mathbf{X}_{m_i\pm}^{(\alpha)}, t_{m_i\pm}, \boldsymbol{\lambda}) = 0 \quad (s_\alpha = 1, 2, \ldots, 2k_\alpha + 1)$$

and the higher order term of $\varepsilon^{2k_\alpha+2}$ will not be considered, the Taylor series expansion gives

$$V(\mathbf{X}_{m_i\pm\varepsilon}^{(\alpha)}, t_{m_i\pm\varepsilon}, \boldsymbol{\lambda}) - V(\mathbf{X}_{m_i\pm}^{(\alpha)}, t_{m_i\pm}, \boldsymbol{\lambda}) = V^{(2k_\alpha+2)}(\mathbf{X}_{m_i\pm}^{(\alpha)}, t_{m_i\pm}, \boldsymbol{\lambda})\varepsilon^{2k_\alpha+2}.$$

From Definition 5.27, the onset and vanishing conditions of the $(2k_\alpha : 2k_\beta)$-synchronization of the two dynamical systems with a corresponding penetration on the constraint boundary $\partial\Omega_{\alpha\beta}$ are

$$V(\mathbf{X}_{m_i\mp\varepsilon}^{(\alpha)}, t_{m_i\mp\varepsilon}, \boldsymbol{\lambda}) - V(\mathbf{X}_{m_i\mp}^{(\alpha)}, t_{m_i\mp}, \boldsymbol{\lambda}) > 0,$$

$$V(\mathbf{X}_{m_i-}^{(\beta)}, t_{m_i-}, \boldsymbol{\lambda}) - V(\mathbf{X}_{m_i-\varepsilon}^{(\beta)}, t_{m_i-\varepsilon}, \boldsymbol{\lambda}) < 0,$$

with

$$V^{(s_\alpha)}(\mathbf{X}_{m_i\mp}^{(\alpha)}, t_{m_i\mp}, \boldsymbol{\lambda}) = 0 \quad (s_\alpha = 1, 2, \ldots, 2k_\alpha + 1),$$

$$V^{(s_\beta)}(\mathbf{X}_{m_i-}^{(\beta)}, t_{m_i-}, \boldsymbol{\lambda}) = 0 \quad (s_\beta = 1, 2, \ldots, 2k_\beta).$$

Thus, one gets

$$V^{(2k_\alpha+2)}(\mathbf{X}^{(\alpha)}_{m_i\mp}, t_{m_i\mp}, \boldsymbol{\lambda}) > 0 \text{ and } V^{(2k_\beta+1)}(\mathbf{X}^{(\beta)}_{m_i-}, t_{m_i-}, \boldsymbol{\lambda}) < 0.$$

In other words, Equation (5.93) is obtained. If Eq.(5.93) holds, the conditions in Definition 5.27 can be obtained for the onset and vanishing condition for synchronization from the penetration.

If the $(2k_\alpha : 2k_\beta)$-synchronization of two dynamical systems to a specific constraint vanishes and appears with a $(2k_\alpha : 2k_\beta)$-desynchronization, the following conditions are required

$$V(\mathbf{X}^{(\alpha)}_{m_i\mp\varepsilon}, t_{m_i\mp\varepsilon}, \boldsymbol{\lambda}) - V(\mathbf{X}^{(\alpha)}_{m_i\mp}, t_{m_i\mp}, \boldsymbol{\lambda}) > 0,$$

$$V(\mathbf{X}^{(\beta)}_{m_i\mp\varepsilon}, t_{m_i\mp\varepsilon}, \boldsymbol{\lambda}) - V(\mathbf{X}^{(\beta)}_{m_i\mp}, t_{m_i\mp}, \boldsymbol{\lambda}) > 0,$$

with the singularity conditions

$$V^{(s_\alpha)}(\mathbf{X}^{(\alpha)}_{m_i\mp}, t_{m_i\mp}, \boldsymbol{\lambda}) = 0 \quad (s_\alpha = 1, 2, \ldots, 2k_\alpha + 1),$$

$$V^{(s_\beta)}(\mathbf{X}^{(\beta)}_{m_i\mp}, t_{m_i\mp}, \boldsymbol{\lambda}) = 0 \quad (s_\beta = 1, 2, \ldots, 2k_\beta + 1).$$

So one obtains

$$V^{(2k_\alpha+2)}(\mathbf{X}^{(\alpha)}_{m_i\mp}, t_{m_i\mp}, \boldsymbol{\lambda}) > 0 \text{ and } V^{(2k_\beta+2)}(\mathbf{X}^{(\beta)}_{m_i\mp}, t_{m_i\mp}, \boldsymbol{\lambda}) > 0,$$

i.e., Equation (5.94) is obtained, vice versa. Therefore, this theorem is proved. ■

In the foregoing theorem, the *onset and vanishing* conditions of the $(2k_\alpha : 2k_\beta)$-synchronization in Eqs.(5.93) and (5.94) for time $t = t_{m_i}$ $(i = 1, 2)$ are also the *vanishing and onset* conditions of the $(2k_\alpha : 2k_\beta)$-penetration and the $(2k_\alpha : 2k_\beta)$-desynchronization, respectively. To explain the synchronization of the two dynamical systems under the condition in Eq.(5.3) in the previous two theorems, such synchronization is sketched in Fig.5.8. On the constraint boundary, any point for synchronization is expressed by $(\mathbf{X}^{(0)}_m, t_m)$. In the two domains, any flows in the vicinity of the boundary are expressed by $(\mathbf{X}^{(\alpha)}_{m-\varepsilon}, t_{m-\varepsilon})$ $(\alpha = 1, 2)$. The onset and vanishing points are $(\mathbf{X}^{(0)}_{m_1}, t_{m_1})$ and $(\mathbf{X}^{(0)}_{m_2}, t_{m_2})$ with red and blue circular symbols. Both of the two points belong to a sub-manifold on the boundary in the $(n+\tilde{n})$-dimensional phase space. Once a flow of the resultant system of two dynamical systems from domain Ω_1 comes to any point of the sub-region on the constraint boundary, the synchronization of the two dynamical systems to the constraint occurs until the point $(\mathbf{X}^{(0)}_{m_2}, t_{m_2})$ is reached. If $t_{m_2} \to \infty$, such synchronization will not disappear forever. For $t_m > t_{m_1}$, once the resultant flows are on the constraint boundary, the synchronization of the two dynamical system to the constraint will keep forever.

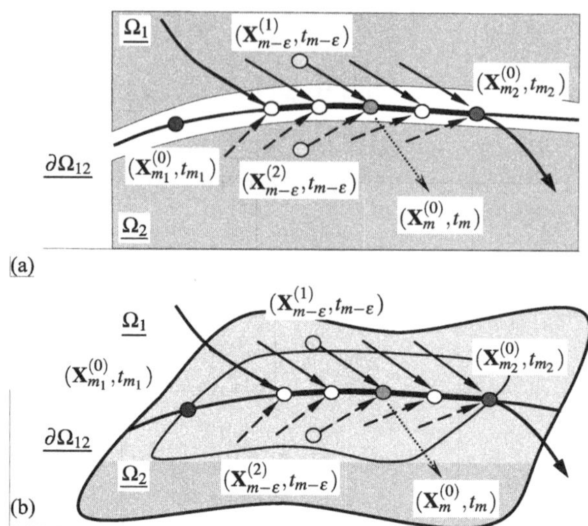

Fig. 5.8 (a) A cross-section view and (b) A 3-D view of the synchronization of resultant flows in vicinity of the constraint boundary $\partial\Omega_{12}$ in $(n+\tilde{n})$-dimensional state space. On the constraint boundary, any point for synchronization is expressed by $(\mathbf{X}^{(0)}_m, t_m)$. In two domains, the resultant flows in the vicinity of the constraint boundary are expressed by $(\mathbf{X}^{(\alpha)}_{m-\varepsilon}, t_{m-\varepsilon})$ $(\alpha = 1,2)$. The onset and vanishing points are $(\mathbf{X}^{(0)}_{m_1}, t_{m_1})$ and $(\mathbf{X}^{(0)}_{m_2}, t_{m_2})$ with red and blue circular symbols.

5.3.6 Desynchronization to constraint

In the previous four theorems, the synchronization for two dynamical systems to the constraint in Eq.(5.3) is discussed. Next, the desynchronization of two dynamical systems will be similarly discussed. The desynchronization is another phenomenon opposite to the synchronization. If $\mathbb{F}^{(\alpha)}(\mathbf{X}^{(\alpha)}, t, \pi^{(\alpha)}) = \mathbb{F}^{(0)}(\mathbf{X}^{(0)}, t, \boldsymbol{\lambda})$ on the constraint boundary, the desynchronization will be discussed first, and the desynchronization for $\mathbb{F}^{(\alpha)}(\mathbf{X}^{(\alpha)}, t, \pi^{(\alpha)}) \neq \mathbb{F}^{(0)}(\mathbf{X}^{(0)}, t, \boldsymbol{\lambda})$ on the constraint boundary will be addressed. The theorems for desynchronization with $\mathbb{F}^{(\alpha)}(\mathbf{X}^{(\alpha)}, t, \pi^{(\alpha)}) = \mathbb{F}^{(0)}(\mathbf{X}^{(0)}, t, \boldsymbol{\lambda})$ are presented as follows.

Theorem 5.5. For two dynamical systems in Eqs.(5.1) and (5.2) with constraint in Eq.(5.3), there is a metric functional of $V(\mathbf{X}, t, \boldsymbol{\lambda}) = f(\varphi(\mathbf{X}, t, \boldsymbol{\lambda}))$ in Eq.(5.29). For $\mathbf{X}^{(\alpha)}_m \in \Omega_\alpha$ $(\alpha \in \{1,2\})$ and $\mathbf{X}^{(0)}_m \in \partial\Omega_{12}$ at time t_m, $\mathbf{X}^{(\alpha)}_m = \mathbf{X}^{(0)}_m$. For any small $\varepsilon > 0$, there is a time interval $[t_{m-\varepsilon}, t_m)$ or $(t_m, t_{m+\varepsilon}]$. For $\mathbf{X}^{(\alpha)} \in \Omega^{\pm\varepsilon}_\alpha$ at $t \in [t_{m-\varepsilon}, t_m)$ or $(t_m, t_{m+\varepsilon}]$, the functional $V(\mathbf{X}^{(\alpha)}, t, \boldsymbol{\lambda})$ is C^{r_α}-continuous and $|V^{(r_\alpha+2)}(\mathbf{X}^{(\alpha)}, t, \boldsymbol{\lambda})| < \infty$ $(r_\alpha \geq k_\alpha + 1)$. For $\mathbf{X}^{(\alpha)} \in \Omega_\alpha$ and $\mathbf{X}^{(0)} \in \partial\Omega_{12}$, suppose $D^{(s_\alpha)}\mathbb{F}^{(\alpha)}(\mathbf{X}^{(\alpha)}, t, \pi^{(\alpha)}) = D^{(s_\alpha)}\mathbb{F}^{(0)}(\mathbf{X}^{(0)}, t, \boldsymbol{\lambda})$ $(s_\alpha = 0, 1, \ldots)$ for $\mathbf{X}^{(\alpha)} = \mathbf{X}^{(0)}$ $(\alpha \in \{1,2\})$. The two dynamical systems in Eqs.(5.1) and (5.2) to constrain in Eq.(5.3) are desynchronized for time $t \in [t_{m_1}, t_{m_2}]$ in the sense of the metric fucntional if and only if

(i) for $\mathbf{X}^{(\alpha)}_m \in \Omega_\alpha$ and $\mathbf{X}^{(0)}_m \in \partial\Omega_{12}$ with any time t_m

$$\mathbf{X}_m^{(\alpha)} = \mathbf{X}_m^{(0)},\ V^{(r_\alpha)}(\mathbf{X}_m^{(\alpha)}, t_m, \lambda) = 0, \\ \text{for } \alpha = 1,2 \text{ and } r_\alpha = 0,1,2,\ldots \tag{5.95}$$

(ii) for $\mathbf{X}_\kappa^{(\alpha)} \in \Omega_\alpha^{+\varepsilon}$ at time $t_\kappa^+ \in (t_m, t_{m+\varepsilon}]$ and $\mathbf{X}_m^{(0)} \in \partial\Omega_{12}$ with $t_m \in (t_{m_1}, t_{m_2})$

$$\mathbf{X}_\kappa^{(\alpha)} \neq \mathbf{X}_m^{(0)},\ V^{(1)}(\mathbf{X}_\kappa^{(\alpha)}, t_\kappa^+, \boldsymbol{\lambda}) > 0 \text{ and} \\ \lim_{t_\kappa^+ \to t_m} V^{(1)}(\mathbf{X}_\kappa^{(\alpha)}, t_\kappa^+, \boldsymbol{\lambda}) = 0 \text{ for } \alpha = 1,2; \tag{5.96}$$

(iii) for $\mathbf{X}_\kappa^{(\alpha)} \in \Omega_\alpha^{-\varepsilon}$ at time $t_\kappa^- \in [t_{m-\varepsilon}, t_m)$ and $\mathbf{X}_m^{(0)} \in \partial\Omega_{12}$ with $t_m \notin [t_{m_1}, t_{m_2}]$

$$\mathbf{X}_\kappa^{(\alpha)} \neq \mathbf{X}_m^{(0)},\ V^{(1)}(\mathbf{X}_\kappa^{(\alpha)}, t_\kappa^-, \boldsymbol{\lambda}) < 0 \text{ and} \\ \lim_{t_\kappa^- \to t_m} V^{(1)}(\mathbf{X}_\kappa^{(\alpha)}, t_\kappa^-, \boldsymbol{\lambda}) = 0 \text{ for } \alpha = 1,2; \tag{5.97}$$

(iv) for $\mathbf{X}_\kappa^{(\alpha)} \in \Omega_\alpha^{\pm\varepsilon}$ at time $t_\kappa^- \in [t_{m-\varepsilon}, t_m)$ or $t_\kappa^+ \in (t_m, t_{m+\varepsilon}]$ and $\mathbf{X}_m^{(0)} \in \partial\Omega_{12}$ with $t_m = t_{m_1}$ and t_{m_2}

$$\mathbf{X}_\kappa^{(\alpha)} \neq \mathbf{X}_m^{(0)},\ \lim_{t_\kappa^\pm \to t_{m\pm}} V^{(1)}(\mathbf{X}_\kappa^{(\alpha)}, t_\kappa^\pm, \boldsymbol{\lambda}) = 0, \\ \lim_{t_\kappa^\pm \to t_{m\pm}} V^{(2)}(\mathbf{X}_\kappa^{(\alpha)}, t_\kappa^\pm, \boldsymbol{\lambda}) < 0 \text{ for } \alpha = 1,2. \tag{5.98}$$

Proof: Once Definitions 5.15, 5.16, 5.19 and 5.20 are used, the proof of this theorem is similar to the proof of Theorem 5.1. ■

Theorem 5.6. For two dynamical systems in Eqs.(5.1) and (5.2) with constraint in Eq.(5.3), there is a metric functional of $V(\mathbf{X}, t, \boldsymbol{\lambda}) = f(\varphi(\mathbf{X}, t, \boldsymbol{\lambda}))$ in Eq.(5.29). For $\mathbf{X}_m^{(\alpha)} \in \Omega_\alpha$ ($\alpha \in \{1,2\}$) and $\mathbf{X}_m^{(0)} \in \partial\Omega_{12}$ at time t_m, $\mathbf{X}_m^{(\alpha)} = \mathbf{X}_m^{(0)}$. For any small $\varepsilon > 0$, there is a time interval $[t_{m-\varepsilon}, t_m)$ or $(t_m, t_{m+\varepsilon}]$. For $\mathbf{X}^{(\alpha)} \in \Omega_\alpha^{\pm\varepsilon}$ at $t \in [t_{m-\varepsilon}, t_m)$ or $(t_m, t_{m+\varepsilon}]$, the functional $V(\mathbf{X}^{(\alpha)}, t, \boldsymbol{\lambda})$ is C^{r_α}-continuous and $|V^{(r_\alpha+2)}(\mathbf{X}^{(\alpha)}, t, \boldsymbol{\lambda})| < \infty (r_\alpha \geq k_\alpha + 1)$. For $\mathbf{X}^{(\alpha)} \in \Omega_\alpha$ and $\mathbf{X}^{(0)} \in \partial\Omega_{12}$, suppose $D^{(s_\alpha)}\mathbb{F}^{(\alpha)}(\mathbf{X}^{(\alpha)}, t, \boldsymbol{\pi}^{(\alpha)}) = D^{(s_\alpha)}\mathbb{F}^{(0)}(\mathbf{X}^{(0)}, t, \boldsymbol{\lambda})$ ($s_\alpha = 0,1,\ldots$) for $\mathbf{X}^{(\alpha)} = \mathbf{X}^{(0)}$ ($\alpha \in \{1,2\}$). The two dynamical systems in Eqs.(5.1) and (5.2) to constraint in Eq.(5.3) are desynchronized for time $t \in [t_{m_1}, t_{m_2}]$ in the sense of the metric fucntional if and only if

(i) for $\mathbf{X}_m^{(\alpha)} \in \Omega_\alpha$ and $\mathbf{X}_m^{(0)} \in \partial\Omega_{12}$ with $t_m \in (t_{m_1}, t_{m_2})$

$$\mathbf{X}_m^{(\alpha)} = \mathbf{X}_m^{(0)},\ V^{(r_\alpha)}(X_m^{(\alpha)}, t_m, \lambda) = 0, \\ \text{for } \alpha = 1,2 \text{ and } r_\alpha = 0,1,2,\ldots \tag{5.99}$$

(ii) for $\mathbf{X}_\kappa^{(\alpha)} \in \Omega_\alpha^{+\varepsilon}$ at time $t_\kappa^+ \in (t_m, t_{m+\varepsilon}]$ and $\mathbf{X}_m^{(0)} \in \partial\Omega_{12}$ with $t_m \in (t_{m_1}, t_{m_2})$

$$\mathbf{X}_{\kappa}^{(\alpha)} \neq \mathbf{X}_{m}^{(0)},\ \lim_{t_{\kappa}^{+} \to t_{m+}} V^{(s_\alpha)}(\mathbf{X}_{\kappa}^{(\alpha)}, t_{\kappa}^{+}, \boldsymbol{\lambda}) = 0 \text{ for } s_\alpha = 1,2,\ldots,2k_\alpha;$$
$$V^{(2k_\alpha+1)}(\mathbf{X}_{\kappa}^{(\alpha)}, t_{\kappa}^{+}, \boldsymbol{\lambda}) > 0 \text{ and } \lim_{t_{\kappa}^{+} \to t_{m}} V^{(2k_\alpha+1)}(\mathbf{X}_{\kappa}^{(\alpha)}, t_{\kappa}^{+}, \boldsymbol{\lambda}) = 0 \text{ for } \alpha = 1,2; \tag{5.100}$$

(iii) for $\mathbf{X}_{\kappa}^{(\alpha)} \in \Omega_{\alpha}^{-\varepsilon}$ at time $t_{\kappa}^{-} \in [t_{m-\varepsilon}, t_m)$ and $\mathbf{X}_{m}^{(0)} \in \partial\Omega_{12}$ with $t_m \notin [t_{m_1}, t_{m_2}]$

$$\mathbf{X}_{\kappa}^{(\alpha)} \neq \mathbf{X}_{m}^{(0)},\ \lim_{t_{\kappa}^{-} \to t_{m-}} V^{(s_\alpha)}(\mathbf{X}_{\kappa}^{(\alpha)}, t_{\kappa}^{-}, \boldsymbol{\lambda}) = 0 \text{ for } s_\alpha = 1,2,\ldots,2k_\alpha;$$
$$V^{(2k_\alpha+1)}(\mathbf{X}_{\kappa}^{(\alpha)}, t_{\kappa}^{-}, \boldsymbol{\lambda}) < 0 \text{ and } \lim_{t_{\kappa}^{-} \to t_{m}} V^{(2k_\alpha+1)}(\mathbf{X}_{\kappa}^{(\alpha)}, t_{\kappa}^{-}, \boldsymbol{\lambda}) = 0 \text{ for } \alpha = 1,2; \tag{5.101}$$

(iv) for $\mathbf{X}_{\kappa}^{(\alpha)} \in \Omega_{\alpha}^{\pm\varepsilon}$ at time $t_{\kappa}^{-} \in [t_{m-\varepsilon}, t_{m-})$ or $t_{\kappa}^{+} \in (t_{m+}, t_{m+\varepsilon}]$ and $\mathbf{X}_{m}^{(0)} \in \partial\Omega_{12}$ with $t_m = t_{m_1}$ and t_{m_2}

$$\mathbf{X}_{\kappa}^{(\alpha)} \neq \mathbf{X}_{m}^{(0)},\ \lim_{t_{\kappa}^{\pm} \to t_{m\pm}} V^{(s_\alpha)}(\mathbf{X}_{\kappa}^{(\alpha)}, t_{\kappa}^{\pm}, \boldsymbol{\lambda}) = 0 \text{ for } s_\alpha = 1,2,\ldots,2k_\alpha+1;$$
$$\lim_{t_{\kappa}^{\pm} \to t_{m\pm}} V^{(2k_\alpha+2)}(\mathbf{X}_{\kappa}^{(\alpha)}, t_{\kappa}^{\pm}, \boldsymbol{\lambda}) < 0 \text{ for } \alpha = 1,2. \tag{5.102}$$

Proof: Once Definitions 5.23, 5.24, 5.27 and 5.28 are used, the proof of this theorem is similar to the proof of Theorem 5.2. ■

From the two foregoing theorems, the conditions for the desynchronization are similar to the conditions for the synchronization. If $t_{m_1} \to -\infty$ and $t_{m_2} \to \infty$, such a desynchronization of two dynamical systems to constraint in Eq.(5.3) is absolute. Once the resultant flows on the constraint boundary are repelled, such a desynchronization can keep forever. To explain the two foregoing theorems, the desynchronization of two dynamical systems to a specific constraint are sketched in Fig. 5.9 through the resultant flows in the vicinity of the constraint boundary $\partial\Omega_{12}$. Any point for desynchronization on the constraint boundary is expressed by $(\mathbf{X}_{m}^{(0)}, t_m)$. In the two domains, the resultant flows in the vicinity of the boundary are expressed by $(\mathbf{X}_{m+\varepsilon}^{(\alpha)}, t_{m+\varepsilon})$ $(\alpha = 1,2)$. The onset point for the desynchronization is denoted by $(\mathbf{X}_{m_1}^{(0)}, t_{m_1})$. For $t_m > t_{m_1}$ and $t_m \to \infty$, all the resultant flows leave from the constraint boundary, which means the desynchronization exists forever. However, if $t_{m_2} > t_{m_1}$ is finite, such desynchronization to the constraint will disappear at a point $(\mathbf{X}_{m_2}^{(0)}, t_{m_2})$. The desynchronization of two dynamical systems with $\mathbb{F}^{(\alpha)}(\mathbf{X}^{(\alpha)}, t, \boldsymbol{\pi}^{(\alpha)}) = \mathbb{F}^{(0)}(\mathbf{X}^{(0)}, t, \boldsymbol{\lambda})$ to a specific constraint are different from those with $\mathbb{F}^{(\alpha)}(\mathbf{X}^{(\alpha)}, t, \boldsymbol{\pi}^{(\alpha)}) \neq \mathbb{F}^{(0)}(\mathbf{X}^{(0)}, t, \boldsymbol{\lambda})$. Therefore, the conditions for the desynchronization of two dynamical systems with discontinuous vector fields are discussed in the following two theorems.

Theorem 5.7. For two dynamical systems in Eqs.(5.1) and (5.2) with constraint in Eq.(5.3), there is a metric functional of $V(\mathbf{X}, t, \boldsymbol{\lambda}) = f(\varphi(\mathbf{X}, t, \boldsymbol{\lambda}))$ in Eq.(5.29). For $\mathbf{X}_{m}^{(\alpha)} \in \Omega_\alpha$ $(\alpha \in \{1,2\})$ and $\mathbf{X}_{m}^{(0)} \in \partial\Omega_{12}$ at time t_m, $\mathbf{X}_{m}^{(\alpha)} = \mathbf{X}_{m}^{(0)}$. For any small $\varepsilon > 0$, there is a time interval $[t_{m-\varepsilon}, t_m)$ or $(t_m, t_{m+\varepsilon}]$. For $\mathbf{X}^{(\alpha)} \in \Omega_{\alpha}^{\pm\varepsilon}$ at

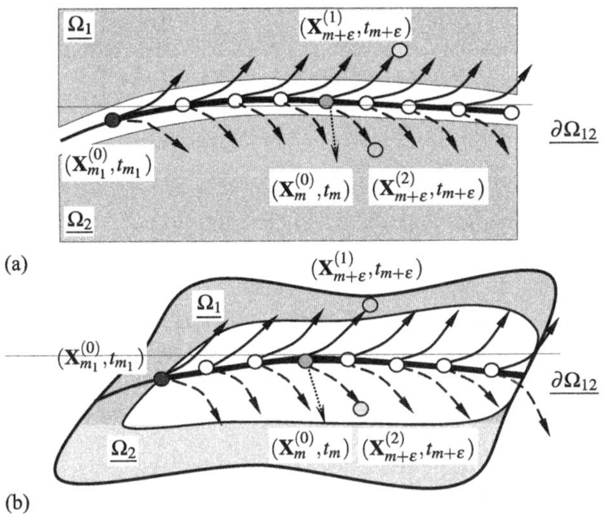

Fig. 5.9 (a) Cross section view and (b) 3-D view for the desynchronization of slave and master flows in vicinity of the boundary $\partial\Omega_{12}$ in $(n+\tilde{n})$-dimensional state space. On the boundary, any point for desynchronization is expressed by $(\mathbf{X}_m^{(0)}, t_m)$. In the two domains, the flows in the vicinity of the boundary are expressed by $(\mathbf{X}_{m+\varepsilon}^{(\alpha)}, t_{m+\varepsilon})$ $(\alpha = 1,2)$. The onset point is $(\mathbf{X}_{m_1}^{(0)}, t_{m_1})$, depicted by a red circular symbol.

$t \in [t_{m-\varepsilon}, t_m)$ or $(t_m, t_{m+\varepsilon}]$, suppose $V(\mathbf{X}^{(\alpha)}, t, \boldsymbol{\lambda})$ is C^{r_α}-continuous $(r_\alpha > 3)$ and $|V^{(r_\alpha+1)}(\mathbf{X}^{(\alpha)}, t, \boldsymbol{\lambda})| < \infty$. For $\mathbf{X}^{(\alpha)} \in \Omega_\alpha$ and $\mathbf{X}^{(0)} \in \partial\Omega_{12}$, $\mathbb{F}^{(\alpha)}(\mathbf{X}^{(\alpha)}, t, \boldsymbol{\pi}^{(\alpha)}) \neq \mathbb{F}^{(0)}(\mathbf{X}^{(0)}, t, \boldsymbol{\lambda})$ at $\mathbf{X}^{(\alpha)} = \mathbf{X}^{(0)}$ $(\alpha \in \{1,2\})$. The two dynamical systems in Eqs.(5.1) and (5.2) to constraint in Eq.(5.3) is desynchronized for time $t \in [t_{m_1}, t_{m_2}]$ in the sense of the metric fucntional if and only if

(i) for $\mathbf{X}_m^{(\alpha)} = \mathbf{X}_m^{(0)}$ and $\mathbf{X}^{(\alpha)}(t) \in \Omega_\alpha$ $(\alpha \in \{1,2\})$ at time $t = t_m \in [t_{m_1}, t_{m_2}]$

$$V(\mathbf{X}_{m+}^{(\alpha)}, t_{m+}, \boldsymbol{\lambda}) = V(\mathbf{X}_m^{(0)}, t_m, \boldsymbol{\lambda}) = 0. \tag{5.103}$$

(ii) for time $t_m \in [t_{m_1}, t_{m_2})$,

$$\mathbf{X}_{m+}^{(\alpha)} = \mathbf{X}_m^{(0)} \text{ and } V^{(1)}(\mathbf{X}_{m+}^{(\alpha)}, t_{m+}, \boldsymbol{\lambda}) > 0 \text{ for } \alpha = 1,2, \tag{5.104}$$

(iii) with a penetration for time $t = t_{m_i}$, $\mathbf{X}_{m_i}^{(\alpha)} = \mathbf{X}_{m_i}^{(0)}$ $(i = 1,2)$,

$$\left.\begin{aligned} &V^{(1)}(\mathbf{X}_{m_i\mp}^{(\alpha)}, t_{m_i\mp}, \boldsymbol{\lambda}) = 0 \text{ and } V^{(2)}(\mathbf{X}_{m_i\mp}^{(\alpha)}, t_{m_i\mp}, \boldsymbol{\lambda}) > 0 \text{ for } \alpha \in \{1,2\}, \\ &V^{(1)}(\mathbf{X}_{m_i+}^{(\beta)}, t_{m_i+}, \boldsymbol{\lambda}) > 0 \text{ for } \beta \in \{1,2\} \text{ and } \beta \neq \alpha, \end{aligned}\right\} \tag{5.105}$$

or with a synchronization for time $t = t_{m_i}$, $\mathbf{X}_{m_i}^{(\alpha)} = \mathbf{X}_{m_i}^{(0)}$,

$$V^{(1)}(\mathbf{X}_{m_i\mp}^{(\alpha)},t_{m_i\mp},\boldsymbol{\lambda})=0 \text{ and } V^{(2)}(\mathbf{X}_{m_i\mp}^{(\alpha)},t_{m_i\mp},\boldsymbol{\lambda})>0 \text{ for } \alpha\in\{1,2\},$$
$$V^{(1)}(\mathbf{X}_{m_i\mp}^{(\beta)},t_{m_i\mp},\boldsymbol{\lambda})=0 \text{ and } V^{(2)}(\mathbf{X}_{m_i\mp}^{(\beta)},t_{m_i\mp},\boldsymbol{\lambda})>0 \text{ for } \beta\in\{1,2\} \text{ and } \beta\neq\alpha. \tag{5.106}$$

Proof: By using Definitions 5.15, 5.19-5.21, the proof of this theorem is similar to the proof of Theorem 5.3. ∎

From the foregoing theorem, the desynchronization of two dynamical systems to a specific constraint requires that the first-order derivative of the metric functional be greater than zero. In addition, the *onset and vanishing* conditions of desynchronization in Eqs.(5.105) and (5.106) are the *vanishing and onset* conditions for onset of the penetration and synchronization with the desynchronization, respectively. The following theorem will give the corresponding conditions for the desynchronization of two dynamical systems to a specific constraint with the higher-order singularity.

Theorem 5.8. For two dynamical systems in Eqs.(5.1) and (5.2) with constraint in Eq.(5.3), there is a metric functional of $V(\mathbf{X},t,\boldsymbol{\lambda})=f(\varphi(\mathbf{X},t,\boldsymbol{\lambda}))$ in Eq.(5.29). For $\mathbf{X}_m^{(\alpha)}\in\Omega_\alpha$ $(\alpha\in\{1,2\})$ and $\mathbf{X}_m^{(0)}\in\partial\Omega_{12}$ at time t_m, $\mathbf{X}_m^{(\alpha)}=\mathbf{X}_m^{(0)}$. For any small $\varepsilon>0$, there is a time interval $[t_{m-\varepsilon},t_m)$ or $(t_m,t_{m+\varepsilon}]$. For $\mathbf{X}^{(\alpha)}\in\Omega_\alpha^{\pm\varepsilon}$ at $t\in[t_{m-\varepsilon},t_m)$ or $(t_m,t_{m+\varepsilon}]$, suppose $V(\mathbf{X}^{(\alpha)},t,\boldsymbol{\lambda})$ is C^{r_α}-continuous $(r_\alpha\geq 2k_\alpha+1)$ and $|V^{(r_\alpha+1)}(\mathbf{X}^{(\alpha)},t,\boldsymbol{\lambda})|<\infty$. $\mathbb{F}^{(\alpha)}(\mathbf{X}^{(\alpha)},t,\boldsymbol{\pi}^{(\alpha)})\neq\mathbb{F}^{(0)}(\mathbf{X}^{(0)},t,\boldsymbol{\lambda})$ at $\mathbf{X}^{(\alpha)}=\mathbf{X}^{(0)}$ for $\mathbf{X}^{(\alpha)}\in\Omega_\alpha$ and $\mathbf{X}^{(0)}\in\partial\Omega_{12}$ $(\alpha\in\{1,2\})$. The two dynamical systems in Eqs.(5.1) and (5.2) to constraint in Eq.(5.3) is desynchronized of the $(2k_1:2k_2)$-type for time $t\in[t_{m_1},t_{m_2}]$ in the sense of the metric fucntional if and only if

(i) for $\mathbf{X}_m^{(\alpha)}=\mathbf{X}_m^{(0)}$ and $\mathbf{X}^{(\alpha)}(t)\in\Omega_\alpha$ $(\alpha=1,2)$ at time $t=t_m\in[t_{m_1},t_{m_2}]$,

$$V(\mathbf{X}_{m+}^{(\alpha)},t_{m+},\boldsymbol{\lambda})=V(\mathbf{X}_m^{(0)},t_m,\boldsymbol{\lambda})=0, \tag{5.107}$$

(ii) for time $t_m\in(t_{m_1},t_{m_2})$,

$$\begin{aligned}&\mathbf{X}_{m+}^{(\alpha)}=\mathbf{X}_m^{(0)} \text{ and } V^{(r_\alpha)}(\mathbf{X}_{m+}^{(\alpha)},t_{m+},\boldsymbol{\lambda})=0(r_\alpha=1,2,\ldots,2k_\alpha),\\&V^{(2k_\alpha+1)}(\mathbf{X}_{m+}^{(\alpha)},t_{m+},\boldsymbol{\lambda})<0 \text{ for } \alpha\in\{1,2\};\\&\mathbf{X}_{m+}^{(\beta)}=\mathbf{X}_m^{(0)} \text{ and } V^{(r_\beta)}(\mathbf{X}_{m+}^{(\beta)},t_{m+},\boldsymbol{\lambda})=0(r_\beta=1,2,\ldots,2k_\beta),\\&V^{(2k_\beta+1)}(\mathbf{X}_{m+}^{(\beta)},t_{m+},\boldsymbol{\lambda})<0 \text{ for } \beta\in\{1,2\} \text{ and } \beta\neq\alpha.\end{aligned} \tag{5.108}$$

(iii) with a $(2k_\alpha:2k_\beta)$-penetration flow for time $t=t_{m_i}$, $\mathbf{X}_{m_i}^{(\alpha)}=\mathbf{X}_{m_i}^{(0)}$,

$$\begin{aligned}&V^{(s_\alpha)}(\mathbf{X}_{m_i\pm}^{(\alpha)},t_{m_i\pm},\boldsymbol{\lambda})=0\ (s_\alpha=1,2,\ldots,2k_\alpha+1)\\&\text{and } V^{(2k_\alpha+2)}(\mathbf{X}_{m_i\pm}^{(\alpha)},t_{m_i\pm},\boldsymbol{\lambda})>0 \text{ for } \alpha\in\{1,2\};\\&V^{(s_\beta)}(\mathbf{X}_{m_i+}^{(\beta)},t_{m_i+},\boldsymbol{\lambda})=0\ (s_\beta=1,2,\ldots,2k_\beta),\\&V^{(2k_\beta+1)}(\mathbf{X}_{m_i+}^{(\beta)},t_{m_i+},\boldsymbol{\lambda})>0 \text{ for } \beta\in\{1,2\} \text{ and } \beta\neq\alpha.\end{aligned} \tag{5.109}$$

or with a $(2k_1 : 2k_2)$-synchronization for time $t = t_{m_i}$, $\mathbf{X}_{m_i}^{(\alpha)} = \mathbf{X}_{m_i}^{(0)}$,

$$\left.\begin{aligned} & V^{(s_\alpha)}(\mathbf{X}_{m_i\pm}^{(\alpha)}, t_{m_i\pm}, \boldsymbol{\lambda}) = 0 \ (s_\alpha = 1,2,\ldots,2k_\alpha + 1) \\ & \text{and } V^{(2k_\alpha+2)}(\mathbf{X}_{m_i\pm}^{(\alpha)}, t_{m_i\pm}, \boldsymbol{\lambda}) > 0 \text{ for } \alpha \in \{1,2\}; \\ & V^{(s_\beta)}(\mathbf{X}_{m_i\pm}^{(\beta)}, t_{m_i\pm}, \boldsymbol{\lambda}) = 0 \ (s_\beta = 1,2,\ldots,2k_\beta + 1), \\ & V^{(2k_\beta+2)}(\mathbf{X}_{m_i\pm}^{(\beta)}, t_{m_i\pm}, \boldsymbol{\lambda}) > 0 \text{ for } \beta \in \{1,2\} \text{ and } \beta \neq \alpha. \end{aligned}\right\} \tag{5.110}$$

Proof: Using Definitions 5.25, 5.27–5.29, the proof of this theorem is similar to Theorem 5.4. ■

In the foregoing theorem, the *onset and vanishing* conditions of the $(2k_1 : 2k_2)$-desynchronization in Eqs.(5.109) and (5.110) are also the *vanishing and onsets* of the $(2k_\alpha : 2k_\beta)$-penetration and the $(2k_1 : 2k_2)$-synchronization, respectively. The $(2k_1 : 2k_2)$-desynchronization requires that the $(2k_1+1 : 2k_2+1)$-order derivatives of the metric function should be greater than zero. The desynchronization of two dynamical systems to a specific constraint is presented in the previous two theorems, as sketched in Fig.5.10 through the resultant flows in the vicinity of the constraint boundary. On the constraint boundary, any point relative to desynchronization is expressed by $(\mathbf{X}_m^{(0)}, t_m)$. In the two domains, the flows in the vicinity of the constraint boundary are expressed by $(\mathbf{X}_{m+\varepsilon}^{(\alpha)}, t_{m+\varepsilon})$ $(\alpha = 1,2)$. The *onset and vanishing* points are $(\mathbf{X}_{m_1}^{(0)}, t_{m_1})$ and $(\mathbf{X}_{m_2}^{(0)}, t_{m_2})$ with red and green circular symbols, which are generated by the two penetration. Both of them belong to a sub-manifold on the constraint boundary in the $(n+\tilde{n})$-dimensional state space. The points $(\mathbf{X}_{m_1}^{(0)}, t_{m_1})$ and $(\mathbf{X}_{m_2}^{(0)}, t_{m_2})$ are starting and vanishing points of the resultant flow relative to desynchronization. If $t_{m_2} \to \infty$, once the desynchronization exists, no any synchronization for such two systems to a specific constraint can be achieved. The desynchronization for $\mathbb{F}^{(\alpha)}(\mathbf{X}^{(\alpha)}, t, \boldsymbol{\pi}^{(\alpha)}) \neq \mathbb{F}^{(\beta)}(\mathbf{X}^{(\beta)}, t, \boldsymbol{\pi}^{(\beta)})$ can be investigated through the two foregoing theorems. From the previous discussion, the penetration of two dynamical systems to a specific constraint is also very important for the onset and vanishing of synchronization and desynchronization.

5.3.7 Penetration to constraint

The synchronization and desynchronization of two dynamical systems to a specific constraint have been discussed. Another important phenomenon is the penetration of two dynamical systems to a specific constraint. The penetration of two dynamical systems with $\mathbb{F}^{(\alpha)}(\mathbf{X}^{(\alpha)}, t, \boldsymbol{\pi}^{(\alpha)}) = \mathbb{F}^{(0)}(\mathbf{X}^{(0)}, t, \boldsymbol{\lambda})(\alpha = 1,2)$ to a specific constraint cannot exist. However, if two dynamical systems to a specific constraint possess discontinuous vector fields, the penetration can occur at the constraint boundary. The corresponding theorems are presented as follows.

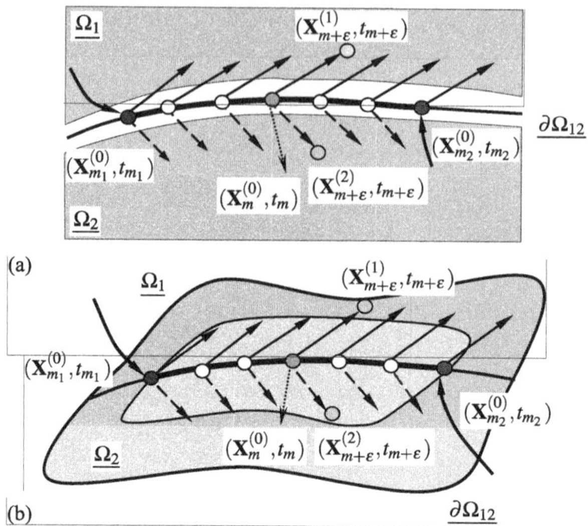

Fig. 5.10 (a) A cross-section view and (b) a 3-D view of the desynchronization of resultant flows in vicinity of the constraint boundary $\partial\Omega_{12}$ in $(n+\tilde{n})$-dimensional state space. On the constant boundary, any point for desynchronization is expressed by $(\mathbf{X}_m^{(0)}, t_m)$. In two domains, the resultant flows in the vicinity of the constant boundary are expressed by $(\mathbf{X}_{m-\varepsilon}^{(\alpha)}, t_{m-\varepsilon})$ $(\alpha = 1,2)$. The onset and vanishing points are $(\mathbf{X}_{m_1}^{(0)}, t_{m_1})$ and $(\mathbf{X}_{m_2}^{(0)}, t_{m_2})$ with red and green circular symbols.

Theorem 5.9. For two dynamical systems in Eqs.(5.1) and (5.2) with constraint in Eq.(5.3), there is a metric functional of $V(\mathbf{X},t,\boldsymbol{\lambda}) = f(\varphi(\mathbf{X},t,\boldsymbol{\lambda}))$ in Eq.(5.29). For $\mathbf{X}_m^{(\alpha)} \in \Omega_\alpha$ $(\alpha \in \{1,2\})$ and $\mathbf{X}_m^{(0)} \in \partial\Omega_{12}$ at time t_m, $\mathbf{X}_m^{(\alpha)} = \mathbf{X}_m^{(0)}$. For any small $\varepsilon > 0$, there is a time interval $[t_{m-\varepsilon}, t_m)$ or $(t_m, t_{m+\varepsilon}]$. For $\mathbf{X}^{(\alpha)} \in \Omega_\alpha^{\pm\varepsilon}$ at $t \in [t_{m-\varepsilon}, t_m)$ or $(t_m, t_{m+\varepsilon}]$, $V(\mathbf{X}^{(\alpha)}, t, \boldsymbol{\lambda})$ is C^{r_α}-continuous $(r_\alpha > 2)$ and $|V^{(r_\alpha+1)}(\mathbf{X}^{(\alpha)}, t, \boldsymbol{\lambda})| < \infty$. $\mathbb{F}^{(\alpha)}(\mathbf{X}^{(\alpha)}, t, \boldsymbol{\pi}^{(\alpha)}) \neq \mathbb{F}^{(0)}(\mathbf{X}^{(0)}, t, \boldsymbol{\lambda})$ at $\mathbf{X}^{(\alpha)} = \mathbf{X}^{(0)}$ for $\mathbf{X}^{(\alpha)} \in \Omega_\alpha$ and $\mathbf{X}^{(0)} \in \partial\Omega_{12}$ $(\alpha \in \{1,2\})$. The two dynamical systems in Eqs.(5.1) and (5.2) to the constraint in Eq.(5.3) is penetrated at time $t \in [t_{m_1}, t_{m_2}]$ in the sense of the metric fucntional if and only if

(i) at time $t = t_m \in (t_{m_1}, t_{m_2})$, $\mathbf{X}_{m-}^{(\alpha)} = \mathbf{X}_m^{(0)} = \mathbf{X}_{m+}^{(\beta)}$ for $\alpha, \beta = 1,2$ and $\alpha \neq \beta$,

$$V^{(1)}(\mathbf{X}_{m-}^{(\alpha)}, t_{m-}, \boldsymbol{\lambda}) < 0 \text{ and } V^{(1)}(\mathbf{X}_{m+}^{(\beta)}, t_{m+}, \boldsymbol{\lambda}) > 0, \tag{5.111}$$

(ii) with a synchronization at time $t = t_{m_i}$, $\mathbf{X}_{m_i-}^{(\alpha)} = \mathbf{X}_{m_i}^{(0)} = \mathbf{X}_{m_i\pm}^{(\beta)}$ $(i = 1,2)$,

$$\begin{aligned} &V^{(1)}(\mathbf{X}_{m_i-}^{(\alpha)}, t_{m_i-}, \boldsymbol{\lambda}) < 0 \text{ for } \alpha \in \{1,2\}, \\ &V^{(1)}(\mathbf{X}_{m_i\pm}^{(\beta)}, t_{m_i\pm}, \boldsymbol{\lambda}) = 0 \text{ and } V^{(2)}(\mathbf{X}_{m_i\pm}^{(\beta)}, t_{m_i\pm}, \boldsymbol{\lambda}) > 0 \\ &\quad \text{for } \beta \in \{1,2\} \text{ and } \beta \neq \alpha, \end{aligned} \tag{5.112}$$

or with a desynchronization at time $t = t_{m_i}$, $\mathbf{X}^{(\alpha)}_{m_i\mp} = \mathbf{X}^{(0)}_{m_i} = \mathbf{X}^{(\beta)}_{m_i+}$ $(i = 1,2)$,

$$\begin{aligned} &V^{(1)}(\mathbf{X}^{(\alpha)}_{m_i\mp}, t_{m_i\mp}, \boldsymbol{\lambda}) = 0 \text{ and } V^{(2)}(\mathbf{X}^{(\alpha)}_{m_i\mp}, t_{m_i\mp}, \boldsymbol{\lambda}) > 0 \text{ for} \alpha \in \{1,2\}, \\ &V^{(1)}(\mathbf{X}^{(\beta)}_{m_i+}, t_{m_i+}, \boldsymbol{\lambda}) > 0 \text{ for } \beta \in \{1,2\} \text{ and } \beta \neq \alpha, \end{aligned} \tag{5.113}$$

or with a switching penetration at time $t = t_{m_i}$, $\mathbf{X}^{(\alpha)}_{m_i\mp} = \mathbf{X}^{(0)}_{m_i} = \mathbf{X}^{(\beta)}_{m_i\pm}$ $(i = 1,2)$,

$$\begin{aligned} &V^{(1)}(\mathbf{X}^{(\alpha)}_{m_i\mp}, t_{m_i\mp}, \boldsymbol{\lambda}) = 0 \text{ and } V^{(2)}(\mathbf{X}^{(\alpha)}_{m_i\mp}, t_{m_i\mp}, \boldsymbol{\lambda}) > 0 \text{ for } \alpha \in \{1,2\}, \\ &V^{(1)}(\mathbf{X}^{(\beta)}_{m_i\pm}, t_{m_i\pm}, \boldsymbol{\lambda}) = 0 \text{ and } V^{(2)}(\mathbf{X}^{(\beta)}_{m_i\pm}, t_{m_i\pm}, \boldsymbol{\lambda}) > 0 \\ &\quad \text{for } \beta \in \{1,2\} \text{ and } \beta \neq \alpha \end{aligned} \tag{5.114}$$

Proof: By using Definitions 5.17, 5.19, 5.20 and 5.22, the proof of this theorem is similar to the proof of Theorem 5.3. ■

Theorem 5.10. For two dynamical systems in Eqs.(5.1) and (5.2) with constraint in Eq.(5.3), there is a metric functional of $V(\mathbf{X}, t, \boldsymbol{\lambda}) = f(\varphi(\mathbf{X}, t, \boldsymbol{\lambda}))$ in Eq.(5.29). For $\mathbf{X}^{(\alpha)}_m \in \Omega_\alpha$ $(\alpha \in \{1,2\})$ and $\mathbf{X}^{(0)}_m \in \partial\Omega_{12}$ at time t_m, $\mathbf{X}^{(\alpha)}_m = \mathbf{X}^{(0)}_m$. For any small $\varepsilon > 0$, there is a time interval $[t_{m-\varepsilon}, t_m)$ or $(t_m, t_{m+\varepsilon}]$. For $\mathbf{X}^{(\alpha)} \in \Omega^{\pm\varepsilon}_\alpha$ at $t \in [t_{m-\varepsilon}, t_m)$ or $(t_m, t_{m+\varepsilon}]$, suppose $V(\mathbf{X}^{(\alpha)}, t, \boldsymbol{\lambda})$ is C^{r_α}-continuous $(r_\alpha \geq 2k_\alpha + 1)$ and $|V^{(r_\alpha+1)}(\mathbf{X}^{(\alpha)}, t, \boldsymbol{\lambda})| < \infty$. For $\mathbf{X}^{(\alpha)} \in \Omega_\alpha$ and $\mathbf{X}^{(0)} \in \partial\Omega_{12}$, $\mathbb{F}^{(\alpha)}(\mathbf{X}^{(\alpha)}, t, \boldsymbol{\pi}^{(\alpha)}) \neq \mathbb{F}^{(0)}(\mathbf{X}^{(0)}, t, \boldsymbol{\lambda})$ for $\mathbf{X}^{(\alpha)} = \mathbf{X}^{(0)}$ $(\alpha \in \{1,2\})$. The two dynamical systems in Eqs.(5.1) and (5.2) to the constraint in Eq.(5.3) is penetrated of the $(2k_1 : 2k_2)$-type for time $t \in [t_{m_1}, t_{m_2}]$ in the sense of the metric fucntional if and only if

(i) for time $t = t_m \in (t_{m_1}, t_{m_2})$, $\mathbf{X}^{(\alpha)}_{m-} = \mathbf{X}^{(0)}_m = \mathbf{X}^{(\beta)}_{m+}$,

$$\begin{aligned} &V^{(2k_\alpha)}(\mathbf{X}^{(\alpha)}_{m-}, t_{m-}, \boldsymbol{\lambda}) = 0 \text{ and } V^{(2k_\alpha+1)}(\mathbf{X}^{(\alpha)}_{m-}, t_{m-}, \boldsymbol{\lambda}) < 0; \\ &V^{(2k_\beta)}(\mathbf{X}^{(\beta)}_{m+}, t_{m+}, \boldsymbol{\lambda}) = 0 \text{ and } V^{(2k_\beta+1)}(\mathbf{X}^{(\beta)}_{m+}, t_{m+}, \boldsymbol{\lambda}) > 0; \\ &\alpha, \beta \in \{1,2\} \text{ and } \alpha \neq \beta. \end{aligned} \tag{5.115}$$

(ii) with a $(2k_1 : 2k_2)$-synchronization at time $t = t_{m_i}$, $\mathbf{X}^{(\alpha)}_{m_i-} = \mathbf{X}^{(0)}_{m_i} = \mathbf{X}^{(\beta)}_{m_i\pm}$ $(i = 1,2)$,

$$\begin{aligned} &V^{(2k_\alpha)}(\mathbf{X}^{(\alpha)}_{m_i-}, t_{m_i-}, \boldsymbol{\lambda}) = 0 \text{ and } V^{(2k_\alpha+1)}(\mathbf{X}^{(\alpha)}_{m_i-}, t_{m_i-}, \boldsymbol{\lambda}) < 0; \\ &V^{(2k_\beta+1)}(\mathbf{X}^{(\beta)}_{m_i\pm}, t_{m_i\pm}, \boldsymbol{\lambda}) = 0 \text{ and } V^{(2k_\alpha+2)}(\mathbf{X}^{(\beta)}_{m_i\pm}, t_{m_i\pm}, \boldsymbol{\lambda}) > 0 \\ &\text{for } \alpha, \beta \in \{1,2\} \text{ and } \beta \neq \alpha, \end{aligned} \tag{5.116}$$

or with a $(2k_1 : 2k_2)$-desynchronization at $t = t_{m_i}$, $\mathbf{X}^{(\alpha)}_{m_i\mp} = \mathbf{X}^{(0)}_{m_i} = \mathbf{X}^{(\beta)}_{m_i+}$ $(i = 1,2)$,

$$
\begin{aligned}
&V^{(2k_\alpha+1)}(\mathbf{X}_{m_i\mp}^{(\alpha)},t_{m_i\mp},\boldsymbol{\lambda})=0 \text{ and } V^{(2k_\alpha+2)}(\mathbf{X}_{m_i\mp}^{(\alpha)},t_{m_i\mp},\boldsymbol{\lambda})>0;\\
&V^{(2k_\beta)}(\mathbf{X}_{m_i+}^{(\beta)},t_{m_i+},\boldsymbol{\lambda})=0 \text{ and } V^{(2k_\beta+1)}(\mathbf{X}_{m_i+}^{(\beta)},t_{m_i+},\boldsymbol{\lambda})>0\\
&\text{for } \alpha,\beta\in\{1,2\} \text{ and } \beta\neq\alpha,
\end{aligned} \tag{5.117}
$$

or with a $(2k_\beta : 2k_\alpha)$ penetration at time $t=t_{m_i}$, $\mathbf{X}_{m_i\mp}^{(\alpha)}=\mathbf{X}_{m_i}^{(0)}=\mathbf{X}_{m_i+}^{(\beta)}$ $(i=1,2)$,

$$
\begin{aligned}
&V^{(2k_\alpha+1)}(\mathbf{X}_{m_i\mp}^{(\alpha)},t_{m_i\mp},\boldsymbol{\lambda})=0 \text{ and } V^{(2k_\alpha+2)}(\mathbf{X}_{m_i\mp}^{(\alpha)},t_{m_i\mp},\boldsymbol{\lambda})>0;\\
&V^{(2k_\beta+1)}(\mathbf{X}_{m_i\pm}^{(\beta)},t_{m_i\pm},\boldsymbol{\lambda})=0 \text{ and } V^{(2k_\beta+2)}(\mathbf{X}_{m_i\pm}^{(\beta)},t_{m_i\pm},\boldsymbol{\lambda})>0\\
&\text{for } \alpha,\beta\in\{1,2\} \text{ and } \beta\neq\alpha.
\end{aligned} \tag{5.118}
$$

Proof: Using Definitions 5.25, 5.27, 5.28 and 5.30, the proof of this theorem is similar to the proof of Theorem 5.4. ■

In the foregoing theorem, the *onset and vanishing* conditions of the $(2k_\alpha : 2k_\beta)$-penetration of the $t\in[t_{m_1},t_{m_2}]$ to a specific constraint in Eqs.(5.116)–(5.118) are also the *vanishing and onset* conditions of the $(2k_1 : 2k_2)$-synchronization, the $(2k_1 : 2k_2)$-desynchronization and the $(2k_\beta : 2k_\alpha)$-penetration, respectively. The penetration of the two dynamical systems to a specific constraint is sketched in Fig.5.11. On the constraint boundary, any point for penetration is expressed by

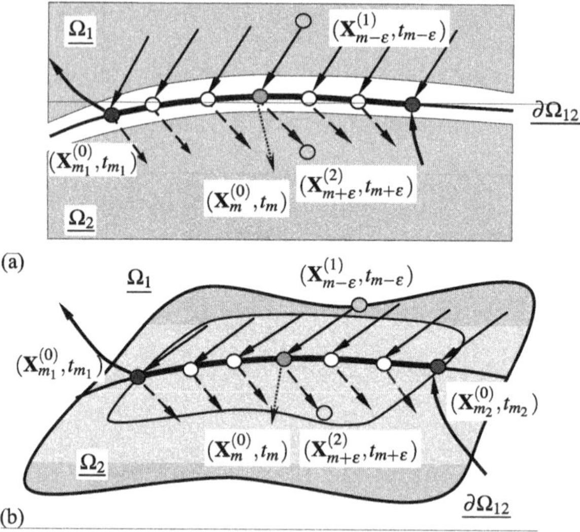

Fig. 5.11 (a) A cross-section view and (b) a 3-D view of the penetration of resultant flows in vicinity of the constraint boundary $\partial\Omega_{12}$ in $(n+\tilde{n})$-dimensional state space. On the constraint boundary, any point for penetration is expressed by $(\mathbf{X}_m^{(0)},t_m)$. In two domains, the resultant flows in the vicinity of the constraint boundary are expressed by $(\mathbf{X}_{m-\varepsilon}^{(\alpha)},t_{m-\varepsilon})$ $(\alpha=1,2)$. The onset and vanishing points are $(\mathbf{X}_{m_1}^{(0)},t_{m_1})$ and $(\mathbf{X}_{m_2}^{(0)},t_{m_2})$ with red and blue circular symbols.

$(\mathbf{X}_m^{(0)}, t_m)$. In two domains, the incoming and outgoing resultant flows in the vicinity of the constraint boundary are expressed by $(\mathbf{X}_{m-\varepsilon}^{(\alpha)}, t_{m-\varepsilon})$ and $(\mathbf{X}_{m+\varepsilon}^{(\beta)}, t_{m+\varepsilon})$ $(\alpha, \beta \in \{1,2\}$ and $\alpha \neq \beta)$. The *onset and vanishing* points are $(\mathbf{X}_{m_1}^{(0)}, t_{m_1})$ and $(\mathbf{X}_{m_2}^{(0)}, t_{m_2})$ with red and blue circular symbols.

5.4 Multiple-constraint synchronization

In this section, the synchronization of two dynamical systems to multiple constraints will be discussed. Following the discussion of the synchronicity of two dynamical systems to a single constraint, the synchronicity of two dynamical systems with multiple constraints can be investigated. Based on the metric functions of multiple constraints, the definitions relative to the synchronicity of two dynamical systems to multiple constraints will be defined. From the corresponding definitions, the corresponding theorems for the synchronicity of two dynamical systems to the constraints are presented.

5.4.1 Synchronicity to multiple-constraints

The ε-domain in the vicinity of the intersected, constraint boundary will be defined through the ε-domain of the jth-constraint boundary. Based on such ε-domain and the intersected constraint boundary, the synchronicity of two dynamical systems to multiple constraints will be discussed in this section.

Definition 5.31. For $\mathbf{X}_m^{(\alpha_j,j)} \in \Omega_{(\alpha_j,j)}$ $(\alpha_j \in \mathbb{I}$ and $j \in \mathbb{L}$ with $\mathbb{I} = \{1,2\}$ and $\mathbb{L} = \{1,2,\ldots,l\})$ and $\mathbf{X}_m^{(0,j)} \in \partial\Omega_{12(j)}$ at time $t_m, \mathbf{X}_m^{(\alpha_j,j)} = \mathbf{X}_m^{(0,j)}$. For any small $\varepsilon > 0$, there is a time interval $[t_{m-\varepsilon}, t_m)$ or $(t_m, t_{m+\varepsilon}]$. The neighborhood of the jth-constraint boundary is defined as

$$
\begin{aligned}
\Omega_{(\alpha_j,j)}^{-\varepsilon} &= \left\{ \mathbf{X}^{(\alpha_j,j)} \middle| \, \|\mathbf{X}^{(\alpha_j,j)}(t) - \mathbf{X}_m^{(0,j)}\| \leq \delta_{(\alpha_j,j)}, \delta_{(\alpha_j,j)} > 0, t \in [t_{m-\varepsilon}, t_m) \right\}, \\
\Omega_{(\alpha_j,j)}^{+\varepsilon} &= \left\{ \mathbf{X}^{(\alpha_j,j)} \middle| \, \|\mathbf{X}^{(\alpha_j,j)}(t) - \mathbf{X}_m^{(0,j)}\| \leq \delta_{(\alpha_j,j)}, \delta_{(\alpha_j,j)} > 0, t \in (t_m, t_{m+\varepsilon}] \right\}.
\end{aligned}
\tag{5.119}
$$

The sub-domains and the intersected boundary are defined as

$$
\Omega_{\alpha} \equiv \Omega_{\alpha_1 \alpha_2 \ldots \alpha_l} = \bigcap_{j=1}^{l} \Omega_{(\alpha_j,j)} \text{ and } \partial\Omega_{12(\mathbf{j})} \equiv \partial\Omega_{12(12\cdots l)} = \bigcap_{j=1}^{l} \partial\Omega_{(\alpha_j \beta_j,j)} \tag{5.120}
$$

For $\mathbf{X}_m^{(\alpha)} \in \Omega_{\alpha}(\alpha = \alpha_1 \alpha_2 \cdots \alpha_l, \alpha_j \in \mathbb{I}$ and $j \in \mathbb{L})$ and $\mathbf{X}_m^{(0)} \in \partial\Omega_{12(\mathbf{j})}$ $(\mathbf{j} = (12\cdots l))$ at time $t_m, \mathbf{X}_m^{(\alpha)} = \mathbf{X}_m^{(0)}$. For any small $\varepsilon > 0$, there is a time interval $[t_{m-\varepsilon}, t_m)$ or

$(t_m, t_{m+\varepsilon}]$. The neighborhood of the intersected constraint boundary $\partial\Omega_{12(j)}$ is defined as

$$\Omega_{\alpha}^{-\varepsilon} = \left\{ \mathbf{X}^{(\alpha)} \middle| \|\mathbf{X}^{(\alpha)}(t) - \mathbf{X}_m^{(0)}\| \leq \delta, \delta > 0, t \in [t_{m-\varepsilon}, t_m) \right\},$$
$$\Omega_{\alpha}^{+\varepsilon} = \left\{ \mathbf{X}^{(\alpha)} \middle| \|\mathbf{X}^{(\alpha)}(t) - \mathbf{X}_m^{(0)}\| \leq \delta, \delta > 0, t \in (t_m, t_{m+\varepsilon}] \right\}. \quad (5.121)$$

where $\delta = \min_{j\in\mathbb{L}, \alpha_j\in\mathbb{I}} (\delta_{(\alpha_j, j)})$ with $\mathbb{L} = \{1, 2, \ldots, l\}$ and $\mathbb{I} = \{1, 2\}$.

Definition 5.32. Three index sets are defined as

$$\mathbb{L} = \bigcup_{i=1}^{3} \mathbb{L}_i \text{ and } \bigcap_{i=1}^{3} \mathbb{L}_i = \varnothing, \quad (5.122)$$

$$\mathbb{L}_i = \varnothing \bigcup \{k_1^{(i)}, k_2^{(i)}, \ldots, k_{l_i}^{(i)}\} \subseteq \mathbb{L} \text{ and } l_1 + l_2 + l_3 = l. \quad (5.123)$$

Definition 5.33. For two dynamical systems in Eqs.(5.1) and (5.2) with constraints in Eq.(5.4), there are $(l+1)$-metric functionals in Eq.(5.37). For $\mathbf{X}_m^{(\alpha_j, j)} \in \Omega_{(\alpha_j, j)}$ ($\alpha_j \in \mathbb{I}$ and $j \in \mathbb{L}$ with $\mathbb{I} = \{1, 2\}$ and $\mathbb{L} = \{1, 2, \ldots, l\}$) and $\mathbf{X}_m^{(0, j)} \in \partial\Omega_{12(j)}$ at time $t_m, \mathbf{X}_m^{(\alpha_j, j)} = \mathbf{X}_m^{(0, j)}$. For any small $\varepsilon > 0$, there is a time interval $[t_{m-\varepsilon}, t_m)$ or $(t_m, t_{m+\varepsilon}]$. The systems in Eqs.(5.1) and (5.2) with constraints in Eq.(5.4) are called an l_1-dimensional synchronization, l_2-dimensional desynchronization and l_3-dimensional penetration for time $t_m \in [t_{m_1}, t_{m_2}]$ in the sense of metric fucntionals

(i) if for $\alpha_j = 1, 2$ and $j \in \mathbb{L}_1$,

$$^{(n:\tilde{n})}V_j(\mathbf{X}_{m-}^{(\alpha_j, j)}, t_{m-}, \boldsymbol{\lambda}_j) = {}^{(n:\tilde{n})}V_j(\mathbf{X}_m^{(0, j)}, t_m, \boldsymbol{\lambda}_j) = 0;$$
$$^{(n:\tilde{n})}V_j(\mathbf{X}_{m-\varepsilon}^{(\alpha_j, j)}, t_{m-\varepsilon}, \boldsymbol{\lambda}_j) - {}^{(n:\tilde{n})}V_j(\mathbf{X}_{m-}^{(\alpha_j, j)}, t_{m-}, \boldsymbol{\lambda}_j) > 0; \quad (5.124)$$

(ii) if for $\alpha_j = 1, 2$ and $j \in \mathbb{L}_2$,

$$^{(n:\tilde{n})}V_j(\mathbf{X}_{m+}^{(\alpha_j, j)}, t_{m+}, \boldsymbol{\lambda}_j) = {}^{(n:\tilde{n})}V_j(\mathbf{X}_m^{(0, j)}, t_m, \boldsymbol{\lambda}_j) = 0;$$
$$^{(n:\tilde{n})}V_j(\mathbf{X}_{m+\varepsilon}^{(\alpha_j, j)}, t_{m+\varepsilon}, \boldsymbol{\lambda}_j) - {}^{(n:\tilde{n})}V_j(\mathbf{X}_{m+}^{(\alpha_j, j)}, t_{m+}, \boldsymbol{\lambda}_j) > 0; \quad (5.125)$$

(iii) if for $\alpha_j, \beta_j \in \{1, 2\}, \alpha_j \neq \beta_j$ and $j \in \mathbb{L}_3$,

$$^{(n:\tilde{n})}V_j(\mathbf{X}_{m-}^{(\alpha_j, j)}, t_{m-}, \boldsymbol{\lambda}_j) = {}^{(n:\tilde{n})}V_j(\mathbf{X}_m^{(0, j)}, t_m, \boldsymbol{\lambda}_j) = 0;$$
$$^{(n:\tilde{n})}V_j(\mathbf{X}_{m-\varepsilon}^{(\alpha_j, j)}, t_{m-\varepsilon}, \boldsymbol{\lambda}_j) - {}^{(n:\tilde{n})}V_j(\mathbf{X}_{m-}^{(\alpha_j, j)}, t_{m-}, \boldsymbol{\lambda}_j) > 0; \quad (5.126)$$

$$^{(n:\tilde{n})}V_j(\mathbf{X}_{m+}^{(\beta_j, j)}, t_{m+}, \boldsymbol{\lambda}_j) = {}^{(n:\tilde{n})}V_j(\mathbf{X}_m^{(0, j)}, t_m, \boldsymbol{\lambda}_j) = 0;$$
$$^{(n:\tilde{n})}V_j(\mathbf{X}_{m+\varepsilon}^{(\beta_j, j)}, t_{m+\varepsilon}, \boldsymbol{\lambda}_j) - {}^{(n:\tilde{n})}V_j(\mathbf{X}_{m+}^{(\beta_j, j)}, t_{m+}, \boldsymbol{\lambda}_j) > 0. \quad (5.127)$$

From the previous definition, among the l-constraints in Eq.(5.4), (i) there are l_1-constraints to make the two dynamical systems be synchronized in the normal directions of the corresponding constraints; (ii) there are l_2-constraints to make the two dynamical systems be desynchronized in the normal directions of the corresponding constraints; and (iii) there are l_3-constraints to make the two dynamical systems be penetrated in the normal directions of the corresponding constraints. If $l_2 = l_3 = 0$ and $l_1 = l$, the two dynamical systems to all the l-constraints are synchronized. If $l_3 = 0$ and $l_1 + l_2 = l$, the two dynamical systems to all the l-constraints are to be synchronized with l_1-constraints and desynchronized with l_2-constraints. If $l_1 = 0$ and $l_2 + l_3 = l$, the two dynamical systems to all the l-constraints are to be desynchronized with l_2-constraints and to be penetrated with l_3-constraints. For this case, the two dynamical systems cannot be synchronized any more for all the l-constraints. If one of three types of synchronicity has changed the current state, the synchronicity of the two dynamical systems will be changed. The three special cases are useful. Therefore, three definitions for the three special cases will be given as follows:

Definition 5.34. For two dynamical systems in Eqs.(5.1) and (5.2) with constraints in Eq.(5.4), there are $(l+1)$-metric functionals in Eq.(5.37). For $\mathbf{X}_m^{(\alpha_j,j)} \in \Omega_{(\alpha_j,j)}$ ($\alpha_j \in \mathbb{I}$ and $j \in \mathbb{L}$ with $\mathbb{I} = \{1,2\}$ and $\mathbb{L} = \{1,2,\ldots,l\}$) and $\mathbf{X}_m^{(0,j)} \in \partial\Omega_{12(j)}$ at time $t_m, \mathbf{X}_m^{(\alpha_j,j)} = \mathbf{X}_m^{(0,j)}$. For any small $\varepsilon > 0$, there is a time interval $[t_{m-\varepsilon}, t_m)$. The two dynamical systems in Eqs.(5.1) and (5.2) with constraints in Eq.(5.4) are called an l-dimensional synchronization for time $t_m \in [t_{m_1}, t_{m_2}]$ in the sense of metric functionals if for $\alpha_j = 1,2$ and $j \in \mathbb{L}$,

$$\begin{aligned} &{}^{(n:\tilde{n})}V_j(\mathbf{X}_{m-}^{(\alpha_j,j)}, t_{m-}, \boldsymbol{\lambda}_j) = {}^{(n:\tilde{n})}V_j(\mathbf{X}_m^{(0,j)}, t_m, \boldsymbol{\lambda}_j) = 0; \\ &{}^{(n:\tilde{n})}V_j(\mathbf{X}_{m-\varepsilon}^{(\alpha_j,j)}, t_{m-\varepsilon}, \boldsymbol{\lambda}_j) - {}^{(n:\tilde{n})}V_j(\mathbf{X}_{m-}^{(\alpha_j,j)}, t_{m-}, \boldsymbol{\lambda}_j) > 0. \end{aligned} \tag{5.128}$$

Definition 5.35. For two dynamical systems in Eqs.(5.1) and (5.2) with constraints in Eq.(5.4), there are $(l+1)$-metric functionals in Eq.(5.37). For $\mathbf{X}_m^{(\alpha_j,j)} \in \Omega_{(\alpha_j,j)}$ ($\alpha_j \in \mathbb{I}$ and $j \in \mathbb{L}$ with $\mathbb{I} = \{1,2\}$ and $\mathbb{L} = \{1,2,\ldots,l\}$) and $\mathbf{X}_m^{(0,j)} \in \partial\Omega_{12(j)}$ at time $t_m, \mathbf{X}_m^{(\alpha_j,j)} = \mathbf{X}_m^{(0,j)}$. For any small $\varepsilon > 0$, there is a time interval $(t_m, t_{m+\varepsilon}]$. The two dynamical systems in Eqs.(5.1) and (5.2) with constraints in Eq.(5.4) are called an l-dimensional desynchronization for time $t_m \in [t_{m_1}, t_{m_2}]$ in the sense of metric functionals if for $\alpha_j \in \mathbb{I}$ and $j \in \mathbb{L}$,

$$\begin{aligned} &{}^{(n:\tilde{n})}V_j(\mathbf{X}_{m+}^{(\alpha_j,j)}, t_{m+}, \boldsymbol{\lambda}_j) = {}^{(n:\tilde{n})}V_j(\mathbf{X}_m^{(0,j)}, t_m, \boldsymbol{\lambda}_j) = 0; \\ &{}^{(n:\tilde{n})}V_j(\mathbf{X}_{m+\varepsilon}^{(\alpha_j,j)}, t_{m+\varepsilon}, \boldsymbol{\lambda}_j) - {}^{(n:\tilde{n})}V_j(\mathbf{X}_{m+}^{(\alpha_j,j)}, t_{m+}, \boldsymbol{\lambda}_j) > 0. \end{aligned} \tag{5.129}$$

Definition 5.36. For two dynamical systems in Eqs.(5.1) and (5.2) with constraints in Eq.(5.4), there are $(l+1)$-metric functionals in Eq.(5.37). For $\mathbf{X}_m^{(\alpha_j,j)} \in \Omega_{(\alpha_j,j)}$

($\alpha_j \in \mathbb{I}$ and $j \in \mathbb{L}$ with $\mathbb{I} = \{1,2\}$ and $\mathbb{L} = \{1,2,\ldots,l\}$) and $\mathbf{X}_m^{(0,j)} \in \partial\Omega_{12(j)}$ at time $t_m, \mathbf{X}_m^{(\alpha_j,j)} = \mathbf{X}_m^{(0,j)}$. For any small $\varepsilon > 0$, there is a time interval $[t_{m-\varepsilon}, t_m)$ or $(t_m, t_{m+\varepsilon}]$. The two dynamical systems in Eqs.(5.1) and (5.2) with constraints in Eq.(5.4) are called an l-dimensional penetration for time $t_m \in [t_{m_1}, t_{m_2}]$ in the sense of metric functionals if for $\alpha_j, \beta_j \in \mathbb{I}$ and $j \in \mathbb{L}$ with $\alpha_j \neq \beta_j$,

$$\begin{aligned}
&{}^{(n:\tilde{n})}V_j(\mathbf{X}_{m-}^{(\alpha_j,j)}, t_{m-}, \boldsymbol{\lambda}_j) = {}^{(n:\tilde{n})}V_j(\mathbf{X}_m^{(0,j)}, t_m, \boldsymbol{\lambda}_j) = 0,\\
&{}^{(n:\tilde{n})}V_j(\mathbf{X}_{m-\varepsilon}^{(\alpha_j,j)}, t_{m-\varepsilon}, \boldsymbol{\lambda}_j) - {}^{(n:\tilde{n})}V_j(\mathbf{X}_{m-}^{(\alpha_j,j)}, t_{m-}, \boldsymbol{\lambda}_j) > 0;
\end{aligned} \tag{5.130}$$

$$\begin{aligned}
&{}^{(n:\tilde{n})}V_j(\mathbf{X}_{m+}^{(\beta_j,j)}, t_{m+}, \boldsymbol{\lambda}_j) = {}^{(n:\tilde{n})}V_j(\mathbf{X}_m^{(0,j)}, t_m, \boldsymbol{\lambda}_j) = 0,\\
&{}^{(n:\tilde{n})}V_j(\mathbf{X}_{m+\varepsilon}^{(\beta_j,j)}, t_{m+\varepsilon}, \boldsymbol{\lambda}_j) - {}^{(n:\tilde{n})}V_j(\mathbf{X}_{m+}^{(\beta_j,j)}, t_{m+}, \boldsymbol{\lambda}_j) > 0.
\end{aligned} \tag{5.131}$$

5.4.2 Singularity to constraints

As discussed in the synchronization of two dynamical systems to the single constraint, the singularity for a flow of the two dynamical systems in Eqs.(5.1) and (5.2) to one of the constraints in Eq.(5.4) can be described. The tangency of a resultant flow to one of l-constraint boundaries is presented first, and then the vanishing and onset of the synchronization of two dynamical systems to the jth-constraint boundary of the l-constraint boundaries will be presented.

Definition 5.37. For two dynamical systems in Eqs.(5.1) and (5.2) with constraints in Eq.(5.4), there are $(l+1)$-metric functionals in Eq.(5.37). For $\mathbf{X}_m^{(\alpha_j,j)} \in \Omega_{(\alpha_j,j)}$ ($\alpha_j \in \mathbb{I}$ and $j \in \mathbb{L}$ with $\mathbb{I} = \{1,2\}$ and $\mathbb{L} = \{1,2,\ldots,l\}$) and $\mathbf{X}_m^{(0,j)} \in \partial\Omega_{12(j)}$ at time $t_m, \mathbf{X}_m^{(\alpha_j,j)} = \mathbf{X}_m^{(0,j)}$. For any small $\varepsilon > 0$, there is a time interval $[t_{m-\varepsilon}, t_m)$ or $(t_m, t_{m+\varepsilon}]$. At $\mathbf{X}^{(\alpha_j,j)} \in \Omega_{(\alpha_j,j)}^{\pm\varepsilon}$ for time $t \in [t_{m-\varepsilon}, t_m)$ or $(t_m, t_{m+\varepsilon}]$, the functional ${}^{(n:\tilde{n})}V_j(\mathbf{X}^{(\alpha_j,j)}, t, \boldsymbol{\lambda}_j)$ is $C^{r_{\alpha_j}}$-continuous and $|{}^{(n:\tilde{n})}V_j^{(r_{\alpha_j}+1)}(\mathbf{X}^{(\alpha_j,j)}, t, \boldsymbol{\lambda}_j)| < \infty$ ($r_{\alpha_j} \geq 2$). A flow of the resultant system for two dynamical systems in Eqs.(5.1) and (5.2) with l-constraints in Eq.(5.4) is called *to be tangential to* the jth -constraint boundary for time $t_m \in [t_{m_1}, t_{m_2}]$ in the sense of metric functionals if all $j \in \mathbb{L}$ and $\alpha_j \in \mathbb{I}$,

$$\begin{aligned}
&{}^{(n:\tilde{n})}V_j(\mathbf{X}_{m\mp}^{(\alpha_j,j)}, t_{m\mp}, \boldsymbol{\lambda}_j) = {}^{(n:\tilde{n})}V_j(\mathbf{X}_m^{(0,j)}, t_m, \boldsymbol{\lambda}_j) = 0;\\
&{}^{(n:\tilde{n})}V_j^{(1)}(\mathbf{X}_{m\mp}^{(\alpha_j,j)}, t_{m\mp}, \boldsymbol{\lambda}_j) = 0;\\
&{}^{(n:\tilde{n})}V_j(\mathbf{X}_{m\mp\varepsilon}^{(\alpha_j,j)}, t_{m\mp\varepsilon}, \boldsymbol{\lambda}_j) - {}^{(n:\tilde{n})}V_j(\mathbf{X}_{m\mp}^{(\alpha_j,j)}, t_{m\mp}, \boldsymbol{\lambda}_j) > 0.
\end{aligned} \tag{5.132}$$

Definition 5.38. For two dynamical systems in Eqs.(5.1) and (5.2) with constraints in Eq.(5.4), there are $(l+1)$-metric functionals in Eq.(5.37). For $\mathbf{X}_m^{(\alpha_j,j)} \in \Omega_{(\alpha_j,j)}$

($\alpha_j \in \mathbb{I}$ and $j \in \mathbb{L}$ with $\mathbb{I} = \{1,2\}$ and $\mathbb{L} = \{1,2,\ldots,l\}$) and $\mathbf{X}_m^{(0,j)} \in \partial\Omega_{12(j)}$ at time $t_m, \mathbf{X}_m^{(\alpha_j,j)} = \mathbf{X}_m^{(0,j)}$. For any small $\varepsilon > 0$, there is a time interval $[t_{m-\varepsilon}, t_m)$ or $(t_m, t_{m+\varepsilon}]$. At $\mathbf{X}^{(\alpha_j,j)} \in \Omega_{(\alpha_j,j)}^{\pm\varepsilon}$ for $t \in [t_{m-\varepsilon}, t_m)$ or $(t_m, t_{m+\varepsilon}]$, the functional ${}^{(n:\tilde{n})}V_j(\mathbf{X}^{(\alpha_j,j)}, t, \boldsymbol{\lambda}_j)$ is $C^{r_{\alpha_j}}$-continuous and $|{}^{(n:\tilde{n})}V_j^{(r_{\alpha_j}+1)}(\mathbf{X}^{(\alpha_j,j)}, t, \boldsymbol{\lambda}_j)| < \infty$ ($r_{\alpha_j} \geq 2$). (i) The synchronization of the two dynamical systems in Eqs.(5.1) and (5.2) with the jth-constraint in Eq.(5.4) is called to be *vanished* to form a penetration on the jth -constraint boundary at time t_m in the sense of metric functionals if for $\alpha_j, \beta_j \in \mathbb{I}$ and $\alpha \neq \beta$ with $j \in \mathbb{L}$,

$$
\begin{aligned}
&{}^{(n:\tilde{n})}V_j(\mathbf{X}_{m\mp}^{(\alpha_j,j)}, t_{m\mp}, \boldsymbol{\lambda}_j) = {}^{(n:\tilde{n})}V_j(\mathbf{X}_{m-}^{(\beta_j,j)}, t_{m-}, \boldsymbol{\lambda}_j)\\
&\qquad = {}^{(n:\tilde{n})}V_j(\mathbf{X}_m^{(0,j)}, t_m, \boldsymbol{\lambda}_j) = 0;\\
&{}^{(n:\tilde{n})}V_j^{(1)}(\mathbf{X}_{m\mp}^{(\alpha_j,j)}, t_{m\mp}, \boldsymbol{\lambda}_j) = 0;\\
&{}^{(n:\tilde{n})}V_j(\mathbf{X}_{m\mp\varepsilon}^{(\alpha_j,j)}, t_{m\mp\varepsilon}, \boldsymbol{\lambda}_j) - {}^{(n:\tilde{n})}V_j(\mathbf{X}_{m\mp}^{(\alpha_j,j)}, t_{m\mp}, \boldsymbol{\lambda}_j) > 0;\\
&{}^{(n:\tilde{n})}V_j(\mathbf{X}_{m-\varepsilon}^{(\beta_j,j)}, t_{m-\varepsilon}, \boldsymbol{\lambda}_j) - {}^{(n:\tilde{n})}V_j(\mathbf{X}_{m-}^{(\beta_j,j)}, t_{m-}, \boldsymbol{\lambda}_j) > 0.
\end{aligned}
\tag{5.133}
$$

(ii) The synchronization of the two dynamical systems in Eqs.(5.1) and (5.2) with the jth-constraint in Eq.(5.4) is called to be *onset* from the penetration on the jth-constraint boundary at time t_m in the sense of metric functionals if for $\alpha_j, \beta_j \in \mathbb{I}$ and $\alpha \neq \beta$ with $j \in \mathbb{L}$,

$$
\begin{aligned}
&{}^{(n:\tilde{n})}V_j(\mathbf{X}_{m-}^{(\alpha)}, t_{m-}, \boldsymbol{\lambda}) = {}^{(n:\tilde{n})}V_j(\mathbf{X}_{m\pm}^{(\beta)}, t_{m\pm}, \boldsymbol{\lambda})\\
&\qquad = {}^{(n:\tilde{n})}V_j(\mathbf{X}_m^{(0)}, t_m, \boldsymbol{\lambda}) = 0;\\
&{}^{(n:\tilde{n})}V_j^{(1)}(\mathbf{X}_{m\pm}^{(\beta)}, t_{m\pm}, \boldsymbol{\lambda}) = 0;\\
&{}^{(n:\tilde{n})}V_j(\mathbf{X}_{m-\varepsilon}^{(\alpha)}, t_{m-\varepsilon}, \boldsymbol{\lambda}) - {}^{(n:\tilde{n})}V_j(\mathbf{X}_{m-}^{(\alpha)}, t_{m-}, \boldsymbol{\lambda}) > 0;\\
&{}^{(n:\tilde{n})}V_j(\mathbf{X}_{m\pm\varepsilon}^{(\beta)}, t_{m\pm\varepsilon}, \boldsymbol{\lambda}) - {}^{(n:\tilde{n})}V_j(\mathbf{X}_{m\pm}^{(\beta)}, t_{m\pm}, \boldsymbol{\lambda}) > 0.
\end{aligned}
\tag{5.134}
$$

Definition 5.39. For two dynamical systems in Eqs.(5.1) and (5.2) with constraints in Eq.(5.4), there are $(l+1)$-metric functionals in Eq.(5.37). For $\mathbf{X}_m^{(\alpha_j,j)} \in \Omega_{(\alpha_j,j)}$ ($\alpha_j \in \mathbb{I}$ and $j \in \mathbb{L}$ with $\mathbb{I} = \{1,2\}$ and $\mathbb{L} = \{1,2,\ldots,l\}$) and $\mathbf{X}_m^{(0,j)} \in \partial\Omega_{12(j)}$ at time $t_m, \mathbf{X}_m^{(\alpha_j,j)} = \mathbf{X}_m^{(0,j)}$. For any small $\varepsilon > 0$, there is a time interval $[t_{m-\varepsilon}, t_m)$ or $(t_m, t_{m+\varepsilon}]$. At $\mathbf{X}^{(\alpha_j,j)} \in \Omega_{(\alpha_j,j)}^{\pm\varepsilon}$ for $t \in [t_{m-\varepsilon}, t_m)$ or $(t_m, t_{m+\varepsilon}]$, the functional ${}^{(n:\tilde{n})}V_j(\mathbf{X}^{(\alpha_j,j)}, t, \boldsymbol{\lambda}_j)$ is $C^{r_{\alpha_j}}$-continuous and $|{}^{(n:\tilde{n})}V_j^{(r_{\alpha_j}+1)}(\mathbf{X}^{(\alpha_j,j)}, t, \boldsymbol{\lambda}_j)| < \infty$ ($r_{\alpha_j} \geq 2$). (i) The synchronization of the two dynamical systems in Eqs.(5.1) and (5.2) with the jth-constraint in Eq.(5.4) are called to be *onset* from the desynchronization on the jth- constraint boundary at time t_m in the sense of metric functionals if for $\alpha_j, \beta_j \in \mathbb{I}$ and $\alpha \neq \beta$ with $j \in \mathbb{L}$,

$$\begin{aligned}
&{}^{(n:\tilde{n})}V_j(\mathbf{X}_{m\pm}^{(\alpha_j,j)}, t_{m\pm}, \boldsymbol{\lambda}_j) = {}^{(n:\tilde{n})}V_j(\mathbf{X}_{m\pm}^{(\beta_j,j)}, t_{m\pm}, \boldsymbol{\lambda}_j) \\
&\quad = {}^{(n:\tilde{n})}V_j(\mathbf{X}_{m}^{(0,j)}, t_m, \boldsymbol{\lambda}) = 0; \\
&{}^{(n:\tilde{n})}V_j^{(1)}(\mathbf{X}_{m\pm}^{(\alpha_j,j)}, t_{m\pm}, \boldsymbol{\lambda}_j) = 0; \\
&{}^{(n:\tilde{n})}V_j(\mathbf{X}_{m\pm\varepsilon}^{(\alpha_j,j)}, t_{m\pm\varepsilon}, \boldsymbol{\lambda}_j) - {}^{(n:\tilde{n})}V_j(\mathbf{X}_{m\pm}^{(\alpha_j,j)}, t_{m\pm}, \boldsymbol{\lambda}_j) > 0; \\
&{}^{(n:\tilde{n})}V_j(\mathbf{X}_{m\pm\varepsilon}^{(\beta_j,j)}, t_{m\pm\varepsilon}, \boldsymbol{\lambda}_j) - {}^{(n:\tilde{n})}V_j(\mathbf{X}_{m\pm}^{(\beta_j,j)}, t_{m\pm}, \boldsymbol{\lambda}_j) > 0.
\end{aligned} \tag{5.135}$$

(ii) The synchronization of the two dynamical systems in Eqs.(5.1) and (5.2) with the jth-constraint in Eq.(5.4) is called to be *vanished* to form the desynchronization on the jth-constraint boundary at time t_m in the sense of metric functionals if for $\alpha_j, \beta_j \in \mathbb{I}$ and $\alpha \neq \beta$ with $j \in \mathbb{L}$,

$$\begin{aligned}
&{}^{(n:\tilde{n})}V_j(\mathbf{X}_{m\mp}^{(\alpha_j,j)}, t_{m\mp}, \boldsymbol{\lambda}_j) = {}^{(n:\tilde{n})}V_j(\mathbf{X}_{m\mp}^{(\beta_j,j)}, t_{m\mp}, \boldsymbol{\lambda}_j) \\
&\quad = {}^{(n:\tilde{n})}V_j(\mathbf{X}_{m}^{(0,j)}, t_m, \boldsymbol{\lambda}_j) = 0; \\
&{}^{(n:\tilde{n})}V_j^{(1)}(\mathbf{X}_{m\mp}^{(\alpha_j,j)}, t_{m\mp}, \boldsymbol{\lambda}_j) = 0; \\
&{}^{(n:\tilde{n})}V_j(\mathbf{X}_{m\mp\varepsilon}^{(\alpha_j,j)}, t_{m\mp\varepsilon}, \boldsymbol{\lambda}_j) - {}^{(n:\tilde{n})}V_j(\mathbf{X}_{m\mp}^{(\alpha_j,j)}, t_{m\mp}, \boldsymbol{\lambda}_j) > 0; \\
&{}^{(n:\tilde{n})}V_j(\mathbf{X}_{m\mp\varepsilon}^{(\beta_j,j)}, t_{m\mp\varepsilon}, \boldsymbol{\lambda}_j) - {}^{(n:\tilde{n})}V_j(\mathbf{X}_{m\mp}^{(\beta_j,j)}, t_{m\mp}, \boldsymbol{\lambda}_j) > 0.
\end{aligned} \tag{5.136}$$

Definition 5.40. For two dynamical systems in Eqs.(5.1) and (5.2) with constraints in Eq.(5.4), there are $(l+1)$-metric functionals in Eq.(5.37). For $\mathbf{X}_m^{(\alpha_j,j)} \in \Omega_{(\alpha_j,j)}$ ($\alpha_j \in \mathbb{I}$ and $j \in \mathbb{L}$ with $\mathbb{I} = \{1,2\}$ and $\mathbb{L} = \{1,2,\ldots,l\}$) and $\mathbf{X}_m^{(0,j)} \in \partial\Omega_{12(j)}$ at time $t_m, \mathbf{X}_m^{(\alpha_j,j)} = \mathbf{X}_m^{(0,j)}$. For any small $\varepsilon > 0$, there is a time interval $[t_{m-\varepsilon}, t_m)$ or $(t_m, t_{m+\varepsilon}]$. At $\mathbf{X}^{(\alpha_j,j)} \in \Omega_{(\alpha_j,j)}^{\pm\varepsilon}$ for $t \in [t_{m-\varepsilon}, t_m)$ or $(t_m, t_{m+\varepsilon}]$, the functional ${}^{(n:\tilde{n})}V_j(\mathbf{X}^{(\alpha_j,j)}, t, \boldsymbol{\lambda}_j)$ is $C^{r_{\alpha_j}}$-continuous and $|{}^{(n:\tilde{n})}V_j^{(r_{\alpha_j}+1)}(\mathbf{X}^{(\alpha_j,j)}, t, \boldsymbol{\lambda}_j)| < \infty$ ($r_{\alpha_j} \geq 2$). (i) The desynchronization of the two dynamical systems in Eqs.(5.1) and (5.2) with the jth-constraint in Eq.(5.4) is called to be *vanished* to form a penetration on the jth-constraint boundary at time t_m if for $\alpha_j, \beta_j \in \mathbb{I}$ and $\alpha \neq \beta$ with $j \in \mathbb{L}$,

$$\begin{aligned}
&{}^{(n:\tilde{n})}V_j(\mathbf{X}_{m\pm}^{(\alpha_j,j)}, t_{m\pm}, \boldsymbol{\lambda}_j) = {}^{(n:\tilde{n})}V_j(\mathbf{X}_{m+}^{(\beta_j,j)}, t_{m+}, \boldsymbol{\lambda}_j) \\
&\quad = {}^{(n:\tilde{n})}V_j(\mathbf{X}_{m}^{(0,j)}, t_m, \boldsymbol{\lambda}_j) = 0; \\
&{}^{(n:\tilde{n})}V_j^{(1)}(\mathbf{X}_{m\pm}^{(\alpha_j,j)}, t_{m\pm}, \boldsymbol{\lambda}_j) = 0; \\
&{}^{(n:\tilde{n})}V_j(\mathbf{X}_{m\pm\varepsilon}^{(\alpha_j,j)}, t_{m\pm\varepsilon}, \boldsymbol{\lambda}_j) - {}^{(n:\tilde{n})}V_j(\mathbf{X}_{m\pm}^{(\alpha_j,j)}, t_{m\pm}, \boldsymbol{\lambda}_j) > 0; \\
&{}^{(n:\tilde{n})}V_j(\mathbf{X}_{m+\varepsilon}^{(\beta_j,j)}, t_{m+\varepsilon}, \boldsymbol{\lambda}_j) - {}^{(n:\tilde{n})}V_j(\mathbf{X}_{m+}^{(\beta_j,j)}, t_{m+}, \boldsymbol{\lambda}_j) > 0.
\end{aligned} \tag{5.137}$$

(ii) The desynchronization of the two dynamical systems in Eqs.(5.1) and (5.2) with the jth-constraint in Eq.(5.4) is called to be *onset* from the jth-penetration flow on the jth-constraint boundary at time t_m in the sense of metric functionals if for $\alpha_j, \beta_j \in \mathbb{I}$ and $\alpha \neq \beta$ with $j \in \mathbb{L}$,

$$\begin{aligned}
&{}^{(n:\tilde{n})}V_j(\mathbf{X}_{m\mp}^{(\alpha)}, t_{m\mp}, \boldsymbol{\lambda}) = {}^{(n:\tilde{n})}V_j(\mathbf{X}_{m+}^{(\beta)}, t_{m+}, \boldsymbol{\lambda}) \\
&\quad = {}^{(n:\tilde{n})}V_j(\mathbf{X}_{m}^{(0)}, t_{m}, \boldsymbol{\lambda}) = 0; \\
&{}^{(n:\tilde{n})}V_j^{(1)}(\mathbf{X}_{m+}^{(\beta)}, t_{m+}, \boldsymbol{\lambda}) = 0; \\
&{}^{(n:\tilde{n})}V_j(\mathbf{X}_{m\mp\varepsilon}^{(\alpha)}, t_{m\mp\varepsilon}, \boldsymbol{\lambda}) - {}^{(n:\tilde{n})}V_j(\mathbf{X}_{m\mp}^{(\alpha)}, t_{m\mp}, \boldsymbol{\lambda}) > 0; \\
&{}^{(n:\tilde{n})}V_j(\mathbf{X}_{m+\varepsilon}^{(\beta)}, t_{m+\varepsilon}, \boldsymbol{\lambda}) - {}^{(n:\tilde{n})}V_j(\mathbf{X}_{m+}^{(\beta)}, t_{m+}, \boldsymbol{\lambda}) > 0.
\end{aligned} \tag{5.138}$$

Definition 5.41. For two dynamical systems in Eqs.(5.1) and (5.2) with constraints in Eq.(5.4), there are $(l+1)$-metric functionals in Eq.(5.37). For $\mathbf{X}_m^{(\alpha_j,j)} \in \Omega_{(\alpha_j,j)}$ ($\alpha_j \in \mathbb{I}$ and $j \in \mathbb{L}$ with $\mathbb{I} = \{1,2\}$ and $\mathbb{L} = \{1,2,\ldots,l\}$) and $\mathbf{X}_m^{(0,j)} \in \partial\Omega_{12(j)}$ at time $t_m, \mathbf{X}_m^{(\alpha_j,j)} = \mathbf{X}_m^{(0,j)}$. For any small $\varepsilon > 0$, there is a time interval $[t_{m-\varepsilon}, t_m)$ or $(t_m, t_{m+\varepsilon}]$. At $\mathbf{X}^{(\alpha_j,j)} \in \Omega_{(\alpha_j,j)}^{\pm\varepsilon}$ for $t \in [t_{m-\varepsilon}, t_m)$ or $(t_m, t_{m+\varepsilon}]$, the functional ${}^{(n:\tilde{n})}V_j(\mathbf{X}^{(\alpha_j,j)}, t, \boldsymbol{\lambda}_j)$ is $C^{r_{\alpha_j}}$-continuous and $|{}^{(n:\tilde{n})}V_j^{(r_{\alpha_j}+1)}(\mathbf{X}^{(\alpha_j,j)}, t, \boldsymbol{\lambda}_j)| < \infty (r_{\alpha_j} \geq 2)$. The penetration of the two dynamical systems in Eqs.(5.1) and (5.2) with the jth-constraint in Eq.(5.4) is called to be *switched* to form a new penetration on the jth-constraint boundary at time t_m in the sense of metric functionals if for $\alpha_j, \beta_j \in \mathbb{I}$ and $\alpha \neq \beta$ with $j \in \mathbb{L}$,

$$\begin{aligned}
&{}^{(n:\tilde{n})}V_j(\mathbf{X}_{m\mp}^{(\alpha_j,j)}, t_{m\mp}, \boldsymbol{\lambda}_j) = {}^{(n:\tilde{n})}V_j(\mathbf{X}_{m\pm}^{(\beta_j,j)}, t_{m\pm}, \boldsymbol{\lambda}_j) \\
&\quad = {}^{(n:\tilde{n})}V_j(\mathbf{X}_{m}^{(0,j)}, t_{m}, \boldsymbol{\lambda}_j) = 0; \\
&{}^{(n:\tilde{n})}V_j^{(1)}(\mathbf{X}_{m\mp}^{(\alpha_j,j)}, t_{m\mp}, \boldsymbol{\lambda}_j) = {}^{(n:\tilde{n})}V_j^{(1)}(\mathbf{X}_{m\pm}^{(\beta_j,j)}, t_{m\pm}, \boldsymbol{\lambda}_j) = 0; \\
&{}^{(n:\tilde{n})}V_j(\mathbf{X}_{m\mp\varepsilon}^{(\alpha_j,j)}, t_{m\mp\varepsilon}, \boldsymbol{\lambda}_j) - {}^{(n:\tilde{n})}V_j(\mathbf{X}_{m\mp}^{(\alpha_j,j)}, t_{m\mp}, \boldsymbol{\lambda}_j) > 0; \\
&{}^{(n:\tilde{n})}V_j(\mathbf{X}_{m\pm\varepsilon}^{(\beta_j,j)}, t_{m\pm\varepsilon}, \boldsymbol{\lambda}_j) - {}^{(n:\tilde{n})}V_j(\mathbf{X}_{m\pm}^{(\beta_j,j)}, t_{m\pm}, \boldsymbol{\lambda}_j) > 0.
\end{aligned} \tag{5.139}$$

5.4.3 Synchronicity with singularity to multiple constraints

As discussed in a single constraint, the synchronicity of two dynamical systems to multiple constraints with higher-order singularity can be described through the following definitions.

Definition 5.42. For two dynamical systems in Eqs.(5.1) and (5.2) with constraints in Eq.(5.4), there are $(l+1)$-metric functionals in Eq.(5.37). For $\mathbf{X}_m^{(\alpha_j,j)} \in \Omega_{(\alpha_j,j)}$

($\alpha_j \in \mathbb{I}$ and $j \in \mathbb{L}$ with $\mathbb{I} = \{1,2\}$ and $\mathbb{L} = \{1,2,\ldots,l\}, \mathbb{L} = \mathbb{L}_1 \cup \mathbb{L}_2 \cup \mathbb{L}_3$) and $\mathbf{X}_m^{(0,j)} \in \partial\Omega_{12(j)}$ at time $t_m, \mathbf{X}_m^{(\alpha_j,j)} = \mathbf{X}_m^{(0,j)}$. For any small $\varepsilon > 0$, there is a time interval $[t_{m-\varepsilon}, t_m)$ or $(t_m, t_{m+\varepsilon}]$. At $\mathbf{X}^{(\alpha_j,j)} \in \Omega^{\pm\varepsilon}_{(\alpha_j,j)}$ for $t \in [t_{m-\varepsilon}, t_m)$ or$(t_m, t_{m+\varepsilon}]$, the functional ${}^{(n:\tilde{n})}V_j(\mathbf{X}^{(\alpha_j,j)}, t, \boldsymbol{\lambda}_j)$ is $C^{r_{\alpha_j}}$-continuous and $|{}^{(n:\tilde{n})}V_j^{(r_{\alpha_j}+1)}(\mathbf{X}^{(\alpha_j,j)}, t, \boldsymbol{\lambda}_j)| < \infty$ ($r_{\alpha_j} \geq 2k_{\alpha_j} + 1$). The two dynamical systems in Eqs.(5.1) and (5.2) with constraints in Eq.(5.4) is called an l_1-dimensional synchronization with the $(2k_{\alpha_j} : 2k_{\beta_j})$-order singularity for all $j \in \mathbb{L}_1$, l_2-dimensional desynchronization with the $(2k_{\alpha_j} : 2k_{\beta_j})$-order singularity for all $j \in \mathbb{L}_2$, and l_3-dimensional penetration with the $(2k_{\alpha_j} : 2k_{\beta_j})$-order singularity for all $j \in \mathbb{L}_3$ for time $t_m \in [t_{m_1}, t_{m_2}]$ in the sense of metric functionals

(i) if for $\alpha_j = 1,2$ and $j \in \mathbb{L}_1$,

$$\begin{aligned}
&{}^{(n:\tilde{n})}V_j(\mathbf{X}_{m-}^{(\alpha_j,j)}, t_{m-}, \boldsymbol{\lambda}_j) = {}^{(n:\tilde{n})}V_j(\mathbf{X}_m^{(0,j)}, t_m, \boldsymbol{\lambda}_j) = 0;\\
&{}^{(n:\tilde{n})}V_j^{(s_{\alpha_j})}(\mathbf{X}_{m-}^{(\alpha_j,j)}, t_{m-}, \boldsymbol{\lambda}_j) = 0, \text{ for } s_{\alpha_j} = 0,1,2,\ldots,2k_{\alpha_j};\\
&{}^{(n:\tilde{n})}V_j(\mathbf{X}_{m-\varepsilon}^{(\alpha_j,j)}, t_{m-\varepsilon}, \boldsymbol{\lambda}_j) - {}^{(n:\tilde{n})}V_j(\mathbf{X}_{m-}^{(\alpha_j,j)}, t_{m-}, \boldsymbol{\lambda}_j) > 0;
\end{aligned} \tag{5.140}$$

(ii) if for $\alpha_j = 1,2$ and $j \in \mathbb{L}_2$,

$$\begin{aligned}
&{}^{(n:\tilde{n})}V_j(\mathbf{X}_{m+}^{(\alpha_j,j)}, t_{m+}, \boldsymbol{\lambda}_j) = {}^{(n:\tilde{n})}V_j(\mathbf{X}_m^{(0,j)}, t_m, \boldsymbol{\lambda}_j) = 0;\\
&{}^{(n:\tilde{n})}V_j^{(s_{\alpha_j})}(\mathbf{X}_{m+}^{(\alpha_j,j)}, t_{m+}, \boldsymbol{\lambda}_j) = 0 \text{ for } s_{\alpha_j} = 0,1,2,\ldots,2k_{\alpha_j};\\
&{}^{(n:\tilde{n})}V_j(\mathbf{X}_{m+\varepsilon}^{(\alpha_j,j)}, t_{m+\varepsilon}, \boldsymbol{\lambda}_j) - {}^{(n:\tilde{n})}V_j(\mathbf{X}_{m+}^{(\alpha_j,j)}, t_{m+}, \boldsymbol{\lambda}_j) > 0;
\end{aligned} \tag{5.141}$$

(iii) if for $\alpha_j, \beta_j \in 1,2$ and $j \in \mathbb{L}_3$ with $\alpha_j \neq \beta_j$,

$$\begin{aligned}
&{}^{(n:\tilde{n})}V_j(\mathbf{X}_{m-}^{(\alpha_j,j)}, t_{m-}, \boldsymbol{\lambda}_j) = {}^{(n:\tilde{n})}V_j(\mathbf{X}_m^{(0,j)}, t_m, \boldsymbol{\lambda}_j) = 0;\\
&{}^{(n:\tilde{n})}V_j^{(s_{\alpha_j})}(\mathbf{X}_{m-}^{(\alpha_j,j)}, t_{m-}, \boldsymbol{\lambda}_j) = 0, \text{ for } s_{\alpha_j} = 0,1,2,\ldots,2k_{\alpha_j};\\
&{}^{(n:\tilde{n})}V_j(\mathbf{X}_{m-\varepsilon}^{(\alpha_j,j)}, t_{m-\varepsilon}, \boldsymbol{\lambda}_j) - {}^{(n:\tilde{n})}V_j(\mathbf{X}_{m-}^{(\alpha_j,j)}, t_{m-}, \boldsymbol{\lambda}_j) > 0;
\end{aligned} \tag{5.142}$$

$$\begin{aligned}
&{}^{(n:\tilde{n})}V_j(\mathbf{X}_{m+}^{(\beta_j,j)}, t_{m+}, \boldsymbol{\lambda}_j) = {}^{(n:\tilde{n})}V_j(\mathbf{X}_m^{(0,j)}, t_m, \boldsymbol{\lambda}_j) = 0;\\
&{}^{(n:\tilde{n})}V_j^{(s_{\beta_j})}(\mathbf{X}_{m+}^{(\beta_j,j)}, t_{m+}, \boldsymbol{\lambda}_j) = 0, \text{ for } s_{\beta_j} = 1,2,\ldots,2k_{\beta_j};\\
&{}^{(n:\tilde{n})}V_j(\mathbf{X}_{m+\varepsilon}^{(\beta_j,j)}, t_{m+\varepsilon}, \boldsymbol{\lambda}_j) - {}^{(n:\tilde{n})}V_j(\mathbf{X}_{m+}^{(\beta_j,j)}, t_{m+}, \boldsymbol{\lambda}_j) > 0.
\end{aligned} \tag{5.143}$$

Consider two dynamical systems with l-constraints to be synchronized, or desynchronized or penetrated with higher-order singularity. The corresponding descriptions for such synchronicity will be given as follows.

Definition 5.43. For two dynamical systems in Eqs.(5.1) and (5.2) with constraints in Eq.(5.4), there are $(l+1)$-metric functionals in Eq.(5.37). For $\mathbf{X}_m^{(\alpha_j,j)} \in \Omega_{(\alpha_j,j)}$ ($\alpha_j \in \mathbb{I}$ and $j \in \mathbb{L}$ with $\mathbb{I} = \{1,2\}$ and $\mathbb{L} = \{1,2,\ldots,l\}$) and $\mathbf{X}_m^{(0,j)} \in \partial\Omega_{12(j)}$ at time $t_m, \mathbf{X}_m^{(\alpha_j,j)} = \mathbf{X}_m^{(0,j)}$. For any small $\varepsilon > 0$, there is a time interval $[t_{m-\varepsilon}, t_m)$ or $(t_m, t_{m+\varepsilon}]$. At $\mathbf{X}^{(\alpha_j,j)} \in \Omega_{(\alpha_j,j)}^{\pm\varepsilon}$ for $t \in [t_{m-\varepsilon}, t_m)$ or$(t_m, t_{m+\varepsilon}]$, the functional ${}^{(n:\tilde{n})}V_j(\mathbf{X}^{(\alpha_j,j)}, t, \boldsymbol{\lambda}_j)$ is $C^{r_{\alpha_j}}$-continuous and $|{}^{(n:\tilde{n})}V_j^{(r_{\alpha_j}+1)}(\mathbf{X}^{(\alpha_j,j)}, t, \boldsymbol{\lambda}_j)| < \infty$ $(r_{\alpha_j} \geq 2k_{\alpha_j}+1)$. The two dynamical systems in Eqs.(5.1) and (5.2) with constraints in Eq.(5.4) is called an l-dimensional synchronization with the $(2k_{\alpha_j} : 2k_{\beta_j})$-order singularity for all $j \in \mathbb{L}$ for time $t_m \in [t_{m_1}, t_{m_2}]$ in the sense of metric functionals if for $\alpha_j = 1,2$ and $j \in \mathbb{L}$,

$$
\begin{aligned}
&{}^{(n:\tilde{n})}V_j(\mathbf{X}_{m-}^{(\alpha_j,j)}, t_{m-}, \boldsymbol{\lambda}_j) = {}^{(n:\tilde{n})}V_j(\mathbf{X}_m^{(0,j)}, t_m, \boldsymbol{\lambda}_j) = 0;\\
&{}^{(n:\tilde{n})}V_j^{(s_{\alpha_j})}(\mathbf{X}_{m-}^{(\alpha_j,j)}, t_{m-}, \boldsymbol{\lambda}_j) = 0, \text{ for } s_{\alpha_j} = 0,1,2,\ldots,2k_{\alpha_j};\\
&{}^{(n:\tilde{n})}V_j(\mathbf{X}_{m-\varepsilon}^{(\alpha_j,j)}, t_{m-\varepsilon}, \boldsymbol{\lambda}_j) - {}^{(n:\tilde{n})}V_j(\mathbf{X}_{m-}^{(\alpha_j,j)}, t_{m-}, \boldsymbol{\lambda}_j) > 0.
\end{aligned}
\tag{5.144}
$$

Definition 5.44. For two dynamical systems in Eqs.(5.1) and (5.2) with constraints in Eq.(5.4), there are $(l+1)$-metric functionals in Eq.(5.37). For $\mathbf{X}_m^{(\alpha_j,j)} \in \Omega_{(\alpha_j,j)}$ ($\alpha_j \in \mathbb{I}$ and $j \in \mathbb{L}$ with $\mathbb{I} = \{1,2\}$ and $\mathbb{L} = \{1,2,\ldots,l\}$) and $\mathbf{X}_m^{(0,j)} \in \partial\Omega_{12(j)}$ at time $t_m, \mathbf{X}_m^{(\alpha_j,j)} = \mathbf{X}_m^{(0,j)}$. For any small $\varepsilon > 0$, there is a time interval $[t_{m-\varepsilon}, t_m)$ or $(t_m, t_{m+\varepsilon}]$. At $\mathbf{X}^{(\alpha_j,j)} \in \Omega_{(\alpha_j,j)}^{\pm\varepsilon}$ for $t \in [t_{m-\varepsilon}, t_m)$ or $(t_m, t_{m+\varepsilon}]$, the functional ${}^{(n:\tilde{n})}V_j(\mathbf{X}^{(\alpha_j,j)}, t, \boldsymbol{\lambda}_j)$ is $C^{r_{\alpha_j}}$-continuous and $|{}^{(n:\tilde{n})}V_j^{(r_{\alpha_j}+1)}(\mathbf{X}^{(\alpha_j,j)}, t, \boldsymbol{\lambda}_j)| < \infty$ $(r_{\alpha_j} \geq 2k_{\alpha_j}+1)$. The two dynamical systems in Eqs.(5.1) and (5.2) with constraints in Eq.(5.4) is called an l-dimensional desynchronization with the $(2k_{\alpha_j} : 2k_{\beta_j})$-order singularity for all $j \in \mathbb{L}$ for time $t_m \in [t_{m_1}, t_{m_2}]$ in the sense of metric functionals if for $\alpha_j = 1,2$ and $j \in \mathbb{L}$,

$$
\begin{aligned}
&{}^{(n:\tilde{n})}V_j(\mathbf{X}_{m+}^{(\alpha_j,j)}, t_{m+}, \boldsymbol{\lambda}_j) = {}^{(n:\tilde{n})}V_j(\mathbf{X}_m^{(0,j)}, t_m, \boldsymbol{\lambda}_j) = 0;\\
&{}^{(n:\tilde{n})}V_j^{(s_{\alpha_j})}(\mathbf{X}_{m+}^{(\alpha_j,j)}, t_{m+}, \boldsymbol{\lambda}_j) = 0 \text{ for } s_{\alpha_j} = 0,1,2,\ldots,2k_{\alpha_j};\\
&{}^{(n:\tilde{n})}V_j(\mathbf{X}_{m+\varepsilon}^{(\alpha_j,j)}, t_{m+\varepsilon}, \boldsymbol{\lambda}_j) - {}^{(n:\tilde{n})}V_j(\mathbf{X}_{m+}^{(\alpha_j,j)}, t_{m+}, \boldsymbol{\lambda}_j) > 0.
\end{aligned}
\tag{5.145}
$$

Definition 5.45. For two dynamical systems in Eqs.(5.1) and (5.2) with constraints in Eq.(5.4), there are $(l+1)$-metric functionals in Eq.(5.37). For $\mathbf{X}_m^{(\alpha_j,j)} \in \Omega_{(\alpha_j,j)}$

($\alpha_j \in \mathbb{I}$ and $j \in \mathbb{L}$ with $\mathbb{I} = \{1,2\}$ and $\mathbb{L} = \{1,2,\ldots,l\}$) and $\mathbf{X}_m^{(0,j)} \in \partial\Omega_{12(j)}$ at time $t_m, \mathbf{X}_m^{(\alpha_j,j)} = \mathbf{X}_m^{(0,j)}$. For any small $\varepsilon > 0$, there is a time interval $[t_{m-\varepsilon}, t_m)$ or $(t_m, t_{m+\varepsilon}]$. At $\mathbf{X}^{(\alpha_j,j)} \in \Omega_{(\alpha_j,j)}^{\pm\varepsilon}$ for $t \in [t_{m-\varepsilon}, t_m)$ or $(t_m, t_{m+\varepsilon}]$, the functional ${}^{(n:\tilde{n})}V_j(\mathbf{X}^{(\alpha_j,j)}, t, \boldsymbol{\lambda}_j)$ is $C^{r_{\alpha_j}}$-continuous and $|{}^{(n:\tilde{n})}V_j^{(r_{\alpha_j}+1)}(\mathbf{X}^{(\alpha_j,j)}, t, \boldsymbol{\lambda}_j)| < \infty$ ($r_{\alpha_j} \geq 2k_{\alpha_j} + 1$). The two dynamical systems in Eqs.(5.1) and (5.2) with constraints in Eq.(5.4) is called an l-dimensional penetration with the $(2k_{\alpha_j} : 2k_{\beta_j})$-order singularity for all $j \in \mathbb{L}$ for time $t_m \in [t_{m_1}, t_{m_2}]$ in the sense of metric functionals if for $\alpha_j, \beta_j \in 1,2$ and $j \in \mathbb{L}$ with $\alpha_j \neq \beta_j$.

$$\left.\begin{aligned}
&{}^{(n:\tilde{n})}V_j(\mathbf{X}_{m-}^{(\alpha_j,j)}, t_{m-}, \boldsymbol{\lambda}_j) = {}^{(n:\tilde{n})}V_j(\mathbf{X}_m^{(0,j)}, t_m, \boldsymbol{\lambda}_j) = 0;\\
&{}^{(n:\tilde{n})}V_j^{(s_{\alpha_j})}(\mathbf{X}_{m-}^{(\alpha_j,j)}, t_{m-}, \boldsymbol{\lambda}_j) = 0, \text{ for } s_{\alpha_j} = 0,1,2,\ldots,2k_{\alpha_j};\\
&{}^{(n:\tilde{n})}V_j(\mathbf{X}_{m-\varepsilon}^{(\alpha_j,j)}, t_{m-\varepsilon}, \boldsymbol{\lambda}_j) - {}^{(n:\tilde{n})}V_j(\mathbf{X}_{m-}^{(\alpha_j,j)}, t_{m-}, \boldsymbol{\lambda}_j) > 0;
\end{aligned}\right\} \tag{5.146}$$

$$\left.\begin{aligned}
&{}^{(n:\tilde{n})}V_j(\mathbf{X}_{m+}^{(\beta_j,j)}, t_{m+}, \boldsymbol{\lambda}_j) = {}^{(n:\tilde{n})}V_j(\mathbf{X}_m^{(0,j)}, t_m, \boldsymbol{\lambda}_j) = 0;\\
&{}^{(n:\tilde{n})}V_j^{(s_{\beta_j})}(\mathbf{X}_{m+}^{(\beta_j,j)}, t_{m+}, \boldsymbol{\lambda}_j) = 0, \text{ for } s_{\beta_j} = 0,1,2,\ldots,2k_{\beta_j};\\
&{}^{(n:\tilde{n})}V_j(\mathbf{X}_{m+\varepsilon}^{(\beta_j,j)}, t_{m+\varepsilon}, \boldsymbol{\lambda}_j) - {}^{(n:\tilde{n})}V_j(\mathbf{X}_{m+}^{(\beta_j,j)}, t_{m+}, \boldsymbol{\lambda}_j) > 0.
\end{aligned}\right\} \tag{5.147}$$

5.4.4 Higher-order singularity to constraints

Since a resultant flow of two dynamical systems to one of l-constraints possesses the higher-order singularity, the synchronicity of the two dynamical systems to the l-constraints will be changed. In this section, the higher-order singularity of a resultant flow of two dynamical systems to the jth-constraint boundary from the l-constraints will be presented herein.

Definition 5.46. For two dynamical systems in Eqs.(5.1) and (5.2) with constraints in Eq.(5.4), there are $(l+1)$-metric functionals in Eq.(5.37). For $\mathbf{X}_m^{(\alpha_j,j)} \in \Omega_{(\alpha_j,j)}$ ($\alpha_j \in \mathbb{I}$ and $j \in \mathbb{L}$ with $\mathbb{I} = \{1,2\}$ and $\mathbb{L} = \{1,2,\cdots,l\}$) and $\mathbf{X}_m^{(0,j)} \in \partial\Omega_{12(j)}$ at time $t_m, \mathbf{X}_m^{(\alpha_j,j)} = \mathbf{X}_m^{(0,j)}$. For any small $\varepsilon > 0$, there is a time interval $[t_{m-\varepsilon}, t_m)$ or $(t_m, t_{m+\varepsilon}]$. At $\mathbf{X}^{(\alpha_j,j)} \in \Omega_{(\alpha_j,j)}^{\pm\varepsilon}$ for $t \in [t_{m-\varepsilon}, t_m)$ or $(t_m, t_{m+\varepsilon}]$, the functional ${}^{(n:\tilde{n})}V_j(\mathbf{X}^{(\alpha_j,j)}, t, \boldsymbol{\lambda}_j)$ is $C^{r_{\alpha_j}}$-continuous and $|{}^{(n:\tilde{n})}V_j^{(r_{\alpha_j}+1)}(\mathbf{X}^{(\alpha_j,j)}, t, \boldsymbol{\lambda}_j)| < \infty$ ($r_{\alpha_j} \geq 2k_{\alpha_j}$). A flow of the resultant system of two dynamical systems in Eqs.(5.1) and (5.2) with l-constraints in Eq.(5.4) is called to be *tangential* to the jth-constraint boundary with *the* $(2k_{\alpha_j} - 1)^{\text{th}}$*-order* for time $t_m \in [t_{m_1}, t_{m_2}]$ in the sense of metric functionals if for $j \in \mathbb{L}$ and $\alpha_j \in \mathbb{I}$,

$$\begin{aligned}
&{}^{(n:\tilde{n})}V_j(\mathbf{X}_{m\mp}^{(\alpha_j,j)},t_{m\mp},\boldsymbol{\lambda}_j) = {}^{(n:\tilde{n})}V_j(\mathbf{X}_m^{(0,j)},t_m,\boldsymbol{\lambda}_j) = 0;\\
&{}^{(n:\tilde{n})}V_j^{(s_{\alpha_j})}(\mathbf{X}_{m\mp}^{(\alpha_j,j)},t_{m\mp},\boldsymbol{\lambda}_j) = 0 \text{ for } s_{\alpha_j} = 1,2,\ldots,2k_{\alpha_j}-1;\\
&{}^{(n:\tilde{n})}V_j(\mathbf{X}_{m\mp\varepsilon}^{(\alpha_j,j)},t_{m\mp\varepsilon},\boldsymbol{\lambda}_j) - {}^{(n:\tilde{n})}V_j(\mathbf{X}_{m\mp}^{(\alpha_j,j)},t_{m\mp},\boldsymbol{\lambda}_j) > 0.
\end{aligned} \tag{5.148}$$

Definition 5.47. For two dynamical systems in Eqs.(5.1) and (5.2) with constraints in Eq.(5.4), there are $(l+1)$-metric functionals in Eq.(5.37). For $\mathbf{X}_m^{(\alpha_j,j)} \in \Omega_{(\alpha_j,j)}$ ($\alpha_j \in \mathbb{I}$ and $j \in \mathbb{L}$ with $\mathbb{I} = \{1,2\}$ and $\mathbb{L} = \{1,2,\ldots,l\}$) and $\mathbf{X}_m^{(0,j)} \in \partial\Omega_{12(j)}$ at time $t_m, \mathbf{X}_m^{(\alpha_j,j)} = \mathbf{X}_m^{(0,j)}$. For any small $\varepsilon > 0$, there is a time interval $[t_{m-\varepsilon}, t_m)$ or $(t_m, t_{m+\varepsilon}]$. At $\mathbf{X}^{(\alpha_j,j)} \in \Omega_{(\alpha_j,j)}^{\pm\varepsilon}$ for $t \in [t_{m-\varepsilon}, t_m)$ or $(t_m, t_{m+\varepsilon}]$, the functional ${}^{(n:\tilde{n})}V_j(\mathbf{X}^{(\alpha_j,j)},t,\boldsymbol{\lambda}_j)$ is $C^{r_{\alpha_j}}$-continuous and $|{}^{(n:\tilde{n})}V_j^{(r_{\alpha_j}+1)}(\mathbf{X}^{(\alpha_j,j)},t,\boldsymbol{\lambda}_j)| < \infty$ $(r_{\alpha_j} \geq 2k_{\alpha_j}+1)$.

(i) The synchronization of the $(2k_{\alpha_j} : 2k_{\beta_j})$-order of the two dynamical systems in Eqs.(5.1) and (5.2) with the jth-constraint in Eq.(5.4) is said to *be vanished* to form the penetration on the jth-constraint boundary from domain $\Omega_{(\alpha_j j)}$ to $\Omega_{(\beta_j j)}$ at time t_m in the sense of metric functionals if for $\alpha_j, \beta_j \in \mathbb{I}$ and $\alpha \neq \beta$ with $j \in \mathbb{L}$,

$$\begin{aligned}
&{}^{(n:\tilde{n})}V_j(\mathbf{X}_{m-}^{(\alpha_j,j)},t_{m-},\boldsymbol{\lambda}_j) = {}^{(n:\tilde{n})}V_j(\mathbf{X}_{m\mp}^{(\beta_j,j)},t_{m\mp},\boldsymbol{\lambda}_j)\\
&\qquad = {}^{(n:\tilde{n})}V_j(\mathbf{X}_m^{(0,j)},t_m,\boldsymbol{\lambda}_j) = 0;\\
&{}^{(n:\tilde{n})}V_j^{(s_{\alpha_j})}(\mathbf{X}_{m-}^{(\alpha_j,j)},t_{m-},\boldsymbol{\lambda}_j) = 0 \text{ for } s_{\alpha_j} = 1,2,\ldots,2k_{\alpha_j};\\
&{}^{(n:\tilde{n})}V_j^{(s_{\beta_j})}(\mathbf{X}_{m\mp}^{(\beta_j,j)},t_{m\mp},\boldsymbol{\lambda}_j) = 0 \text{ for } s_{\beta_j} = 1,2,\ldots,2k_{\beta_j}+1;\\
&{}^{(n:\tilde{n})}V_j(\mathbf{X}_{m-\varepsilon}^{(\alpha_j,j)},t_{m-\varepsilon},\boldsymbol{\lambda}_j) - {}^{(n:\tilde{n})}V_j(\mathbf{X}_{m-}^{(\alpha_j,j)},t_{m-},\boldsymbol{\lambda}_j) > 0;\\
&{}^{(n:\tilde{n})}V_j(\mathbf{X}_{m\mp\varepsilon}^{(\beta_j,j)},t_{m\mp\varepsilon},\boldsymbol{\lambda}_j) - {}^{(n:\tilde{n})}V_j(\mathbf{X}_{m\mp}^{(\beta_j,j)},t_{m\mp},\boldsymbol{\lambda}_j) > 0.
\end{aligned} \tag{5.149}$$

(ii) The synchronization of the $(2k_{\alpha_j} : 2k_{\beta_j})$-order of the two dynamical systems in Eqs.(5.1) and (5.2) with the jth constraint in Eq.(5.4) is said to be onset from the penetration on the jth constraint boundary from domain $\Omega_{(\alpha_j,j)}$ to $\Omega_{(\beta_j,j)}$ at time t_m in the sense of metric functionals if for $\alpha_j, \beta_j \in \mathbb{I}$ and $\alpha \neq \beta$ with $j \in \mathbb{L}$,

$$\begin{aligned}
&{}^{(n:\tilde{n})}V_j(\mathbf{X}_{m-}^{(\alpha)},t_{m-},\boldsymbol{\lambda}) = {}^{(n:\tilde{n})}V_j(\mathbf{X}_{m\pm}^{(\beta)},t_{m\pm},\boldsymbol{\lambda})\\
&\qquad = {}^{(n:\tilde{n})}V_j(\mathbf{X}_m^{(0)},t_m,\boldsymbol{\lambda}) = 0;
\end{aligned} \tag{5.150a}$$

$$ {}^{(n:\tilde{n})}V_j^{(s_{\alpha_j})}(\mathbf{X}_{m-}^{(\alpha_j,j)},t_{m-},\boldsymbol{\lambda}_j) = 0 \text{ for } s_{\alpha_j} = 1,2,\cdots,2k_{\alpha_j};$$

$$
\begin{aligned}
&{}^{(n:\tilde{n})}V_j^{(s_{\beta_j})}(\mathbf{X}_{m\mp}^{(\beta_j,j)},t_{m\mp},\boldsymbol{\lambda}_j)=0 \text{ for } s_{\beta_j}=1,2,\cdots,2k_{\beta_j}+1;\\
&{}^{(n:\tilde{n})}V_j(\mathbf{X}_{m-\varepsilon}^{(\alpha)},t_{m-\varepsilon},\boldsymbol{\lambda})-{}^{(n:\tilde{n})}V_j(\mathbf{X}_{m-}^{(\alpha)},t_{m-},\boldsymbol{\lambda})>0;\\
&{}^{(n:\tilde{n})}V_j(\mathbf{X}_{m\pm\varepsilon}^{(\beta)},t_{m\pm\varepsilon},\boldsymbol{\lambda})-{}^{(n:\tilde{n})}V_j(\mathbf{X}_{m\pm}^{(\beta)},t_{m\pm},\boldsymbol{\lambda})>0.
\end{aligned}
\tag{5.150b}
$$

Definition 5.48. For two dynamical systems in Eqs.(5.1) and (5.2) with constraints in Eq.(5.4), there are $(l+1)$-metric functionals in Eq.(5.37). For $\mathbf{X}_m^{(\alpha_j,j)}\in\Omega_{(\alpha_j,j)}$ ($\alpha_j\in\mathbb{I}$ and $j\in\mathbb{L}$ with $\mathbb{I}=\{1,2\}$ and $\mathbb{L}=\{1,2,\ldots,l\}$) and $\mathbf{X}_m^{(0,j)}\in\partial\Omega_{12(j)}$ at time $t_m,\mathbf{X}_m^{(\alpha_j,j)}=\mathbf{X}_m^{(0,j)}$. For any small $\varepsilon>0$, there is a time interval $[t_{m-\varepsilon},t_m)$ or $(t_m,t_{m+\varepsilon}]$. At $\mathbf{X}^{(\alpha_j,j)}\in\Omega_{(\alpha_j,j)}^{\pm\varepsilon}$ for $t\in[t_{m-\varepsilon},t_m)$ or $(t_m,t_{m+\varepsilon}]$, the functional ${}^{(n:\tilde{n})}V_j(\mathbf{X}^{(\alpha_j,j)},t,\boldsymbol{\lambda}_j)$ is $C^{r_{\alpha_j}}$-continuous and $|{}^{(n:\tilde{n})}V_j^{(r_{\alpha_j}+1)}(\mathbf{X}^{(\alpha_j,j)},t,\boldsymbol{\lambda}_j)|<\infty$ ($r_{\alpha_j}\geq 2k_{\alpha_j}+1$).

(i) The $(2k_{\alpha_j}:2k_{\beta_j})$-synchronization of the two dynamical systems in Eqs.(5.1) and (5.2) with the jth-constraint in Eq.(5.4) is called to be *onset* from the desynchronization on the jth-constraint boundary at time t_m in the sense of metric functionals if for $\alpha_j,\beta_j\in\mathbb{I}$ and $\alpha_j\neq\beta_j$ with $j\in\mathbb{L}$,

$$
\begin{aligned}
&{}^{(n:\tilde{n})}V_j(\mathbf{X}_{m\pm}^{(\alpha_j,j)},t_{m\pm},\boldsymbol{\lambda}_j)={}^{(n:\tilde{n})}V_j(\mathbf{X}_{m\pm}^{(\beta_j,j)},t_{m\pm},\boldsymbol{\lambda}_j)\\
&\quad={}^{(n:\tilde{n})}V_j(\mathbf{X}_m^{(0,j)},t_m,\boldsymbol{\lambda}_j)=0;\\
&{}^{(n:\tilde{n})}V_j^{(s_{\alpha_j})}(\mathbf{X}_{m\pm}^{(\alpha_j,j)},t_{m\pm},\boldsymbol{\lambda}_j)=0 \text{ for } s_{\alpha_j}=1,2,\cdots,2k_{\alpha_j}+1;\\
&{}^{(n:\tilde{n})}V_j^{(s_{\beta_j})}(\mathbf{X}_{m\pm}^{(\beta_j,j)},t_{m\pm},\boldsymbol{\lambda}_j)=0 \text{ for } s_{\beta_j}=1,2,\cdots,2k_{\beta_j}+1;\\
&{}^{(n:\tilde{n})}V_j(\mathbf{X}_{m\pm\varepsilon}^{(\alpha_j,j)},t_{m\pm\varepsilon},\boldsymbol{\lambda}_j)-{}^{(n:\tilde{n})}V_j(\mathbf{X}_{m\pm}^{(\alpha_j,j)},t_{m\pm},\boldsymbol{\lambda}_j)>0;\\
&{}^{(n:\tilde{n})}V_j(\mathbf{X}_{m\pm\varepsilon}^{(\beta_j,j)},t_{m\pm\varepsilon},\boldsymbol{\lambda}_j)-{}^{(n:\tilde{n})}V_j(\mathbf{X}_{m\pm}^{(\beta_j,j)},t_{m\pm},\boldsymbol{\lambda}_j)>0.
\end{aligned}
\tag{5.151}
$$

(ii) The $(2k_{\alpha_j}:2k_{\beta_j})$-synchronization of the two dynamical systems in Eqs.(5.1) and (5.2) with the jth-constraint in Eq.(5.4) is said to be *vanishing* to form the desynchronization on the jth-constraint boundary at time t_m in the sense of metric functionals if for $\alpha_j,\beta_j\in\mathbb{I}$ and $\alpha\neq\beta$ with $j\in\mathbb{L}$,

$$
\begin{aligned}
&{}^{(n:\tilde{n})}V_j\left(\mathbf{X}_{m\mp}^{(\alpha_j,j)},t_{m\mp},\boldsymbol{\lambda}_j\right)={}^{(n:\tilde{n})}V_j\left(\mathbf{X}_{m\mp}^{(\beta_j,j)},t_{m\mp},\boldsymbol{\lambda}_j\right)\\
&\quad={}^{(n:\tilde{n})}V_j\left(\mathbf{X}_m^{(0,j)},t_m,\lambda_j\right)=0;\\
&{}^{(n:\tilde{n})}V_j^{(s_{\alpha_j})}(\mathbf{X}_{m\mp}^{(\alpha_j,j)},t_{m\mp},\boldsymbol{\lambda}_j)=0 \text{ for } s_{\alpha_j}=1,2,\ldots,2k_{\alpha_j}+1;\\
&{}^{(n:\tilde{n})}V_j^{(s_{\beta_j})}(\mathbf{X}_{m\mp}^{(\beta_j,j)},t_{m\mp},\boldsymbol{\lambda}_j)=0 \text{ for } s_{\beta_j}=1,2,\ldots,2k_{\beta_j}+1;\\
&{}^{(n:\tilde{n})}V_j(\mathbf{X}_{m\mp\varepsilon}^{(\alpha_j,j)},t_{m\mp\varepsilon},\boldsymbol{\lambda}_j)-{}^{(n:\tilde{n})}V_j(\mathbf{X}_{m\mp}^{(\alpha_j,j)},t_{m\mp},\boldsymbol{\lambda}_j)>0;\\
&{}^{(n:\tilde{n})}V_j(\mathbf{X}_{m\mp\varepsilon}^{(\beta_j,j)},t_{m\mp\varepsilon},\boldsymbol{\lambda}_j)-{}^{(n:\tilde{n})}V_j(\mathbf{X}_{m\mp}^{(\beta_j,j)},t_{m\mp},\boldsymbol{\lambda}_j)>0.
\end{aligned}
\tag{5.152}
$$

Definition 5.49. For two dynamical systems in Eqs.(5.1) and (5.2) with constraints in Eq.(5.4), there are $(l+1)$-metric functionals in Eq.(5.37). For $\mathbf{X}_m^{(\alpha_j,j)} \in \Omega_{(\alpha_j,j)}$ ($\alpha_j \in \mathbb{I}$ and $j \in \mathbb{L}$ with $\mathbb{I} = \{1,2\}$ and $\mathbb{L} = \{1,2,\ldots,l\}$) and $\mathbf{X}_m^{(0,j)} \in \partial\Omega_{12(j)}$ at time $t_m, \mathbf{X}_m^{(\alpha_j,j)} = \mathbf{X}_m^{(0,j)}$. For any small $\varepsilon > 0$, there is a time interval $[t_{m-\varepsilon}, t_m)$ or $(t_m, t_{m+\varepsilon}]$. At $\mathbf{X}^{(\alpha_j,j)} \in \Omega_{(\alpha_j,j)}^{\pm\varepsilon}$ for $t \in [t_{m-\varepsilon}, t_m)$ or $(t_m, t_{m+\varepsilon}]$, the functional ${}^{(n:\tilde{n})}V_j(\mathbf{X}^{(\alpha_j,j)}, t, \boldsymbol{\lambda}_j)$ is $C^{r_{\alpha_j}}$-continuous and $|{}^{(n:\tilde{n})}V_j^{(r_{\alpha_j}+1)}(\mathbf{X}^{(\alpha_j,j)}, t, \boldsymbol{\lambda}_j)| < \infty$ ($r_{\alpha_j} \geq 2k_{\alpha_j}+1$).

(i) The desynchronization of the $(2k_{\alpha_j} : 2k_{\beta_j})$-order of the two dynamical systems in Eqs.(5.1) and (5.2) with the jth-constraint in Eq.(5.4) is called *to be vanished* to form the penetration on the jth-constraint boundary from domain $\Omega_{(\alpha_j,j)}$ to $\Omega_{(\beta_j,j)}$ at time t_m in the sense of metric functionals if for $\alpha_j, \beta_j \in \mathbb{I}$ and $\alpha \neq \beta$ with $j \in \mathbb{L}$,

$$\left.\begin{aligned}
&{}^{(n:\tilde{n})}V_j(\mathbf{X}_{m\pm}^{(\alpha_j,j)}, t_{m\pm}, \boldsymbol{\lambda}_j) = {}^{(n:\tilde{n})}V_j(\mathbf{X}_{m+}^{(\beta_j,j)}, t_{m+}, \boldsymbol{\lambda}_j)\\
&\quad = {}^{(n:\tilde{n})}V_j(\mathbf{X}_m^{(0,j)}, t_m, \boldsymbol{\lambda}_j) = 0;\\
&{}^{(n:\tilde{n})}V_j^{(s_{\alpha_j})}(\mathbf{X}_{m\pm}^{(\alpha_j,j)}, t_{m\pm}, \boldsymbol{\lambda}_j) = 0 \text{ for } s_{\alpha_j} = 1,2,\ldots,2k_{\alpha_j}+1;\\
&{}^{(n:\tilde{n})}V_j^{(s_{\beta_j})}(\mathbf{X}_{m+}^{(\beta_j,j)}, t_{m+}, \boldsymbol{\lambda}_j) = 0 \text{ for } s_{\beta_j} = 1,2,\ldots,2k_{\beta_j};\\
&{}^{(n:\tilde{n})}V_j(\mathbf{X}_{m\pm\varepsilon}^{(\alpha_j,j)}, t_{m\pm\varepsilon}, \boldsymbol{\lambda}_j) - {}^{(n:\tilde{n})}V_j(\mathbf{X}_{m\pm}^{(\alpha_j,j)}, t_{m\pm}, \boldsymbol{\lambda}_j) > 0;\\
&{}^{(n:\tilde{n})}V_j(\mathbf{X}_{m+\varepsilon}^{(\beta_j,j)}, t_{m+\varepsilon}, \boldsymbol{\lambda}_j) - {}^{(n:\tilde{n})}V_j(\mathbf{X}_{m+}^{(\beta_j,j)}, t_{m+}, \boldsymbol{\lambda}_j) > 0.
\end{aligned}\right\} \tag{5.153}$$

(ii) The desynchronization of the $(2k_{\alpha_j} : 2k_{\beta_j})$-order of the two dynamical systems in Eqs.(5.1) and (5.2) with the jth-constraint in Eq.(5.4) is said to be *onset* from the penetration on the jth-constraint boundary from domain $\Omega_{(\alpha_j,j)}$ to $\Omega_{(\beta_j,j)}$ at time t_m in the sense of metric functionals if for $\alpha_j, \beta_j \in \mathbb{I}$ and $\alpha \neq \beta$ with $j \in \mathbb{L}$,

$$\left.\begin{aligned}
&{}^{(n:\tilde{n})}V_j(\mathbf{X}_{m\mp}^{(\alpha)}, t_{m\mp}, \boldsymbol{\lambda}) = {}^{(n:\tilde{n})}V_j(\mathbf{X}_{m+}^{(\beta)}, t_{m+}, \boldsymbol{\lambda})\\
&\quad = {}^{(n:\tilde{n})}V_j(\mathbf{X}_m^{(0)}, t_m, \boldsymbol{\lambda}) = 0;\\
&{}^{(n:\tilde{n})}V_j^{(s_{\alpha_j})}(\mathbf{X}_{m\mp}^{(\alpha_j,j)}, t_{m\mp}, \boldsymbol{\lambda}_j) = 0 \text{ for } s_{\alpha_j} = 1,2,\ldots,2k_{\alpha_j}+1;\\
&{}^{(n:\tilde{n})}V_j^{(s_{\beta_j})}(\mathbf{X}_{m+}^{(\beta_j,j)}, t_{m+}, \boldsymbol{\lambda}_j) = 0 \text{ for } s_{\beta_j} = 1,2,\ldots,2k_{\beta_j};\\
&{}^{(n:\tilde{n})}V_j(\mathbf{X}_{m\mp\varepsilon}^{(\alpha)}, t_{m\mp\varepsilon}, \boldsymbol{\lambda}) - {}^{(n:\tilde{n})}V_j(\mathbf{X}_{m\mp}^{(\alpha)}, t_{m\mp}, \boldsymbol{\lambda}) > 0;\\
&{}^{(n:\tilde{n})}V_j(\mathbf{X}_{m+\varepsilon}^{(\beta)}, t_{m+\varepsilon}, \boldsymbol{\lambda}) - {}^{(n:\tilde{n})}V_j(\mathbf{X}_{m+}^{(\beta)}, t_{m+}, \boldsymbol{\lambda}) > 0.
\end{aligned}\right\} \tag{5.154}$$

Definition 5.50. For two dynamical systems in Eqs.(5.1) and (5.2) with constraints in Eq.(5.4), there are $(l+1)$-metric functionals in Eq.(5.37). For $\mathbf{X}_m^{(\alpha_j,j)} \in \Omega_{(\alpha_j,j)}$ ($\alpha_j \in \mathbb{I}$ and $j \in \mathbb{L}$ with $\mathbb{I}=\{1,2\}$ and $\mathbb{L}=\{1,2,\ldots,l\}$) and $\mathbf{X}_m^{(0,j)} \in \partial\Omega_{12(j)}$ at time $t_m, \mathbf{X}_m^{(\alpha_j,j)} = \mathbf{X}_m^{(0,j)}$. For any small $\varepsilon > 0$, there is a time interval $[t_{m-\varepsilon}, t_m)$ or $(t_m, t_{m+\varepsilon}]$. At $\mathbf{X}^{(\alpha_j,j)} \in \Omega_{(\alpha_j,j)}^{\pm\varepsilon}$ for $t \in [t_{m-\varepsilon}, t_m)$ or $(t_m, t_{m+\varepsilon}]$, the functional ${}^{(n:\tilde{n})}V_j(\mathbf{X}^{(\alpha_j,j)}, t, \boldsymbol{\lambda}_j)$ is $C^{r_{\alpha_j}}$-continuous and $|{}^{(n:\tilde{n})}V_j^{(r_{\alpha_j}+1)}(\mathbf{X}^{(\alpha_j,j)}, t, \boldsymbol{\lambda}_j)| < \infty$ ($r_{\alpha_j} \geq 2k_{\alpha_j}+1$). The $(2k_{\alpha_j} : 2k_{\beta_j})$-penetration of the two dynamical systems in Eqs.(5.1) and (5.2) with the jth-constraint in Eq.(5.4) is called to be *switched* from the $(2k_{\beta_j} : 2k_{\alpha_j})$-penetration on the jth-constraint boundary at time t_m in the sense of metric functionals if for $\alpha_j, \beta_j \in \mathbb{I}$ and $\alpha_j \neq \beta_j$ with $j \in \mathbb{L}$,

$$\begin{aligned}
&{}^{(n:\tilde{n})}V_j(\mathbf{X}_{m\mp}^{(\alpha_j,j)}, t_{m\mp}, \boldsymbol{\lambda}_j) = {}^{(n:\tilde{n})}V_j(\mathbf{X}_{m\pm}^{(\beta_j,j)}, t_{m\pm}, \boldsymbol{\lambda}_j)\\
&\qquad = {}^{(n:\tilde{n})}V_j(\mathbf{X}_m^{(0,j)}, t_m, \boldsymbol{\lambda}_j) = 0;\\
&{}^{(n:\tilde{n})}V_j^{(s_{\alpha_j})}(\mathbf{X}_{m\mp}^{(\alpha_j,j)}, t_{m\mp}, \boldsymbol{\lambda}_j) = 0 \text{ for } s_{\alpha_j} = 1,2,\ldots,2k_{\alpha_j}+1;\\
&{}^{(n:\tilde{n})}V_j^{(s_{\beta_j})}(\mathbf{X}_{m\pm}^{(\beta_j,j)}, t_{m\pm}, \boldsymbol{\lambda}_j) = 0 \text{ for } s_{\beta_j} = 1,2,\ldots,2k_{\beta_j}+1;\\
&{}^{(n:\tilde{n})}V_j(\mathbf{X}_{m\mp\varepsilon}^{(\alpha_j,j)}, t_{m\mp\varepsilon}, \boldsymbol{\lambda}_j) - {}^{(n:\tilde{n})}V_j(\mathbf{X}_{m\mp}^{(\alpha_j,j)}, t_{m\mp}, \boldsymbol{\lambda}_j) > 0;\\
&{}^{(n:\tilde{n})}V_j(\mathbf{X}_{m\pm\varepsilon}^{(\beta_j,j)}, t_{m\pm\varepsilon}, \boldsymbol{\lambda}_j) - {}^{(n:\tilde{n})}V_j(\mathbf{X}_{m\pm}^{(\beta_j,j)}, t_{m\pm}, \boldsymbol{\lambda}_j) > 0.
\end{aligned} \tag{5.155}$$

5.4.5 Synchronization to all constraints

In this section, from the definitions for the synchronicity of two dynamical systems with multiple constraints, the necessary and sufficient conditions for such synchronicity of the two dynamical systems to multi-constraints will be discussed. Because of many constraints for two dynamical systems, the synchronicity for each one of constraints should be discussed.

Theorem 5.11. For two dynamical systems in Eqs.(5.1) and (5.2) with constraints in Eq.(5.4), there are $(l+1)$-metric functionals in Eq.(5.37). For $\mathbf{X}_m^{(\alpha_j,j)} \in \Omega_{(\alpha_j,j)}$ ($\alpha_j \in \mathbb{I}$ and $j \in \mathbb{L}$ with $\mathbb{I}=\{1,2\}$ and $\mathbb{L}=\{1,2,\ldots,l\}$) and $\mathbf{X}_m^{(0,j)} \in \partial\Omega_{12(j)}$ at time $t_m, \mathbf{X}_m^{(\alpha_j,j)} = \mathbf{X}_m^{(0,j)}$. For any small $\varepsilon > 0$, there is a time interval $[t_{m-\varepsilon}, t_m)$ or $(t_m, t_{m+\varepsilon}]$. For $\mathbf{X}^{(\alpha_j,j)} \in \Omega_{(\alpha_j,j)}$ and $\mathbf{X}^{(0,j)} \in \partial\Omega_{12(j)}$, at $\mathbf{X}^{(\alpha_j,j)} = \mathbf{X}^{(0,j)}$, the condition $D^{(s_{\alpha_j})}\mathbb{F}^{(\alpha_j,j)}(\mathbf{X}^{(\alpha_j,j)}, t, \boldsymbol{\pi}^{(\alpha_j,j)}) = D^{(s_{\alpha_j})}\mathbb{F}^{(0,j)}(\mathbf{X}^{(0,j)}, t, \boldsymbol{\lambda}_j)$ ($s_{\alpha_j} = 0,1,2,\ldots$)

holds . The two dynamical systems in Eqs.(5.1) and (5.2) to l-constraints in Eq.(5.4) are synchronized with l-dimensions for time $t \in [t_{m_1}, t_{m_2}]$ in the sense of metric functionals if and only if

(i) for all $j \in \mathbb{L}, \mathbf{X}_m^{(\alpha_j,j)} \in \Omega_{(\alpha_j,j)}$ and $\mathbf{X}_m^{(0,j)} \in \partial\Omega_{12(j)}$ for any time t_m,

$$\begin{aligned} &\mathbf{X}_m^{(\alpha_j,j)} = \mathbf{X}_m^{(0,j)}, {}^{(n:\tilde{n})}V_j^{(s_{\alpha_j})}(\mathbf{X}_m^{(\alpha_j,j)}, t_m, \boldsymbol{\lambda}_j) = 0 \\ &\text{for } \alpha_j = 1,2 \text{ and } s_{\alpha_j} = 0,1,2,\ldots \end{aligned} \tag{5.156}$$

(ii) for all $j \in \mathbb{L}, \mathbf{X}_\kappa^{(\alpha_j,j)} \in \Omega_{(\alpha_j,j)}^{-\varepsilon}$ at time $t_\kappa^- \in [t_{m-\varepsilon}, t_m)$ and $\mathbf{X}_m^{(0,j)} \in \partial\Omega_{12(j)}$ with $t_m \in (t_{m_1}, t_{m_2})$,

$$\begin{aligned} &\mathbf{X}_\kappa^{(\alpha_j,j)} \neq \mathbf{X}_m^{(0,j)}, {}^{(n:\tilde{n})}V_j^{(1)}(\mathbf{X}_\kappa^{(\alpha_j,j)}, t_\kappa^-, \boldsymbol{\lambda}_j) < 0 \text{ and} \\ &\lim_{t_\kappa^- \to t_m} {}^{(n:\tilde{n})}V_j^{(1)}(\mathbf{X}_\kappa^{(\alpha_j,j)}, t_\kappa^-, \boldsymbol{\lambda}_j) = 0 \text{ for } \alpha_j = 1,2; \end{aligned} \tag{5.157}$$

(iii) for the jth-constraint $(j \in \mathbb{L}), \mathbf{X}_\kappa^{(\alpha_j,j)} \in \Omega_{(\alpha_j,j)}^{+\varepsilon}$ at time $t_\kappa^+ \in (t_m, t_{m+\varepsilon}]$ and $\mathbf{X}_m^{(0,j)} \in \partial\Omega_{12(j)}$ with $t_m \notin [t_{m_1}, t_{m_2}]$,

$$\begin{aligned} &\mathbf{X}_\kappa^{(\alpha_j,j)} \neq \mathbf{X}_m^{(0,j)}, {}^{(n:\tilde{n})}V_j^{(1)}(\mathbf{X}_\kappa^{(\alpha_j,j)}, t_\kappa^+, \boldsymbol{\lambda}_j) > 0 \text{ and} \\ &\lim_{t_\kappa^+ \to t_m} {}^{(n:\tilde{n})}V_j^{(1)}(\mathbf{X}_\kappa^{(\alpha_j,j)}, t_\kappa^+, \boldsymbol{\lambda}_j) = 0 \text{ for } \alpha_j = 1,2; \end{aligned} \tag{5.158}$$

(iv) for the jth-constraint $(j \in \mathbb{L}), \mathbf{X}_\kappa^{(\alpha_j,j)} \in \Omega_{(\alpha_j,j)}^{+\varepsilon}$ at time $t_\kappa^- \in [t_{m-\varepsilon}, t_{m-}), t_\kappa^+ \in (t_{m+}, t_{m+\varepsilon}]$ and $\mathbf{X}_m^{(0,j)} \in \partial\Omega_{12(j)}$ with $t_m = t_{m_1}$ and t_{m_2},

$$\begin{aligned} &\mathbf{X}_\kappa^{(\alpha)} \neq \mathbf{X}_m^{(0)}, \lim_{t_\kappa^\pm \to t_{m\pm}} {}^{(n:\tilde{n})}V_j^{(1)}(\mathbf{X}_\kappa^{(\alpha)}, t_\kappa^\pm, \boldsymbol{\lambda}) = 0, \\ &\lim_{t_\kappa^\pm \to t_{m\pm}} {}^{(n:\tilde{n})}V_j^{(2)}(\mathbf{X}_\kappa^{(\alpha)}, t_\kappa^\pm, \boldsymbol{\lambda}) < 0 \text{ for } \alpha_j = 1,2. \end{aligned} \tag{5.159}$$

Proof: The proof is similar to the proof of Theorem 5.1 for each $j \in \mathbb{L}$. For all $j \in \mathbb{L}$, if the conditions in Eqs.(5.156) and (5.157) are satisfied, from Definition 5.34, the two dynamical systems in Eqs.(5.1) and (5.2) are synchronized for time $t \in (t_{m_1}, t_{m_2})$ in the sense of Eq.(5.4), vice versa. If the onset and vanishing conditions in Eqs.(5.158) and (5.159) hold, from Definition 5.37, the synchronization of two dynamical systems will start to form and to vanish, vice versa. This theorem is proved. ■

Theorem 5.12. For two dynamical systems in Eqs.(5.1) and (5.2) with constraints in Eq.(5.4), there are $(l+1)$-metric functionals in Eq.(5.37). For $\mathbf{X}_m^{(\alpha_j,j)} \in \Omega_{(\alpha_j,j)}$

($\alpha_j \in \mathbb{I}$ and $j \in \mathbb{L}$ with $\mathbb{I} = \{1,2\}$ and $\mathbb{L} = \{1,2,\ldots,l\}$) and $\mathbf{X}_m^{(0,j)} \in \partial\Omega_{12(j)}$ at time $t_m, \mathbf{X}_m^{(\alpha_j,j)} = \mathbf{X}_m^{(0,j)}$. For any small $\varepsilon > 0$, there is a time interval $[t_{m-\varepsilon}, t_m)$ or $(t_m, t_{m+\varepsilon}]$. For $\mathbf{X}^{(\alpha_j,j)} \in \Omega_{(\alpha_j,j)}$ and $\mathbf{X}^{(0,j)} \in \partial\Omega_{12(j)}$, at $\mathbf{X}^{(\alpha_j,j)} = \mathbf{X}^{(0,j)}$, the relation $D^{(s_{\alpha_j})}\mathbb{F}^{(\alpha_j,j)}(\mathbf{X}^{(\alpha_j,j)}, t, \pi^{(\alpha_j,j)}) = D^{(s_{\alpha_j})}\mathbb{F}^{(0,j)}(\mathbf{X}^{(0,j)}, t, \boldsymbol{\lambda}_j)$ $(s_{\alpha_j} = 0,1,2,\ldots)$ holds. The two dynamical systems in Eqs.(5.1) and (5.2) to l-constraints in Eq.(5.4) are synchronized with l-dimensions for time $t \in [t_{m_1}, t_{m_2}]$ in the sense of metric functionals if and only if

(i) for all $j \in \mathbb{L}, \mathbf{X}_m^{(\alpha_j,j)} \in \Omega_{(\alpha_j,j)}$ and $\mathbf{X}_m^{(0,j)} \in \partial\Omega_{12(j)}$ with $t_m \in (t_{m_1}, t_{m_2})$,

$$\begin{aligned} &\mathbf{X}_m^{(\alpha_j,j)} = \mathbf{X}_m^{(0,j)},\ {}^{(n:\tilde{n})}V_j^{(s_{\alpha_j})}(\mathbf{X}_m^{(\alpha_j,j)}, t_m, \boldsymbol{\lambda}_j) = 0 \\ &\text{for } \alpha_j = 1,2 \text{ and } s_{\alpha_j} = 0,1,2,\ldots \end{aligned} \tag{5.160}$$

(ii) for all $j \in \mathbb{L}, \mathbf{X}_\kappa^{(\alpha_j,j)} \in \Omega_{(\alpha_j,j)}^{-\varepsilon}$ at time $t_\kappa^- \in [t_{m-\varepsilon}, t_m)$ and $\mathbf{X}_m^{(0,j)} \in \partial\Omega_{12(j)}$ with $t_m \in (t_{m_1}, t_{m_2})$,

$$\mathbf{X}_\kappa^{(\alpha_j,j)} \neq \mathbf{X}_m^{(0)},$$

$$\begin{aligned} &\lim_{t_\kappa^- \to t_{m-}} {}^{(n:\tilde{n})}V_j^{(s_{\alpha_j})}(\mathbf{X}_\kappa^{(\alpha_j,j)}, t_\kappa^-, \boldsymbol{\lambda}_j) = 0 \text{ for } s_{\alpha_j} = 1,2,\ldots,2k_{\alpha_j}; \\ &{}^{(n:\tilde{n})}V_j^{(2k_{\alpha_j}+1)}(\mathbf{X}_\kappa^{(\alpha)}, t_\kappa^-, \boldsymbol{\lambda}) < 0, \end{aligned} \tag{5.161}$$

$$\lim_{t_\kappa^- \to t_m} {}^{(n:\tilde{n})}V_j^{(2k_{\alpha_j}+1)}(\mathbf{X}_\kappa^{(\alpha_j,j)}, t_\kappa^-, \boldsymbol{\lambda}) = 0 \text{ for } \alpha_j = 1,2;$$

(iii) for the jth-constraint $(j \in \mathbb{L}), \mathbf{X}_\kappa^{(\alpha_j,j)} \in \Omega_{(\alpha_j,j)}^{+\varepsilon}$ at time $t_\kappa^+ \in (t_m, t_{m+\varepsilon}]$ and $\mathbf{X}_m^{(0,j)} \in \partial\Omega_{12(j)}$ with $t_m \notin [t_{m_1}, t_{m_2}]$,

$$\mathbf{X}_\kappa^{(\alpha_j,j)} \neq \mathbf{X}_m^{(0)},$$

$$\begin{aligned} &\lim_{t_\kappa^- \to t_{m-}} {}^{(n:\tilde{n})}V_j^{(s_{\alpha_j})}(\mathbf{X}_\kappa^{(\alpha_j,j)}, t_\kappa^+, \boldsymbol{\lambda}_j) = 0 \text{ for } s_{\alpha_j} = 1,2,\ldots,2k_{\alpha_j}; \\ &{}^{(n:\tilde{n})}V_j^{(2k_{\alpha_j}+1)}(\mathbf{X}_\kappa^{(\alpha)}, t_\kappa^+, \boldsymbol{\lambda}) < 0, \end{aligned} \tag{5.162}$$

$$\lim_{t_\kappa^+ \to t_m} {}^{(n:\tilde{n})}V_j^{(2k_{\alpha_j}+1)}(\mathbf{X}_\kappa^{(\alpha_j,j)}, t_\kappa^+, \boldsymbol{\lambda}) = 0 \text{ for } \alpha_j = 1,2;$$

(iv) for the jth-constraint $(j \in \mathbb{L}), \mathbf{X}_\kappa^{(\alpha_j,j)} \in \Omega_{(\alpha_j,j)}^{\pm\varepsilon}$ at time $t_\kappa^- \in [t_{m-\varepsilon}, t_{m-})$ and $t_\kappa^+ \in (t_{m+}, t_{m+\varepsilon}]$, and $\mathbf{X}_m^{(0,j)} \in \partial\Omega_{12(j)}$ with $t_m = t_{m_1}$ and t_{m_2}

$$\mathbf{X}_{\kappa}^{(\alpha)} \neq \mathbf{X}_{m}^{(0)},$$

$$\lim_{t_{\kappa}^{\pm} \to t_{m\pm}} {}^{(n:\tilde{n})}V_{j}^{(s_{\alpha_j})}(\mathbf{X}_{\kappa}^{(\alpha_j,j)}, t_{\kappa}^{\pm}, \boldsymbol{\lambda}_j) = 0 \text{ for } s_{\alpha_j} = 1,2,\ldots,2k_{\alpha_j}+1, \tag{5.163}$$

$$\lim_{t_{\kappa}^{\pm} \to t_{m\pm}} {}^{(n:\tilde{n})}V_{j}^{(2k_{\alpha_j}+2)}(\mathbf{X}_{\kappa}^{(\alpha_j,j)}, t_{\kappa}^{\pm}, \boldsymbol{\lambda}_j) < 0 \text{ for } \alpha = 1,2.$$

Proof: The proof is similar to the proof of Theorem 5.2 for each $j \in \mathbb{L}$. For all $j \in \mathbb{L}$, if the conditions in Eqs.(5.160) and (5.161) are satisfied, from Definition 5.38, the slave and master systems in Eqs.(5.1) and (5.2) are synchronized for time $t \in (t_{m_1}, t_{m_2})$ in the sense of Eq.(5.4), vice versa. If the onset and vanishing conditions in Eqs.(5.162) and (5.163) hold, from Definition 5.41, the synchronization will start to form and to vanish, vice versa. The proof is completed. ■

Theorem 5.13. For two dynamical systems in Eqs.(5.1) and (5.2) with constraints in Eq.(5.4), there are $(l+1)$-metric functionals in Eq.(5.37). For $\mathbf{X}_m^{(\alpha_j,j)} \in \Omega_{(\alpha_j,j)}$ ($\alpha_j \in \mathbb{I}$ and $j \in \mathbb{L}$ with $\mathbb{I} = \{1,2\}$ and $\mathbb{L} = \{1,2,\ldots,l\}$) and $\mathbf{X}_m^{(0,j)} \in \partial\Omega_{12(j)}$ at time t_m, $\mathbf{X}_m^{(\alpha_j,j)} = \mathbf{X}_m^{(0,j)}$. For any small $\varepsilon > 0$, there is a time interval $[t_{m-\varepsilon}, t_m)$ or $(t_m, t_{m+\varepsilon}]$. At $\mathbf{X}^{(\alpha_j,j)} \in \Omega_{(\alpha_j,j)}^{\pm\varepsilon}$ for time $t \in [t_{m-\varepsilon}, t_m)$ or $(t_m, t_{m+\varepsilon}]$, the functional ${}^{(n:\tilde{n})}V_j(\mathbf{X}^{(\alpha_j,j)}, t, \boldsymbol{\lambda}_j)$ is $C^{r_{\alpha_j}}$-continuous and $|{}^{(n:\tilde{n})}V_j^{(r_{\alpha_j}+1)}(\mathbf{X}^{(\alpha_j,j)}, t, \boldsymbol{\lambda}_j)| < \infty (r_{\alpha_j} \geq 3)$. For $\mathbf{X}^{(\alpha_j,j)} \in \Omega_{(\alpha_j,j)}$ and $\mathbf{X}^{(0,j)} \in \partial\Omega_{12(j)}$, at $\mathbf{X}^{(\alpha_j,j)} = \mathbf{X}^{(0,j)}$, $\mathbb{F}^{(\alpha_j,j)}(\mathbf{X}^{(\alpha_j,j)}, t, \boldsymbol{\pi}^{(\alpha_j,j)}) \neq \mathbb{F}^{(0,j)}(\mathbf{X}^{(0,j)}, t, \boldsymbol{\lambda}_j)$. The two dynamical systems in Eqs.(5.1) and (5.2) to l-constraints in Eq.(5.4) are synchronized with l-dimensions for time $t \in [t_{m_1}, t_{m_2}]$ in the sense of metric functionals if and only if

(i) for all $j \in \mathbb{L}$, $\mathbf{X}_{m-}^{(\alpha_j,j)} = \mathbf{X}_m^{(0,j)}$ and $\mathbf{X}_m^{(\alpha_j,j)} \in \Omega_{(\alpha_j,j)}$ $(\alpha_j \in \mathbb{I})$ at $t_m \in [t_{m_1}, t_{m_2}]$,

$${}^{(n:\tilde{n})}V_j(\mathbf{X}_{m-}^{(\alpha_j,j)}, t_{m-}, \boldsymbol{\lambda}_j) = {}^{(n:\tilde{n})}V_j(\mathbf{X}_m^{(0,j)}, t_m, \boldsymbol{\lambda}_j) = 0; \tag{5.164}$$

(ii) for all $j \in \mathbb{L}$, time $t_m \in (t_{m_1}, t_{m_2})$,

$$\mathbf{X}_{m-}^{(\alpha_j,j)} = \mathbf{X}_m^{(0,j)} \text{ and } {}^{(n:\tilde{n})}V_j^{(1)}(\mathbf{X}_{m-}^{(\alpha_j,j)}, t_{m-}, \boldsymbol{\lambda}_j) < 0 \text{ for } \alpha_j = 1,2; \tag{5.165}$$

(iii) with the jth-penetration for time $t = t_{m_i}$, $\mathbf{X}_{m_i}^{(\alpha)} = \mathbf{X}_{m_i}^{(0)}$ $(i = 1,2)$ for $\alpha_j, \beta_j \in \mathbb{I}$ and $\beta_j \neq \alpha_j$,

$$\begin{gathered} {}^{(n:\tilde{n})}V_j^{(1)}(\mathbf{X}_{m_i\pm}^{(\alpha_j,j)}, t_{m_i\pm}, \boldsymbol{\lambda}_j) = 0 \text{ and } {}^{(n:\tilde{n})}V_j^{(2)}(X_{m_i\pm}^{(\alpha)}, t_{m_i\pm}, \boldsymbol{\lambda}_j) > 0, \\ {}^{(n:\tilde{n})}V_j^{(1)}(\mathbf{X}_{m_i-}^{(\beta_j,)}, t_{m_i-}, \boldsymbol{\lambda}_j) < 0; \end{gathered} \tag{5.166}$$

or with the jth-desynchronization for time $t = t_{m_i}$, $\mathbf{X}_{m_i}^{(\alpha)} = \mathbf{X}_{m_i}^{(0)}$ $(i = 1,2)$ for $\alpha_j, \beta_j \in \mathbb{I}$ and $\beta_j \neq \alpha_j$,

$$
\begin{aligned}
&{}^{(n:\tilde{n})}V_j^{(1)}(\mathbf{X}_{m_i\pm}^{(\alpha_j,j)},t_{m_i\pm},\boldsymbol{\lambda}_j)=0 \text{ and } {}^{(n:\tilde{n})}V_j^{(2)}(\mathbf{X}_{m_i\pm}^{(\alpha_j,j)},t_{m_i\pm},\boldsymbol{\lambda}_j)>0,\\
&{}^{(n:\tilde{n})}V_j^{(1)}(\mathbf{X}_{m_i\pm}^{(\beta_j,j)},t_{m_i\pm},\boldsymbol{\lambda}_j)=0 \text{ and } {}^{(n:\tilde{n})}V_j^{(2)}(\mathbf{X}_{m_i\pm}^{(\beta_j,j)},t_{m_i\pm},\boldsymbol{\lambda}_j)>0.
\end{aligned}
\tag{5.167}
$$

Proof: The proof is similar to the proof of Theorem 5.3 for each $j\in\mathbb{L}$. For all $j\in\mathbb{L}$, if the conditions in Eqs.(5.164) and (5.165) are satisfied, from Definition 5.34, the two dynamical systems in Eqs.(5.1) and (5.2) to l-constraints in Eq.(5.4) are synchronized for time $t\in[t_{m_1},t_{m_2}]$, vice versa. If the onset and vanishing conditions in Eqs.(5.166) and (5.167) hold, from Definition 5.38 or 5.39, the synchronization of the two dynamical systems to l-constraints in Eq.(5.4) will start to form and to vanish, vice versa. The proof is completed. ■

In the foregoing theorem, the synchronization of the two systems is without any singularity except for the onset and vanishing conditions on the boundaries of the constraints. If the synchronization of two dyanmical systems with higher-order singularity, the corresponding theorems can be presented as follows.

Theorem 5.14. For two dynamical systems in Eqs.(5.1) and (5.2) with constraints in Eq.(5.4), there are $(l+1)$-metric functionals in Eq.(5.37). For $\mathbf{X}_m^{(\alpha_j,j)}\in\Omega_{(\alpha_j,j)}$ ($\alpha_j\in\mathbb{I}$ and $j\in\mathbb{L}$ with $\mathbb{I}=\{1,2\}$ and $\mathbb{L}=\{1,2,\ldots,l\}$) and $\mathbf{X}_m^{(0,j)}\in\partial\Omega_{12(j)}$ at time $t_m,\mathbf{X}_m^{(\alpha_j,j)}=\mathbf{X}_m^{(0,j)}$. For any small $\varepsilon>0$, there is a time interval $[t_{m-\varepsilon},t_m)$ or $(t_m,t_{m+\varepsilon}]$. At $\mathbf{X}^{(\alpha_j,j)}\in\Omega_{(\alpha_j,j)}^{\pm\varepsilon}$ for time $t\in[t_{m-\varepsilon},t_m)$ or $(t_m,t_{m+\varepsilon}]$, the functional ${}^{(n:\tilde{n})}V_j(\mathbf{X}^{(\alpha_j,j)},t,\boldsymbol{\lambda}_j)$ is $C^{r_{\alpha_j}}$-continuous and $|{}^{(n:\tilde{n})}V_j^{(r_{\alpha_j}+1)}(\mathbf{X}^{(\alpha_j,j)},t,\boldsymbol{\lambda}_j)|<\infty$ ($r_{\alpha_j}\geq 2k_{\alpha_j}+1$). For $\mathbf{X}^{(\alpha_j,j)}\in\Omega_{(\alpha_j,j)}$ and $\mathbf{X}^{(0,j)}\in\partial\Omega_{12(j)}$, at $\mathbf{X}^{(\alpha_j,j)}=\mathbf{X}^{(0,j)}$, $\mathbb{F}^{(\alpha_j,j)}(\mathbf{X}^{(\alpha_j,j)},t,\pi^{(\alpha_j,j)})\neq\mathbb{F}^{(0,j)}(\mathbf{X}^{(0,j)},t,\boldsymbol{\lambda}_j)$. The two dynamical systems in Eqs.(5.1) and (5.2) to l-constraints in Eq.(5.4) are synchronized of the $(2k_{\alpha_j}:2k_{\beta_j})$-type with l-dimensions for time $t\in[t_{m_1},t_{m_2}]$ in the sense of metric functionals if and only if

(i) for all $j\in\mathbb{L},\mathbf{X}_{m-}^{(\alpha_j,j)}=\mathbf{X}_m^{(0,j)}$ and $\mathbf{X}^{(\alpha_j,j)}(t)\in\Omega_{(\alpha_j,j)}$ ($\alpha_j\in\mathbb{I}$) at time $t=t_m\in[t_{m_1},t_{m_2}]$,

$$
{}^{(n:\tilde{n})}V_j(\mathbf{X}_{m-}^{(\alpha_j,j)},t_{m-},\boldsymbol{\lambda}_j)={}^{(n:\tilde{n})}V_j(\mathbf{X}_m^{(0,j)},t_m,\boldsymbol{\lambda}_j)=0;
\tag{5.168}
$$

(ii) for all $j\in\mathbb{L}$ and time $t_m\in(t_{m_1},t_{m_2})$,

$$
\begin{aligned}
&\mathbf{X}_{m-}^{(\alpha_j,j)}=\mathbf{X}_m^{(0,j)},\\
&{}^{(n:\tilde{n})}V_j^{(s_{\alpha_j})}(\mathbf{X}_{m-}^{(\alpha_j,j)},t_{m-},\boldsymbol{\lambda}_j)=0 \text{ for } s_{\alpha_j}=1,2,\ldots,2k_{\alpha_j},\\
&{}^{(n:\tilde{n})}V_j^{(2k_{\alpha_j}+1)}(\mathbf{X}_{m-}^{(\alpha_j,j)},t_{m-},\boldsymbol{\lambda}_j)<0 \text{ for } \alpha_j=1,2;
\end{aligned}
\tag{5.169}
$$

(iii) with the jth-penetration of the $(2k_{\alpha_j}:2k_{\beta_j})$-type for time $t=t_{m_i},\mathbf{X}_{m_i}^{(\alpha_j,j)}=\mathbf{X}_{m_i}^{(0,j)}$ $(i=1,2)$,

$$
\begin{aligned}
&{}^{(n:\tilde{n})}V_j^{(s_{\alpha_j})}(\mathbf{X}_{m_i\pm}^{(\alpha_j,j)},t_{m_i\pm},\boldsymbol{\lambda}_j)=0\ (s_{\alpha_j}=1,2,\ldots,2k_{\alpha_j}+1),\\
&\text{and } V^{(2k_{\alpha_j}+2)}(\mathbf{X}_{m_i\pm}^{(\alpha_j,j)},t_{m_i\pm},\boldsymbol{\lambda}_j)>0 \text{ for } \alpha_j\in\{1,2\};\\
&{}^{(n:\tilde{n})}V_j^{(s_{\beta_j})}(\mathbf{X}_{m_i-}^{(\beta_j,j)},t_{m_i-},\boldsymbol{\lambda}_j)=0(s_{\beta_j}=1,2,\ldots,2k_{\beta_j}),\\
&{}^{(n:\tilde{n})}V_j^{(2k_{\beta}+1)}(\mathbf{X}_{m_i-}^{(\beta)},t_{m_i-},\boldsymbol{\lambda}_j)<0 \text{ for } \beta_j\in\{1,2\} \text{ and } \beta_j\neq\alpha_j;
\end{aligned}
\tag{5.170}
$$

or with the jth-desynchronization of the $(2k_{\alpha_j}:2k_{\beta_j})$-type for time $t=t_{m_i},\mathbf{X}_{m_i}^{(\alpha_j,j)}=\mathbf{X}_{m_i}^{(0)}$ $(i=1,2)$,

$$
\begin{aligned}
&{}^{(n:\tilde{n})}V_j^{(s_{\alpha_j})}\left(\mathbf{X}_{m_i\pm}^{(\alpha_j,j)},t_{m_i\pm},\boldsymbol{\lambda}_j\right)=0\ (s_{\alpha_j}=1,2,\ldots,2k_{\alpha_j}+1),\\
&\text{and } V^{(2k_{\alpha_j}+2)}\left(\mathbf{X}_{m_i\pm}^{(\alpha_j,j)},t_{m_i\pm},\boldsymbol{\lambda}_j\right)>0 \text{ or } \alpha_j\in\{1,2\};\\
&{}^{(n:\tilde{n})}V_j^{(s_{\beta_j})}\left(\mathbf{X}_{m_i\pm}^{(\beta_j,j)},t_{m_i\pm},\boldsymbol{\lambda}_j\right)=0(s_{\beta_j}=1,2,\ldots,2k_{\beta_j}+1),\\
&{}^{(n:\tilde{n})}V_j^{(2k_{\beta_j}+2)}\left(\mathbf{X}_{m_i\pm}^{(\beta_j,j)},t_{m_i\pm},\boldsymbol{\lambda}_j\right)>0 \text{ for } \beta_j\in\{1,2\} \text{ and } \beta_j\neq\alpha_j.
\end{aligned}
\tag{5.171}
$$

Proof: The proof is similar to the proof of Theorem 5.4 for each $j\in\mathbb{L}$. For all $j\in\mathbb{L}$, if the conditions in Eqs.(5.168) and (5.169) are satisfied, from Definition 5.38, the slave and master systems in Eqs.(5.1) and (5.2) are synchronized with the $(2k_{\alpha_j}:2k_{\beta_j})$-type for time $t\in[t_{m_1},t_{m_2}]$ in the sense of Eq.(5.4), vice versa. If the onset and vanishing conditions in Eqs.(5.170) and (5.171) hold, from Definition 5.42 or 5.43, the synchronization will start to form and to vanish, vice versa. The proof is completed. ■

5.4.6 Desynchronization to all constraints

In this section, from the definitions for the desynchronicity of two dynamical systems to multiple constraints, the necessary and sufficient conditions for such desynchronicity will be discussed. Because of many constraints for two dynamical systems, the synchronicity for each single one of constraints should be discussed.

Theorem 5.15. For two dynamical systems in Eqs.(5.1) and (5.2) with constraints in Eq.(5.4), there are $(l+1)$-metric functionals in Eq.(5.37). For $\mathbf{X}_m^{(\alpha_j,j)}\in\Omega_{(\alpha_j,j)}$ ($\alpha_j\in\mathbb{I}$ and $j\in\mathbb{L}$ with $\mathbb{I}=\{1,2\}$ and $\mathbb{L}=\{1,2,\ldots,l\}$) and $\mathbf{X}_m^{(0,j)}\in\partial\Omega_{12(j)}$ at time $t_m,\mathbf{X}_m^{(\alpha_j,j)}=\mathbf{X}_m^{(0,j)}$. For any small $\varepsilon>0$, there is a time interval $[t_{m-\varepsilon},t_m)$ or $(t_m,t_{m+\varepsilon}]$. For $\mathbf{X}^{(\alpha_j,j)}\in\Omega_{(\alpha_j,j)}$ and $\mathbf{X}^{(0,j)}\in\partial\Omega_{12(j)}$, at $\mathbf{X}^{(\alpha_j,j)}=\mathbf{X}^{(0,j)}$, $D^{(s_{\alpha_j})}\mathbb{F}^{(\alpha_j,j)}$

$(\mathbf{X}^{(\alpha_j,j)},t,\pi^{(\alpha_j,j)}) = D^{(s_{\alpha_j})}\mathbb{F}^{(0,j)}(\mathbf{X}^{(0,j)},t,\boldsymbol{\lambda}_j)$ $(s_{\alpha_j}=0,1,2,\cdots)$. The two dynamical systems in Eqs.(5.1) and (5.2) to l-constraints in Eq.(5.4) are desynchronized with l-dimensions for time $t\in[t_{m_1},t_{m_2}]$ in the sense of metric functionals if and only if

(i) for all $j\in\mathbb{L}, \mathbf{X}_m^{(\alpha_j,j)}\in\Omega_{(\alpha_j,j)}$ and $\mathbf{X}_m^{(0,j)}\in\partial\Omega_{12(j)}$ for any time t_m,

$$\begin{aligned}&\mathbf{X}_m^{(\alpha_j,j)} = \mathbf{X}_m^{(0,j)},\ {}^{(n:\tilde{n})}V_j^{(s_{\alpha_j})}(\mathbf{X}_m^{(\alpha_j,j)},t_m,\boldsymbol{\lambda}_j)=0,\\ &\text{for } \alpha_j=1,2 \text{ and } s_{\alpha_j}=0,1,2,\ldots\end{aligned}\tag{5.172}$$

(ii) for all $j\in\mathbb{L}, \mathbf{X}_\kappa^{(\alpha_j,j)}\in\Omega_{(\alpha_j,j)}^{+\varepsilon}$ at time $t_\kappa^+\in(t_m,t_{m+\varepsilon}]$ and $\mathbf{X}_m^{(0,j)}\in\partial\Omega_{12(j)}$ with $t_m\in(t_{m_1},t_{m_2})$,

$$\begin{aligned}&\mathbf{X}_\kappa^{(\alpha_j,j)}\neq\mathbf{X}_m^{(0,j)},\ {}^{(n:\tilde{n})}V_j^{(1)}(\mathbf{X}_\kappa^{(\alpha_j,j)},t_\kappa^+,\boldsymbol{\lambda}_j)>0 \text{ and}\\ &\lim_{t_\kappa^+\to t_m}{}^{(n:\tilde{n})}V_j^{(1)}(\mathbf{X}_\kappa^{(\alpha_j,j)},t_\kappa^+,\boldsymbol{\lambda}_j)=0 \text{ for } \alpha_j=1,2;\end{aligned}\tag{5.173}$$

(iii) for the jth-constraint $(j\in\mathbb{L}), \mathbf{X}_\kappa^{(\alpha_j,j)}\in\Omega_{(\alpha_j,j)}^{-\varepsilon}$ at time $t_\kappa^-\in[t_{m-\varepsilon},t_m)$ and $\mathbf{X}_m^{(0,j)}\in\partial\Omega_{12(j)}$ with $t_m\notin[t_{m_1},t_{m_2}]$,

$$\begin{aligned}&\mathbf{X}_\kappa^{(\alpha_j,j)}\neq\mathbf{X}_m^{(0,j)},\ {}^{(n:\tilde{n})}V_j^{(1)}(\mathbf{X}_\kappa^{(\alpha_j,j)},t_\kappa^-,\boldsymbol{\lambda}_j)<0 \text{ and}\\ &\lim_{t_\kappa^-\to t_m}{}^{(n:\tilde{n})}V_j^{(1)}(\mathbf{X}_\kappa^{(\alpha_j,j)},t_\kappa^-,\boldsymbol{\lambda}_j)=0 \text{ for } \alpha_j=1,2;\end{aligned}\tag{5.174}$$

(iv) for the jth-constraint $(j\in\mathbb{L}), \mathbf{X}_\kappa^{(\alpha_j,j)}\in\Omega_{(\alpha_j,j)}^{+\varepsilon}$ at time $t_\kappa^-\in[t_{m-\varepsilon},t_{m-}), t_\kappa^+\in(t_{m+},t_{m+\varepsilon}]$ and $\mathbf{X}_m^{(0,j)}\in\partial\Omega_{12(j)}$ with $t_m=t_{m_1}$ and t_{m_2},

$$\begin{aligned}&\mathbf{X}_\kappa^{(\alpha)}\neq\mathbf{X}_m^{(0)},\ \lim_{t_\kappa^\pm\to t_{m\pm}}{}^{(n:\tilde{n})}V_j^{(1)}\left(\mathbf{X}_\kappa^{(\alpha)},t_\kappa^\pm,\boldsymbol{\lambda}_j\right)=0,\\ &\lim_{t_\kappa^\pm\to t_{m\pm}}{}^{(n:\tilde{n})}V_j^{(2)}\left(\mathbf{X}_\kappa^{(\alpha)},t_\kappa^\pm,\boldsymbol{\lambda}\right)<0 \text{ for } \alpha_j=1,2.\end{aligned}\tag{5.175}$$

Proof: The proof is similar to the proof of Theorem 5.1 for each $j\in\mathbb{L}$. For all $j\in\mathbb{L}$, if the conditions in Eqs.(5.172) and (5.173) are satisfied, from Definition 5.35, the two dynamical systems in Eqs.(5.1) and (5.2) to constraints in Eq.(5.4) are desynchronized for time $t\in(t_{m_1},t_{m_2})$, vice versa. If the onset and vanishing conditions in Eqs.(5.174) and (5.175) hold, from Definition 5.37, the desynchronization will start to form and to vanish, vice versa. This theorem is proved. ■

Theorem 5.16. For two dynamical systems in Eqs.(5.1) and (5.2) with constraints in Eq.(5.4), there are $(l+1)$-metric functionals in Eq.(5.37). For $\mathbf{X}_m^{(\alpha_j,j)}\in\Omega_{(\alpha_j,j)}$

($\alpha_j \in \mathbb{I}$ and $j \in \mathbb{L}$ with $\mathbb{I} = \{1,2\}$ and $\mathbb{L} = \{1,2,\ldots,l\}$) and $\mathbf{X}_m^{(0,j)} \in \partial\Omega_{12(j)}$ at time $t_m, \mathbf{X}_m^{(\alpha_j,j)} = \mathbf{X}_m^{(0,j)}$. For any small $\varepsilon > 0$, there is a time interval $[t_{m-\varepsilon}, t_m)$ or $(t_m, t_{m+\varepsilon}]$. For $\mathbf{X}^{(\alpha_j,j)} \in \Omega_{(\alpha_j,j)}$ and $\mathbf{X}^{(0,j)} \in \partial\Omega_{12(j)}$, at $\mathbf{X}^{(\alpha_j,j)} = \mathbf{X}^{(0,j)}, D^{s_{\alpha_j}}\mathbb{F}^{(\alpha_j,j)}(\mathbf{X}^{(\alpha_j,j)}, t, \pi^{(\alpha_j,j)}) = D^{s_{\alpha_j}}\mathbb{F}^{(0,j)}(\mathbf{X}^{(0,j)}, t, \boldsymbol{\lambda}_j)$ $(s_{\alpha_j} = 0,1,2,\ldots)$. The two dynamical systems in Eqs.(5.1) and (5.2) to l-constraints in Eq.(5.4) are desynchronized with l-dimensions for time $t \in [t_{m_1}, \infty)$ in the sense of metric functionals if and only if

(i) for all $j \in \mathbb{L}, \mathbf{X}_m^{(\alpha_j,j)} \in \Omega_{(\alpha_j,j)}$ and $\mathbf{X}_m^{(0,j)} \in \partial\Omega_{12(j)}$ with$t_m \in (t_{m_1}, t_{m_2})$,

$$\begin{aligned} &\mathbf{X}_m^{(\alpha_j,j)} = \mathbf{X}_m^{(0,j)}, \ {}^{(n:\tilde{n})}V_j^{(s_{\alpha_j})}(\mathbf{X}_m^{(\alpha_j,j)}, t_m, \boldsymbol{\lambda}_j) = 0, \\ &\text{for } \alpha_j = 1,2 \text{ and } s_{\alpha_j} = 0,1,2,\ldots \end{aligned} \tag{5.176}$$

(ii) for all $j \in \mathbb{L}, \mathbf{X}_\kappa^{(\alpha_j,j)} \in \Omega_{(\alpha_j,j)}^{+\varepsilon}$ at time $t_\kappa^+ \in (t_m, t_{m+\varepsilon}]$ and $\mathbf{X}_m^{(0,j)} \in \partial\Omega_{12(j)}$ with $t_m \in (t_{m_1}, t_{m_2})$,

$$\mathbf{X}_\kappa^{(\alpha_j,j)} \neq \mathbf{X}_m^{(0)},$$

$$\begin{aligned} &\lim_{t_\kappa^- \to t_{m-}} {}^{(n:\tilde{n})}V_j^{(s_{\alpha_j})}(\mathbf{X}_\kappa^{(\alpha_j,j)}, t_\kappa^+, \boldsymbol{\lambda}_j) = 0 \text{ for } s_{\alpha_j} = 1,2,\ldots,2k_{\alpha_j}; \\ &{}^{(n:\tilde{n})}V_j^{(2k_{\alpha_j}+1)}(\mathbf{X}_\kappa^{(\alpha_j,j)}, t_\kappa^+, \boldsymbol{\lambda}_j) < 0 \text{ and} \end{aligned} \tag{5.177}$$

$$\lim_{t_\kappa^+ \to t_m} {}^{(n:\tilde{n})}V_j^{(2k_{\alpha_j}+1)}(\mathbf{X}_\kappa^{(\alpha_j,j)}, t_\kappa^+, \boldsymbol{\lambda}_j) = 0 \text{ for } \alpha_j = 1,2;$$

(iii) for the jth-constraint $(j \in \mathbb{L}), \mathbf{X}_\kappa^{(\alpha_j,j)} \in \Omega_{(\alpha_j,j)}^{-\varepsilon}$ at time $t_\kappa^- \in [t_{m-\varepsilon}, t_m)$ and $\mathbf{X}_m^{(0,j)} \in \partial\Omega_{12(j)}$ with $t_m \notin [t_{m_1}, t_{m_2}]$,

$$\mathbf{X}_\kappa^{(\alpha_j,j)} \neq \mathbf{X}_m^{(0)},$$

$$\begin{aligned} &\lim_{t_\kappa^- \to t_{m-}} {}^{(n:\tilde{n})}V_j^{(s_{\alpha_j})}(\mathbf{X}_\kappa^{(\alpha_j,j)}, t_\kappa^-, \boldsymbol{\lambda}_j) = 0 \text{ for } s_{\alpha_j} = 1,2,\ldots,2k_{\alpha_j}; \\ &{}^{(n:\tilde{n})}V_j^{(2k_{\alpha_j}+1)}(\mathbf{X}_\kappa^{(\alpha)}, t_\kappa^-, \boldsymbol{\lambda}_j) < 0 \text{ and} \end{aligned} \tag{5.178}$$

$$\lim_{t_\kappa^- \to t_m} {}^{(n:\tilde{n})}V_j^{(2k_{\alpha_j}+1)}(\mathbf{X}_\kappa^{(\alpha_j,j)}, t_\kappa^-, \boldsymbol{\lambda}_j) = 0 \text{ for } \alpha_j = 1,2;$$

(iv) for the jth-constraint $(j \in \mathbb{L}), \mathbf{X}_\kappa^{(\alpha_j,j)} \in \Omega_{(\alpha_j,j)}^{\pm\varepsilon}$ at time $t_\kappa^- \in [t_{m-\varepsilon}, t_{m-})$ and $t_\kappa^+ \in (t_{m+}, t_{m+\varepsilon}]$ and $\mathbf{X}_m^{(0,j)} \in \partial\Omega_{12(j)}$ with $t_m = t_{m_1}$ and t_{m_2},

$$\mathbf{X}_{\kappa}^{(\alpha)} \neq \mathbf{X}_{m}^{(0)},$$

$$\lim_{t_{\kappa}^{\pm} \to t_{m\pm}} {}^{(n:\tilde{n})}V_{j}^{(s_{\alpha_j})}(\mathbf{X}_{\kappa}^{(\alpha_j,j)}, t_{\kappa}^{\pm}, \boldsymbol{\lambda}_j) = 0 \text{ for } s_{\alpha_j} = 1,2,\ldots,2k_{\alpha_j}+1; \tag{5.179}$$

$$\lim_{t_{\kappa}^{\pm} \to t_{m\pm}} {}^{(n:\tilde{n})}V_{j}^{(2k_{\alpha_j}+2)}(\mathbf{X}_{\kappa}^{(\alpha_j,j)}, t_{\kappa}^{\pm}, \boldsymbol{\lambda}_j) < 0 \text{ for } \alpha = 1,2.$$

Proof: The proof is similar to the proof of Theorem 5.2 for each $j \in \mathbb{L}$. For all $j \in \mathbb{L}$, if the conditions in Eqs.(5.176) and (5.177) are satisfied, from Definition 5.44, the two dynamical systems in Eqs.(5.1) and (5.2) to constraints in Eq.(5.4) are synchronized for time $t \in (t_{m_1}, t_{m_2})$, vice versa. If the onset and vanishing conditions in Eqs.(5.178) and (5.179) hold, from Definition 5.47, the synchronization will start to form and to vanish, vice versa. The proof is completed. ■

From the foregoing theorem, the desynchronization requires all the higher-order derivatives of the metric functions in Eq.(5.37) should be zero on the constraint surfaces and the the highest-order derivative of the metric functions in domain should be greater than zero. In practical applications, such a condition is too strong for one to control the desynchronization of the slave and master systems. Therefore, such a condition can be released through a discontinuous vector fileds to the slave and master systems. Therefore, the following theorem for the desynchronization will be stated.

Theorem 5.17. For two dynamical systems in Eqs.(5.1) and (5.2) with constraints in Eq.(5.4), there are $(l+1)$-metric functionals in Eq.(5.37). For $\mathbf{X}_{m}^{(\alpha_j,j)} \in \Omega_{(\alpha_j,j)}$ ($\alpha_j \in \mathbb{I}$ and $j \in \mathbb{L}$ with $\mathbb{I} = \{1,2\}$ and $\mathbb{L} = \{1,2,\ldots,l\}$) and $\mathbf{X}_{m}^{(0,j)} \in \partial\Omega_{12(j)}$ at time $t_m, \mathbf{X}_{m}^{(\alpha_j,j)} = \mathbf{X}_{m}^{(0,j)}$. For any small$\varepsilon > 0$, there is a time interval $[t_{m-\varepsilon}, t_m)$ or $(t_m, t_{m+\varepsilon}]$. At $\mathbf{X}^{(\alpha_j,j)} \in \Omega_{(\alpha_j,j)}^{\pm\varepsilon}$ for time $t \in [t_{m-\varepsilon}, t_m)$ or $(t_m, t_{m+\varepsilon}]$, the functional ${}^{(n:\tilde{n})}V_j(\mathbf{X}^{(\alpha_j,j)}, t, \boldsymbol{\lambda}_j)$ is $C^{r_{\alpha_j}}$-continuous and $|{}^{(n:\tilde{n})}V_j^{(r_{\alpha_j}+1)}(\mathbf{X}^{(\alpha_j,j)}, t, \boldsymbol{\lambda}_j)| < \infty (r_{\alpha_j} \geq 3)$. For $\mathbf{X}^{(\alpha_j,j)} \in \Omega_{(\alpha_j,j)}$ and $\mathbf{X}^{(0,j)} \in \partial\Omega_{12(j)}$, at $\mathbf{X}^{(\alpha_j,j)} = \mathbf{X}^{(0,j)}, \mathbb{F}^{(\alpha_j,j)}(\mathbf{X}^{(\alpha_j,j)}, t, \boldsymbol{\pi}^{(\alpha_j,j)}) \neq \mathbb{F}^{(0,j)}(\mathbf{X}^{(0,j)}, t, \boldsymbol{\lambda}_j)$. The two dynamical systems in Eqs.(5.1) and (5.2) to l-constraints in Eq.(5.4) are desynchronized with l-dimensions for time $t \in [t_{m_1}, t_{m_2}]$ in the sense of metric functionals if and only if

(i) for all $j \in \mathbb{L}, \mathbf{X}_{m-}^{(\alpha_j,j)} = \mathbf{X}_{m}^{(0,j)}$and $\mathbf{X}^{(\alpha_j,j)}(t) \in \Omega_{(\alpha_j,j)}$ ($\alpha_j \in \mathbb{I}$) at$t_m \in [t_{m_1}, t_{m_2}]$,

$${}^{(n:\tilde{n})}V_j(\mathbf{X}_{m+}^{(\alpha_j,j)}, t_{m+}, \boldsymbol{\lambda}_j) = {}^{(n:\tilde{n})}V_j(\mathbf{X}_{m}^{(0,j)}, t_m, \boldsymbol{\lambda}_j) = 0; \tag{5.180}$$

(ii) for all $j \in \mathbb{L}$ and time $t_m \in (t_{m_1}, t_{m_2})$,

$$\mathbf{X}_{m+}^{(\alpha_j,j)} = \mathbf{X}_{m}^{(0,j)} \text{ and } {}^{(n:\tilde{n})}V_j^{(1)}(\mathbf{X}_{m+}^{(\alpha_j,j)}, t_{m+}, \boldsymbol{\lambda}_j) > 0 \text{ for } \alpha_j = 1,2; \tag{5.181}$$

(iii) with the jth-penetration for time $t = t_{m_i}, \mathbf{X}_{m_i}^{(\alpha)} = \mathbf{X}_{m_i}^{(0)}$ $(i = 1,2)$ for $\alpha_j \in \mathbb{I}$,

$$
\begin{aligned}
&{}^{(n:\tilde{n})}V_j^{(1)}(\mathbf{X}_{m_i\pm}^{(\alpha_j,j)},t_{m_i\pm},\boldsymbol{\lambda}_j)=0 \text{ and } {}^{(n:\tilde{n})}V_j^{(2)}(\mathbf{X}_{m_i\pm}^{(\alpha_j,j)},t_{m_i\pm},\boldsymbol{\lambda}_j)>0,\\
&{}^{(n:\tilde{n})}V_j^{(1)}(\mathbf{X}_{m_i+}^{(\beta)},t_{m_i+},\boldsymbol{\lambda}_j)>0 \text{ for } \beta_j\in\mathbb{I} \text{ and } \beta_j\neq\alpha_j;
\end{aligned}
\tag{5.182}
$$

or with the jth-synchronization for time $t=t_{m_i}, \mathbf{X}_{m_i}^{(\alpha)}=\mathbf{X}_{m_i}^{(0)}$ $(i=1,2)$ for $\alpha_j,\beta_j\in\mathbb{I}$ and $\beta_j\neq\alpha_j$,

$$
\begin{aligned}
&{}^{(n:\tilde{n})}V_j^{(1)}(\mathbf{X}_{m_i\pm}^{(\alpha_j,j)},t_{m_i\pm},\boldsymbol{\lambda}_j)=0 \text{ and } {}^{(n:\tilde{n})}V_j^{(2)}(\mathbf{X}_{m_i\pm}^{(\alpha_j,j)},t_{m_i\pm},\boldsymbol{\lambda}_j)>0,\\
&{}^{(n:\tilde{n})}V_j^{(1)}(\mathbf{X}_{m_i\pm}^{(\beta_j,j)},t_{m_i\pm},\boldsymbol{\lambda}_j)=0 \text{ and } {}^{(n:\tilde{n})}V_j^{(2)}(\mathbf{X}_{m_i\pm}^{(\beta_j,j)},t_{m_i\pm},\boldsymbol{\lambda}_j)>0.
\end{aligned}
\tag{5.183}
$$

Proof: The proof is similar to the proof of Theorem 5.3 for each $j\in\mathbb{L}$. For all $j\in\mathbb{L}$, if the conditions in Eqs.(5.180) and (5.181) are satisfied, from Definition 5.35, the two dynamical systems in Eqs.(5.1) and (5.2) to constraints in Eq.(5.4) are desynchronized for time $t\in[t_{m_1},t_{m_2}]$, vice versa. If the onset and vanishing conditions in Eqs.(5.182) and (5.183) hold, from Definition 5.39 or 5.40, the desynchronization will start to form and to vanish, vice versa. The proof is completed. ■

In the foregoing theorem, the desynchronization of two dynamical systems to multiple constraints is without any singularity except for the onset and vanishing condition. If the desynchronization of two dynamical systems to multiple constraints possesses higher-order singularity, the following theorem is presented.

Theorem 5.18. For two dynamical systems in Eqs.(5.1) and (5.2) with constraints in Eq.(5.4), there are $(l+1)$-metric functionals in Eq.(5.37). For $\mathbf{X}_m^{(\alpha_j,j)}\in\Omega_{(\alpha_j,j)}$ ($\alpha_j\in\mathbb{I}$ and $j\in\mathbb{L}$ with $\mathbb{I}=\{1,2\}$ and $\mathbb{L}=\{1,2,\ldots,l\}$) and $\mathbf{X}_m^{(0,j)}\in\partial\Omega_{12(j)}$ at time $t_m, \mathbf{X}_m^{(\alpha_j,j)}=\mathbf{X}_m^{(0,j)}$. For any small $\varepsilon>0$, there is a time interval $[t_{m-\varepsilon},t_m)$ or $(t_m,t_{m+\varepsilon}]$. At $\mathbf{X}^{(\alpha_j,j)}\in\Omega_{(\alpha_j,j)}^{\pm\varepsilon}$ for time $t\in[t_{m-\varepsilon},t_m)$ or $(t_m,t_{m+\varepsilon}]$, the functional ${}^{(n:\tilde{n})}V_j(\mathbf{X}^{(\alpha_j,j)},t,\boldsymbol{\lambda}_j)$ is $C^{r_{\alpha_j}}$-continuous and $|{}^{(n:\tilde{n})}V_j^{(r_{\alpha_j}+1)}(\mathbf{X}^{(\alpha_j,j)},t,\boldsymbol{\lambda}_j)|<\infty$ $(r_{\alpha_j}\geq 2k_{\alpha_j}+1)$. For $\mathbf{X}^{(\alpha_j,j)}\in\Omega_{(\alpha_j,j)}$ and $\mathbf{X}^{(0,j)}\in\partial\Omega_{12(j)}$, at $\mathbf{X}^{(\alpha_j,j)}=\mathbf{X}^{(0,j)}$, $\mathbb{F}^{(\alpha_j,j)}(\mathbf{X}^{(\alpha_j,j)},t,\pi^{(\alpha_j,j)})\neq\mathbb{F}^{(0,j)}(\mathbf{X}^{(0,j)},t,\boldsymbol{\lambda}_j)$. The two dynamical systems in Eqs.(5.1) and (5.2) to l-constraints in Eq.(5.4) are desynchronized of the $(2k_{\alpha_j}:2k_{\beta_j})$-type with l-dimensions for time $t\in[t_{m_1},t_{m_2}]$ in the sense of metric functionals if and only if

(i) for all $j\in\mathbb{L}, \mathbf{X}_{m+}^{(\alpha_j,j)}=\mathbf{X}_m^{(0,j)}$ and $\mathbf{X}_m^{(\alpha_j,j)}\in\Omega_{(\alpha_j,j)}$ $(\alpha_j\in\mathbb{I})$ at $t_m\in[t_{m_1},t_{m_2}]$,

$$
{}^{(n:\tilde{n})}V_j(\mathbf{X}_{m+}^{(\alpha_j,j)},t_{m+},\boldsymbol{\lambda}_j)={}^{(n:\tilde{n})}V_j(\mathbf{X}_m^{(0,j)},t_m,\boldsymbol{\lambda}_j)=0;
\tag{5.184}
$$

(ii) for all $j\in\mathbb{L}$ and time $t_m\in(t_{m_1},t_{m_2})$,

$$\mathbf{X}_{m+}^{(\alpha_j,j)} = \mathbf{X}_{m}^{(0,j)},$$

$$ {}^{(n:\tilde{n})}V_j^{(s_{\alpha_j})}(\mathbf{X}_{m+}^{(\alpha_j,j)},t_{m+},\boldsymbol{\lambda}_j) = 0 \ s_{\alpha_j} = 1,2,\ldots,2k_{\alpha_j}, \tag{5.185}$$

$$ {}^{(n:\tilde{n})}V_j^{(2k_{\alpha_j}+1)}(\mathbf{X}_{m+}^{(\alpha_j,j)},t_{m+},\boldsymbol{\lambda}_j) < 0 \text{ for } \alpha_j = 1,2;$$

(iii) with the jth-penetration of the $(2k_{\alpha_j} : 2k_{\beta_j})$-type for time $t = t_{m_i}, \mathbf{X}_{m_i}^{(\alpha_j,j)} = \mathbf{X}_{m_i}^{(0,j)}$ $(i = 1,2)$,

$$\begin{aligned}
&{}^{(n:\tilde{n})}V_j^{(s_{\alpha_j})}(\mathbf{X}_{m_i\pm}^{(\alpha_j,j)},t_{m_i\pm},\boldsymbol{\lambda}_j) = 0 \ (s_{\alpha_j} = 1,2,\ldots,2k_{\alpha_j}+1)\\
&\text{and } V^{(2k_{\alpha_j}+2)}(\mathbf{X}_{m_i\pm}^{(\alpha_j,j)},t_{m_i\pm},\boldsymbol{\lambda}_j) > 0 \text{ for } \alpha_j \in \{1,2\};\\
&{}^{(n:\tilde{n})}V_j^{(s_{\beta_j})}(\mathbf{X}_{m_i+}^{(\beta_j,j)},t_{m_i+},\boldsymbol{\lambda}_j) = 0(s_{\beta_j} = 1,2,\ldots,2k_{\beta_j})\\
&{}^{(n:\tilde{n})}V_j^{(2k_{\beta}+1)}(\mathbf{X}_{m_i+}^{(\beta)},t_{m_i+},\boldsymbol{\lambda}_j) > 0 \text{ for } \beta_j \in \{1,2\} \text{ and } \beta_j \neq \alpha_j;
\end{aligned} \tag{5.186}$$

or with the jth-desynchronization of the $(2k_{\alpha_j} : 2k_{\beta_j})$-type for time $t = t_{m_i}, \mathbf{X}_{m_i}^{(\alpha_j,j)} = \mathbf{X}_{m_i}^{(0)}$ $(i = 1,2)$,

$$\begin{aligned}
&{}^{(n:\tilde{n})}V_j^{(s_{\alpha_j})}(\mathbf{X}_{m_i\pm}^{(\alpha_j,j)},t_{m_i\pm},\boldsymbol{\lambda}_j) = 0 \ (s_{\alpha_j} = 1,2,\ldots,2k_{\alpha_j}+1)\\
&\text{and } V^{(2k_{\alpha_j}+2)}(\mathbf{X}_{m_i\pm}^{(\alpha_j,j)},t_{m_i\pm},\boldsymbol{\lambda}_j) > 0 \text{ for } \alpha_j \in \{1,2\};\\
&{}^{(n:\tilde{n})}V_j^{(s_{\beta_j})}(\mathbf{X}_{m_i\pm}^{(\beta_j,j)},t_{m_i\pm},\boldsymbol{\lambda}_j) = 0 \ (s_{\beta_j} = 1,2,\ldots,2k_{\beta_j}+1),\\
&{}^{(n:\tilde{n})}V_j^{(2k_{\beta_j}+2)}(\mathbf{X}_{m_i\pm}^{(\beta_j,j)},t_{m_i\pm},\boldsymbol{\lambda}_j) > 0 \text{ for } \beta_j \in \{1,2\} \text{ and} \beta_j \neq \alpha_j.
\end{aligned} \tag{5.187}$$

Proof: The proof is similar to the proof of Theorem 5.4 for each $j \in \mathbb{L}$. For all$j \in \mathbb{L}$, if the conditions in Eqs.(5.184) and (5.185) are satisfied, from Definition 5.44, the two dynamical systems in Eqs.(5.1) and (5.2) to constraints in Eq.(5.4) are synchronized with the $(2k_{\alpha_j} : 2k_{\beta_j})$-type for time $t \in [t_{m_1}, t_{m_2}]$, vice versa. If the onset and vanishing conditions in Eqs.(5.186) and (5.187) hold, from Definition 5.47 or 5.48, the desynchronization will start to form and to vanish, vice versa. The proof is completed. ■

5.4.7 Penetration to all constraints

If $\mathbb{F}^{(\alpha_j,j)}(\mathbf{X}^{(\alpha_j,j)},t,\pi^{(\alpha_j,j)}) = \mathbb{F}^{(0,j)}(\mathbf{X}^{(0,j)},t,\boldsymbol{\lambda}_j)$ for $\mathbf{X}^{(\alpha_j,j)} = \mathbf{X}^{(0,j)}$ and $\alpha_j \in \{1,2\}$ with all $j \in \mathbb{L}$, the two dynamical systems to multiple constraints do not have any penetration. For such penetration, $\mathbb{F}^{(\alpha_j,j)}(\mathbf{X}^{(\alpha_j,j)},t,\boldsymbol{\pi}^{(\alpha_j,j)}) \neq \mathbb{F}^{(0,j)}(\mathbf{X}^{(0,j)},t,\boldsymbol{\lambda}_j)$ should hold. Thus, the corresponding conditions for the l-dimensional penetration of two dynamical systems with l-constraints are presented through the following theorems.

Theorem 5.19. For two dynamical systems in Eqs.(5.1) and (5.2) with constraints in Eq.(5.4), there are $(l+1)$-metric functionals in Eq.(5.37). For $\mathbf{X}_m^{(\alpha_j,j)} \in \Omega_{(\alpha_j,j)}$ ($\alpha_j \in \mathbb{I}$ and $j \in \mathbb{L}$ with $\mathbb{I} = \{1,2\}$ and $\mathbb{L} = \{1,2,\ldots,l\}$) and $\mathbf{X}_m^{(0,j)} \in \partial\Omega_{12(j)}$ at time $t_m, \mathbf{X}_m^{(\alpha_j,j)} = \mathbf{X}_m^{(0,j)}$. For any small $\varepsilon > 0$, there is a time interval $[t_{m-\varepsilon}, t_m)$ or $(t_m, t_{m+\varepsilon}]$. At $\mathbf{X}^{(\alpha_j,j)} \in \Omega_{(\alpha_j,j)}^{\pm\varepsilon}$ for time $t \in [t_{m-\varepsilon}, t_m)$ or $(t_m, t_{m+\varepsilon}]$, the functional ${}^{(n:\tilde{n})}V_j(\mathbf{X}^{(\alpha_j,j)},t,\boldsymbol{\lambda}_j)$ is $C^{r_{\alpha_j}}$-continuous and $|{}^{(n:\tilde{n})}V_j^{(r_{\alpha_j}+1)}(\mathbf{X}^{(\alpha_j,j)},t,\boldsymbol{\lambda}_j)| < \infty (r_{\alpha_j} \geq 3)$. For $\mathbf{X}^{(\alpha_j,j)} \in \Omega_{(\alpha_j,j)}$ and $\mathbf{X}^{(0,j)} \in \partial\Omega_{12(j)}$, at $\mathbf{X}^{(\alpha_j,j)} = \mathbf{X}^{(0,j)}, \mathbb{F}^{(\alpha_j,j)}(\mathbf{X}^{(\alpha_j,j)},t,\boldsymbol{\pi}^{(\alpha_j,j)}) \neq \mathbb{F}^{(0,j)}(\mathbf{X}^{(0,j)},t,\boldsymbol{\lambda}_j)$. The two dynamical systems in Eqs.(5.1) and (5.2) to constraints in Eq.(5.4) is penetrated with l-dimensions for time $t \in (t_{m_1}, t_{m_2})$ in the sense of metric functionals if and only if

(i) for all $j \in \mathbb{L}$, time $t = t_m \in (t_{m_1}, t_{m_2})$, $\mathbf{X}_{m-}^{(\alpha_j,j)} = \mathbf{X}_m^{(0)} = \mathbf{X}_{m+}^{(\beta_j,j)}$,

$$ {}^{(n:\tilde{n})}V_j^{(1)}(\mathbf{X}_{m-}^{(\alpha_j,j)},t_{m-},\boldsymbol{\lambda}_j) < 0 \text{ and } {}^{(n:\tilde{n})}V_j^{(1)}(\mathbf{X}_{m+}^{(\beta_j,j)},t_{m+},\boldsymbol{\lambda}_j) > 0; \tag{5.188} $$

(ii) with the synchronization to the jth-constraint for time $t = t_{m_i}, \mathbf{X}_{m_i-}^{(\alpha_j,j)} = \mathbf{X}_{m_i}^{(0,j)} = \mathbf{X}_{m_i\pm}^{(\beta_j,j)}$ $(i = 1,2)$,

$$ \begin{gathered} {}^{(n:\tilde{n})}V_j^{(1)}(\mathbf{X}_{m_i-}^{(\alpha_j,j)},t_{m_i-},\boldsymbol{\lambda}_j) < 0, \\ {}^{(n:\tilde{n})}V_j^{(1)}(\mathbf{X}_{m_i\pm}^{(\beta_j,j)},t_{m_i\pm},\boldsymbol{\lambda}_j) = 0 \text{ and } {}^{(n:\tilde{n})}V_j^{(2)}(\mathbf{X}_{m_i\pm}^{(\beta_j,j)},t_{m_i\pm},\boldsymbol{\lambda}_j) > 0; \end{gathered} \tag{5.189} $$

or with the desynchronization to the jth-constraint only for time$t = t_{m_i}, \mathbf{X}_{m_i\mp}^{(\alpha)} = \mathbf{X}_{m_i}^{(0)} = \mathbf{X}_{m_i+}^{(\beta)}$ $(i = 1,2)$,

$$ \begin{gathered} {}^{(n:\tilde{n})}V_j^{(1)}(\mathbf{X}_{m_i\mp}^{(\alpha_j,j)},t_{m_i\mp},\boldsymbol{\lambda}_j) = 0 \text{ and } {}^{(n:\tilde{n})}V_j^{(2)}(\mathbf{X}_{m_i\mp}^{(\alpha_j,j)},t_{m_i\mp},\boldsymbol{\lambda}_j) > 0, \\ {}^{(n:\tilde{n})}V_j^{(1)}(\mathbf{X}_{m_i+}^{(\beta_j,j)},t_{m_i+},\boldsymbol{\lambda}_j) > 0; \end{gathered} \tag{5.190} $$

or with the switching penetration to the jth-constraint only for time$t = t_{m_i}, \mathbf{X}_{m_i\mp}^{(\alpha)} = \mathbf{X}_{m_i}^{(0)} = \mathbf{X}_{m_i\pm}^{(\beta)}$ $(i = 1,2)$,

$$ \begin{gathered} {}^{(n:\tilde{n})}V_j^{(1)}(\mathbf{X}_{m_i\mp}^{(\alpha_j,j)},t_{m_i\mp},\boldsymbol{\lambda}_j) = 0 \text{ and } {}^{(n:\tilde{n})}V_j^{(2)}(\mathbf{X}_{m_i\mp}^{(\alpha_j,j)},t_{m_i\mp},\boldsymbol{\lambda}_j) > 0, \\ {}^{(n:\tilde{n})}V_j^{(1)}(\mathbf{X}_{m_i\pm}^{(\beta_j,j)},t_{m_i\pm},\boldsymbol{\lambda}_j) = 0 \text{ and } {}^{(n:\tilde{n})}V_j^{(2)}(\mathbf{X}_{m_i\pm}^{(\beta_j,j)},t_{m_i\pm},\boldsymbol{\lambda}_j) > 0. \end{gathered} \tag{5.191} $$

Proof: The proof is similar to the proof of Theorem 5.3 for each $j \in \mathbb{L}$. For all $j \in \mathbb{L}$, if the conditions in Eq.(5.188) are satisfied, the two dynamical systems in Eqs.(5.1) and (5.2) to constraints in Eq.(5.4) are penetrated with l-dimensions for time $t \in [t_{m_1}, t_{m_2}]$, vice versa. If the onset and vanishing conditions in Eqs.(5.189)–(5.191) are satisfied, the penetration of the two dynamical systems with l- constraints will start to be formed or to vanish, vice versa. This theorem is proved. ■

Theorem 5.20. For two dynamical systems in Eqs.(5.1) and (5.2) with constraints in Eq.(5.4), there are $(l+1)$-metric functionals in Eq.(5.37). For $\mathbf{X}_m^{(\alpha_j,j)} \in \Omega_{(\alpha_j,j)}$ ($\alpha_j \in \mathbb{I}$ and $j \in \mathbb{L}$ with $\mathbb{I} = \{1,2\}$ and $\mathbb{L} = \{1,2,\ldots,l\}$) and $\mathbf{X}_m^{(0,j)} \in \partial\Omega_{12(j)}$ at time $t_m, \mathbf{X}_m^{(\alpha_j,j)} = \mathbf{X}_m^{(0,j)}$. For any small $\varepsilon > 0$, there is a time interval $[t_{m-\varepsilon}, t_m)$ or $(t_m, t_{m+\varepsilon}]$. At $\mathbf{X}^{(\alpha_j,j)} \in \Omega^{\pm\varepsilon}_{(\alpha_j,j)}$ for time $t \in [t_{m-\varepsilon}, t_m)$ or $(t_m, t_{m+\varepsilon}]$, the functional ${}^{(n:\tilde{n})}V_j(\mathbf{X}^{(\alpha_j,j)}, t, \boldsymbol{\lambda}_j)$ is $C^{r_{\alpha_j}}$-continuous and $|{}^{(n:\tilde{n})}V_j^{(r_{\alpha_j}+1)}(\mathbf{X}^{(\alpha_j,j)}, t, \boldsymbol{\lambda}_j)| < \infty$ ($r_{\alpha_j} \geq 2k_{\alpha_j}+1$). For $\mathbf{X}^{(\alpha_j,j)} \in \Omega_{(\alpha_j,j)}$ and $\mathbf{X}^{(0,j)} \in \partial\Omega_{12(j)}$, at $\mathbf{X}^{(\alpha_j,j)} = \mathbf{X}^{(0,j)}$, $\mathbb{F}^{(\alpha_j,j)}(\mathbf{X}^{(\alpha_j,j)}, t, \pi^{(\alpha_j,j)}) \neq \mathbb{F}^{(0,j)}(\mathbf{X}^{(0,j)}, t, \boldsymbol{\lambda}_j)$. The two dynamical systems in Eqs.(5.1) and (5.2) to l-constraints in Eq.(5.4) is penetrated of the $(2k_1 : 2k_2)$-type with l-dimensions for time $t \in [t_{m_1}, t_{m_2})$ in the sense of metric functionals if and only if

(i) for all $j \in \mathbb{L}$, time $t = t_m \in (t_{m_1}, t_{m_2})$, $\mathbf{X}_{m-}^{(\alpha)} = \mathbf{X}_m^{(0)} = \mathbf{X}_{m+}^{(\beta)}$,

$$\begin{aligned}
&{}^{(n:\tilde{n})}V_j^{(s_{\alpha_j})}(\mathbf{X}_{m-}^{(\alpha_j,j)}, t_{m-}, \boldsymbol{\lambda}_j) = 0 \text{ for } s_{\alpha_j} = 1,2,\ldots,2k_{\alpha_j}\\
&\quad\text{and } {}^{(n:\tilde{n})}V_j^{(2k_{\alpha_j}+1)}(\mathbf{X}_{m-}^{(\alpha_j,j)}, t_{m-}, \boldsymbol{\lambda}_j) < 0;\\
&{}^{(n:\tilde{n})}V_j^{(2k_{\beta_j})}(\mathbf{X}_{m+}^{(\beta_j,j)}, t_{m+}, \boldsymbol{\lambda}_j) = 0 \text{ for } s_{\beta_j} = 1,2,\ldots,2k_{\beta_j}\\
&\quad\text{and } {}^{(n:\tilde{n})}V_j^{(2k_{\beta_j}+1)}(\mathbf{X}_{m+}^{(\beta_j,j)}, t_{m+}, \boldsymbol{\lambda}_j) > 0;
\end{aligned} \tag{5.192}$$

(ii) with the synchronization of the $(2k_{\alpha_j} : 2k_{\beta_j})$-type to the jth-constraint for time $t = t_{m_i}, \mathbf{X}_{m_i-}^{(\alpha)} = \mathbf{X}_{m_i}^{(0)} = \mathbf{X}_{m_i\pm}^{(\beta)}$ $(i = 1,2)$,

$$\begin{aligned}
&{}^{(n:\tilde{n})}V_j^{(s_{\alpha_j})}(\mathbf{X}_{m-}^{(\alpha_j,j)}, t_{m-}, \boldsymbol{\lambda}_j) = 0 \text{ for } s_{\alpha_j} = 1,2,\ldots,2k_{\alpha_j}\\
&\quad\text{and } {}^{(n:\tilde{n})}V_j^{(2k_{\alpha_j}+1)}(\mathbf{X}_{m-}^{(\alpha_j,j)}, t_{m-}, \boldsymbol{\lambda}_j) < 0;\\
&{}^{(n:\tilde{n})}V_j^{(s_{\beta_j})}(\mathbf{X}_{m\pm}^{(\beta_j,j)}, t_{m\pm}, \boldsymbol{\lambda}_j) = 0 \text{ for } s_{\beta_j} = 1,2,\ldots,2k_{\beta_j}+1\\
&\quad\text{and } {}^{(n:\tilde{n})}V_j^{(2k_{\beta_j}+2)}(\mathbf{X}_{m\pm}^{(\beta_j,j)}, t_{m\pm}, \boldsymbol{\lambda}_j) > 0;
\end{aligned} \tag{5.193}$$

or with the desynchronization of the $(2k_{\alpha_j}:2k_{\beta_j})$-type to the jth-constraint only for time $t=t_{m_i},\mathbf{X}_{m_i\mp}^{(\alpha)}=\mathbf{X}_{m_i}^{(0)}=\mathbf{X}_{m_i+}^{(\beta)}$ $(i=1,2)$,

$$\left.\begin{aligned}&{}^{(n:\tilde{n})}V_j^{(s_{\alpha_j})}(\mathbf{X}_{m\mp}^{(\alpha_j,j)},t_{m\mp},\boldsymbol{\lambda}_j)=0 \text{ for } s_{\alpha_j}=1,2,\ldots,2k_{\alpha_j}+1\\&\text{and } {}^{(n:\tilde{n})}V_j^{(2k_{\alpha_j}+2)}(\mathbf{X}_{m\mp}^{(\alpha_j,j)},t_{m\mp},\boldsymbol{\lambda}_j)>0;\\&{}^{(n:\tilde{n})}V_j^{(s_{\beta_j})}(\mathbf{X}_{m+}^{(\beta_j,j)},t_{m+},\boldsymbol{\lambda}_j)=0 \text{ for } s_{\beta_j}=1,2,\ldots,2k_{\beta_j}\\&\text{and } {}^{(n:\tilde{n})}V_j^{(2k_{\beta_j}+1)}(\mathbf{X}_{m+}^{(\beta_j,j)},t_{m+},\boldsymbol{\lambda}_j)>0;\end{aligned}\right\}\tag{5.194}$$

or with the penetration of the $(2k_\beta:2k_\alpha)$-type to the jth-constraint only for time $t=t_{m_i},\mathbf{X}_{m_i\mp}^{(\alpha)}=\mathbf{X}_{m_i}^{(0)}=\mathbf{X}_{m_i+}^{(\beta)}$ $(i=1,2)$,

$$\left.\begin{aligned}&{}^{(n:\tilde{n})}V_j^{(s_{\alpha_j})}(\mathbf{X}_{m\mp}^{(\alpha_j,j)},t_{m\mp},\boldsymbol{\lambda}_j)=0 \text{ for } s_{\alpha_j}=1,2,\ldots,2k_{\alpha_j}+1\\&\text{and } {}^{(n:\tilde{n})}V_j^{(2k_{\alpha_j}+2)}(\mathbf{X}_{m\mp}^{(\alpha_j,j)},t_{m\mp},\boldsymbol{\lambda}_j)>0;\\&{}^{(n:\tilde{n})}V_j^{(s_{\beta_j})}(\mathbf{X}_{m\pm}^{(\beta_j,j)},t_{m\pm},\boldsymbol{\lambda}_j)=0 \text{ for } s_{\beta_j}=1,2,\ldots,2k_{\beta_j}+1\\&\text{and } {}^{(n:\tilde{n})}V_j^{(2k_{\beta_j}+2)}(\mathbf{X}_{m\pm}^{(\beta_j,j)},t_{m\pm},\boldsymbol{\lambda}_j)>0.\end{aligned}\right\}\tag{5.195}$$

Proof: The proof is similar to the proof of Theorem 5.4 for each $j\in\mathbb{L}$. For all $j\in\mathbb{L}$, if the conditions are satisfied in Eq.(5.192), the two dynamical systems in Eqs.(5.1) and (5.2) to constraints in Eq.(5.4) are penetrated of the $(2k_{\alpha_j}:2k_{\beta_j})$-type with l-dimensions for $t\in(t_{m_1},t_{m_2})$, vice versa. If the switching conditions for the synchronization-penetration, desynchronization-penetration, penetration-penetration in Eqs.(5.193)–(5.195) are satisfied, the onset and vanishing of the $(2k_{\alpha_j}:2k_{\beta_j})$-penetration with l-dimensions occur, vice versa. This theorem is proved. ■

5.4.8 Synchronization-desynchronization-penetration

In this section, the mixture of the synchronization, desynchronization and penetration to multiple constraints is discussed.

Theorem 5.21. For two dynamical systems in Eqs.(5.1) and (5.2) with constraints in Eq.(5.4), there are $(l+1)$-metric functionals in Eq.(5.37). For $\mathbf{X}_m^{(\alpha_j,j)}\in\Omega_{(\alpha_j,j)}$ ($\alpha_j\in\mathbb{I}$ and $j\in\mathbb{L}$ with $\mathbb{I}=\{1,2\}$ and $\mathbb{L}=\{1,2,\ldots,l\},\mathbb{L}=\mathbb{L}_1\cup\mathbb{L}_2$) and $\mathbf{X}_m^{(0,j)}\in\partial\Omega_{12(j)}$ at time $t_m,\mathbf{X}_m^{(\alpha_j,j)}=\mathbf{X}_m^{(0,j)}$. For any small $\varepsilon>0$, there is a time interval $[t_{m-\varepsilon},t_m)$ or

$(t_m, t_{m+\varepsilon}]$. For $\mathbf{X}^{(\alpha_j,j)} \in \Omega_{(\alpha_j,j)}$ and $\mathbf{X}^{(0,j)} \in \partial\Omega_{12(j)}$, at $\mathbf{X}^{(\alpha_j,j)} = \mathbf{X}^{(0,j)} D^{(s_{\alpha_j})}\mathbb{F}^{(\alpha_j,j)}(\mathbf{X}^{(\alpha_j,j)}, t, \pi^{(\alpha_j,j)}) = D^{(s_{\alpha_j})}\mathbb{F}^{(0,j)}(\mathbf{X}^{(0,j)}, t, \boldsymbol{\lambda}_j)$ $(s_{\alpha_j} = 0,1,2,\ldots)$. The two dynamical systems in Eqs.(5.1) and (5.2) to l-constraints in Eq.(5.4) are of the (l_1,l_2)-synchronization and desynchronization for time $[t_{m_1}, t_{m_2}]$ in the sense of metric functionals if and only if

(i) for $\mathbf{X}_m^{(\alpha_j,j)} \in \Omega_{(\alpha_j,j)}$ and $\mathbf{X}_m^{(0,j)} \in \partial\Omega_{12(j)}$ for any time t_m,

$$\begin{gathered}\mathbf{X}_m^{(\alpha_j,j)} = \mathbf{X}_m^{(0,j)},\ {}^{(n:\tilde{n})}V_j^{(s_{\alpha_j})}(\mathbf{X}_m^{(\alpha_j,j)}, t_m, \boldsymbol{\lambda}_j) = 0 \\ \text{for } \alpha_j = 1,2 \text{ and } s_{\alpha_j} = 0,1,2,\ldots\end{gathered} \tag{5.196}$$

(ii) for all $j \in \mathbb{L}_1$ and $\alpha_j = 1,2$,

$$\begin{aligned}&\mathbf{X}_\kappa^{(\alpha_j,j)} \neq \mathbf{X}_m^{(0,j)},\ {}^{(n:\tilde{n})}V_j^{(1)}(\mathbf{X}_\kappa^{(\alpha_j,j)}, t_\kappa^-, \boldsymbol{\lambda}_j) < 0 \text{ and} \\ &\lim_{t_\kappa^- \to t_m} {}^{(n:\tilde{n})}V_j^{(1)}(\mathbf{X}_\kappa^{(\alpha_j,j)}, t_\kappa^-, \boldsymbol{\lambda}_j) = 0 \text{ for } \alpha_j = 1,2,\end{aligned} \tag{5.197}$$

with $\mathbf{X}_{\kappa_1}^{(\alpha_j,j)} \in \Omega_{(\alpha_j,j)}^{-\varepsilon}$ at time $t_\kappa^- \in [t_{m-\varepsilon}, t_m)$ and $\mathbf{X}_m^{(0,j)} \in \partial\Omega_{12(j)}$ for $t_m \in (t_{m_1}, t_{m_2})$;

(iii) for all $j \in \mathbb{L}_2$ and $\alpha_j = 1,2$,

$$\begin{aligned}&\mathbf{X}_\kappa^{(\alpha_j,j)} \neq \mathbf{X}_m^{(0,j)},\ {}^{(n:\tilde{n})}V_j^{(1)}(\mathbf{X}_\kappa^{(\alpha_j,j)}, t_\kappa^+, \boldsymbol{\lambda}_j) > 0 \text{ and} \\ &\lim_{t_\kappa^+ \to t_m} {}^{(n:\tilde{n})}V_j^{(1)}(\mathbf{X}_\kappa^{(\alpha_j,j)}, t_\kappa^+, \boldsymbol{\lambda}_j) = 0 \text{ for } \alpha_j = 1,2\end{aligned} \tag{5.198}$$

with $\mathbf{X}_\kappa^{(\alpha_j,j)} \in \Omega_{(\alpha_j,j)}^{+\varepsilon}$ at time $t_\kappa^+ \in (t_m, t_{m+\varepsilon}]$ and $\mathbf{X}_m^{(0,j)} \in \partial\Omega_{12(j)}$ for $t_m \notin (t_{m_1}, t_{m_2})$;

(iv) for some $j \in \mathbb{L}_1$,

$$\begin{aligned}&\mathbf{X}_\kappa^{(\alpha_j,j)} \neq \mathbf{X}_m^{(0,j)},\ {}^{(n:\tilde{n})}V_j^{(1)}(\mathbf{X}_\kappa^{(\alpha_j,j)}, t_\kappa^+, \boldsymbol{\lambda}_j) > 0 \text{ and} \\ &\lim_{t_\kappa^+ \to t_m} {}^{(n:\tilde{n})}V_j^{(1)}(\mathbf{X}_\kappa^{(\alpha_j,j)}, t_\kappa^+, \boldsymbol{\lambda}_j) = 0 \text{ for } \alpha_j = 1,2\end{aligned} \tag{5.199}$$

with $\mathbf{X}_\kappa^{(\alpha_j,j)} \in \Omega_{(\alpha_j,j)}^{+\varepsilon}$ at time $t_\kappa \in (t_m, t_{m+\varepsilon}]$ and $\mathbf{X}_m^{(0,j)} \in \partial\Omega_{12(j)}$ for $t_m \notin (t_{m_1}, t_{m_2})$;

or for some $j \in \mathbb{L}_2$,

$$\begin{aligned}&\mathbf{X}_\kappa^{(\alpha_j,j)} \neq \mathbf{X}_m^{(0,j)},\ {}^{(n:\tilde{n})}V_j^{(1)}(\mathbf{X}_\kappa^{(\alpha_j,j)}, t_\kappa^-, \boldsymbol{\lambda}_j) < 0 \text{ and} \\ &\lim_{t_\kappa^- \to t_m} {}^{(n:\tilde{n})}V_j^{(1)}(\mathbf{X}_\kappa^{(\alpha_j,j)}, t_\kappa^-, \boldsymbol{\lambda}_j) = 0 \text{ for } \alpha_j = 1,2\end{aligned} \tag{5.200}$$

with $\mathbf{X}_\kappa^{(\alpha_j,j)} \in \Omega_{(\alpha_j,j)}^{-\varepsilon}$ at time $t_\kappa \in [t_{m-\varepsilon}, t_m)$ and $\mathbf{X}_m^{(0,j)} \in \partial\Omega_{12(j)}$ for $t_m \in (t_{m_1}, t_{m_2})$;

(v) for $j \in \mathbb{L}$ in (iv) and $\alpha_j = 1,2$,

$$\begin{aligned}&\mathbf{X}_{\kappa}^{(\alpha)} \neq \mathbf{X}_m^{(0)}, \lim_{t_\kappa^{\pm} \to t_{m\pm}} {}^{(n:\tilde{n})}V_j^{(1)}(\mathbf{X}_\kappa^{(\alpha)}, t_\kappa^{\pm}, \boldsymbol{\lambda}_j) = 0,\\ &\lim_{t_\kappa^{\pm} \to t_{m\pm}} {}^{(n:\tilde{n})}V_j^{(2)}(\mathbf{X}_\kappa^{(\alpha)}, t_\kappa^{\pm}, \boldsymbol{\lambda}) < 0 \text{ for } \alpha_j = 1,2,\end{aligned} \tag{5.201}$$

with $\mathbf{X}_\kappa^{(\alpha_j,j)} \in \Omega_{(\alpha_j,j)}^{\pm\varepsilon}$ at time $t_\kappa^- \in [t_{m-\varepsilon}, t_{m-}), t_\kappa^+ \in (t_{m+}, t_{m+\varepsilon}]$ and $\mathbf{X}_m^{(0,j)} \in \partial\Omega_{12(j)}$ for $t_m = t_{m_1}$ and t_{m_2}.

Proof: The proof is similar to the proof of Theorem 5.1 for each $j \in \mathbb{L}$. For all $j \in \mathbb{L}$, if the conditions are satisfied, the two dynamical systems in Eqs.(5.1) and (5.2) to l-constraints in Eq.(5.4) are of the (l_1, l_2)-synchronization and desynchronization for time $[t_{m_1}, t_{m_2}]$, vice versa. This theorem is proved. ■

Theorem 5.22. For two dynamical systems in Eqs.(5.1) and (5.2) with constraints in Eq.(5.4), there are $(l+1)$-metric functionals in Eq.(5.37). For $\mathbf{X}_m^{(\alpha_j,j)} \in \Omega_{(\alpha_j,j)}$ ($\alpha_j \in \mathbb{I}$ and $j \in \mathbb{L}$ with $\mathbb{I} = \{1,2\}$ and $\mathbb{L} = \{1,2,\ldots,l\}$, $\mathbb{L} = \mathbb{L}_1 \cup \mathbb{L}_2$) and $\mathbf{X}_m^{(0,j)} \in \partial\Omega_{12(j)}$ at time $t_m, \mathbf{X}_m^{(\alpha_j,j)} = \mathbf{X}_m^{(0,j)}$. For any small $\varepsilon > 0$, there is a time interval $[t_{m-\varepsilon}, t_m)$ or $(t_m, t_{m+\varepsilon}]$. For $\mathbf{X}^{(\alpha_j,j)} \in \Omega_{(\alpha_j,j)}$ and $\mathbf{X}^{(0,j)} \in \partial\Omega_{12(j)}$, at $\mathbf{X}^{(\alpha_j,j)} = \mathbf{X}^{(0,j)}, D^{(s_{\alpha_j})}\mathbb{F}^{(\alpha_j,j)}(\mathbf{X}^{(\alpha_j,j)}, t, \boldsymbol{\pi}^{(\alpha_j,j)}) = D^{(s_{\alpha_j})}\mathbb{F}^{(0,j)}(\mathbf{X}^{(0,j)}, t, \boldsymbol{\lambda}_j)$ $(s_{\alpha_j} = 0,1,2,\ldots)$. The two dynamical systems in Eqs.(5.1) and (5.2) to l-constraints in Eq.(5.4) are of an $(l_1,l_2) - (2k_{\alpha_j}+1)^{\text{th}}$-synchronization and $(2k_{\alpha_j}+1)^{\text{th}}$-desynchronization for time $[t_{m_1}, t_{m_2}]$ in the sense of metric functionals if and only if

(i) for $\mathbf{X}_m^{(\alpha_j,j)} \in \Omega_{(\alpha_j,j)}$ and $\mathbf{X}_m^{(0,j)} \in \partial\Omega_{12(j)}$ for any time t_m,

$$\begin{aligned}&\mathbf{X}_m^{(\alpha_j,j)} = \mathbf{X}_m^{(0,j)}, {}^{(n:\tilde{n})}V_j^{(s_{\alpha_j})}(\mathbf{X}_m^{(\alpha_j,j)}, t_m, \boldsymbol{\lambda}_j) = 0\\ &\text{for } \alpha_j = 1,2 \text{ and } s_{\alpha_j} = 0,1,2,\ldots\end{aligned} \tag{5.202}$$

(ii) for all $j \in \mathbb{L}_1$ and $\alpha_j = 1,2$,

$$\begin{aligned}&\mathbf{X}_\kappa^{(\alpha_j,j)} \neq \mathbf{X}_m^{(0,j)},\\ &\lim_{t_\kappa^- \to t_{m-}} {}^{(n:\tilde{n})}V_j^{(s_{\alpha_j})}(\mathbf{X}_\kappa^{(\alpha_j,j)}, t_\kappa^-, \boldsymbol{\lambda}_j) = 0 \text{ for } s_{\alpha_j} = 1,2,\ldots,2k_{\alpha_j};\\ &{}^{(n:\tilde{n})}V_j^{(2k_{\alpha_j}+1)}(\mathbf{X}_\kappa^{(\alpha)}, t_\kappa^-, \boldsymbol{\lambda}_j) < 0 \text{ and}\end{aligned} \tag{5.203}$$

$$\lim_{t_\kappa^- \to t_m} {}^{(n:\tilde{n})}V_j^{(2k_{\alpha_j}+1)}(\mathbf{X}_\kappa^{(\alpha_j,j)}, t_\kappa^-, \boldsymbol{\lambda}_j) = 0 \text{ for } \alpha_j = 1,2;$$

with $\mathbf{X}_{\kappa_1}^{(\alpha_j,j)} \in \Omega_{(\alpha_j,j)}^{-\varepsilon}$ at time $t_\kappa^- \in [t_{m-\varepsilon}, t_m)$ and $\mathbf{X}_m^{(0,j)} \in \partial\Omega_{12(j)}$ for $t_m \in (t_{m_1}, t_{m_2})$;

(iii) for all $j \in \mathbb{L}_2$ and $\alpha_j = 1,2$,

$$\begin{aligned}
&\mathbf{X}_\kappa^{(\alpha_j,j)} \neq \mathbf{X}_m^{(0,j)},\\
&\lim_{t_\kappa^+ \to t_{m+}} {}^{(n:\tilde{n})}V_j^{(s_{\alpha_j})}(\mathbf{X}_\kappa^{(\alpha_j,j)}, t_\kappa^+, \boldsymbol{\lambda}_j) = 0 \text{ for } s_{\alpha_j} = 1,2,\ldots,2k_{\alpha_j};\\
&{}^{(n:\tilde{n})}V_j^{(2k_{\alpha_j}+1)}(\mathbf{X}_\kappa^{(\alpha_j,j)}, t_\kappa^+, \boldsymbol{\lambda}_j) > 0 \text{ and}\\
&\lim_{t_\kappa^+ \to t_m} {}^{(n:\tilde{n})}V_j^{(2k_{\alpha_j}+1)}(\mathbf{X}_\kappa^{(\alpha_j,j)}, t_\kappa^+, \boldsymbol{\lambda}_j) = 0 \text{ for } \alpha_j = 1,2;
\end{aligned} \tag{5.204}$$

with $\mathbf{X}_\kappa^{(\alpha_j,j)} \in \Omega_{(\alpha_j,j)}^{+\varepsilon}$ at time $t_\kappa^+ \in (t_m, t_{m+\varepsilon}]$ and $\mathbf{X}_m^{(0,j)} \in \partial\Omega_{12(j)}$ for $t_m \notin (t_{m_1}, t_{m_2})$;

(iv) for some $j \in \mathbb{L}_1$,

$$\begin{aligned}
&\mathbf{X}_\kappa^{(\alpha_j,j)} \neq \mathbf{X}_m^{(0,j)},\\
&\lim_{t_\kappa^+ \to t_{m+}} {}^{(n:\tilde{n})}V_j^{(s_{\alpha_j})}(\mathbf{X}_\kappa^{(\alpha_j,j)}, t_\kappa^+, \boldsymbol{\lambda}_j) = 0 \text{ for } s_{\alpha_j} = 1,2,\ldots,2k_{\alpha_j};\\
&{}^{(n:\tilde{n})}V_j^{(2k_{\alpha_j}+1)}(\mathbf{X}_\kappa^{(\alpha)}, t_\kappa^+, \boldsymbol{\lambda}_j) > 0 \text{ and}\\
&\lim_{t_\kappa^+ \to t_m} {}^{(n:\tilde{n})}V_j^{(2k_{\alpha_j}+1)}(\mathbf{X}_\kappa^{(\alpha_j,j)}, t_\kappa^+, \boldsymbol{\lambda}_j) = 0 \text{ for } \alpha_j = 1,2;
\end{aligned} \tag{5.205}$$

with $\mathbf{X}_\kappa^{(\alpha_j,j)} \in \Omega_{(\alpha_j,j)}^{+\varepsilon}$ at time $t_\kappa \in (t_m, t_{m+\varepsilon}]$ and $\mathbf{X}_m^{(0,j)} \in \partial\Omega_{12(j)}$ for $t_m \notin (t_{m_1}, t_{m_2})$;
or for some $j \in \mathbb{L}_2$,

$$\begin{aligned}
&\mathbf{X}_\kappa^{(\alpha_j,j)} \neq \mathbf{X}_m^{(0,j)},\\
&\lim_{t_\kappa^- \to t_{m+}} {}^{(n:\tilde{n})}V_j^{(s_{\alpha_j})}(\mathbf{X}_\kappa^{(\alpha_j,j)}, t_\kappa^-, \boldsymbol{\lambda}_j) = 0 \text{ for } s_{\alpha_j} = 1,2,\ldots,2k_{\alpha_j};\\
&{}^{(n:\tilde{n})}V_j^{(2k_{\alpha_j}+1)}(\mathbf{X}_\kappa^{(\alpha_j,j)}, t_\kappa^-, \boldsymbol{\lambda}_j) < 0 \text{ and}
\end{aligned} \tag{5.206}$$

with $\mathbf{X}_K^{(\alpha_{j_2},j_2)} \in \Omega_{(\alpha_{j_2},j_2)}^{+\varepsilon}$ at time $t_k \in (t_m, t_{m+\varepsilon}]$ and $\mathbf{X}_m^{(0,j_1)} \in \partial\Omega_{12(j_1)}$ for $t_m \notin [t_{m_1}, t_{m_2}]$;

$$\lim_{t_\kappa^- \to t_m} {}^{(n:\tilde{n})}V_j^{(2k_{\alpha_j}+1)}(\mathbf{X}_\kappa^{(\alpha_j,j)}, t_\kappa^-, \boldsymbol{\lambda}_j) = 0 \text{ for } \alpha_j = 1,2;$$

(v) for $j \in \mathbb{L}$ in (iv) and $\alpha_j = 1,2$,

$$\mathbf{X}_\kappa^{(\alpha)} \neq \mathbf{X}_m^{(0)},$$

$$\lim_{t_\kappa^\pm \to t_{m\pm}} {}^{(n:\tilde{n})}V_j^{(s_{\alpha_j})}(\mathbf{X}_\kappa^{(\alpha_j,j)}, t_\kappa^\pm, \boldsymbol{\lambda}_j) = 0 \text{ for } s_{\alpha_j} = 1,2,\ldots,2k_{\alpha_j}; \tag{5.207}$$

$$\lim_{t_\kappa^\pm \to t_{m\pm}} {}^{(n:\tilde{n})}V_j^{(2k_{\alpha_j}+1)}(\mathbf{X}_\kappa^{(\alpha_j,j)}, t_\kappa^\pm, \boldsymbol{\lambda}_j) < 0 \text{ for } \alpha = 1,2$$

with $\mathbf{X}_\kappa^{(\alpha_j,j)} \in \Omega_{(\alpha_j,j)}^{\pm\varepsilon}$ at time $t_\kappa^- \in [t_{m-\varepsilon}, t_{m-})$ or $t_\kappa^+ \in (t_{m+}, t_{m+\varepsilon}]$ and $\mathbf{X}_m^{(0,j)} \in \partial\Omega_{12(j)}$ for $t_m = t_{m_1}$ and t_{m_2}.

Proof: The proof is similar to the proof of Theorem 5.2 for each $j \in \mathbb{L}$. For all $j \in \mathbb{L}$, if the conditions are satisfied, the two dynamical systems in Eqs.(5.1) and (5.2) to constraints in Eq.(5.4) arc of an $(l_1,l_2)-(2k_{\alpha_j}+1)^{\text{th}}$-synchronization and $(2k_{\alpha_j}+1)^{\text{th}}$-desynchronization, vice versa. This theorem is proved. ■

Theorem 5.23. For two dynamical systems in Eqs.(5.1) and (5.2) with constraints in Eq.(5.4), there are $(l+1)$-metric functionals in Eq.(5.37). For $\mathbf{X}_m^{(\alpha_j,j)} \in \Omega_{(\alpha_j,j)}$ ($\alpha_j \in \mathbb{I}$ and $j \in \mathbb{L}$ with $\mathbb{I} = \{1,2\}$ and $\mathbb{L} = \{1,2,\ldots,l\}$) and $\mathbf{X}_m^{(0,j)} \in \partial\Omega_{12(j)}$ at time $t_m, \mathbf{X}_m^{(\alpha_j,j)} = \mathbf{X}_m^{(0,j)}$. For any small $\varepsilon > 0$, there is a time interval $[t_{m-\varepsilon}, t_m)$ or $(t_m, t_{m+\varepsilon}]$. At $\mathbf{X}^{(\alpha_j,j)} \in \Omega_{(\alpha_j,j)}^{\pm\varepsilon}$ for time $t \in [t_{m-\varepsilon}, t_m)$ or $(t_m, t_{m+\varepsilon}]$, the functional ${}^{(n:\tilde{n})}V_j(\mathbf{X}^{(\alpha_j,j)}, t, \boldsymbol{\lambda}_j)$is $C^{r_{\alpha_j}}$-continuous and $|{}^{(n:\tilde{n})}V_j^{(\alpha_j+1)}(\mathbf{X}^{(\alpha_j,j)}, t, \boldsymbol{\lambda}_j)| < \infty$ $(r_{\alpha_j} \geq 3)$. For $\mathbf{X}^{(\alpha_j,j)} \in \Omega_{(\alpha_j,j)}$ and $\mathbf{X}^{(0,j)} \in \partial\Omega_{12(j)}$, at $\mathbf{X}^{(\alpha_j,j)} = \mathbf{X}^{(0,j)}, \mathbb{F}^{(\alpha_j,j)}(\mathbf{X}^{(\alpha_j,j)}, t, \pi^{(\alpha_j,j)}) \neq \mathbb{F}^{(0,j)}(\mathbf{X}^{(0,j)}, t, \boldsymbol{\lambda}_j)$. The two dynamical systems in Eqs.(5.1) and (5.2) to l-constraints in Eq.(5.4) are of the (l_1,l_2,l_3)-synchnonization, desynchronization and penetration for time $t \in [t_{m_1}, t_{m_2}]$ in the sense of metric functionals if and only if

(i) for $\mathbf{X}_{m+}^{(\alpha_j,j)} = \mathbf{X}_m^{(0,j)}$ and $\mathbf{X}^{(\alpha_j,j)}(t) \in \Omega_{(\alpha_j,j)}$ $(\alpha_j \in \mathbb{I})$, at time $t = t_m \in [t_{m_1}, t_{m_2}]$,

$${}^{(n:\tilde{n})}V_j(\mathbf{X}_{m+}^{(\alpha_j,j)}, t_{m+}, \boldsymbol{\lambda}_j) = {}^{(n:\tilde{n})}V_j(\mathbf{X}_m^{(0,j)}, t_m, \boldsymbol{\lambda}_j) = 0 \tag{5.208}$$

with all $j \in \mathbb{L} = \mathbb{L}_1 \cup \mathbb{L}_2 \cup \mathbb{L}_3$;
(ii) for time $t_m \in (t_{m_1}, t_{m_2})$ for all $j \in \mathbb{L}_1$,

$$\left.\begin{aligned} &\mathbf{X}_{m-}^{(\alpha_j,j)} = \mathbf{X}_m^{(0,j)} \text{ and } {}^{(n:\tilde{n})}V_j^{(1)}(\mathbf{X}_{m-}^{(\alpha_j,j)}, t_{m-}, \boldsymbol{\lambda}_j) < 0 \text{ for } \alpha_j \in \mathbb{I}, \\ &\mathbf{X}_{m-}^{(\beta_j,j)} = \mathbf{X}_m^{(0,j)} \text{ and } {}^{(n:\tilde{n})}V_j^{(1)}(\mathbf{X}_{m-}^{(\beta_j,j)}, t_{m-}, \boldsymbol{\lambda}_j) < 0 \text{ for } \beta_j \in \mathbb{I}; \end{aligned}\right\} \tag{5.209}$$

(iii) for time $t_m \in (t_{m_1}, t_{m_2})$ for all $j \in \mathbb{L}_2$,

$$\mathbf{X}_{m+}^{(\alpha_j,j)} = \mathbf{X}_m^{(0,j)} \text{ and } {}^{(n:\tilde{n})}V_j^{(1)}(\mathbf{X}_{m+}^{(\alpha_j,j)}, t_{m+}, \boldsymbol{\lambda}_j) > 0 \text{ for } \alpha_j \in \mathbb{I},$$
$$\mathbf{X}_{m+}^{(\beta_j,j)} = \mathbf{X}_m^{(0,j)} \text{ and } {}^{(n:\tilde{n})}V_j^{(1)}(\mathbf{X}_{m+}^{(\beta_j,j)}, t_{m+}, \boldsymbol{\lambda}_j) > 0 \text{ for } \beta_j \in \mathbb{I}; \tag{5.210}$$

(vi) for time $t_m \in (t_{m_1}, t_{m_2})$ for all $j \in \mathbb{L}_3$,

$$\left.\begin{array}{l} \mathbf{X}_{m-}^{(\alpha_j,j)} = \mathbf{X}_m^{(0,j)} \text{ and} \\ {}^{(n:\tilde{n})}V_j^{(1)}(\mathbf{X}_{m-}^{(\alpha_j,j)}, t_{m-}, \boldsymbol{\lambda}_j) < 0 \end{array}\right\} \text{ for } \alpha_j \in \mathbb{I},$$
$$\left.\begin{array}{l} \mathbf{X}_{m+}^{(\beta_j,j)} = \mathbf{X}_m^{(0,j)} \text{ and} \\ {}^{(n:\tilde{n})}V_j^{(1)}(\mathbf{X}_{m+}^{(\beta_j,j)}, t_{m+}, \boldsymbol{\lambda}_j) > 0 \end{array}\right\} \text{ for } \beta_j \in \mathbb{I} \text{ and } \beta_j \neq \alpha_j; \tag{5.211}$$

(v) for $j \in \mathbb{L}_k, k \in \{1,2,3\}$ with time $t = t_{m_i}, \mathbf{X}_{m_i}^{(\alpha_j,j)} = \mathbf{X}_{m_i}^{(0,j)}$ $(i = 1,2)$,

$$\left.\begin{array}{l} {}^{(n:\tilde{n})}V_j^{(1)}(\mathbf{X}_{m_i\pm}^{(\alpha_j,j)}, t_{m_i\pm}, \boldsymbol{\lambda}_j) = 0 \\ {}^{(n:\tilde{n})}V_j^{(2)}(\mathbf{X}_{m_i\pm}^{(\alpha_j,j)}, t_{m_i\pm}, \boldsymbol{\lambda}_j) > 0 \end{array}\right\} \text{ for } \alpha_j \in \{1,2\}; \tag{5.212}$$

and/or

$$\left.\begin{array}{l} {}^{(n:\tilde{n})}V_j^{(1)}(\mathbf{X}_{m_i\pm}^{(\beta_j,j)}, t_{m_i\pm}, \boldsymbol{\lambda}_j) = 0 \\ {}^{(n:\tilde{n})}V_j^{(2)}(\mathbf{X}_{m_i\pm}^{(\beta_j,j)}, t_{m_i\pm}, \boldsymbol{\lambda}_j) > 0 \end{array}\right\} \text{ for } \beta_j \in \{1,2\}. \tag{5.213}$$

Proof: The proof is similar to the proof of Theorem 5.3 for each $j \in \mathbb{L}$. For all $j \in \mathbb{L}$, if the conditions are satisfied, of the (l_1, l_2, l_3)-synchnonization, desynchronization and penetration for time$t \in [t_{m_1}, t_{m_2}]$, vice versa. This theorem is proved. ■

Theorem 5.24. For two dynamical systems in Eqs.(5.1) and (5.2) with constraints in Eq.(5.4), there are $(l+1)$-metric functionals in Eq.(5.37). For $\mathbf{X}_m^{(\alpha_j,j)} \in \Omega_{(\alpha_j,j)}$ ($\alpha_j \in \mathbb{I}$ and $j \in \mathbb{L}$ with $\mathbb{I} = \{1,2\}$ and $\mathbb{L} = \{1,2,\ldots,l\}$) and $\mathbf{X}_m^{(0,j)} \in \partial\Omega_{12(j)}$ at time $t_m, \mathbf{X}_m^{(\alpha_j,j)} = \mathbf{X}_m^{(0,j)}$. For any small $\varepsilon > 0$, there is a time interval $[t_{m-\varepsilon}, t_m)$ or $(t_m, t_{m+\varepsilon}]$. At $\mathbf{X}^{(\alpha_j,j)} \in \Omega_{(\alpha_j,j)}^{\pm\varepsilon}$ for time $t \in [t_{m-\varepsilon}, t_m)$ or $(t_m, t_{m+\varepsilon}]$, the functional ${}^{(n:\tilde{n})}V_j(\mathbf{X}^{(\alpha_j,j)}, t, \boldsymbol{\lambda}_j)$ is $C^{r_{\alpha_j}}$-continuous and $|{}^{(n:\tilde{n})}V_j^{(r_{\alpha_j}+1)}(\mathbf{X}^{(\alpha_j,j)}, t, \boldsymbol{\lambda}_j)| < \infty$ $(r_{\alpha_j} \geq 2k_{\alpha_j} + 1)$. For $\mathbf{X}^{(\alpha_j,j)} \in \Omega_{(\alpha_j,j)}$ and $\mathbf{X}^{(0,j)} \in \partial\Omega_{12(j)}$, at $\mathbf{X}^{(\alpha_j,j)} = \mathbf{X}^{(0,j)}$, $\mathbb{F}^{(\alpha_j,j)}(\mathbf{X}^{(\alpha_j,j)}, t, \pi^{(\alpha_j,j)}) \neq \mathbb{F}^{(0,j)}(\mathbf{X}^{(0,j)}, t, \boldsymbol{\lambda}_j)$. The two dynamical systems in Eqs.(5.1) and (5.2) to l-constraints in Eq.(5.4) are of the (l_1, l_2, l_3)-synchnonization, desynchronization and penetration of the $(2k_{\alpha_j} : 2k_{\beta_j})$-type for all $j \in \mathbb{L} = \mathbb{L}_1 \cup \mathbb{L}_2 \cup \mathbb{L}_3$ for time $t \in [t_{m_1}, t_{m_2}]$ in the sense of metric functionals if and only if

(i) for $\mathbf{X}_{m+}^{(\alpha_j,j)} = \mathbf{X}_m^{(0,j)}$ and $\mathbf{X}^{(\alpha_j,j)}(t) \in \Omega_{(\alpha_j,j)}$ $(\alpha_j \in \mathbb{I})$, at time $t = t_m \in [t_{m_1}, t_{m_2}]$,

$$ {}^{(n:\tilde{n})}V_j(\mathbf{X}_{m+}^{(\alpha_j,j)}, t_{m+}, \boldsymbol{\lambda}_j) = {}^{(n:\tilde{n})}V_j(\mathbf{X}_m^{(0,j)}, t_m, \boldsymbol{\lambda}_j) = 0 \tag{5.214} $$

with all $j \in \mathbb{L} = \mathbb{L}_1 \cup \mathbb{L}_2 \cup \mathbb{L}_3$;
(ii) for time $t_m \in (t_{m_1}, t_{m_2})$ for all $j \in \mathbb{L}_1$,

$$ \left.\begin{aligned} &\mathbf{X}_{m-}^{(\alpha_j,j)} = \mathbf{X}_m^{(0,j)}, \\ &{}^{(n:\tilde{n})}V_j^{(s_{\alpha_j})}(\mathbf{X}_{m-}^{(\alpha_j,j)}, t_{m-}, \boldsymbol{\lambda}_j) = 0 \text{ for } s_{\alpha_j} = 1,2,\ldots,2k_{\alpha_j}, \\ &{}^{(n:\tilde{n})}V_j^{(2k_{\alpha_j}+1)}(\mathbf{X}_{m-}^{(\alpha_j,j)}, t_{m-}, \boldsymbol{\lambda}_j) < 0 \text{ for } \alpha_j = 1,2; \end{aligned}\right\} \tag{5.215} $$

(iii) for time $t_m \in (t_{m_1}, t_{m_2})$ for all $j \in \mathbb{L}_2$,

$$ \left.\begin{aligned} &\mathbf{X}_{m+}^{(\alpha_j,j)} = \mathbf{X}_m^{(0,j)}, \\ &{}^{(n:\tilde{n})}V_j^{(s_{\alpha_j})}(\mathbf{X}_{m+}^{(\alpha_j,j)}, t_{m+}, \boldsymbol{\lambda}_j) = 0 \text{ for } s_{\alpha_j} = 1,2,\ldots,2k_{\alpha_j}, \\ &{}^{(n:\tilde{n})}V_j^{(2k_{\alpha_j}+1)}(\mathbf{X}_{m+}^{(\alpha_j,j)}, t_{m+}, \boldsymbol{\lambda}_j) > 0 \text{ for } \alpha_j = 1,2; \end{aligned}\right\} \tag{5.216} $$

(vi) for time $t_m \in (t_{m_1}, t_{m_2})$ for all $j \in \mathbb{L}_3$,

$$ \left.\begin{aligned} &\mathbf{X}_{m-}^{(\alpha_j,j)} = \mathbf{X}_m^{(0,j)}, \\ &{}^{(n:\tilde{n})}V_j^{(s_{\alpha_j})}(\mathbf{X}_{m-}^{(\alpha_j,j)}, t_{m-}, \boldsymbol{\lambda}_j) = 0 \text{ for } s_{\alpha_j} = 1,2,\ldots,2k_{\alpha_j}, \\ &{}^{(n:\tilde{n})}V_j^{(2k_{\alpha_j}+1)}(\mathbf{X}_{m-}^{(\alpha_j,j)}, t_{m-}, \boldsymbol{\lambda}_j) < 0 \text{ for } \alpha_j \in \mathbb{I}; \\ &\mathbf{X}_{m+}^{(\beta_j,j)} = \mathbf{X}_m^{(0,j)}, \\ &{}^{(n:\tilde{n})}V_j^{(s_{\beta_j})}(\mathbf{X}_{m+}^{(\beta_j,j)}, t_{m+}, \boldsymbol{\lambda}_j) = 0 \text{ for } s_{\beta_j} = 1,2,\cdots,2k_{\beta_j}, \\ &{}^{(n:\tilde{n})}V_j^{(2k_{\beta_j}+1)}(\mathbf{X}_{m+}^{(\beta_j,j)}, t_{m+}, \boldsymbol{\lambda}_j) > 0 \text{ for } \beta_j \in \mathbb{I} \text{ and } \beta_j \neq \alpha_j; \end{aligned}\right\} \tag{5.217} $$

(v) for one of $j \in \mathbb{L}$ with the $(2k_{\alpha_j} : 2k_{\beta_j})$-singularity for time $t = t_{m_i}$, $\mathbf{X}_{m_i}^{(\alpha_j,j)} = \mathbf{X}_{m_i}^{(0,j)}$ $(i = 1,2)$,

$$ \left.\begin{aligned} &{}^{(n:\tilde{n})}V_j^{(s_{\alpha_j})}(\mathbf{X}_{m_i\pm}^{(\alpha_j,j)}, t_{m_i\pm}, \boldsymbol{\lambda}_j) = 0 \text{ for } s_{\alpha_j} = 1,2,\ldots,2k_{\alpha_j}+1, \\ &V^{(2k_{\alpha_j}+2)}(\mathbf{X}_{m_i\pm}^{(\alpha_j,j)}, t_{m_i\pm}, \boldsymbol{\lambda}_j) > 0 \text{ for } \alpha_j \in \{1,2\}; \end{aligned}\right\} \tag{5.218} $$

and/or

$$
\begin{aligned}
&{}^{(n:\tilde{n})}V_j^{(s_{\beta_j})}(\mathbf{X}_{m_i+}^{(\beta_j,j)}, t_{m_i+}, \boldsymbol{\lambda}_j) = 0 \text{ for } s_{\beta_j} = 1,2,\ldots,2k_{\beta_j}+1,\\
&{}^{(n:\tilde{n})}V_j^{(2k_\beta+2)}(\mathbf{X}_{m_i+}^{(\beta_j,j)}, t_{m_i+}, \boldsymbol{\lambda}_j) > 0 \text{ for } \beta_j \in \{1,2\}.
\end{aligned}
\quad (5.219)
$$

Proof: The proof is similar to the proof of Theorem 5.4 for each $j \in \mathbb{L}$. For all $j \in \mathbb{L}$, if the conditions are satisfied, two dynamical systems in Eqs.(5.1) and (5.2) to l-constraints in Eq.(5.4) are of the (l_1, l_2, l_3)-synchnonization, desynchronization and penetration of the $(2k_{\alpha_j} : 2k_{\beta_j})$-type for time $t \in [t_{m_1}, t_{m_2}]$, vice versa. This theorem is proved. ■

5.5 Conclusions

In this chapter, a theory for synchronization of multiple dynamical systems to specific constraints was presented from a theory of discontinuous dynamical systems. The concepts on synchronization of two or more dynamical systems to specific constraints were systematically presented. Based on specific constraints, metric functionals were porposed to measure the synchronicity of the two or more dynamical systems to such specific constraints. The synchronization, desynchronization and penetration of two or more dynamical systems to specific constraints were discussed through the metric functionals, and the necessary and sufficient conditions for such synchronicity were developed. The synchronization of two dynamical systems to a single specific constraint was first discussed, and further the synchronicity of two dynamical systems to multiple constraints was investigated. The meaning of synchronization for dynamical systems with constraints is extended. This chapter provides a general frame work to control slave systems which can be synchronized with master systems through specific constraints.

References

Abarbanel H.D.I., Korzinov L., Mees A.I. and Rulkov N.F., 1997, Small Force Control of Nonlinear systems to given orbits, *IEEE Transactions of Circuits and Systems-I Fundamental Theory and Applications*, **44**, 1018–1023.

Afraimovich V., Cordonet A. and Rulkov N.F., 2002, Generalized synchronization of chaos in noninvertible maps, *Physical Review E*, **66**, 016208-1∼016208–6.

Barreto E., Josic K., Morales C., Sander E. and So P., 2003, The geometry of chaos synchronization, *Chaos*, **13**, 151–164.

Boccaletti S., 2008, *The Synchronized Dynamics of Complex Systems*, Elsevier, Amsterdam.

Boccaletti S., Kurhts J., Osipov G., Valladars D.L. and Zhou C.S., 2002, The synchronization of chaotic systems, *Physics Reports*, **366**, 1–101.

Boccaletti S., Farini A. and Arecchi F.T., 1997, Adaptive synchronization of chaos for secure communication, *Physical Review E*, **55**, 4979–4981.

Bowong S., Moukam Kakmeni F.M., Dimi J.L. and Koina R., 2006, Synchronizing chaotic dynamics with uncertainties using a predictable synchronization delay design, *Communications in Nonlinear Science and Numerical Simulation*, **11**, 973–987.

Bowong S., Moukam Kakmeni F.M. and Siewe, M., 2007, Secure communication via parameter modulation in a class of chaotic systems, *Communications in Nonlinear Science and Numerical Simulation*, **12**, 397–410.

Carroll T.L. and Pecaora L.M., 1991, Synchronized chaotic circuit, *IEEE Transactions on Circuit and Systems*, **38**, 453–456.

Chen Y., Rangarajan G. and Ding M., 2006, Stability of synchronized dynamics and pattern formation in coupled systems: Review of some recent results, *Communications in Nonlinear Science and Numerical Simulation*, **11**, 934–960.

Campos E. and Urias J., 2004, Multimodal synchronization of chaos, *Chaos*, **14**, 48–53.

Cruz-Hernández C. and Romero-Haros N., 2008, Communicating via synchronized time-delay Chua's circuits, *Communications in Nonlinear Science and Numerical Simulation*, **13**, 645–659.

Ding M., Ding E.-J., Ditto W.L., Gluckman B., In V., Peng J.-H., Spano M.L. and Yang W., 1997, Control and synchronization of chaos in high dimensional systems: Review of some recent results, *Chaos*, **7**, 644–652.

Ding M. and Ott E., 1994, Enhancing synchronization of chaotic systems, *Physical Review E*, **49**, R945–R948.

Enjieu Kadji H.G., Chabi Orou J.B. and Woafo P., 2008, Synchronization dynamics in a ring of four mutually coupled biological systems, *Communications in Nonlinear Science and Numerical Simulation*, **13**, 1361–1372.

Fallahi K., Raoufi R. and Khoshbin H., 2008, An application of Chen system for secure chaotic communication based on extended Kalman filter and multi-shift cipher algorithm, *Communications in Nonlinear Science and Numerical Simulation*, **13**, 763–781.

Feng X.-Q. and Shen K., 2005, phase synchronization and anti-phase synchronization of chaos for degenerate optical parametric oscillator, *Chinese Physics*, **14**, 1526–1532.

Ghosh D., Saha P. and Roy Chowdhury A., 2007, On synchronization of a forced delay dynamical system via the Galerkin approximation, *Communications in Nonlinear Science and Numerical Simulation*, **12**, 928–941.

Hayashi C., 1964, *Nonlinear Oscillations in Physical Systems*, McGraw-Hill, New York.

Hu M., Xu Z. and Zhang R., 2008, Full state hybrid projective synchronization of a general class of chaotic maps, *Communications in Nonlinear Science and Numerical Simulation*, **13**, 782–789.

Hung Y.-C., Ho M.-C., Lih J.-S. and Jiang I-M., 2006, Chaos synchronization of two stochastically coupled random Booleen network, *Physics Letters A*, **356**, 35–43.
Huygens (Hugenii) C., 1673, *Horologium Oscillatorium*. Apud F. Muguet, Parisiis, France. (English Translation, 1986, *The pendulum Clock*, Iowa State University, Ames.)
Jackson E.A., 1991, Controls of dynamic flows with attractors, *Physical Review E*, **44**, 4839–4853.
Kapitaniak T., 1994, Synchronization of chaos using continuous control, *Physical Review E*, **50**, 1642–1644.
Kiani-B A., Fallahi K., Pariz N. and Leung H., 2009, A chaotic secure communication scheme using fractional chaotic systems based on an extended fractional Kalman filter, *Communications in Nonlinear Science and Numerical Simulation*, **14**, 863–879.
Kocarev L. and Parlitz U., 1995, General approach for chaotic synchronization with application to communication, *Physical Review Letters*, **74**, 1642–1644.
Kocarev L. and Parlitz U., 1996, Synchronizing spatiotemporal chaos in coupled nonlinear oscillators, *Physical Review Letters*, **77**, 2206–2209.
Koronovski A.A., Hramov A.E. and Khromova I.A., 2004, Duration of the process of complete synchronization of two completed identical chaotic systems, *Technical Physics Letters*, **30**, 291–294.
Kuramoto Y., 1984, *Chemical Oscillations, Waves, and Turbulence*, Springer-Verlag, Berlin.
Lu J., 2008, Generalized (complete, lag, anticipated) synchronization of discrete-time chaotic systems, *Communications in Nonlinear Science and Numerical Simulation*, **13**, 1851–1859.
Luo A.C.J., 2005, A theory for non-smooth dynamical systems on the connectable domains, *Communications in Nonlinear Science and Numerical Simulation*, **10**, 1–55.
Luo A.C.J., 2006, *Singularity and Dynamics on Discontinuous Vector Fields*, Elsevier, Amsterdam.
Luo A.C.J., 2008, *Global Transversality, Resonance and Chaotic Dynamics*, World Scientific, New Jersey.
Luo A.C.J., 2009, A theory for synchronization of dynamical systems, *Communications in Nonlinear Science and Numerical Simulation*, **14**, 1901–1951.
Mosekilde E., Maistrenko Y. and Postnov D., 2001, *Chaotic Synchronization: Applications to Living Systems*, World Scientific, New Jersey.
Newell T.C., Alsing P.S., Gavrielides A. and Kovanis V., 1994, Synchronization of chaotic diode resonators by occasional proportional feedback, *Physical Review Letters*, **72**, 1647–1650.
Osipov G.V., Kurths J. and Zhou C.S., 2007, *Synchronization in Oscillatory Networks*, Springer, Berlin.
Pareek N.K., Patidar V. and Sud K.K., 2005, Cryptography using multiple one-dimensional chaotic maps, *Communications in Nonlinear Science and Numerical Simulation*, **10**, 715–723.

Pecora L.M. and Carroll T.L., 1990, Synchronization in chaotic systems, *Physical Review Letters*, **64**, 821–824.

Pecora L.M., Carrol T.L., Jonson G. and Mar D., 1997, Volume-preserving and volume-expansion synchronized chaotic systems, *Physical Review E*, **56**, 5090–5100.

Pyragas K., 1998, Properties of generalized synchronization of chaos, *Nonlinear Analysis: Modelling and Control*, Vilnius, IMI, **3**, 1–28.

Peng J.H., Ding E.J., Ding M. and Yang W., 1996, Synchronizing hyperchaos with a scalar transmitted signal, *Physical Review Letters*, **76**, 904–907.

Pikovsky A., Rosenblum M. and Kurths J., 2001, *Synchronization: A Universal Concept in Nonlinear Science*, Cambridge University Press, Cambridge.

Peng Y., Wang J. and Jian Z., 2009, Synchrony of two uncoupled neurons under half wave sine current stimulation, *Communications in Nonlinear Science and Numerical Simulation*, **14**, 1570–1575.

Pyragas K., 1992, Continuous control of chaos by self-controlling feedback, *Physical Letters A*, **170**, 421–428.

Pyragas K., 1996, Weak and strong synchronization of chaos, *Physical Review E*, **54**, R4508–R4511.

Rayleigh J., 1945, *The Theory of Sound*, Dover Publishers, New York.

Rulkov N.F., 2001, Regularization of synchronized chaotic bursts, *Physical Review Letters*, **86**, 183–186.

Rulkov N.F., Sushchik M.M., Tsimring L.S. and Abarbanel H.D., 1995, Generalized synchronization of chaos in directionally coupled chaotic systems, *Physical Review E*, **50**, 1642–1644.

Soto-Crespo J.M. and Akhmediev N., 2005, Soliton as strange attractor: nonlinear synchronization and chaos, *Physical Review Letters*, **95**, 024101–1∼024101–4.

Stocker J.J., 1950, *Nonlinear Vibrations*, Interscience Publishers, New York.

Stojanovski T., Kocarev and Harris R., 1979, Application of symbolic dynamics in chaos synchronization, *IEEE Transactions of Circuits and Systems-I Fundamental Theory and Applications*, **44**, 1014–1018.

Teufel A., Steindl A. and Troger H., 2006, Synchronization of two flow-excited pendula, *Communications in Nonlinear Science and Numerical Simulation*, **11**, 577–594.

van der Pol B., 1927, Forced oscillations in a circuit with resistance. *Philosophical Magazine*, **3**, 64–80.

Wang D., Zhong Y. and Chen S., 2008, Lag synchronizing chaotic system based on a single controller, *Communications in Nonlinear Science and Numerical Simulation*, **13**, 637–644.

Wang H., Lu Q. and Wang Q., 2008, Bursting and synchronization transition in the coupled modified ML neurons, *Communications in Nonlinear Science and Numerical Simulation*, **13**, 1668–1675.

Wang X.-Y., Yu Q., 2009, A block encryption algorithm based on dynamic sequences of multiple chaotic systems, *Communications in Nonlinear Science and Numerical Simulation*, **14**, 574–581.

Xiang T., Wong K. and Liao X., 2008, An improved chaotic cryptosystem with external key, *Communications in Nonlinear Science and Numerical Simulation*, **13**, 1879–1887.

Yamapi R. and Woafo P., 2006, Synchronized states in a ring of four mutually coupled self-sustained electromechanical devices, *Communications in Nonlinear Science and Numerical Simulation*, **11**, 186–202.

Yang T. and Chua L.O., 1999, Generalized synchronization of chaos via linear transformations, *International Journal of Bifurcation and Chaos*, **9**, 215–219.

Zhan M., Wang X. and Gong X., Wei G.W. and Lai C.-H., 2003, Complete synchronization and generalized synchronization of one way coupled time-delay systems, *Physical Review E*, **68**, 036208–1∼036208–5.

Zaks M.A., Park E.-H., Rosenblum M.G. and Kurths J., 1999, Alternating locking ratio in imperfect phase synchronization, *Physical Review Letters*, **82**, 4228–4231.

Chapter 6
The Complexity in Activity of Biological Neurons

Yong Xie, Jian-Xue Xu

Abstract We sum up our work about neurodynamics in this chapter. It is widely considered that the nervous system in man and animals is a rather complicated nonlinear dynamical system. Therefore, it is both necessary and important to understand the behavior occurred in the nervous system from the perspective of nonlinear dynamics. Actually, a great many of novel and puzzling phenomena are just observed in a single neuron, but their physiological or dynamical mechanisms remain open so far. In other words, single neurons are not simple. We show many firing patterns in theoretical neuronal models or neurophysiological experiments of single neurons in rats in this chapter. And then we introduce three representative examples of best known mathematical neuron models. The two types of neuronal excitability are illustrated by the Hodgkin-Huxley mode and the Morris-Lecar model. Especially, it is shown that we can change the types of neuronal excitability using the methods of bifurcation control. Besides, we display bursting and its topological classification, and explain bifurcation, chaos and crisis by the existing neuronal models. We give emphasis to sensitive responsiveness of aperiodic firing neurons to external stimuli, and show experimental phenomena and their underlying nonlinear mechanisms. The synchronization between neurons is remarked simply. We stress a constructive role of noise in the nervous system, and depict the phenomena of stochastic resonance and coherence resonance, and give their dynamical mechanisms. The common analysis methods are presented for the time series of the interspike intervals. Finally, we give two application examples about controlling chaos and stochastic resonance, and draw some conclusions.

MOE Key Laboratory of Strength and Vibration, School of Aerospace, Xi'an Jiaotong University, Xi'an 710049, China.
Emails: yxie@mail.xjtu.edv.cn, jxxu@mail.xjtu.edv.cn

6.1 Complicated firing patterns in biological neurons

6.1.1 Time series of membrane potential

We can obtain the time course of membrane potential (or transmembrane potential) easily in neurophysiological experiments. Membrane potential is the voltage difference (or electrical potential difference) between the interior and exterior of a cell. The cell membrane acts as a barrier which prevents the inside solution (intracellular fluid) from mixing with the outside solution (extracellular fluid). Actually such a membrane surrounds the cell to provide a stable environment for biochemical processes running in the interior of the cell. The membrane potential arises from the action of ion channels, ion pumps, and ion transporters embedded in the cell membrane. Neurons are specialized to use changes in membrane potential for rapid communication with other neurons, muscles and secretory cells. There are two kinds of the membrane potential: the relatively static one and the specific dynamic one. The former is the resting membrane potential, which means the membrane potential of a neuron at rest; the latter include the graded membrane potential and the action potential. An action potential (also known as a nerve impulse or a spike) is a self-regenerating wave of electrochemical activity that allows neurons to carry a signal over a distance. In general, the action potential can be superior to graded membrane potential if the action potential sharpens the temporal structure of neuronal responses by amplifying fast transients of the membrane potential. On the other hand, the graded membrane potential is superior for discrimination between stimuli on a fine time scale (Kretzberg et al., 2001).

Usually neurons depolarizes from the resting potential and produces the action potential, it travels down the axon to the synapses. Surprisingly, many neurons can fire the action potential spontaneously. On reaching a (chemical) synapse, a neurotransmitter is released causing a localized change in potential in the postsynaptic membrane of the target neuron by opening ion channels in its membrane. To our knowledge, the states of individual ion channels are either closed or open, and they are relatively random. The cooperation of a large number of ion channels embedded in the membrane, however, can generate action potentials. Maybe this behavior is a kind of emergent phenomenon in complex adaptive systems.

6.1.2 Firing patterns: spiking and bursting

Many neurons can exhibit repetitive (tonic) spiking and/or bursting (Izhikevich, 2007). Repetitive spiking means a neuron is typically constantly active. Bursting, however, is a dynamic state where a neuron repeatedly fires discrete groups or bursts of spikes. Each such burst is followed by a period of quiescence before the next burst occurs. A burst of two spikes is called a doublet, of three spikes is called a triplet, four - quadruplet, etc. (for detail, see http://www.scholarpedia.org/article/Bursting).

There is an evident difference in the membrane potential for spiking and bursting (Chay et al., 1995). Namely, for spiking there is no clear underlying slow wave while for bursting several spikes ride on the slow wave.

Spiking neurons are classified into two types (Izhikevich, 2007; Rinzel and Ermentrou, 1989; Izhikevich, 2000; Xie et al., 2008a; Tsuji et al., 2007), namely type-I excitability and type-II excitability according to the frequency response characteristics of a neuron to a constant current stimulus. A neuron with type-I excitability is characterized by a continuous FI (the firing frequency versus the applied current) curve that shows oscillations starting with an arbitrarily low frequency. The firing frequency varies continuously from almost zero to a certain value with a wide dynamic range as the applied current changes. In contrast, a neuron with type-II excitability is characterized by a discontinuous FI curve with the oscillations starting with a nonzero frequency, and the response frequency range is narrow. For a neuron with type-I excitability there is an apparent threshold for the appearance of spikes, while there is no true threshold for a neuron with type-II excitability, rigorously speaking (Izhikevich, 2007). Therefore there is a great difference in firing behavior between them. Clearly, the two types of excitability have different neurocomputational properties (Izhikevich, 2007). Type I excitable neurons can smoothly encode the strength of input, e.g., the strength of applied dc-current or the strength of incoming synaptic bombardment, into the frequency of their spiking output. Type II neurons cannot do that. Instead, they can act as threshold elements reporting when the strength of input is above certain value. Both properties are important in neural computations.

Almost every neuron can burst if stimulated or manipulated pharmacologically (for detail, see http://www.scholarpedia.org/article/Bursting). Many burst autonomously due to the interplay of fast ionic currents responsible for spiking activity and slower currents that modulate the activity. Neuronal bursting can play important roles in communication between neurons. In particular, bursting neurons are important for motor pattern generation and synchronization. Rinzel (1987) first proposed a scheme according to the dynamical mechanisms of bursting onset and termination to classify bursting neurons, and identified three types, i.e., square wave bursting, parabolic bursting and elliptic bursting. It was later extended by Bertram et al., who included another type. Izhikevich (2000, 2007) provided the complete classification, identifying all 16 topological types, and foretelling 120 all possible types.

Neuronal firing behavior can be periodic or aperiodic. The periodic firing includes periodic spiking and periodic bursting; while the aperiodic firing consists of irregular spiking and irregular bursting. Now more and more evidence shows that a rather large part of such aperiodic firing is deterministic chaos, namely, chaotic spiking or chaotic bursting (Aihara and Matsumoto, 1986; Mpitsos et al., 1988; Hoffman et al., 1995; Gong et al., 1998, 2002; Xu et al., 1997; Longtin, 1993a; Ren et al., 1997, 2001). Figure 6.1 shows four kinds of firing pattern, and they are obtained by numerical simulation of the Hindmarsh-Rose neuron model (Hindmarsh and Rose, 1984).

In neurophysiological experiments, the series of interspike intervals (ISIs) is usually recorded to display the firing pattern of a neuron. Interestingly, it is widely con-

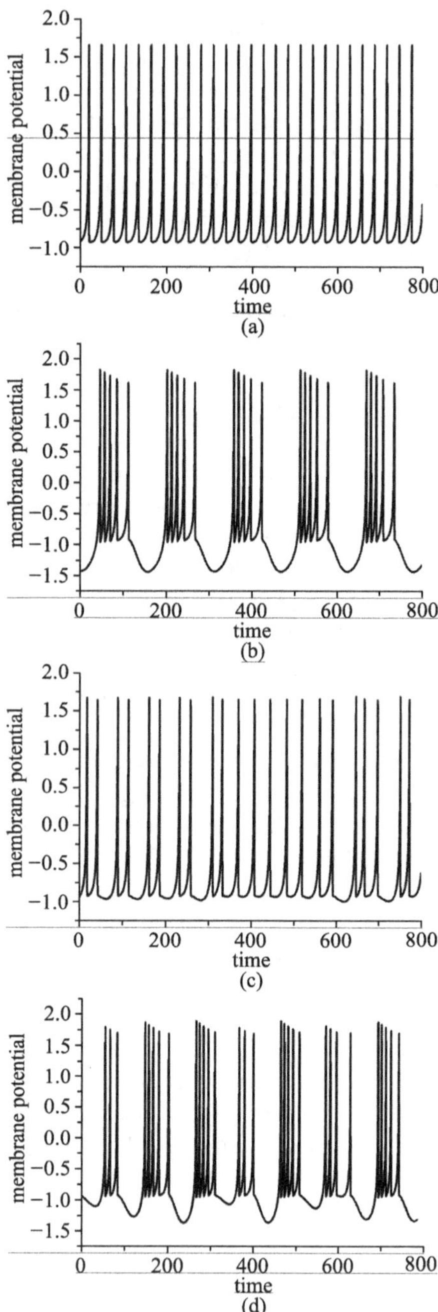

Fig. 6.1 Firing pattern of the Hindmarsh-Rose neuron model with different applied currents. (a) periodic spiking; (b) periodic bursting; (c) chaotic spiking; and (d) chaotic bursting.

sidered that neural information transmitted from presynaptic to postsynaptic neurons is embedded in the series of ISIs (Gong et al., 2002; So et al., 1998; Suzuki et al., 2000). Integer multiple spiking (IMS) (Gu et al., 2001; Yang et al., 2002; Xie et al., 2004a) is seemingly random firing behavior whose interspike interval histogram (ISIH) exhibits multimodal structure with peaks at integer multiples of a basic ISI. Furthermore the amplitude of the peaks decays with increasing ISI except for the first few peaks, and the return map of ISI series has a crystal lattice structure, as seen in Fig. 6.2. The IMS herein is numerically simulated from the Morris-Lecar model. Similarly, integer multiple bursting (IMB) denotes that the inter-burst intervals (IBIs) of the IMB exhibit multi-mode and are approximately integer multiples of a basic IBI. Figure 6.3 shows the firing pattern of the IMB, which is observed in the experiment on an experimental neural pacemaker (Gu et al., 2003). In addition, bursting is a considerable complicated firing pattern, and has various types. Figure 6.4, For example, demonstrates a periodic parabolic bursting in a neural pacemaker after the addtion of 5 μMol/L veratridine (Xie et al., 2003a). The upper trace is the

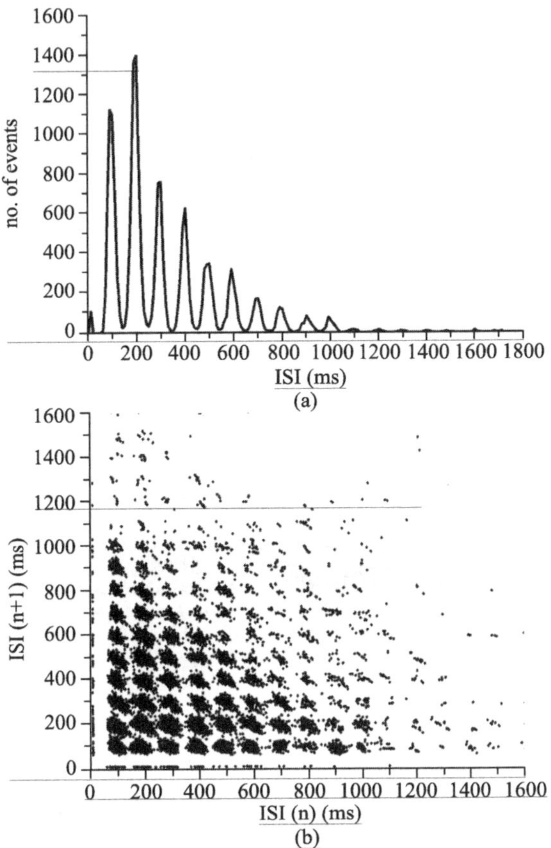

Fig. 6.2 Integer multiple spiking in Ref. [24]. (a) the ISI histogram; and (b) the return map of the ISI series.

time course of membrane potential, and mainly exhibits the active phase of a burst; while the ISI as a function of time is shown in the lower trace. It can be seen clearly that parabolic bursting is characterized by a spike frequency which is low at the beginning, high in the middle, and low again near the end of the active phase. ISI series looks like a family of parabolic curves, as seen in the lower trace.

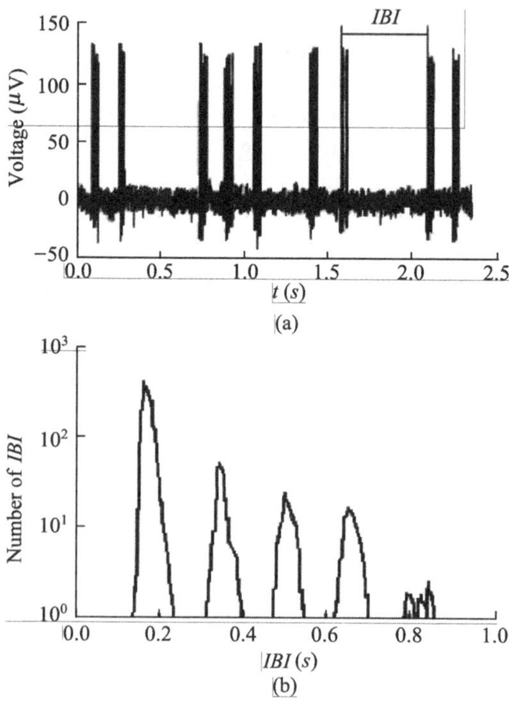

Fig. 6.3 The integer multiple bursting in the experiment in Ref. [25]. (a) The firing trains; and (b) IBI histogram.

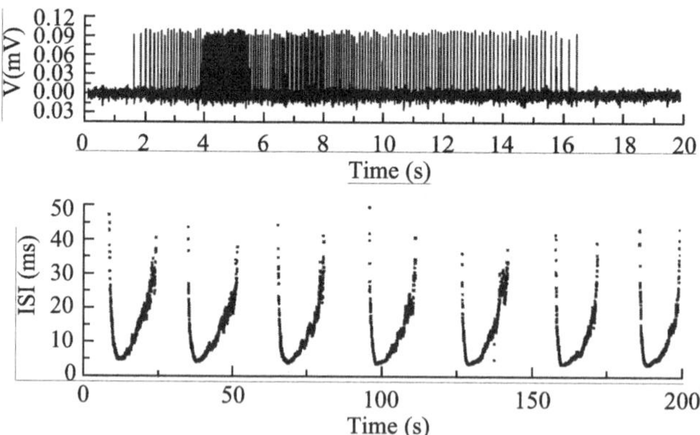

Fig. 6.4 Parabolic bursting induced by Veratridine in rat injured sciatic nerves.

In fact, there are different transition styles between firing patterns (Ren et al., 1997; Mandelblat et al., 2001; Xie et al., 2003b; Li et al., 2003): mainly including period adding and period-doubling cascades, which have been observed in neural pacemakers. Figure 6.5 shows period-adding and period-doubling cascades recorded in the different neural pacemakers, respectively. Note that ISI versus time or ISI versus the number of ISI is not really a bifurcation diagram. However, the action of the drugs is slowly permeable, and tunes up slowly the control parameters of the neural pacemakers, therefore, time or the number of ISI reflects the changes in the control parameters, and thus ISI versus time or ISI versus the number of ISI can be considered to be a bifurcation diagram, roughly speaking.

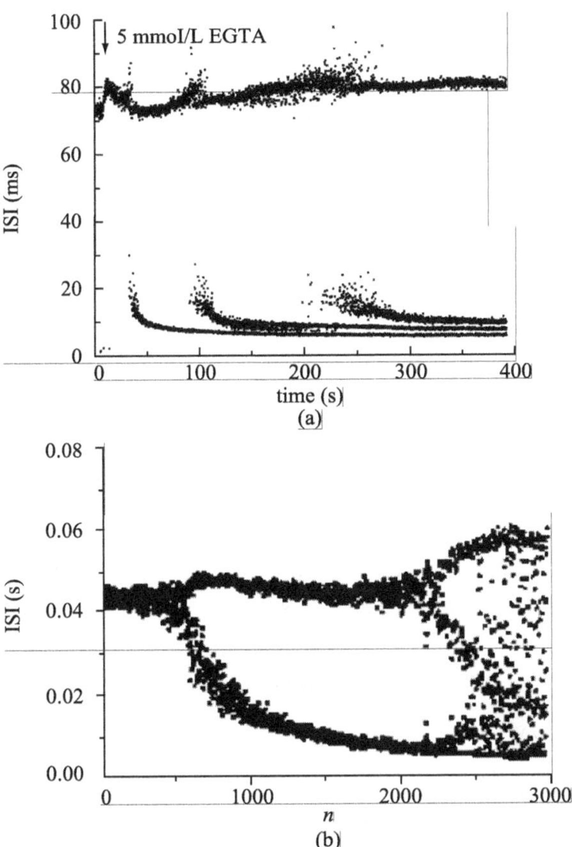

Fig. 6.5 Period-adding and period-doubling cascades. (a) Period-adding cascade in (Xie et al., 2003b). The neural pacemaker exhibited period-one bursting, period-two bursting, period-three bursting and period-four bursting after the addition of 5mMol EGTA, respectively, as time elapsed. (b) Period-doubling cascade route to chaos. The firing cascade of the neural pacemaker is induced by the decrease of $[Ca^{2+}]_o$, see (Li et al., 2003).

6.2 Mathematical models

6.2.1 HH model

The Hodgkin-Huxley (HH) model is one of the most famous neuron models in computational neuroscience. It is the first quantitative model of the electrical excitability of neurons based on a number of experiments in giant squid axons (Hodgkin and Huxley, 1952). This model consists of three ionic currents: fast inward sodium current, time-dependent outward potassium current and time-independent leak current. And the circuit diagram is shown in Fig. 6.6.

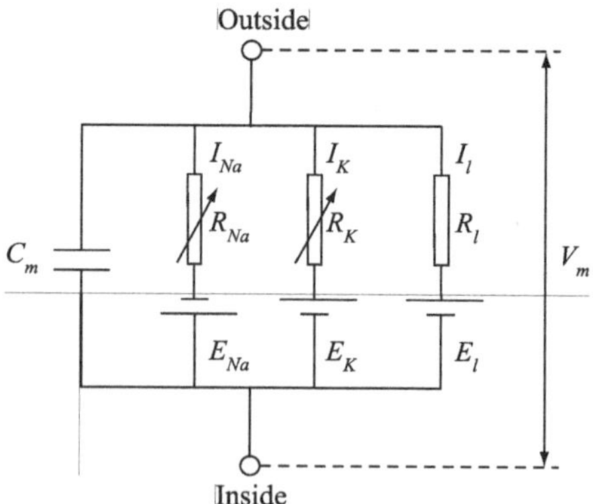

Fig. 6.6 The circuit diagram corresponding to the HH model.

As we know, the HH model describes successfully how the action potentials in neurons are initiated and propagated; and it is expressed by a set of nonlinear ordinary differential equations as follows:

$$\begin{aligned}
\frac{dV}{dt} &= \frac{1}{C_M}(I_{ext} - g_{Na}m^3h(V - V_{Na}) - g_K n^4(V - V_K) - g_L(V - V_L)),\\
\frac{dm}{dt} &= \alpha_m(V)(1-m) - \beta_m(V)m,\\
\frac{dh}{dt} &= \alpha_h(V)(1-h) - \beta_h(V)h,\\
\frac{dn}{dt} &= \alpha_n(V)(1-n) - \beta_n(V)n,
\end{aligned} \tag{6.1}$$

V represents the membrane potential, which is the electrical potential difference (voltage) across the neuronal membrane. m and h are gating variables that represent the activation and inactivation of the sodium current, since h decreases when m increases. n denotes the activation gating variable of the potassium current. Obviously, m, h, and n obey equations of the same form, but with different voltage dependences for their steady state values and time constants. α_m,β_m, α_h, β_h, α_n, and

β_n are functions of V that are defined as follows:

$$\alpha_m(V) = 0.1(25.0 - V)/\exp((25.0 - V)/10.0 - 1.0),$$
$$\beta_m(V) = 4.0\exp(-V/18.0),$$
$$\alpha_h(V) = 0.07\exp(-V/20.0),$$
$$\beta_h(V) = 1.0/(\exp((-V + 30.0)/10.0) + 1.0),$$
$$\alpha_n(V) = 0.01(10.0 - V)/(\exp((10.0 - V)/10.0) - 1.0),$$
$$\beta_n(n) = 0.125\exp(-V/80.0).$$

The HH model includes the following parameters: $V_{Na} = 115.0$ mV, $V_K = -12.0$ mV, and $V_L = 10.599$ mV, representing the equilibrium potentials of the sodium, potassium, and leak currents, respectively. They are determined uniquely by the Nernst equation. Thus, these parameter values are controllable by changing the ionic concentrations within and outside the membrane. $g_{Na} = 120.0$ mS/cm^2, $g_K = 36.0$ mS/cm^2, and $g_L = 0.3$ mS/cm^2 represent the maximum conductance of the corresponding ionic currents. They reflect the ionic-channel density distributed over the membrane. $C_M = 1.0$ μF/cm^2 is the membrane capacitance. I_{ext} represents the externally applied current, and usually serves as a bifurcation parameter of the system. If I_{ext} exceeds a certain threshold value, the HH model neuron can exhibit periodic spiking.

6.2.2 FitzHugh-Nagumo model

The FitzHugh-Nagumo (FHN) model (FitzHugh, 1961) is a simplification of the HH model. As seen above, the HH has four variables. The variables kept in the FHN model are only the excitable variable and the recovery variable which are characterized as being the fast and slow variables respectively. Actually, the FHN model adiabatically eliminates the h and m gates, and retains only the membrane potential V and a slow variable w similar to n. Because of its simple two-variable form and generality, the FHN has been used widely. It is able to reproduce many qualitative electrical characteristics of a neuron, such as the existence of firing threshold, relative and absolute refractory periods, and the generation of action potentials under the action of applied currents. The FHN model is described by the following equations (FitzHugh, 1961):

$$\begin{aligned} \frac{dV}{dt} &= V - \frac{V^3}{3} - w + I_{ext}, \\ \frac{dw}{dt} &= \phi(V + a - bw). \end{aligned} \tag{6.2}$$

Here, parameter are often taken as $a = 0.7$, $b = 0.8$, and $\phi = 0.08$. I_{ext} denotes the externally applied current and is usually considered to be the control parameter. Clearly, the membrane potential variable V has cubic nonlinearity that allows regenerative self-excitation via a positive feedback; and the recovery variable w has a linear dynamics that provides a slower negative feedback. Similarly, if the applied current I_{ext} exceeds a certain threshold value, the FHN can also display a characteristic excursion in the phase plane, before the variables V and w relax back to their resting values. In fact, the dynamical behavior of this model can be nicely described by zapping between the left and right branch of the cubic nullcline in the phase plane.

6.2.3 Hindmarsh-Rose model

In 1982 and 1984, Hindmarsh and Rose (1984) constructed the Hindmarsh-Rose (HR) neuron model to model the synchronization of firing of two snail neurons in a relatively simple way that did not use the full HH equations. Naturally, at the time they modified the FHN model to account for tail current reversal. This point is crucial for the development of the HR model. Now the HR can be used to study the spiking-bursting behavior of the membrane potential observed in experiments made with a single neuron. The HR model has the mathematical form of a system of three nonlinear ordinary differential equations in the following (Hindmarsh and Rose, 1984):

$$
\begin{aligned}
\frac{dx}{dt} &= y - ax^3 + bx^2 - z + I_{ext}, \\
\frac{dy}{dt} &= c - dx^2 - y, \\
\frac{dz}{dt} &= r\left[s(x - x_0) - z\right].
\end{aligned}
\tag{6.3}
$$

Where x denotes the membrane potential, which is written in dimensionless units. The other two variables, y and z, which take into account the transport of ions across the membrane through the ion channels. The transport of sodium and potassium ions is made through fast ion channels and its rate is measured by y, which is called the spiking variable. The transport of other ions is made thorough slow channels, and is taken into account thorough z, which is called the bursting variable. In fact, z is a slow variable and usually regarded as the bifurcation parameter of the fast subsystem $x-y$ when the method of the slow-fast subsystem decomposition is adopted to analyze the dynamics of the HR model.

The HR model has eight parameters: a, b, c, d, r, s, x_0 and I_{ext}. Commonly, we fix some of them and let the other to be control parameters. Frequently, the applied current I_{ext} is considered to be the control parameter. Other parameters take $a = 1.0$, $b = 3.0$, $c = 1.0$, $d = 5.0$, $s = 4.0$ and $x_0 = -1.6$. The parameter r is something of the order of 10e-3, and I_{ext} ranges between -10 and 10.

6.3 Nonlinear mechanisms of firing patterns

6.3.1 Dynamical mechanisms underlying Type I excitability and Type II excitability

The definition of neuronal excitability is first introduced here. In existing textbooks it is that a "subthreshold" synaptic input evokes a small graded postsynaptic potential (PSP), while a "superthreshold" input evokes a large all-or-none action potential. Unfortunately, this definition cannot be adopted to define excitability of dynamical systems because many systems, including neuronal models have neither all-or-none action potentials nor firing thresholds (Kretzberg et al., 2001; Izhikevich, 2000). A purely geometrical definition, therefore, is more reasonable to describe neuronal excitability (Kretzberg et al., 2001). From the geometrical point of view, a dynamical system having a stable equilibrium is excitable if there is a large-amplitude piece of trajectory that starts in a small neighborhood of the equilibrium, leaves the neighborhood, and then returns to the equilibrium.

Despite a large number of biophysical mechanisms, there are only four co-dimension-1 bifurcations of equilibrium underlying neuronal excitability because they have the lowest co-dimension and hence they are the most likely to be seen experimentally, namely, saddle-node on invariant circle (SNIC) bifurcation, saddle-node (off invariant circle) bifurcation, supercritical Hopf bifurcation and subcritical Hopf bifurcation (Izhikevich, 2007). In general, the former one underlies type-I excitability, while the latter three mediate type-II excitability. When the resting state of a neuron is near a SNIC bifurcation, the neuron can fire all-or-none spikes with an arbitrary low frequency, it has a well-defined threshold manifold, and it acts as an integrator (Izhikevich, 2001); i.e. the higher the frequency of incoming pulses, the sooner it fires. In contrast, when the resting state is near a Hopf bifurcation, the neuron fires in a certain frequency range, its spikes are not all-or-none, it does not have a well-defined threshold manifold, it can fire in response to an inhibitory pulse, and it acts as a resonator (Izhikevich, 2001); i.e. it responds preferentially to a certain (resonant) frequency of the input. Increasing the input frequency may actually delay or terminate its firing.

In the context of neurons, the stable equilibrium corresponds to the resting state of a neuron. All trajectories starting in a sufficiently small region of the stable equilibrium converge back to the equilibrium. Such trajectories correspond to subthreshold PSPs. In contrast, the large-excursion trajectory corresponds to firing a spike. Therefore, superthreshold PSPs are those that push the state of the neuron to or near the beginning of the large trajectory thereby initiating the spike. Furthermore, the limit cycle in the phase space corresponds to periodic spiking or periodic bursting or periodic subthreshold oscillation of a neuron.

6.3.2 *Dynamical mechanism for the onset of firing in the HH model*

For the HH model, the bifurcation diagram of the membrane potential V versus the applied current I_{ext} is shown in Fig. 6.7. From Fig. 6.7(a), we can see that the neuron undergoes a Hopf bifurcation (HB) from quiescence to periodic spiking at $I_{ext} = 9.780\ \ \mu\text{A/cm}^2$. Moreover, the amplitude of the periodic oscillation decreases with an increase in the externally applied current, and the periodic oscillation terminates at $I_{ext} = 154.527\ \ \mu\text{A/cm}^2$, where another Hopf bifurcation occurs. Obviously, the left Hopf bifurcation is subcritical from Fig. 6.7(b). The bifurcation diagrams in Fig. 6.7 were produced using the software package XPPAUT (Ermentrout, 2002). The left Hopf bifurcation is considerably important because it is the dynamical mechanism of neuronal excitability from quiescence to periodic spiking

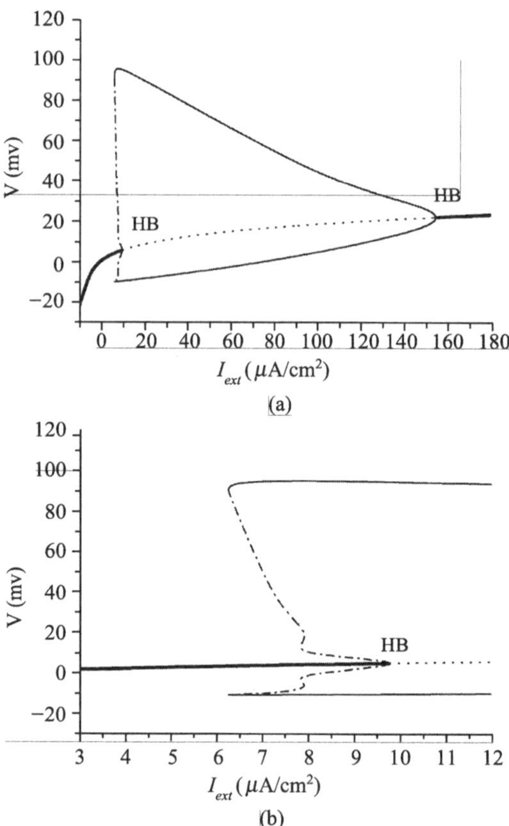

Fig. 6.7 Bifurcation diagram of the HH model (Figure 1 in (Xie et al., 2008b)). The thick solid lines represent stable steady states, while the dotted line represents unstable equilibrium points. The thin lines represent the maximum and minimum values of stable limit cycles, and the dash-dotted lines represent the maximum and minimum values of unstable limit cycles. (b) is the magnification of (a) near the left Hopf bifurcation point.

(Xie et al., 2008b). As to the right Hopf bifurcation, however, the intensity of the external applied current generally exceeds the normal physiological range. The HH model neuron, therefore, is of Type II excitability.

6.3.3 Type I excitability and Type II excitability displayed in the Morris-Lecar model

The Morris-Lecar (ML) model (Morris and Lecar, 1981) is a variation of the Hodgkin-Huxley model. Originally it was postulated in order to describe the various oscillatory response patterns of the Barnacle muscle fiber. The differential equation and V-dependent functions are

$$C\frac{dV}{dt} = -\bar{g}_{Ca} m_\infty(V)(V - V_{Ca}) - \bar{g}_K w(V - V_K) - \bar{g}_L(V - V_L) + I_0,$$
$$\frac{dw}{dt} = \varphi\frac{[w_\infty(V) - w]}{\tau_w(V)}. \tag{6.4}$$

With steady states for the Ca^{2+} and K^+ current fractions:

$$m_\infty(V) = 0.5 * \{1 + \tanh[(V - V_1)/V_2]\},$$
$$w_\infty(V) = 0.5 * \{\tanh[(V - V_3)/V_4]\}.$$

And a transition rate,

$$\tau_w(V) = 1/\cosh[(V - V_3)/(2 * V_4)].$$

Here the parameter C denotes the capacity; $\bar{g}_{Ca}$, $\bar{g}_K$ and $\bar{g}_L$ are the maximal conductance of calcium, potassium and leak, respectively, and V_{Ca}, V_K and V_L are the corresponding reversal potentials. I_0 denotes the total synaptic inputs from the environment vary slowly with time, and ϕ represents the change between slow and fast regions of the neuron. All conductances are in mS/cm^2 and voltages in mV; the capacity C is μF/cm^2 and currents in μA/cm^2. Here, I_0 is generally the only free parameter.

One can see that the equation is a two-dimensional description of neuronal spike dynamics. The first equation describes the evolution of the transmembrane potential V, the second equation the evolution of a slow recovery variable w, which represents the open probability for the potassium channel.

We consider the bifurcation diagram of V versus I_0, and focus our attention on the dynamical mechanisms of excitability. Interestingly, the ML model neuron can undergo a Hopf or SNIC bifurcation depending on two different sets of values of parameters (Rinzel and Ermentrout 1989, Morris and Lecar, 1981). In these two parameter sets, except for V_3, V_4,$\bar{g}_{Ca}$ and ϕ, the rest of parameters is the same, namely, $V_1 = -1.2$, $V_2 = 18$, $\bar{g}_K = 8$, $\bar{g}_L = 2$, $V_K = -84$, $V_L = -60$, $V_{Ca} = 120$ and $C = 20$.

When $V_3 = 2$, $V_4 = 30$, $\bar{g}_{Ca} = 4.4$ and $\phi = 0.04$, the ML model neuron undergoes a subcritical Hopf bifurcation from the resting state to periodic spiking. The bifurcation diagram of V versus I_0 is shown in Fig. 6.8. Evidently, at $I_0 = 93.86$ the subcritical Hopf bifurcation destabilizes the left rest state to fire. A saddle-node bifurcation of limit cycles occurs at $I_0 = 88.29$, that is, a stable limit cycle merges with an unstable limit cycle denoted by the dot line. In particular, at the left and right handsides of Fig. 6.8 the dynamic mechanisms are similar.

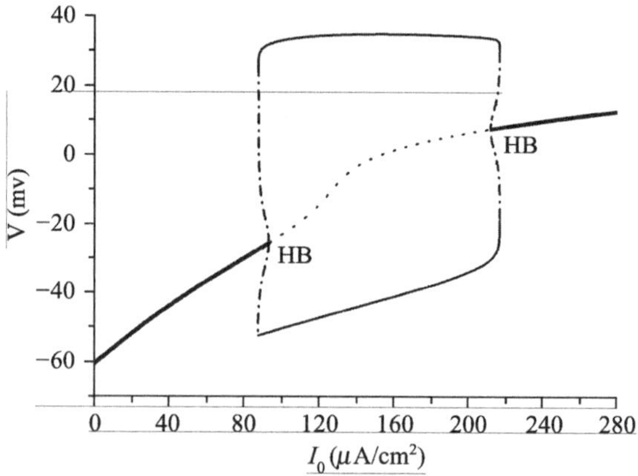

Fig. 6.8 The bifurcation diagram of the membrane potential V versus I_0 (Figure 1 in (Xie et al., 2005)). The bold solid lines represent stable focus points, and the thin lines correspond to the maximum and minimum values for membrane potential of periodic spiking, respectively. The dash-dot line between the two HB bifurcation points is composed of unstable focus points.

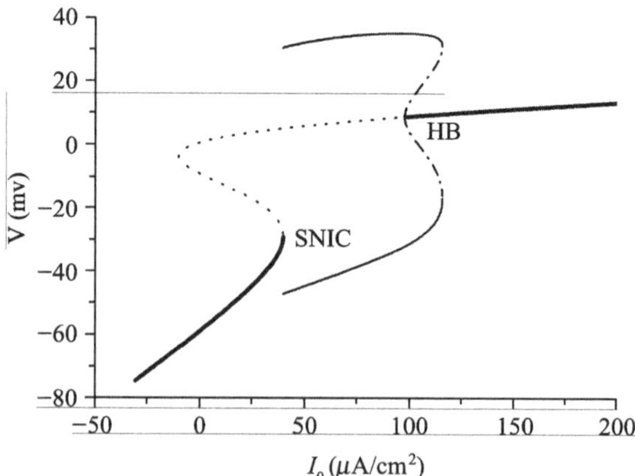

Fig. 6.9 The bifurcation diagram of the membrane potential versus I_0 (Figure 2 in (Xie et al., 2005)). The left bold line consists of stable nodes, while the right bold line stable focus points. The meaning of thin lines is the same as the caption of Fig. 6.8.

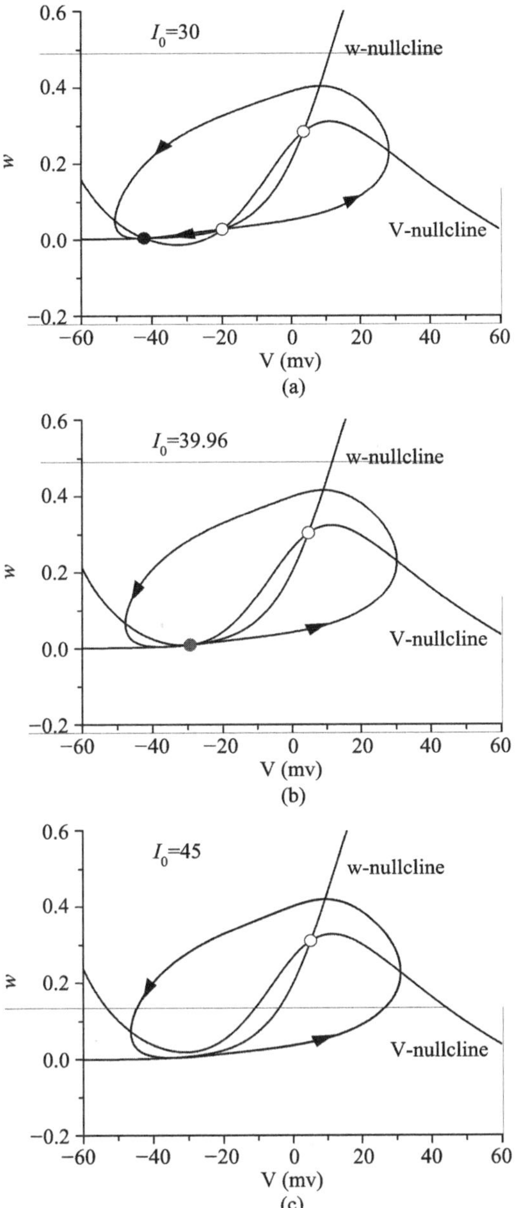

Fig. 6.10 The nullclines in the phase plane of $V - w$ near the SNIC bifurcation (Figure 3 in (Xie et al., 2005)). The open circles are the unstable fixed points, and the solid circle in (a) is stable node. The gray circle in (b) is a saddle-node. (a)The unstable invariant manifold is also drawn before the bifurcation at $I_0 = 30$. (b) At $I_0 = 39.96$ the saddle merges with the node, and the SNIC occurs. (c) A limit cycle corresponding to the periodic spiking is shown when $I_0 = 45$.

When $V_3 = 12$, $V_4 = 17.4$, $\bar{g}_{Ca} = 4$ and $\phi = 1/15$, the model neuron transits a SNIC bifurcation at $I_0 = 39.96$, and the state of the neuron varies from the resting state to periodic spiking. Figure 6.9 shows the bifurcation diagram of V versus I_0. Clearly, a subcritical Hopf bifurcation occurs at $I_0 = 97.79$, and a saddle-node bifurcation of limit cycles at $I_0 = 116.1$. Near the SNIC bifurcation, the change in relative positions of the V-nullcline and w-nullcline are shown in Figs. 6.10(a)-(c). As is known, the intersections of these two nullclines are fixed points. We draw the unstable invariant manifold of the saddle, as seen in Fig. 6.10(a). As to the calculation of the unstable invariant manifold we give a simple description. To begin with, the eigenvector for the positive eigenvalue of the saddle is computed, and then the equations are integrated forward in time with initial conditions that are on the eigenvector and slightly off of the singular point.

Thus we can see that the ML model neuron can undergo Hopf or SNIC bifurcation depending on the values of parameters.

6.3.4 Change in types of neuronal excitability via bifurcation control

As mentioned above, changes in types of neuronal excitability actually imply changes in dynamical mechanisms underlying neuronal excitability, that is, variation in types of bifurcation. Specifically, we convert Type I excitability into Type II excitability by a washout filter-aided dynamic feedback controller (Xie et al., 2008a). In other words, such a controller is adopted to create a Hopf bifurcation before the occurrence of a SNIC bifurcation. It is known that static state feedback does not apply to problems where the dynamics and the targeted operating point are uncertain (Hassouneh et al., 2004). Moreover, static state feedback changes the operating conditions of the open-loop system. This may result in waste of control energy and also induce degradation of system performance. Fortunately, washout filters can overcome these difficulties. In fact, a washout filter is a high pass filter that washes out steady state inputs, while passing transient inputs (Abed et al., 1994). The use of washout filters ensures that all the equilibrium points of an open-loop system are preserved in the closed-loop system; namely, their locations are not changed. In addition, washout filters facilitate automatic following of targeted operating points, which results in vanishing control energy once stabilization is achieved and a steady state is reached.

Here, a two-dimensional Hindmarsh-Rose (HR) type model (Tsuji et al., 2007) is used as a model neuron because it not only can exhibit Type I excitability under appropriate values of parameters but also possesses a set of simple expression formulas. It is described by the following equations:

$$\frac{dx}{dt} = c\left(x - \frac{x^3}{3} - y + z\right),$$
$$\frac{dy}{dt} = \frac{x^2 + dx - by + a}{c}, \tag{6.5}$$

where x and y denote the cell membrane potential and a recovery variable, respectively. a,b,c,d and z are parameters. In particular, z represents the external stimulus. Bifurcation behavior of this model has been explored in detail.

Under a set of parameter values, namely, $a = 0.42$, $b = 1.0$, $c = 3.0$, and $d = 1.8$, the neuron exhibits Type I excitability as the external stimulus z changes, as shown in Fig. 6.11. There is a SNIC bifurcation at $z = 0.3463$, where the neuron model generates the SNIC bifurcation from quiescence to firing. A subcritical Hopf bifurcation occurs at $z = 2.3420$.

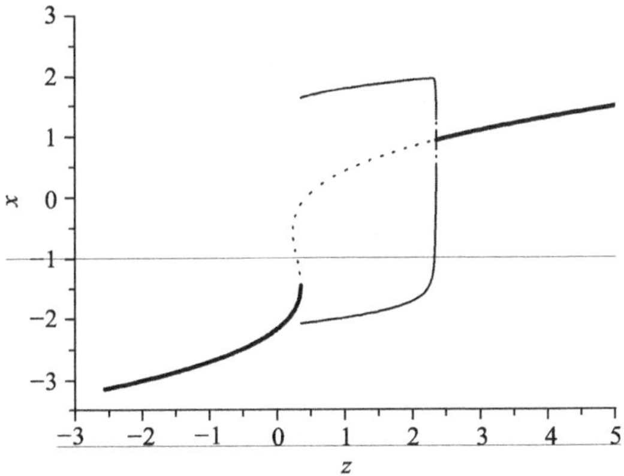

Fig. 6.11 The bifurcation diagram of the HR type model with Type I excitability (Figure 1(a) in (Xie et al., 2008a)). The thick solid lines denote stable steady states, while the dotted line shows unstable equilibrium points. The thin lines represent the maximum and minimum values of stable limit cycles, and the dashed lines are the maximum and minimum values of unstable limit cycles.

We introduce a Hopf bifurcation at $z_0 = -0.5$ via a washout filter-aided dynamic feedback controller. This makes neuronal excitability change from Type I excitability into Type II excitability.

The equations of the two-dimensional HR model with a dynamic feedback controller through a washout filter are given as follows:

$$\frac{dx}{dt} = c(x - \frac{x^3}{3} - y + z) + u,$$
$$\frac{dy}{dt} = \frac{x^2 + dx - by + a}{c}, \tag{6.6}$$
$$\frac{dw}{dt} = x - d_f w,$$
$$u = g(v), \quad v = x - d_f w,$$

where $d_f > 0$ is the reciprocal of the filter time constant, and we set $d_f = 0.1$. v is the output function of the washout filter.

For the above closed-loop system, in addition to the creation of a Hopf bifurcation, our controller can be designed to control the criticality of the bifurcation (see in the following). It is well known that only the quadratic and cubic terms in a nonlinear system generating a Hopf bifurcation influence the bifurcation stability coefficient (Abed and Fu, 1986; Chen et al., 2001). In order to simplify the choice of control parameters, however, we represent our controller in the following simple form with only a linear term and a cubic term:

$$u = K_l(x - d_f w) + K_n(x - d_f w)^3. \tag{6.7}$$

Note that introduction of the washout filter to the two-dimensional HR model does not affect the equilibrium structure of the original system during a control process. As we shall see later, the linear control gain K_l determines two basic critical conditions, but has no effect on the criticality of the bifurcation because of no contribution to the bifurcation stability coefficient; the nonlinear control gain K_n, on the other hand, controls the criticality of the bifurcation, but has no influence on the locations of equilibrium points.

Suppose that a Hopf bifurcation is created at a desired parameter value $z_0 = -0.5$ before the emergence of the SNIC bifurcation. For the closed-loop system it has only one equilibrium point at $z_0 = -0.5$, namely,

$$(x_0, y_0, w_0) = (x_0, y_0, x_0/d_f) = (-2.48104, 2.10968, -24.8104).$$

The Jacobian matrix of the closed-loop system is given as follows.

$$\begin{bmatrix} c(1-x^2) + K_l + 3K_n(x - d_f w)^2 & -c & -K_l d_f - 3K_n(x - d_f w)^2 d_f \\ \dfrac{2x+d}{c} & -\dfrac{b}{c} & 0 \\ 1 & 0 & -d_f \end{bmatrix}. \tag{6.8}$$

It is clear that the nonlinear control term has no influence on the Jacobian matrix at the equilibrium point. Thus, the Jacobian matrix becomes

$$\begin{bmatrix} c(1-x^2) + K_l & -c & -K_l d_f \\ \dfrac{2x+d}{c} & -\dfrac{b}{c} & 0 \\ 1 & 0 & -d_f \end{bmatrix}. \tag{6.9}$$

The corresponding characteristic equation has the following form,

$$p_0\lambda^3 + p_1\lambda^2 + p_2\lambda + p_3 = 0, \tag{6.10}$$

where

$$
\begin{aligned}
p_0 &= 1, \\
p_1 &= d_f + \frac{b}{c} - c + cx^2 - K_l, \\
p_2 &= \frac{bd_f - bK_l}{c} - d_f c + d_f cx^2 + 2x + d - b + bx^2, \\
p_3 &= d_f(2x + d - b + bx^2).
\end{aligned}
$$

If a Hopf bifurcation occurs, the Jacobian matrix of the closed-loop system must satisfy the basic critical conditions (Chen et al., 2001; Guckenheimer and Holmes, 1997). One is the eigenvalue assignment. Namely, the characteristic equation has a pair of pure imaginary eigenvalues $\lambda_1 = \omega_0 i$ and $\lambda_2 = \bar{\lambda}_1 = -\omega_0 i$, and the other eigenvalues have negative real parts at $z_0 = -0.5$. The other is the transversality condition. That is, the eigenvalues λ_1 and λ_2 cross the imaginary axis with some nonzero speed at the Hopf bifurcation point $(x_0, y_0, w_0; z_0)$. To avoid solving directly all eigenvalues, we employ a more convenient and efficient algorithm criterion for detecting the existence of Hopf bifurcations, which is on basis of the Routh-Hurwitz stability criterion and described by the coefficients of the characteristic equation instead of eigenvalues (Liu, 1994).

In this way, the eigenvalues assignment corresponds to the following condtions.

$$
\begin{aligned}
& p_3 > 0, \\
& \Delta_1 = p_1 > 0, \\
& \Delta_2 = \begin{vmatrix} p_1 & p_0 \\ p_3 & p_2 \end{vmatrix} = 0.
\end{aligned}
$$

Substituting the parameter values and the location of the equilibrium point, we can get

$$
\begin{gathered}
p_3 = 0.19935 > 0, \\
K_l < 15.90000, \\
K_l^2 - 26.62042K_l + 169.85670 = 0.
\end{gathered}
$$

There are two solutions for the above equation, namely, $K_l = 16.01299$ and $K_l = 10.60743$. Apparently, only $K_l = 10.60743$ meets the eigenvalue assignment. Next, we examine if $K_l = 10.60743$ satisfies the transversality condtion, which is written as

$$
\left.\frac{\partial \Delta_2}{\partial z}\right| = -62.03510 + 4.72179K_l \neq 0, \text{ namely } K_l \neq 13.13804.
$$

Apparently, $K_l = 10.60743$ satisfies the transversality condition. As a result, we take $K_l = 10.60743$ according to the two basic critical conditions for the occurrence of the Hopf bifurcation. For a while, let $K_n = 0$, then the bifurcation diagram of the closed-loop system is shown in Fig. 6.12. As expected, a Hopf bifurcation is created at $z_0 = -0.5$. At the same time, the right Hopf bifurcation is moved to $z = 9.656$.

Notice that the created Hopf bifurcation here is subcritcial. Thus, we have made the neuronal excitability change from Type I excitability to Type II excitability.

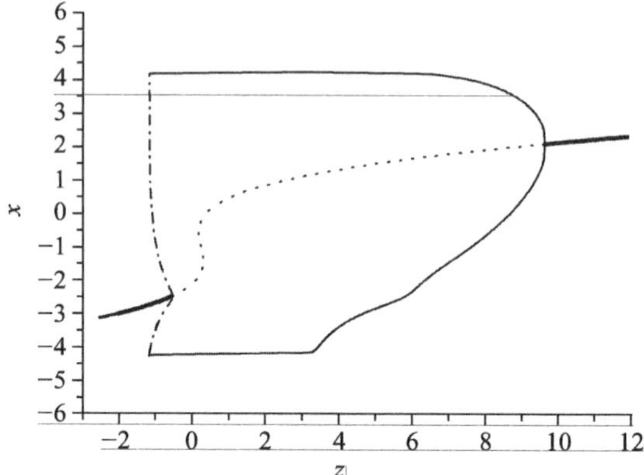

Fig. 6.12 The bifurcation diagram of the closed-loop system with only the linear control term $K_l = 10.60743$.

In what follows, we can control the criticality of the created Hopf bifurcation by the nonlinear control term. At a small neighborhood of a Hopf bifurcation point the bifurcated periodic solution of the limit cycle has the amplitude of $O(\varepsilon)$, here, $\varepsilon = \sqrt{|z - z_0|}$. The asymptotic stability of such a periodic solution is governed by one characteristic exponent given by a real smooth even function $\beta(\varepsilon) = \beta_2\varepsilon^2 + \beta_4\varepsilon^4 + \cdots$. If $\beta(\varepsilon) < 0$, the periodic solution is asymptotically stable, otherwise it is unstable. From the expression of $\beta(\varepsilon)$, typically, it can be seen that the local stability of the bifurcated periodic solution, namely, the criticality of the bifurcation is determined by the sign of β_2, which is called the bifurcation stability coefficient. Here, we apply the center manifold and normal form theory to derive the closed-form analytic expression for β_2.

As seen above, after determining the linear control gain $K_l = 10.60743$ according to the two basic critical conditions for the Hopf bifurcation, the Jacobian matrix of the closed-loop system becomes a constant matrix. Therefore, we can numerically compute all eigenvalues of the matirx and their corresponding eigenvectors. In fact, this is a necessary step in deriving the analytic expression for β_2 with respect to K_n in order to employ the center manifold and normal form theory.

The constant matrix is

$$\begin{bmatrix} -4.85923 & -3.0 & -1.06074 \\ -1.05403 & -0.33333 & 0.0 \\ 1.0 & 0.0 & -0.1 \end{bmatrix}.$$

The eigenvalues and their corresponding eigenvectors are

$$\lambda_1 = -1.16241 \times 10^{-10} + 0.19408i,$$
$$\lambda_2 = -1.16241 \times 10^{-10} - 0.19408i,$$
$$\lambda_3 = -5.29256,$$

and

$$v_1 = \begin{bmatrix} 0.08441 + 0.16381i \\ -0.42457 - 0.27080i \\ 0.84407 \end{bmatrix}, \quad v_2 = \bar{v}_1, \quad v_3 = \begin{bmatrix} -0.96125 \\ -0.20431 \\ 0.18512 \end{bmatrix}.$$

Here, i is the imaginary unit. Due to very small real parts of λ_1 and λ_2, the matrix can be considered to have a pair of pure imaginary eigenvalues. Another is a negative eigevalue. For notational simplicity, let $\omega_0 = Im(\lambda_1) = 0.19408$ and $M = \lambda_3 = -5.29256$. We construct a matrix P as follows:

$$P = (\mathrm{Re}(v_1), -\mathrm{Im}(v_1), v_3).$$

Here, Re and Im mean extracting the real part and the imaginary part of a complex-valued expression, respectively.

That is,

$$P = \begin{bmatrix} 0.08441 & -0.16381 & -0.96124 \\ -0.42457 & 0.27080 & -0.20430 \\ 0.84407 & 0.0 & 0.18512 \end{bmatrix}.$$

Taking the following coordinate transformation,

$$\begin{bmatrix} x \\ y \\ w \end{bmatrix} = \begin{bmatrix} x_0 \\ y_0 \\ w_0 \end{bmatrix} + P \begin{bmatrix} X \\ Y \\ W \end{bmatrix}, \tag{6.11}$$

we can obtain

$$\begin{bmatrix} x \\ y \\ w \end{bmatrix} = \begin{bmatrix} -2.48104 + 0.08441X - 0.16381Y - 0.96124W \\ 2.10968 - 0.42457X + 0.27080Y - 0.20430W \\ -24.81039 + 0.84407X + 01.8512W \end{bmatrix}.$$

Substituting the coordinate transformation into the closed-loop system, and then making the following transformation, we can get a system under a new coordinate system as follows:

$$\begin{bmatrix} \frac{dX}{dt} \\ \frac{dY}{dt} \\ \frac{dW}{dt} \end{bmatrix} = P^{-1} \begin{bmatrix} \frac{dx}{dt} \\ \frac{dy}{dt} \\ \frac{dw}{dt} \end{bmatrix} = \begin{bmatrix} F^1(X,Y,W) \\ F^2(X,Y,W) \\ F^3(X,Y,W) \end{bmatrix}, \tag{6.12}$$

where P^{-1} is the following inverse matrix of P

$$P^{-1} = \begin{bmatrix} 0.20947 & 0.12671 & 1.22753 \\ -0.39215 & 3.45558 & 1.77737 \\ -0.95509 & -0.57777 & -0.19511 \end{bmatrix}.$$

At $(X,Y,W) = (0,0,0)$ the Jacobian matrix of the new system is

$$\begin{bmatrix} -1.0\times 10^{-10} & -0.19408 & 3.0\times 10^{-9} \\ 0.19408 & 0.0 & -1.0\times 10^{-9} \\ 0.0 & 0.0 & -5.29256 \end{bmatrix}.$$

We can regard the Jacobian matrix as the following matrix of the real canonical form by ignoring very small entries,

$$\begin{bmatrix} 0.0 & -0.19408 & 0.0 \\ 0.19408 & 0.0 & 0.0 \\ 0.0 & 0.0 & -5.29256 \end{bmatrix}.$$

As a result, the Jacobian matrix of the new system has the following property:

$$\begin{bmatrix} \frac{\partial F^1}{\partial X} & \frac{\partial F^1}{\partial Y} & \frac{\partial F^1}{\partial W} \\ \frac{\partial F^2}{\partial X} & \frac{\partial F^2}{\partial Y} & \frac{\partial F^2}{\partial W} \\ \frac{\partial F^3}{\partial X} & \frac{\partial F^3}{\partial Y} & \frac{\partial F^3}{\partial W} \end{bmatrix}\Bigg|_{(0,0,0)} = \begin{bmatrix} 0 & -\omega_0 & 0 \\ \omega_0 & 0 & 0 \\ 0 & 0 & M \end{bmatrix}. \tag{6.13}$$

Here, we can apply the center manifold and normal form theory to derive the analytic expression for the bifurcation stability coefficient β_2. In fact, by following the procedures provided in (Hassard, 1981), the bifurcation stability coefficient has a unified expression, regardless of the detailed form of the transformed system with a real canonical form, as follows:

$$\begin{aligned} \beta_2(K_n) = 2\mathrm{Re}\Bigg(& \Big(g_{20}(z_0)g_{11}(z_0) - 2\left|g_{11}(z_0)\right|^2 \\ & -\frac{1}{3}\left|g_{02}(z_0)\right|^2\Big)\frac{i}{2\omega_0} + \frac{g_{21}(z_0,K_n)}{2}\Bigg), \end{aligned} \tag{6.14}$$

where

$$g_{20}(z_0) = \frac{1}{4}\left(\frac{\partial^2 F^1}{\partial X^2} - \frac{\partial^2 F^1}{\partial Y^2} + 2\frac{\partial^2 F^2}{\partial X \partial Y} + i\left(\frac{\partial^2 F^2}{\partial X^2} - \frac{\partial^2 F^2}{\partial Y^2} - 2\frac{\partial^2 F^1}{\partial X \partial Y}\right)\right),$$

$$g_{11}(z_0) = \frac{1}{4}\left(\frac{\partial^2 F^1}{\partial X^2} + \frac{\partial^2 F^1}{\partial Y^2} + i\left(\frac{\partial^2 F^2}{\partial X^2} + \frac{\partial^2 F^2}{\partial Y^2}\right)\right),$$

$$g_{02}(z_0) = \frac{1}{4}\left(\frac{\partial^2 F^1}{\partial X^2} - \frac{\partial^2 F^1}{\partial Y^2} - 2\frac{\partial^2 F^2}{\partial X \partial Y} + i\left(\frac{\partial^2 F^2}{\partial X^2} - \frac{\partial^2 F^2}{\partial Y^2} + 2\frac{\partial^2 F^1}{\partial X \partial Y}\right)\right),$$

$$g_{21}(z_0, K_n) = G_{21}(z_0, K_n) + 2G_{110}s_{11} + G_{101}s_{20},$$

$$G_{21}(z_0, K_n) = \frac{1}{8}\left(\frac{\partial^3 F^1}{\partial X^3} + \frac{\partial^3 F^1}{\partial X \partial Y^2} + \frac{\partial^3 F^2}{\partial X^2 \partial Y} + \frac{\partial^3 F^2}{\partial Y^3} + i\left(\frac{\partial^3 F^2}{\partial X^3} + \frac{\partial^3 F^2}{\partial X \partial Y^2} - \frac{\partial^3 F^1}{\partial X^2 \partial Y} - \frac{\partial^2 F^1}{\partial Y^3}\right)\right),$$

$$G_{110} = \frac{1}{2}\left(\frac{\partial^2 F^1}{\partial X \partial W} + \frac{\partial^2 F^2}{\partial Y \partial W} + i\left(\frac{\partial^2 F^2}{\partial X \partial W} - \frac{\partial^2 F^1}{\partial Y \partial W}\right)\right),$$

$$G_{101} = \frac{1}{2}\left(\frac{\partial^2 F^1}{\partial X \partial W} - \frac{\partial^2 F^2}{\partial Y \partial W} + i\left(\frac{\partial^2 F^1}{\partial Y \partial W} + \frac{\partial^2 F^2}{\partial X \partial W}\right)\right),$$

$$s_{11} = -h_{11}/M,$$

$$s_{20} = -h_{20}/(M - 2i\omega_0),$$

$$h_{11} = \frac{1}{4}\left(\frac{\partial^2 F^3}{\partial X^2} + \frac{\partial^2 F^3}{\partial Y^2}\right),$$

$$h_{20} = \frac{1}{4}\left(\frac{\partial^2 F^3}{\partial X^2} - \frac{\partial^2 F^3}{\partial Y^2} - 2i\frac{\partial^2 F^3}{\partial X \partial Y}\right).$$

As above, all derivatives take their values at $(X, Y, W; z_0) = (0, 0, 0; -0.5)$. In this way, we obtain closed-form analytic expression for β_2 as follows:

$$\beta_2 = 0.22527 \times 10^{-1} + 2\mathrm{Re}(0.64645 \times 10^{-3}K_n + i0.34531 \times 10^{-3}K_n).$$

If K_n is a real number with $K_n < -17.42357$, $\beta_2 < 0$. As a consequence, $K_n < -17.42357$ ensures that the periodic solution bifurcated from the Hopf bifurcation is asymptotically stable, and then makes the Hopf bifurcation change from subcritical into supercritical. In contrast, if $K_n > -17.42357$, then $\beta_2 > 0$, and the created Hopf bifurcation is subcritical.

Let us investigate bifurcation behavior in the case of $K_n = -20$ to verify the accuracy of our analytic expression for β_2. The bifurcation diagram is shown in Figs. 6.13(a) and 6.13(b). Figure 6.13(b) is enlargement of Fig. 6.13(a) near the created Hopf bifurcation point. From Fig. 6.13(b), it is clear that the created Hopf bifurcation is supercritical. Thus, we can make the created Hopf bifurcation supercritical via the nonlinear control term with $K_n = -20$. And we see the structure and locations of the equilibrium points are not changed. Also, the bifurcation points of the left (created) and the right Hopf bifurcations are not varied. The bifurcation value of the left HB is $z = -0.5$, while that of the right HB is $z = 9.656$. In other words, the nonlinear control term only exerts an influence on the criticality of the bifurcations, namely, the bifurcation stability coefficient, but no effect on the structure and locations of the equilibrium points. Actually, these features can be seen from their calculation processes and expressions.

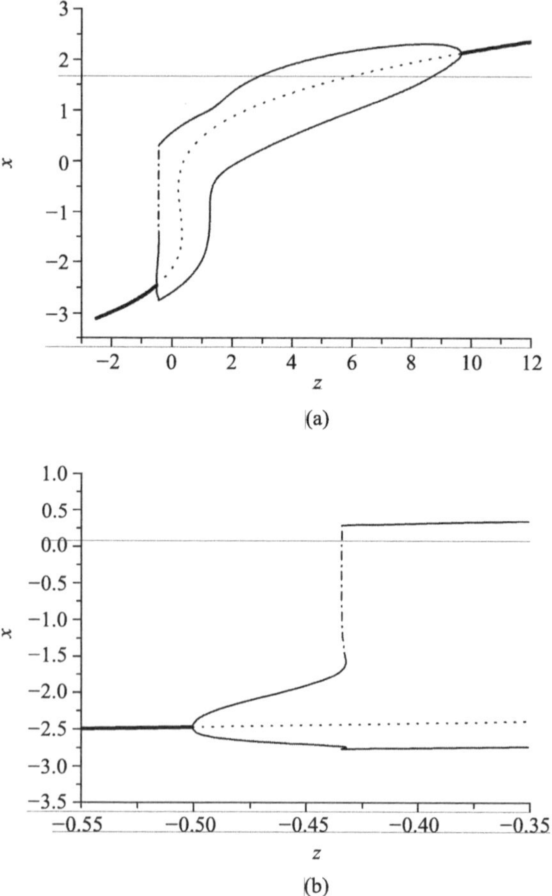

Fig. 6.13 The bifurcation diagram and time series of the closed-loop system with $K_l = 10.60743$ and $K_n = -20.0$. (a) The bifurcation diagram, and (b) is enlargement of (a) near the left Hopf bifurcation point.

6.3.5 Bursting and its topological classification

Different ionic mechanisms of bursting may lead to different dynamical mechanisms, which in turn determine the neuro-computational properties of bursters (Izhikevich, 2007), i.e., how they respond to the input. Therefore, much effort is devoted to studying and classifying the dynamics of bursting.

Most mathematical models of bursting neurons can be written in the slow-fast form [2] as follows:

$$\begin{aligned} \frac{dx}{dt} &= f(x,y), \\ \frac{dy}{dt} &= \mu g(x,y). \end{aligned} \tag{6.15}$$

where x represents the fast subsystem responsible for spiking activity, the modification of spiking attributes to the role of the slow subsystem y, and $\mu << 1$ is the ratio of time scales. Such systems are singularly perturbed system, for which the standard method for analysis is to set $\mu = 0$ and consider the fast and the slow subsystems separately. This method is called dissection of neuronal bursting or slow-fast subsystem decomposition (Rinzel, 1985). Therefore, we can treat y as a vector of slowly changing bifurcation parameters to investigate the dynamical behavior of the fast subsystem.

Usually, the fast subsystem has a limit cycle attractor corresponding to spiking for some values of y and a stable equilibrium attractor corresponding to resting state for other values of y. As y oscillates between the two regions where x exhibits spiking and resting state, respectively, the whole system can burst. Now the key problem is to make y oscillate. In the simple case the slow subsystem has a limit cycle attractor, which is relatively insensitive to the fast variable, and thus the slow variable exhibits an autonomous oscillation that periodically drives the fast subsystem display spiking and the resting state. Such a bursting is called slow wave bursting. Parabolic bursting (Bertram et al., 1995; Ermentrout and Kopell, 1986; Rinzel and Lee, 1987; Baer et al., 1995) belongs to this case, for example. The slow subsystem, however, must be at least tow-dimensional. When the fast subsystem has a bistable range of resting and spiking states, where the stable equilibrium and limit cycle attractors coexist for the same value of y, a hysteresis loop is created for the slow variable. Such a bursting is called hysteresis-loop bursting. Square wave bursting (Rinzel and Ermentrout, 1989;Bertram et al., 1995) is one of this bursting.

Bursting neurons are distinguished qualitatively according to their topological type (Izhikevich, 2007), which is determined by the two bifurcations of the fast subsystem: one is from resting state to spiking, and the other is from spiking to resting state. The former means that the state of the fast subsystem changes from a stable equilibrium into a limit cyle attractor, the latter denotes the state of the fast subsystem transforms from a limit cycle attractor into a stable equilibrium.

As we know, an equilibrium has only 4 possible bifurcations of co-dimension 1: saddle-node bifurcation (fold), saddle-node bifurcation on invariant circle (circle), supercritical Hopf bifurcation (Hopf) and Subcritical Hopf bifurcation (subHopf), while also a limit cycle attractor has only 4 possible bifurcations of co-dimension 1 if the fast subsystem is two-dimensional: saddle-node bifurcation on invariant circle (circle), saddle homoclinic orbit bifurcation (homoclinic), supercirical Hopf bifurcation (Hopf) and fold limit cycle (fold cycle). Thus, there are 16 different bifurcation combinations, resulting in 16 different topological types of fast-slow bursting neurons with 2-dimensional fast subsystems. They can be named after the bifurcations involved. If the constraint that the fast subsystem is two-dimensional is removed, then the topological type has 120 all possible types for bursting neurons (Izhikevich, 2007; Izhikevich, 2000).

The topological types of the three known bursting neurons are as follows.

Square-wave bursting belongs to the fold/homoclinic type. Namely, the fast subsystem undergoes fold (saddle-node off limit cycle) bifurcation resulting in the transition from resting state to spiking. After spiking, the fast subsystem undergoes saddle homoclinic orbit bifurcation resulting in the transition from spiking to resting. As to parabolic bursting, it is the circle/circle type because the transition from spiking to resting and back to spiking of the fast subsystem occurs via saddle-node on invariant circle bifurcation. For elliptic bursting, the fast subsystem undergoes subcritical Hopf bifurcation leading to the transition from resting state to spiking. After that, spiking stops via fold limit cycle bifurcation. Therefore, elliptic bursting is subHopf/fold cycle type.

6.3.6 Bifurcation, chaos and Crisis

Bifurcation structures and chaos phenomena are frequently observed in experimental neural pacemakers and theoretical neuron models. Aihara, et al. (1985) find an alternating periodic-chaotic sequence experimentally observed in periodically forced neural oscillators of giant axons of squids, and demonstrate that such a sequence can be qualitatively described by the HH model. Matsumoto et al. (1987) discover that chaotic potential responses could be evoked in periodically forced squid axons immersed in normal seawater and intermittent chaos appears through a subcritical period-doubling bifurcation. In addition, Takahashi, et al. (1990) find the periodic potential responses appear through either tangent bifurcation or type III bifurcation in the same experimental objects. Ren et al. (1997) claim that they observe period-adding cascades with or without chaos and period doubling cascades many times in the experimental neural pacemakers. Shilnikov and Cymbalyuk (2005) investigate a continuous and reversible transition between periodic tonic spiking and bursting activities in a model of a heart interneuron from the medicinal leech and find that the Blue-Sky Catastrophe is the dynamical mechanism for the transition between Tonic Spiking and Bursting. Also, they show the model can demonstrate co-existence of a periodic tonic spiking with either periodic or chaotic tonic spiking (Cymbalyuk and Shilnikov, 2005). Feudel et al (2002) study global bifurcations of the chaotic attractor in a modified HH model of thermally sensitive neurons, and observe an abrupt increase of the interspike intervals in a certain temperature region, and identify this as a homoclinic bifurcation of a saddle-focus fixed point embedded in the chaotic attractors. Interestingly, Guckenheimer and Oliva (2002) demonstrate the existence of chaotic solutions in the HH model with its original parameters, which is previously unobserved dynamics in such a model. Figure 6.14 shows the bifurcation diagram of ISI versus the time-scale factor r (Xie et al., 2004b). We can see a period-adding cascade with chaos as r is decreased from Fig. 6.14. Here, the firing pattern alternates between periodic bursting and chaotic bursting, which results from chaos appearance by period-doubling cascades and chaos termination by saddle-node bifurcations.

An interior crisis occurs at the transition point between chaotic spiking and chaotic bursting in the HR neuron model, where the change of the attractor size is sudden but continuous (Xie et al., 2004b). Also, Jin et al. (2006) show a crisis of interspike intervals in periodically forced HH model. Figure 6.15 shows an interior

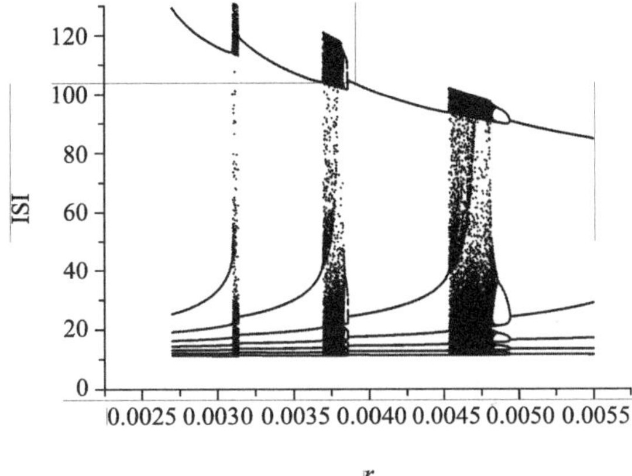

Fig. 6.14 Bifurcation diagram of ISI versus the time-scale factor r (see Fig. 1(b) in (Xie et al., 2004b))

crisis occurs at the transition point between chaotic spiking and chaotic bursting in the modified Chay model (Xie et al., 2004b). By the way, although the Chay model is used to describe the electrical activities of pancreatic β cells, and, recently, it is frequently considered to be a neuron model.

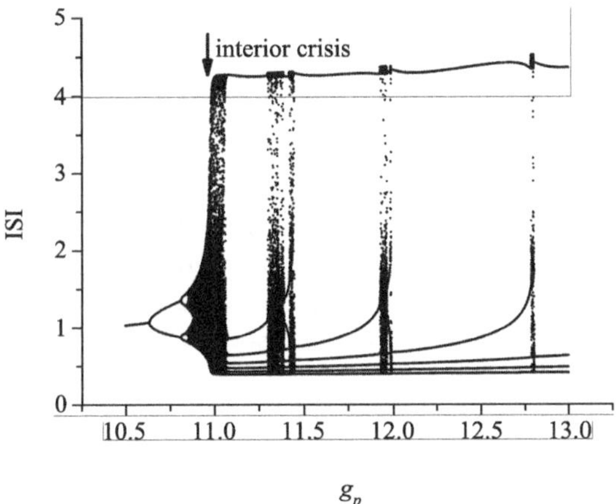

Fig. 6.15 Bifurcation diagram of ISI versus the maximal conductance of the slow K^+ current g_p in the modified Chay model (see Fig. 5(b) in (Xie et al., 2004b)).

6.4 Sensitive responsiveness of aperiodic firing neurons to external stimuli

6.4.1 Experimental phenomena

In a number of neurophysiological experiments about responsiveness of the chronically compressed dorsal root ganglion (DRG) neurons in rats to drugs, neuroscientists have found that aperiodic firing neurons are more sensitive to external stimuli than periodic firing neurons (Hu et al., 2000; Yang et al., 2000). To demonstrate the universality of the sensitive responsiveness, various drugs with different action mechanisms have been applied to the chronic compression DRG neurons, for example, higher Ca^{2+} solution, norepinephrine (NE) and tetraethylammonium (TEA). Interestingly, all of these drugs can induce this phenomenon of the sensitive responsiveness in the experiments.

Let us now give in brief the experimental results on the phenomenon of sensitive responsiveness (Yang et al., 2000). Here, we only show the responses of chronic compression DRG neurons to TEA. Spike series from a total of 91 DRG neurons with spontaneous firing behavior in 25 anesthetized rats were recorded extracellularly. There are 32 periodic firing neurons and 59 aperiodic firing neurons, including 44 with chaotic bursting, 7 with integer multiple firing and 8 with chaotic spiking patterns, in the 91 DRG neurons. After bath application of 2 mmol/L TEA to the DRG neurons for 3 minutes, the percentage of periodic and aperiodic firing neurons exhibiting obvious response are 27.3 and 93.2, respectively. Moreover, the responses of aperiodic firing neurons are more intensive than those of periodic firing neurons. Figure 6.16 shows the responses of injured DRG neurons with different firing patterns to TEA. In Figs. 6.16(a)-(d), top panels, middle panels and bottom panels show the spike trains, the histogram of mean firing rate (bin width 1 second) and the ISI series, respectively.

From the experiments about the responses of periodic firing neurons and aperiodic firing neurons to TEA with different concentration, we observed that the aperiodic firing neurons exhibited a gradually enhanced response as the concentration of TEA was increased from 0.5 mmol/L to 2 mmol/L and 10 mmol/L, while the periodic firing neurons only produced a faint response to 10 mmol/L TEA, as shown in Fig. 6.17.

From a number of experimental results, we can find that aperiodic firing neurons respond more easily and intensively than periodic firing neurons to external stimulation of drugs. What mechanisms govern such a phenomenon? Here, we devote our attention to the study of this problem in terms of dynamical systems theory.

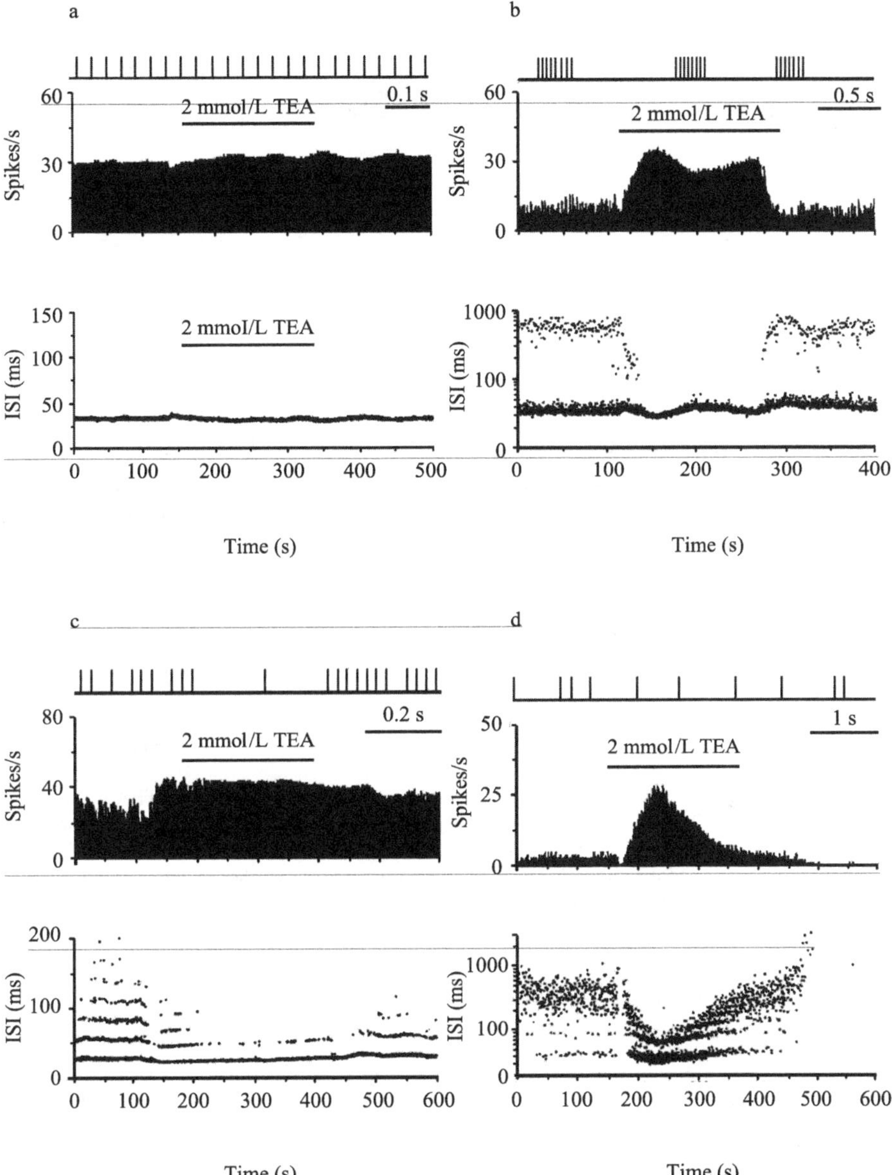

Fig. 6.16 Responses of injured DRG neurons with different firing patterns to 2 mmol/L TEA (Fig. 1 in (Yang et al., 2000)). (a) There was no response induced by TEA in a periodic firing neuron. A significant response was induced by TEA in neurons with chaotic bursting (b), integer multiple firing (c) and chaotic spiking patterns (d).

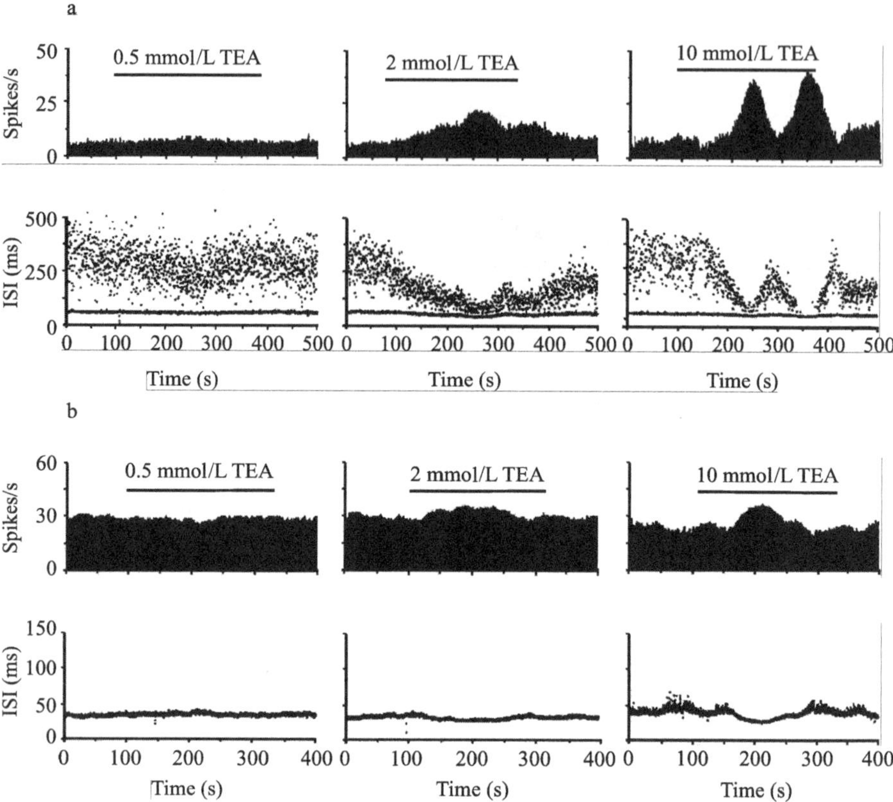

Fig. 6.17 Response of a chaotic bursting neuron and a periodic firing neuron to TEA with different concentration (Fig. 2 in (Yang et al., 2000)). (a) shows a developed evident response of the chaotic bursting neuron to TEA, and the response enhanced gradually as the concentration of TEA was increased from 0.5 mMol/L to 2 mMol/L and 10 mMol/L. (b) shows the periodic firing neuron had just a little response to 10 mMol/L TEA.

6.4.2 Nonlinear mechanisms

Actually, many neuron models are described by a set of first-order ordinary differential equations. Quite frequently some models are chaotic systems under a certain parameter range. As we know, a key development describing chaotic systems is periodic orbit theory. Periodic orbits open a door to the understanding of the chaotic dynamics. It has now been a widely accepted notion that unstable periodic orbits (UPO's) constitute the most fundamental building blocks of a chaotic system (Schmelcher and Diakonos, 1998; Davidchack and Lai, 1999; So et al., 1996, 1997). Theoretically, there are the infinite number of UPO's embedded in a chaotic attractor. Moreover, these UPO's reveal the skeleton of the chaotic attractor because they are dense in it. As a result, the UPO's carry essential information concerning characteristic features of the chaotic system, and allow the calculation of many dynamical invariants of physical interest like Lyapunov exponents, fractal dimensions and entropies of the attractor by knowing their positions and properties (Cvitanovic,1998; Auerbach et al., 1987; Lathrop and Kostelich, 1989). This reflects the importance of the UPO's for the analysis and decoding of the dynamics on the attractor. Therefore, UPO's seem to be the optimal practical tool for the description of chaotic systems. More recently, the theory of UPO's has been applied to understand the mechanism of neural coding and decoding (So et al., 1998). This is the very reason why UPO's is used to characterize the activity of aperiodic firing neurons in this study. We use the recurrence method to compute the distribution of UPO's at every period, which denotes how often a chaotic trajectory visit UPO's with the same period. Although this method was proposed before a decade and more, it is sufficient enough for our purpose.

First, we investigate a celebrated neuron model, the HR model, and reveal the dynamical mechanisms for the sensitive responsiveness of aperiodic firing neurons to external stimuli. And then we turn to the modified Chay model, and find the phenomenon of sensitive responsiveness also occurs in this model. By the way, our goal is to choose these two models to interpret the dynamical mechanisms for the phenomenon of sensitive responsiveness, but not use them to model the injured DRG neurons in rat.

It is well known that the neuronal firing pattern can be changed after the application of a drug. This is because most of drugs affect the function of a neuronal system by varying the efficacy of various ion channels in the membrane. Here, we fix $I = 3.0$ and vary r to reflect the action of external stimulation in the HR model, as seen previously.

The bifurcation diagram of ISI versus r is shown in Fig. 6.18. Figure 6.14 is the enlargement of Fig. 6.18 over the range of $[0.0027, 0.0055]$. From Fig. 6.18, we can see the firing pattern at first undergoes from simple period spiking to chaotic firing via a period-doubling sequence, and then a saddle-node bifurcation abruptly terminates this chaotic firing in the form of intermittency, and reorganizes simultaneously period-3 bursting with the decrease of r. After that, the alternation between period-doubling sequence and saddle-node bifurcation is repeated and at the same time a narrow chaotic firing region occurs between them, as seen in Fig. 6.14. Finally, the chaotic firing region disappears after a certain periodic bursting.

From Fig. 6.18, we see clearly that periodic spiking and period-2 bursting exist in a broad range of r, respectively. Thus, they can retain easily their corresponding firing patterns under a small perturbation of r. As r is reduced, the chaotic firing range and the periodic bursting range appear alternately until the disappearance of the chaotic firing range. In Fig. 6.14, we can find that a saddle-node bifurcation terminates a chaotic regime, and simultaneously reorganizes a new periodic bursting whose number of firing will increase by one. Furthermore, the size of a chaotic regime is much smaller than that of the adjacent periodic regime. The larger the firing number of a periodic bursting is, the smaller the size of the adjacent chaotic regime becomes. Evidently, a period-doubling cascade leads to a chaotic regime, and a saddle-node bifurcation terminates the chaotic regime via intermittency. Namely, there are bifurcation points at the two ends of the chaotic regime.

Therefore, an aperiodic firing neuron located at a chaotic range crosses bifurcation points much more easily than the corresponding periodic firing one under a small perturbation of r.

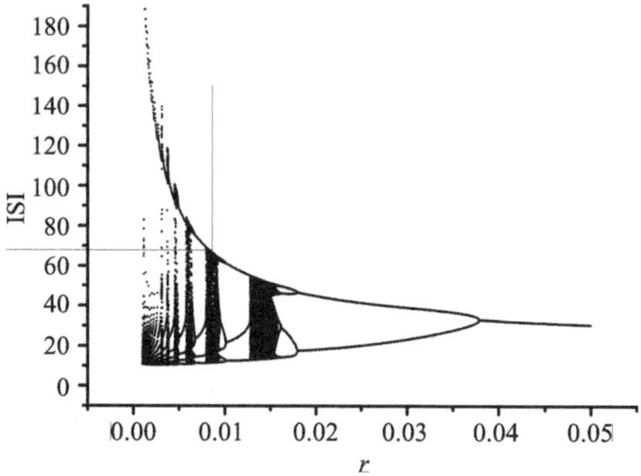

Fig. 6.18 Bifurcation diagram of ISI versus the time-factor r within the range of [0.001, 0.05] (Figure 1(a) in (Xie et al., 2004b)).

It is well known that qualitative variations of dynamical behavior of a system occur at bifurcation points as control parameters are changed. This implies the firing behavior varies qualitatively over bifurcation points for the HR model. Consequently, an aperiodic firing neuron is more sensitive than a periodic firing one under a small perturbation of r. It follows that various bifurcations should be one of dynamical mechanisms for the sensitive responsiveness of aperiodic firing neurons to external stimuli.

To order to demonstrate whether such a phenomenon of sensitive responsiveness of aperiodic firing neurons occurs in biophysical models, let us to investigate a pancreatic β-cell model. Since physiological evidence shows the intracellular calcium concentration Ca^{2+} changes rather quickly during depolarization, it may not be a low dynamic variable. Thus, we adopt the modified Chay model, which is based on the hypothesis of Ca^{2+} -activated K^+ channel (Chay et al., 1995). The equations are listed as follows:

$$
\begin{aligned}
-C_m \frac{dV}{dt} &= g_I m_\infty^3 h_\infty (V - V_I) + g_K n^4 (V - V_K) + g_P p (V - V_K) \\
&\quad + g_L (V - V_L), \\
\frac{dn}{dt} &= \frac{n_\infty - n}{\tau_n}, \\
\frac{dp}{dt} &= \frac{m_\infty^3 h_\infty (V_I - V) - k_C p/(1-p)}{\tau_p} (1-p)^2,
\end{aligned}
\tag{6.16}
$$

where V is the membrane potential, n is the gating variable, p is a slow variable which denotes the fraction of the available Ca^{2+} -sensitive K^+ channels at time

t. The maximal conductance of the slow K^+ current g_p is chosen as the control parameter.

The bifurcation diagram of ISI versus g_p is shown in Fig. 6.19. Figure 6.15 is the enlargement of Fig.6.19 in the range of [10.5, 13.0]. From Figs. 6.19 and 6.15, we can see that dynamical behavior of the Chay model is similar to that of the HR model. As g_p is increased, the firing pattern undergoes a period-doubling cascade and changes from periodic spiking to chaotic spiking, and then becomes suddenly chaotic bursting via an interior crisis. With further increasing g_p, the firing pattern transforms from chaotic bursting into period-7 bursting via an inverse period-doubling cascade. Hereafter, chaotic firing regimes and periodic bursting regimes appear alternately, as shown in Fig. 6.15. Saddle-node bifurcation and period-doubling bifurcation dominate this variation. Moreover, they are at the left and right of chaotic regimes, respectively. Finally, the system terminates the firing behavior at about $g_p = 26.855$. In addition, it is noted that the number of spikes of periodic bursting decreases by one if only the system strides over a chaotic regime.

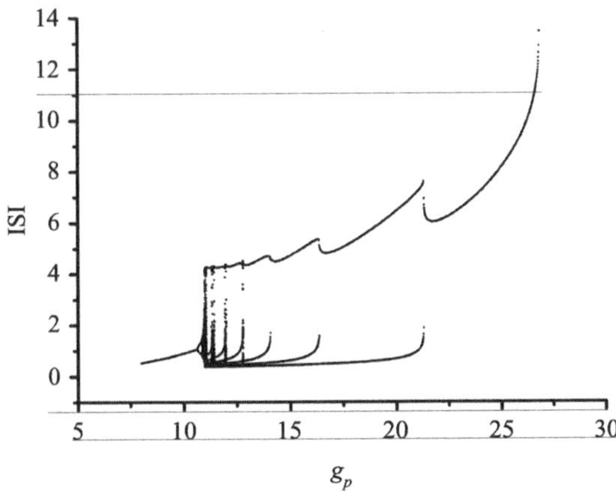

Fig. 6.19 Bifurcation diagrams of ISI versus g_p with $g_p \in [8, 27]$ (Figure 5(a) in (Xie et al., 2004b)).

As seen in Fig. 6.15, the smaller the firing number of a periodic bursting is, the narrower the chaotic regime adjacent to this periodic bursting is. This is contrary to the case of the HR model. The fact holds, however, that is, the size of chaotic regime is much smaller than that of the adjacent periodic regime. Aperiodic firing cells, therefore, get across bifurcation points more easily than periodic firing cells do under a small perturbation of g_p. In other words, aperiodic firing cells exhibit a significant change more easily than periodic firing cells when subjected to a small perturbation of g_p. Hence, various bifurcations remain still one of mechanisms for sensitivity of aperiodic firing cells to external stimulation.

Now let us turn to the case of crisis. In fact, crisis phenomena occur in almost all chaotic systems. There should be crisis phenomena emerging in chaotic neuron models. Generally speaking, there are three different types of crises (Grebogi et al.,

1986, 1987), i.e., attractor merging crisis, interior crisis, and boundary crisis, where a chaotic attractor undergoes a sudden change as a function of the control parameters. For attractor merging crisis, a multi-piece chaotic attractor merges together to increase in size smoothly. Interior crisis means a chaotic attractor increases in size abruptly. If a chaotic attractor suddenly vanishes, it is considered a boundary crisis or an exterior crisis occurs.

In the following we focus our attention on the case of interior crisis. As to exterior crisis, since it is related to abrupt disappearance or appearance of a chaotic attractor, without any doubt, there is a qualitative variation in the dynamical behavior of a system as the control parameter varies. Generally, an attractor merging crisis appears in many systems with symmetries, where two (or more) chaotic attractors merge to form a single chaotic attractor, and it also leads to an abrupt change in dynamical behavior. Thus both of exterior crisis and attractor merging crisis can result in the sensitive responsiveness of aperiodic firing cells to external excitation if they exist.

Now, we investigate the change in the firing pattern of the modified Chay model before and after the interior crisis. It is can be seen from Fig. 6.15 that there is a sudden increase in the size of a chaotic attractor as the parameter g_p passes through a critical value, as denoted by an arrow. The incremental portion comes from a chaotic saddle that already exists for parameter values below the crisis. This chaotic saddle is an invariant and nonattracting set having horseshoe-type dynamics and resembles the new portion of the larger attractor just above the crisis in phase space. When the crisis occurs, the chaotic saddle collides with the chaotic attractor and becomes part of the chaotic attractor at the crisis (Kim and Stringer, 1992). This is the very origin of the abrupt enlargement of the chaotic attractor via the interior crisis.

We take $g_p = 10.9$ below the crisis and $g_p = 11.0$ above the crisis. They correspond to the firing states of chaotic spiking and chaotic bursting, respectively. The membrane potentials are shown in Fig. 6.20. The upper panel is before the crisis, while the lower one is after the crisis. Clearly, there is an evident difference between membrane potentials, namely, for chaotic spiking there is no clear underlying slow wave while for chaotic bursting an unpredictable number of spikes ride on the slow wave. Figure 6.21 shows the first return maps of the ISI series in these two cases, where the sizes and structures of the chaotic attractors are completely different. The big dot and the small dot denote respectively the cases of $g_p = 10.9$ and $g_p = 11.0$. Furthermore, the distributions of UPO's are calculated from 30 000 ISI's before and after the interior crisis, respectively. The embedding dimension is chosen as $d = 3$ and time delay as $\tau = 1$. There is a significant difference in the distribution of UPO's between $g_p = 10.9$ and $g_p = 11.0$, as seen in Fig. 6.22. Interestingly, the number of UPO's increases suddenly at period 7. This is because the ISI series of after the crisis is dense near the period-7 UPO's. Thus, if an aperiodic firing cell resides near an interior crisis, also it can exhibit the sensitive response to external stimulation.

To sum up, various bifurcations and crises are possible mechanisms for sensitive responsiveness of aperiodic firing neurons to external excitation. They all can result in obvious changes in the firing activity of an aperiodic firing neuron to external weak stimulation. It is worth while to say that there are some periodic windows in some chaotic regimes, for those periodic ones are very narrow in scale and do not form the leading pattern of periodic bursting in the HR model, the influence of them is negligible.

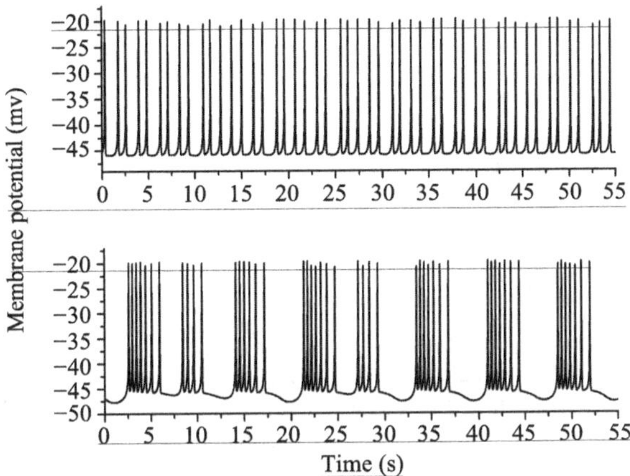

Fig. 6.20 Membrane potentials before and after the interior crisis (Figure 6 in (Xie et al., 2004b)).

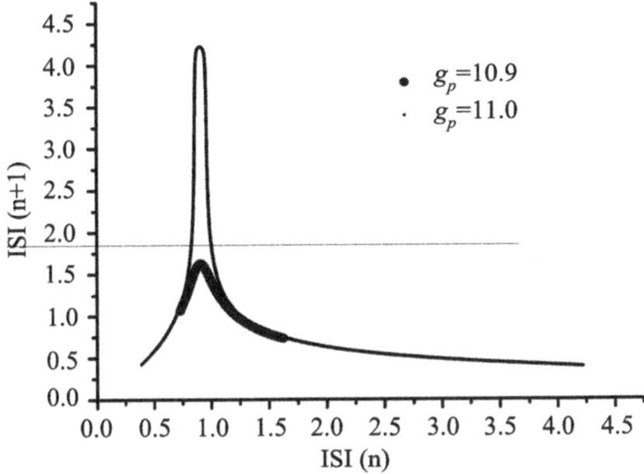

Fig. 6.21 First return maps of ISI before and after the interior crisis (Figure 7 in (Xie et al., 2004b)).

Up to now, the phenomenon of the sensitive responsiveness that aperiodic firing neurons are more sensitive to external stimulation has been investigated using dynamical systems theory, and the dynamical mechanisms underlying such sensitive responsiveness are revealed. Bifurcations and crises are considered as possible mechanisms. We think that the sensitivity of neurons with aperiodic firing activity to external stimulation reflects a universal property of excitable cells with deterministic chaos. The phenomenon of sensitive responsiveness can provide curing some difficult diseases with important directions, for example, the time instant and dose of taking drugs. That is, we can expect to control and cure some diseases with some minimum doses of drugs at appropriate instant of time making use of the sensitive phenomenon.

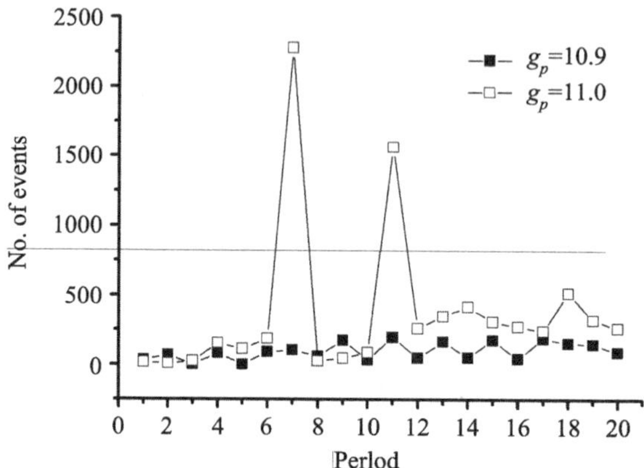

Fig. 6.22 Comparison of the distribution of UPOs computed from 30 000 ISI's before and after the interior crisis, respectively (Figure 8 in (Xie et al., 2004b)).

6.5 Synchronization between neurons

6.5.1 Significance of synchronization in the nervous system

Oscillatory responses in cat visual cortex exhibit inter-columnar synchronization (Gray et al., 1989). Now it is widely considered that the synchronous oscillation plays a role in feature binding, neuronal communication and motor coordination (Singer, 1993; Fries, 2001; Schnitlzer and Gross, 2005). Particularly, feature binding denotes how a large collection of coupled neurons combines external signals with internal memories into new coherent patterns of meaning. An external stimulus spreads over an ensemble of coupled neurons, building up a corresponding collective state. Thus, the synchronization among many coupled neurons is the basis of a coherent perception. The synchronous oscillation, however, can be modulated by task constrains, such as attention. To sum up, the neural synchronization may be the important mechanism of the information integration in the particular cortical area or between different cortical areas.

6.5.2 Coupling: electrical coupling and chemical coupling

Neuron communication is completed via synapses (Michael and Bennett, 2006). The synapse is a unique junction that allows for the transfer of neural information from one neuron to the next. A synapse is usually located between a presynaptic axon and a postsynaptic dendrite. There are two types of synapses, electrical and chemical. Electrical synapses are actually gap junctions, which are clusters of intercellular channels that connect the interiors of coupled neurons via special protein channels (Rozental et al., 2001). This allows for the direct flow of ions from one neuron to another and rapid signal transmission between the neurons. Generally this communication is unidirectional, but bidirectional communication is also possible. We often refer to the coupling scheme of neurons coupled in this way as electrical coupling. These synapses are therefore largely found in smooth and cardiac muscle and in certain regions of the brain. On the other hand, chemical synapses are specialized junctions through which neurons signal to each other and to non-neuronal cells such as those in muscles or glands (more detail see http://en.wikipedia.org/wiki/Chemical_synapse). This type of synapses is crucial to the biological computations that underlie perception and thought. Since a chemical synapse has a synaptic cleft, neurons must utilize a form of the neurotransmitter release to achieve neuron communication. Thus this communication is usually unidirectional. Compared to chemical synapses, electrical synapses conduct nerve impulses faster, but unlike chemical synapses they do not have gain.

Synchronization between neurons coupled by electrical or chemical synapses has been observed experimentally (Neiman and Russell, 2002; Elson, 1998). Now there are a large number of scientific literatures devoted to the study of synchronization mechanisms between coupled neuron models. Actually, there may be no connection between neurons in different brain regions. Many physiological experiments of brain activity, however, show that synchronous oscillation to the same stimulation can appear in different brain regions (Neiman and Russell, 2002). This means that the synchronization can also occur in pairs of noncoupled neurons.

Since synchronization of electrically coupled neurons has been studied extensively, we here show synchronous behavior of chemically coupled HR model neurons (Wu et al., 2005). Specially, we consider two identical HR model neurons with reciprocal synaptic connections. The differential equations of the coupled system are given as

$$\begin{aligned} \frac{dx_i}{dt} &= y_i - ax_i^3 + bx_i^2 - z_i + I_{ext} + e_s\left(\frac{x_i + V_c}{1+\exp\left(\frac{x_j - X_0}{Y_0}\right)}\right), \\ \frac{dy_i}{dt} &= c - dx_i^2 - y_i, \\ \frac{dz_i}{dt} &= r\left[s(x_i - x_0) - z_i\right]. \end{aligned} \tag{6.17}$$

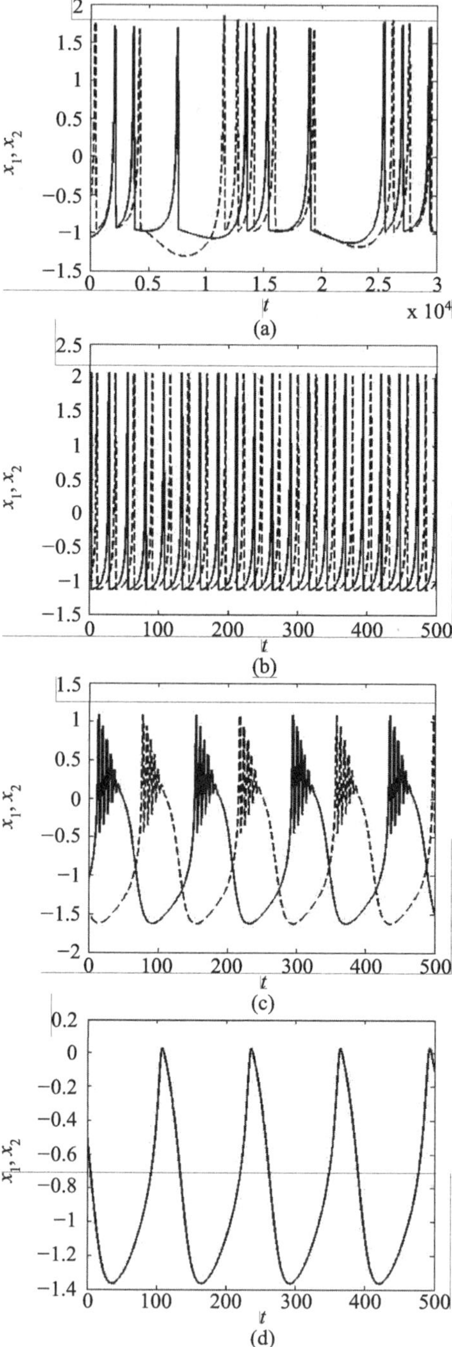

Fig. 6.23 Figs. (a) and (b) are the time courses of membrane potential of two neurons for excitatory synapse. (a) irregular activity for $e_s = 0.03$, (b) period 1 antisynchrony for $e_s = 0.3$, respectively. Figs. (c) and (d) are the time courses of membrane potential of two neurons for inhibitory synapse. (c) full antisynchrony for $e_s = -0.45$, (d) full synchrony for $e_s = -0.9$(Figure 1. in (Wu et al., 2005)).

Here, $i = 1, 2$, $j = 2, 1$, and $i \neq j$. In the numerical simulation, we take $a = 1.0$, $b = 3.0$, $c = 1.0$, $d = 5.0$, $s = 4.0$, $r = 0.006$, $x_0 = -1.56$, and $I_{ext} = 3.0$. e_s is the strength of the synaptic coupling, and $V_c = 1.4$ is the synaptic reverse potential which is selected so that the currents injected into the postsynaptic neuron are always negative for inhibitory synapses and positive for excitatory synapses. Since each neuron must receive an input every time the other neuron produces a spike, we set $Y_0 = 0.01$ and $X_0 = 0.85$.

The chemical synapse is excitatory for $e_s > 0$ and is inhibitory for $e_s < 0$. The results show that two neurons are irregular oscillation with small excitatory coupling strength, and are in full antisynchrony for enough excitatory coupling strength, as shown in Figs. 6.23(a) and (b). Interestingly, these results are contrary to traditional view. For the synchrony course of two coupled neuron with inhibitory synapse, the oscillation of the two neurons is irregular for small coupling intensity, and the phase difference between the two neurons increases gradually with coupling strength increasing till $e_s = -0.45$ at which the phase difference are biggest and the two neurons exhibit full antisynchrony, see Fig. 6.23(c). As the intensity of inhibitory coupling increases, the phase difference between neurons decreases. The two neurons are in full synchronization for $e_s = -0.9$, as seen in Fig. 6.23(d).

Our results show that excitatory synapses can antisynchronize two neurons, and weak or moderate inhibitory synaptic coupling can antisynchronize two neurons too, but strong inhibitory synapse can foster phase synchrony of two neurons. In (Neiman and Russell, 2002), authors show that synchronization of nonidentical neuronal oscillators which are not coupled can still be achieved via a specific mechanism of noise-induced slow dynamics. In addition, note that a common signal including noise can induce full synchronization or phase synchronization in the two uncoupled neurons.

6.6 Role of noise in the nervous system

6.6.1 Constructive role: stochastic resonance and coherence resonance

Noise permeates every level of the nervous system, from the perception of sensory signals to the generation of motor responses, and poses a fundamental problem for information processing (Faisal et al., 2008). In particular, external sensory stimuli are intrinsically noisy because they are either thermodynamic or quantum mechanical in nature. For example, all forms of chemical sensing (including smell and gustation) are affected by thermodynamic noise because molecules arrive at the receptor at random rates owing to diffusion and because receptor proteins are limited in their ability to accurately count the number of signalling molecules. Now it has been widely considered that neurons can use a phenomenon of stochastic resonance (SR) to detect weak signals in information processing (Longtin and Bulsara, 1991;

Chialvo et al., 1997; Douglass et al., 1993; Levin and Miller, 1996; Russell et al., 1999; Longtin, 1993b, 1997; Longtin and Hinzer, 1996 Douglass, 1993; Levin and Miller, 1996; Moss et al., 2004; Braun et al., 1994; Gammaitoni et al., 1998; Wellens et al., 2004).

SR is a phenomenon in which a nonlinear system is subjected to a periodic modulated signal so weak as to be normally undetectable, but it becomes detectable due to resonance between the weak deterministic signal and stochastic noise (Gammaitoni et al., 1998). In other words, stochastic noise enhances a weak input signal. Generally, a system embedded in a noisy environment acquires an enhanced sensitivity towards small external time-dependent forcings by the mechanism of SR, when the noise intensity reaches some finite level. Many scientists, therefore, think that the biological sensory system utilizes this phenomenon to detect emergent signals from outer environment. In fact, the neurophysiological experiments on SR have been conducted, three popular examples of which are the mechanoreceptor cells of crayfish, the sensory hair cells of cricket and human visual perception (more detail see: http://www.scholarpedia.org/article/Stochastic_resonance).

Simultaneously, a phenomenon of coherence resonance (CR) has also attracted much interest in the fields of neuroscience and physics (Pikovsky and Kurths, 1997). CR is sometimes called autonomous stochastic resonance (ASR) (Longtin, 1997), and refers to a phenomenon in which addition of certain amount of noise in excitable system makes its oscillatory responses most coherent without any weak signal. Thus a coherence measure of stochastic oscillations attains an extremum at optimal noise intensity. Theoretically, CR can occur in excitable systems such as FHN model and biophysical neuron models. Actually CR has been demonstrated in an experimental neural pacemaker (Gu et al., 2002).

Moreover, there exists suprathreshold stochastic resonance (SSR), which is a variant of SR that occurs for a specific set of conditions that are somewhat different from those of SR (http://www.scholarpedia.org/article/Suprathreshold_stochastic_resonance). Like SR, SSR describes the observation of noise enhanced behaviour in signal processing systems. Unlike conventional SR, SSR does not disappear when the signal is no longer subthreshold. SSR has also been demonstrated in integrate-and-fire neurons in the context of other noise-based enhancement effects, and saturation. In recent modelling studies SSR has been observed in a model of the electrically stimulated auditary nerve.

6.6.2 *Stochastic resonance: When does it not occur in neuronal models?*

We reported that subthreshold oscillations could hinder the detection of a weak signal via SR in neuronal systems (Gong et al., 1998). Through the FHN model and the Chay model, we have studied their subthreshold oscillations in certain parameter regimes, and we have found that the existence of subthreshold oscillation could hinder the detection of weak signals for neurons. It seems more difficult to say that

SR does not occur than to find it, because one can not check for every one of the combinations of the amplitude and the frequency of a stimulus as well as the noise intensity in the periodic driven stochastic excitable cell model; however, from our definition and calculation, a conclusion can be drawn that SR will not occur when the effect of the subthreshold oscillation is very large compared with that of a periodic stimulus. Consequently, a weak signal whether periodic or aperiodic (finite bandwidth) can not be amplified if it is overwhelmed by the subthreshold oscillation.

We note that the effect of the subthreshold oscillation can easily be neglected, not only because it exists in a very small parameter range with a very small amplitude, it can also bring a gamma-like distribution of spontaneous discharges. Under a periodic stimulus, neurons having the subthreshold oscillation may also exhibit a multimodal structure in the ISIH, a lattice form in the return map, and the evolution of the height of multi-peaks to individual maxima. All these were recorded in neurophysiological experiments without paying attention to it.

The excitable cell model in our study seems robust to external perturbations (stimuli). To get a stable performance, robustness is an essential quality when designing a real weak signal amplifier based on SR. However, this also brings much trouble if the system happens to work in the regime of subthreshold oscillation, which greatly degrades the performance of the amplifier to enlarge weak signals via SR.

The positive role played by subthreshold oscillations in encoding certain stimuli via the mechanism of ASR was highlighted recently where it acted as the periodic forcing. To our knowledge, we are the first to report the negative role played by the subthreshold oscillation on SR phenomenon. As a result, if we make use of it to encode certain stimuli via ASR, the subthreshold oscillation should be strengthened. On the other hand, if we detect weak signals via SR, it should be got rid of.

6.6.3 Global dynamics and stochastic resonance of the forced FitzHugh-Nagumo neuron model

We consider the periodically forced FHN neuron model in the following form:

$$\begin{aligned} \varepsilon\frac{dv}{dt} &= v(v-a)(1-v)-w, \\ \frac{dw}{dt} &= v-dw-b+r\sin(\beta t). \end{aligned} \tag{6.18}$$

The variable v is the fast voltage-like variable and w is the slow recovery variable. We fix the values of the constants to $\varepsilon = 0.005$, $d = 1.0$, and $\beta = 7.5$. Here the slow variable of the neuronal model is driven by the external weak signal; the reason for this is to allow comparison with the results obtained by other scholars.

We show that the small-amplitude subthreshold periodic oscillation and the large-amplitude suprathreshold periodic oscillation coexist commonly in some parameter regions of the forced excitable FHN neuron model (Gong and Xu, 2001), as shown in Fig.6.24 (Figs. 3(a) and (b) in (Gong and Xu, 2001)). To address the question, a white Gaussian noise is added on the second equation. We find that the random transitions induced by noise between the subthreshold oscillation and the suprathreshold oscillation are the essential mechanism underlying stochastic resonance studied by us. The signal-to-noise ratio (SNR) as a function of noise intensity D is shown in Fig. 6.25 (Fig. 7 in (Gong and Xu, 2001)). Clearly, the SNR increases with noise intensity D, reaches a maximum, and decreases again, displaying the typical feature of stochastic resonance. This kind of bistability was also found in the HH neuron model with time-dependent sinusoidal stimulation, but stochastic resonance was discussed only in the region where the periodically forced HH neuron model has one attractor, a stable nonfiring state. It is no doubt that the appearance of such dynamic bistability should exist in other forced excitable neuronal models such as the ML

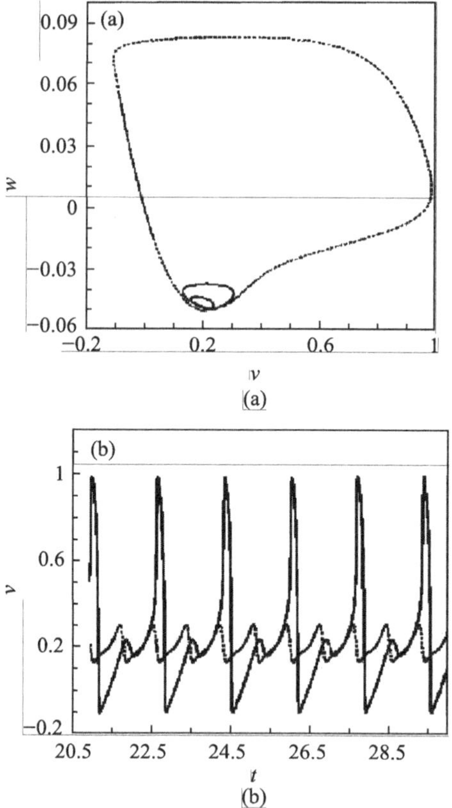

Fig. 6.24 The subthreshold small-amplitude periodic solution with period $T_1 = T_0$ and The suprathreshold large-amplitude periodic solution with period $T_2 = 2T_0$, here $T_0 = 2\pi/\beta$: (a) the coexistent attractors in the state space $v - w$, the dashed line is for the suprathreshold oscillation and the solid line is for the subthreshold oscillation; (b) Membrane potential versus time for the two attractors, the dashed line is for the subthreshold oscillation and the solid line is for the suprathreshold oscillation. The units of the variables are arbitrary.

neuron model and the Chay neuron model. Moreover, such bistability has been observed experimentally in neurons. Therefore our results may help us to understand stochastic resonance in these neuron systems.

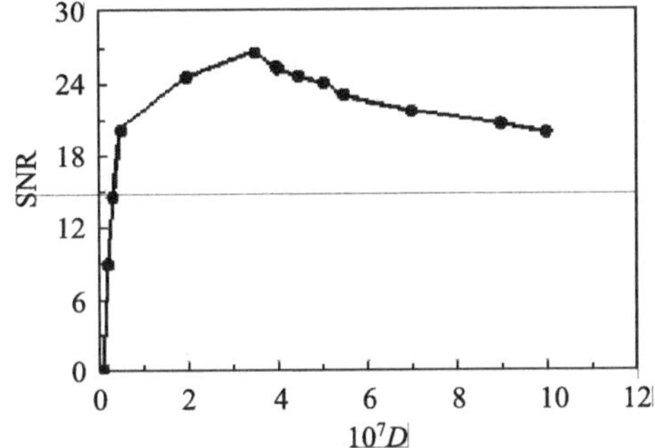

Fig. 6.25 Signal-to-noise ratios versus noise intensity D. The units of the axes are arbitrary.

In comparison with the previous studies about stochastic resonance, our work shows that stochastic resonance of the excitable neuronal model is related to the dynamic bistability. Furthermore, the transitions induced by noise between the two dynamic oscillation are studied by us. The mean of the return time and the mean-to-variance ratio of the pulse number distributions are calculated in our works, the results suggest that these values can serve to distinguish our case from the previous studies about stochastic resonance in the typical bistable nonlinear system and the excitable neuronal model. Moreover, it is interesting to note that for our case the external signal can be regarded not only as a subthreshold stimulation but also a suprathreshold stimulation, thus our studies also extend the classical stochastic resonance which is used to detect a subthreshold signal to a new range that can be used to detect suprathreshold signal for neurons.

Through comparing the stability of the firings of the FHN neuron model with smooth basin boundary and that with fractal basin boundary, we can draw the conclusion that the stability of firings of the forced FHN neuron with fractal basin boundary can be changed easily under the small noise perturbation. Figure 6.26 exhibits that the basin boundary between the two coexistent attractors has a fractal structure, and it is verified by the further calculation about the dimension of the basin boundary. This result suggests that in order to maintain the stability of firing state subjected to random perturbations, the neuron model should be operated in the region where the basin of attraction is smooth. The result also suggests that when we study the dynamic behavior of some typical neuronal models, much attention should be paid to the global dynamics of these systems. As shown in the present studies, the global characteristics may have significant effects on some issues we are interested in.

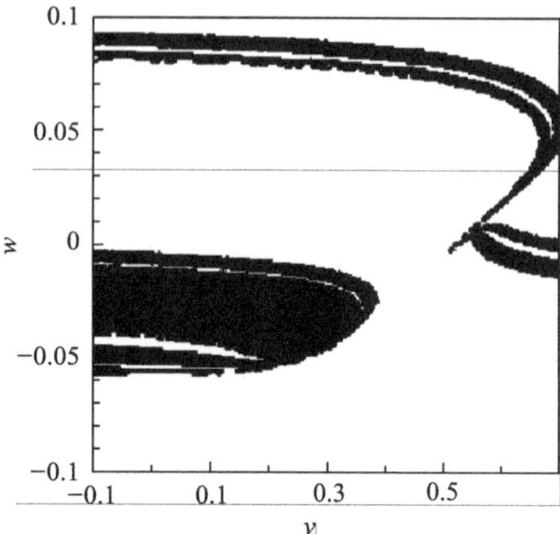

Fig. 6.26 Basin of attraction for the forced FHN neuronal model when $b = 0.259\,92$, $r = 0.016\,3$ (Figure 15 in (Gong and Xu, 2001)). The black dots represent the basin of attraction of the subthreshold oscillation, the other region is the basin of attraction of the suprathreshold oscillation. The units of the v and w are arbitrary.

6.6.4 A novel dynamical mechanism of neural excitability for integer multiple spiking

We show that saddle-node on invariant circle (SNIC) bifurcation is a novel dynamical mechanism for integer multiple spiking (Xie et al., 2004a), and a neuron with integer multiple spiking may employ the phenomenon of stochastic resonance to detect external weak signals and transmit neural information. Integer multiple spiking (IMS), here, is seemingly random firing behavior whose interspike interval histogram (ISIH) exhibits multimodal structure with peaks at integer multiples of a basic interspike interval (ISI). Furthermore the amplitude of the peaks decays with increasing ISI except for the first few peaks, and the return map of ISI series has a crystal lattice structure. Such a phenomenon is referred to as stochastic phase locking or skipping by Longtin (Longtin, 1995), it has been already observed in a variety of neurophysiological experiments. Most recently, Xing and Hu et al. (2001) observed the IMS in spontaneous discharge from injured dorsal root ganglion neurons. Actually, this special kind of firing pattern was found earlier in the auditory fibers of squirrel monkey and of the cat retinal ganglion cells and primary visual cortex, and mechanoreceptor of the macaque monkey and crayfish when subject to periodic stimulus (Gammaitoni et al., 1998).

The dynamical mechanisms of neural excitability for the IMS involved only Hopf bifurcation (including supercritical and subcritical) in the existing studies. Since there exist two major dynamical mechanisms of neural excitability, i.e., SNIC and Hopf bifurcations, we want to investigate whether a model neuron near SNIC bifurcation can exhibit the firing behavior of the IMS.

There are various noise sources for a neuron, such as ionic channel conductance fluctuations, synaptic fluctuations and thermal noise. In general, therefore, the noise component can be described using Gaussian white noise. Since the ML model under a set of parameter values can undergo a SNIC bifurcation from the rest state to repetitive spiking, here, we utilize this model to simulate the IMS when the model neuron is subjected to a subthreshold periodic stimulus and Gaussian white noise.

In the following calculation, we take parameter values: $V_1 = -1.2$, $V_2 = 18$, $V_3 = 12$, $V_4 = 17.4$,$\bar{g}_{Ca} = 4$, $\bar{g}_K = 8$, $\bar{g}_L = 2$, $V_{Ca} = 120$, $V_K = -84$, $V_L = -60$, $\phi = 1/15$and $C = 20$. The bifurcation diagram of the membrane potential versus I_0 is shown in Fig. 6.9. This model neuron transits a SNIC bifurcation at $I_0 = 39.96$, and the state of the neuron changes from the rest state to repetitive spiking. A subcritical Hopf bifurcation (HB) occurs at $I_0 = 97.79$, and a saddle-node bifurcation of limit cycles occurs at $I_0 = 116.1$. Here, we are only interested in the dynamical mechanism of excitability from the rest state to repetitive spiking, namely, the SNIC bifurcation.

As stated above, a subthreshold periodic stimulus $I_1 \sin(2\pi f t)$ and noise $\xi(t)$ are added up to the right hand of the first equation to produce the IMS. Since $I_1 \sin(2\pi f t)$ is subthreshold, it alone is insufficient to evoke firing of the neuron. The $\xi(t)$ is chosen as a Gaussian white noise with the statistical properties as $\langle \xi(t) \rangle = 0$ and $\langle \xi_1(t)\xi_2(t) \rangle = 2D\delta(t_1 - t_2)$, where D is the noise intensity, and δ is the Dirac-function.

We fix the constant current $I_0 = 37\mu$A/cm^2, the frequency of the stimulus $f = 10$ Hz, and the amplitude of the stimulus $I_1 = 8$ μA/cm^2, which is subthreshold one. For a while the intensity of noise is chosen as $D = 130$. The stochastic Runge-Kutta algorithm proposed by Honeycutt is used to integrate the stochastic ML mode with the integration time step of 0.1 msec. A realization of this stochastic model is implemented, and the time course of the membrane potential is shown in Fig. 6.27(a). Clearly, the firing occurs near a preferred phase of the stimulus, but there can be a random number of periods skipped between two successive firings. In other words, noise can produce stochastic phase locking in the ML model with a subthreshold periodic stimulus in the vicinity of the SNIC bifurcation. Figure 6.27(b) shows the series of ISIs. We can see that the ISIs are concentrated at integer multiples of the stimulus period of 100 msec and exhibit a structure of distinct layers. The peaks of the ISIH decay exponentially except for the first two peaks, as seen in Fig. 6.2(a). Furthermore, the width of the ISIH peaks determines the degree of phase locking: sharp peaks correspond to a high degree of phase locking, i.e., a narrow range of phases of the periodic stimulus during which firing preferentially occurs. The return map of the ISI series has a crystal lattice structure, as shown in Fig. 6.2(b). These all exhibit the major features of the IMS observed in the neurophysiological experiments. It follows that the firing behavior of the IMS is simulated successfully by use of the ML under the action of subthreshold periodic stimulus and Gaussian white noise. SNIC bifurcation, therefore, is a novel dynamical mechanism of neural excitability for the IMS.

Evidently, the deterministic ML model is not a bistable system near the SNIC bifurcation. To a certain extent, however, the stochastic ML model is a nearly bistable

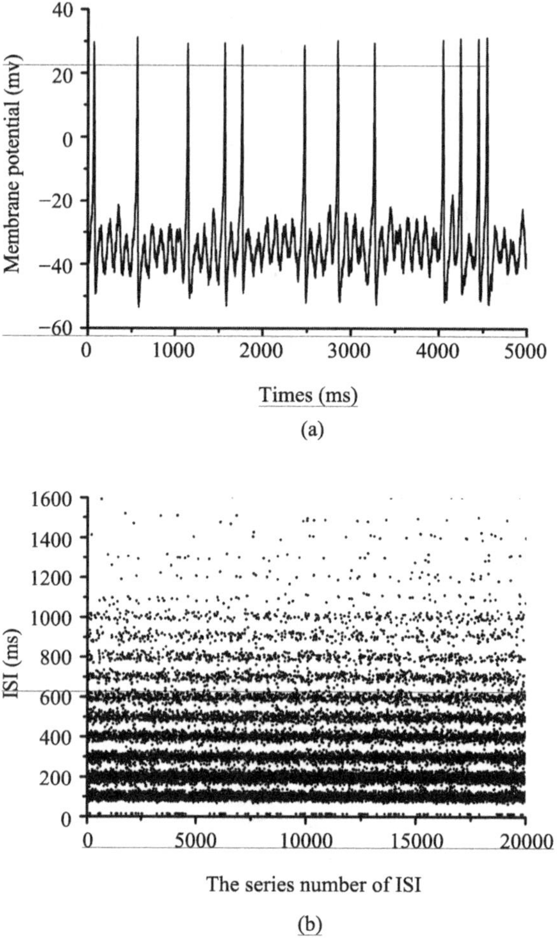

Fig. 6.27 The simulation of Integer multiple spiking in the stochastic ML model (Figure 4 in (Xie et al., 2004a)). The first 200 msec data is discarded to avoid transient. (a) The time series of membrane potential. (b) The ISI versus the series number of ISI.

system, i.e., noise-induced bistablity. It follows that the SR can be characterized by the ISIH, which corresponds to the residence time histogram (RTH) of a bistable system. Figure 6.28 shows the height of the first three peaks in the ISIH computed from 20,000 ISIs as a function of noise intensity D. We can see that a change in the height of the second (or third) peak by varying the noise intensity: from a small peak at low noise intensity, going through a maximum at an intermediate noise intensity and decreasing again at a high noise intensity. An optimal noise intensity can be found at where the maximum is located. This phenomenon is a signature of a resonance (Longtin, 1993b; Wiesenfeld and Moss, 1995), and it is known as SR, which is nonlinear cooperative effect in which a weak periodic stimulus entrains large-scale environmental fluctuations, with the result that the periodic component is greatly enhanced. Therefore, neurons may make use of the SR phenomenon to detect weak signals and transmit neural information, in which the noise plays a constructive role.

In summary, the IMS occurs in the vicinity of a SNIC bifurcation. This point is different from the existing investigations in which the IMS was obtained near a Hopf bifurcation. Hence, the SNIC bifurcation is a novel dynamical mechanism of neural excitability underlying the IMS. Besides, we have investigated stochastic resonance in the ML model neurons near the SNIC bifurcation. Neurons may make use of the special firing behavior of the IMS to detect weak signals and transmit neural information.

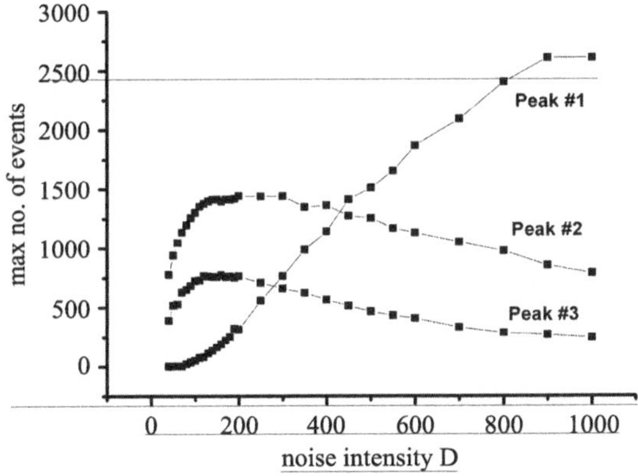

Fig. 6.28 Height of the first three peaks in the ISIH as a function of noise intensity D (Figure 5 in (Xie et al., 2004a)).

6.6.5 A Further Insight into Stochastic Resonance in an Integrate-and-fire Neuron with Noisy Periodic Input

To check whether double SR occurs in an integrate-and-fire (IF) neuron with input noise, a further insight into SR of the IF neuron is made. Since the IF neuron is one of the simplest neuron models, it is frequently used to get qualitative characteristics about SR. As we know, the noise enters the neuron in two ways, i.e. as the input noise, or as the threshold noise. The former causes the membrane voltage to fluctuate, while the latter causes the threshold to fluctuate. For the input noise, Bulsara et al. (1995) studied the cooperative behavior between the noise and the periodic stimulation in a simplified version by the method of mirror image; Shimokawa et al. (1999, 2000), Plesser and Geisel (1999) studied the SR based on numerically calculating the mean first passage time, which is also named as interspike interval (ISI) in the background of neural discharge. As for the threshold noise, Barbi et al. (2003) recently reported the double SR, and they stressed that it is interesting to check whether the double SR can be observed in the case of the input noise. Inspired by them, we are dedicated to investigating whether the double SR occurs in the IF neuron with the input noise, and clarify the comparison between the two cases.

We consider the IF neuron described by the following equation:

$$\begin{aligned} \frac{dV(t)}{dt} &= -\frac{V}{\tau} + \mu + A\sin(2\pi t/T + \varphi_0) + \sigma\xi(t), \quad V(t) \le S_0, \\ V(t^+) &= V_0, \quad \text{if} \quad V(t) = S_0 \end{aligned} \tag{6.19}$$

where V is the membrane potential; S_0 is the constant firing threshold; $V_0 < S_0$ represents the post-discharge resetting potential; τ represents the characteristic membrane charge-discharge time; $\mu\tau$ represents the resting potential; $\xi(\cdot)$is the Gaussian white noise satisfying $< \xi(t+\tau)\xi(t) >= \delta(\tau)$; σ is the noise intensity; and A, T and φ_0 are the amplitude, period, and initial phase, repectively. Once V reaches S_0, it is immediately reset to V_0 and a spike pulse is generated. The output of the neuron consists of the sequence of these pulses. When there is no input signal, namely$A = 0$, Eq. (6.19) is the so-called Ornstein-Uhlenbeck with an absorbing boundary at S_0.

Let us temporarily omit the role of the noise, we have

$$\begin{aligned} \overline{V}(t,\varphi_0) = V_0 \exp\left(-\frac{t-t_0}{\tau}\right) + \tau\mu\left(1-\exp\left(-\frac{t-t_0}{\tau}\right)\right) \\ + \frac{A}{1+\tau^2\Omega^2}(\tau\sin(\Omega t+\varphi_0) - \tau^2\Omega\cos(\Omega t+\varphi_0)) \\ - \frac{A}{1+\tau^2\Omega^2}\left[\tau\sin(\Omega t_0+\varphi_0) - \tau^2\Omega\cos(\Omega t_0+\varphi_0)\right]\exp\left(-\frac{t-t_0}{\tau}\right). \end{aligned} \tag{6.20}$$

If $t \to \infty$, then

$$\overline{V}(t,\varphi_0) \to \tau\mu + \frac{A}{1+\tau^2\Omega^2}\left[\tau\sin(\Omega t+\varphi_0) - \tau^2\Omega\cos(\Omega t+\varphi_0)\right]$$

with the maximum value

$$\overline{V}_{\max} = \tau\mu + \frac{\tau A}{\sqrt{1+\tau^2\Omega^2}}. \tag{6.21}$$

Then from $\overline{V}_{\max} = S_0$ we get the critical amplitude $A_c = \frac{1}{\tau}\sqrt{1+\tau^2\Omega^2}(S_0 - \tau\mu)$. For suprathreshold periodic inputs, i.e. $A \ge A_c$, Eq. (6.19) has complex dynamics. For instance, if during each M input signal periods, just N spikes are generated, the discharge pattern is a $M:N$ phase locking with M and N to be positive integers. If the noise is considered, the counterpart is the $M:N$ stochastic phase locking, which is a statistical definition.

Next we introduce the method for the probability distribution of ISIs (Buonocore et al., 1987). For this purpose, we make a transform $V(t) = X(t) + \overline{V}(t,\varphi_0)$, then Eq. (6.19) turns into the Ornstein-Uhlenbeck with an absorbing boundary at the transformed threshold, namely,

$$\begin{aligned} dX(t) &= -\frac{X}{\tau}dt + \sigma dW(t), \quad X(t) \le S_{\text{mod}}(t,\varphi_0) = S_0 - \overline{V}(t,\varphi_0) \\ X(t^+) &= 0, \quad \text{if} \quad X(t) = S_{\text{mod}}(t,\varphi_0) = S_0 - \overline{V}(t,\varphi_0) \end{aligned} \tag{6.22}$$

If define the first passage time of Eq. (6.22) as

$$\tau_f = \inf\{u : X(u) > S_{\text{mod}}(u,\varphi_0) | X(0) = 0 < S_{\text{mod}}(0,\varphi_0)\}, \tag{6.23}$$

then τ_f is a random variable, and it is the same as an ISI defined by Eq. (6.19). Let $g(S_{\text{mod}}(u,\varphi_0), u|0)$ represent the probability density function of τ_f, then $g(S_{\text{mod}}(u,\varphi_0), u|0)$ satisfies the integral equation (Buonocore et al., 1987).

$$\begin{aligned} g(S_{\text{mod}}(t,\varphi_0),t|0) &= -2\Psi(S_{\text{mod}}(t,\varphi_0),t|0,0) \\ &\quad +2\int_0^t g(S_{\text{mod}}(s,\varphi_0),s|0)\Psi(S_{\text{mod}}(t,\varphi_0),t|S_{\text{mod}}(s,\varphi_0),s)ds, \end{aligned} \tag{6.24}$$

where a nonsingular integral kernel Ψ is defined as

$$\Psi(S_{\text{mod}}(t,\varphi_0),t|S_{\text{mod}}(s,\varphi_0),s) = p(S_{\text{mod}}(t,\theta),t|y,s)H(t,s,y) \tag{6.25}$$

with

$$\begin{aligned} H(t,s,y) &= \frac{1}{2}\left(S'_{\text{mod}}(t,\varphi_0) + \frac{S_{\text{mod}}(t,\varphi_0)}{\tau}\right) \\ &\quad + \frac{\exp((t-s)/\tau)/\tau}{1-\exp(2(t-s)/\tau)}\left[S_{\text{mod}}(t,\varphi_0)\exp((t-s)/\tau) - y\right], \end{aligned} \tag{6.26}$$

$$p(x,t|y,s) = \frac{1}{\sqrt{\pi\sigma^2\tau(1-e^{-2(t-s)/\tau})}}\exp\left(-\frac{(x-ye^{-(t-s)/\tau})^2}{\sigma^2\tau(1-e^{-2(t-s)/\tau})}\right). \tag{6.27}$$

Here $p(x,t|y,s)$ is the probability density function of the Ornstein- Uhlenbeck process with free boundaries. Solving Eq. (6.24) yields the probability density function, namely,

$$\begin{aligned} g(S_{\text{mod}}(\Delta t,\varphi_0),\Delta t|0) &= -2\Psi(S_{\text{mod}}(\Delta t,\varphi_0),\Delta t|0,0), \\ g(S_{\text{mod}}(k\Delta t,\varphi_0),k\Delta t|0) &= -2\Psi(S_{\text{mod}}(k\Delta t,\varphi_0),k\Delta t|0,0) \\ &\quad +2\Delta t\sum_{j=1}^{k-1} g(S_{\text{mod}}(j\Delta t,\varphi_0),j\Delta t|0), \\ &\quad \times\Psi(S_{\text{mod}}(k\Delta t,\varphi_0),k\Delta t|S_{\text{mod}}(j\Delta t,\varphi_0),j\Delta t), \\ &\qquad k = 2,3,\ldots \end{aligned} \tag{6.28}$$

Now we turn to the Markov analysis of SR. Since $g(S_{\text{mod}}(u,\varphi_0), u|0)$ depends only on the initial phase φ_0 at time $t = t_0$, it can be rewritten as $g(u|\varphi_0)$ for brief. If

φ_0 denotes the previous discharge phase and $f(\varphi|\varphi_0)d\varphi$ represents the probability with which the following discharge phase falls within $(\varphi, \varphi + d\varphi)$, then $f(\varphi|\varphi_0)$ reads

$$f(\varphi|\varphi_0) = \frac{1}{\Omega}\sum_{k=0}^{\infty} g\left(kT + \frac{\varphi - \varphi_0}{\Omega}|\varphi_0\right). \tag{6.29}$$

Further, if the initial discharge phase distribution is $h_1(\varphi_0)$, then the successive discharge phase distributions obey a recursive relation $h_{n+1}(\varphi) = \int_0^{2\pi} f(\varphi|\varphi_0) h_n(\varphi_0)d\varphi_0 (n = 0, 1, 2, \ldots)$, and there exists a stationary discharge phase distribution $h_s(\varphi) = \lim_{n\to\infty} h_n(\varphi)$. By approximating the continuous phase axis by discrete phase state, the continuous kernel f is reduced to a finite-order stochastic matrix with elements $f(2\pi j/N|2\pi k/N)2\pi/N,\ j,k = 0, 1, \ldots, N$. Correspondingly, $h_s(\varphi)$ is approximated by the eigenvector belonging to the eigenvalue 1. From $h_s(\varphi)$ the stationary probability density function $g(t)$ of ISI can be calculated according to

$$g(t) = \int_0^{2\pi} g(t|\varphi)h_s(\varphi)d\varphi. \tag{6.30}$$

The importance of $g(t)$ is in that it quantifies the coherence between the input signal and the output signal. It is also possible to compute the statistical properties such as autocorrelation and power spectral density of the spiking train, and finally the signal-to-noise ratio (SNR) (Shimokawa et al., 2000) is

$$S^* = \frac{2\pi}{\langle \tau_f \rangle}\left|\int_0^{2\pi} h_s(\varphi)e^{-i\phi}d\varphi\right|^2, \tag{6.31}$$

where $\langle \tau_f \rangle = \int_0^{\infty} t g(t)dt$ is the mean ISI.

For the IF neuron with the threshold noise (Barbi et al., 2003), when the subthreshold input is strong enough, there exist two different noise intensities at which $g(nT)$(n takes 1 and 2 or 0.5) attain their maximums, respectively, and as a result the output SNR reaches two extremes near the noise intensities, which is the so-called double SR. Now in order to check whether this double SR occurs in the IF neuron with the input noise, let us make a comparison between the SR in the two cases. Seen from Fig.6.29(a), when the input signal is subthreshold, the highest peak of $g(t)$ moves towards the origin, so there exist three different noise intensities such that $g(nT)$(n takes 1 or 2 or 0.5) attains the maximum, respectively, as shown in Fig.6.29(b). But compared with Fig.6.29(c), these noise intensities nearly have no bearing with the optimal noise intensity of SR, and this is completely inconsistent with the case with the threshold noise. Furthermore, comparing Fig. 6.29(c) with Fig. 6.29(d), we see at the optimal intensity $\sigma_{opt} \approx 0.75$ (where the resonant peak is observed) that the mean ISI nearly equals the input period, i.e. $< \tau_f > \approx T$. Therefore, the occurrence of SR here, similar to the case in (Shimokawa et al., 1999) is due to the 1:1 stochastic phase locking, even though there isn't an obvious plateau-like flattening on the curve of $< \tau_f >$ vs. σ.

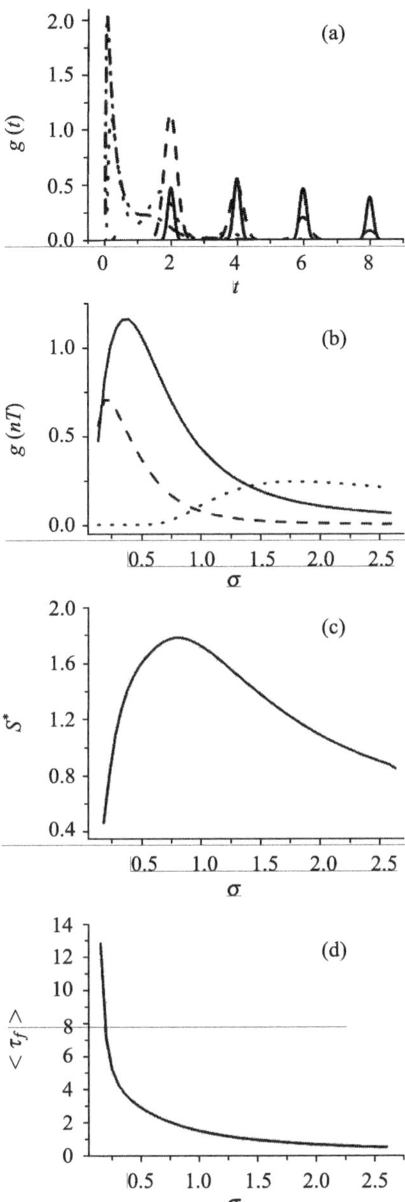

Fig. 6.29 (Figure 1 in (Kang et al., 2005a)). (a) The stationary distribution of ISI with different noise intensity: 0.15 (solid), 0.4 (dash), 1.15 (dot), 1.9 (dash dot); (b) the evolution of $g(nT)$ via noise intensity with n=1(line), n=0.5(dot), n=2(dash); (c) the SNR and (d) mean ISI via the noise intensity. The parameters are $\tau = 1.0, \mu = 0.35, A = 1.0, \Omega = 0.5\pi$ and $S_0 = 1.0$.

Since the evolution of $g(nT)$ does not contain the material information of SR, the subsequent plots only show the dependence of mean ISI and SNR via the noise intensity. To check whether the double SR exists, we choose a "low" frequency and a "high" frequency to give numerical results just as that done in (Barbi et al., 2003), respectively. In the low frequency case (see Fig.6.30), only if the subthreshold input has a amplitude very close to the critical one $A_c = 0.7086$ such as $A = 0.7085$, a hump is observed on the curve of SNR via noise intensity left to the main resonant peak, however this hump is not a secondary resonant peak. For the given $A = 0.7085$, the optimal noise intensity is $\sigma_{opt} = 0.54 \pm 0.01$, where the mean ISI is $< \tau_f >= 4.0 \pm 0.03$, so the occurrence of SR is owing to the 2:5 stochastic phase locking. Noting that not only at the noise level where the hump is observed $< \tau_f >$ is a little larger than T, but there is a plateau-like flattening corresponding to the hump, the occurrence of the hump corresponds to the 1:1 stochastic phase locking. In the high frequency case (see Fig.6.31), even when the input amplitude is exactly the same as the critical one, i.e. $A = A_c = 0.8242$, an obvious hump different from the

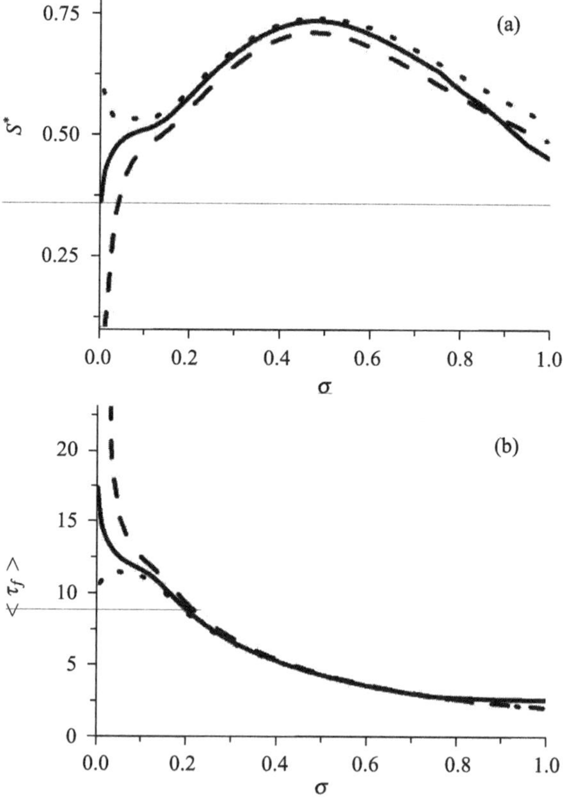

Fig. 6.30 (Figure 2 in (Buonocore et al., 1987)) The evolution of (a) SNR and (b) mean ISI via noise intensity. The parameters $\tau = 1.0$, $S_0 = 1.0$, $T = 10$, $\mu = 0.4$ and A is 0.7085(solid), 0.69(dash), 0.72(dot).

main SR peak is not obvious. The main SR peak is observed at $\sigma_{opt} = 0.21 \pm 0.01$ with $< \tau_f >= 3.68 \pm 0.03$, therefore in this case the occurrence of SR is due to the 2:1 stochastic phase locking. The above analysis shows us there does not exist a double SR in the IF neuron with the input noise, but when the input frequency is low enough, a resonant structure similar to that of a double SR can be observed. Why such a hump is obvious only in the low input frequency case? The explanation should be connected with the most left peak at the stronger noise level in Fig.6.29 (a). In fact, the peak is noise-dominated, and it will soon overwhelm the coherent integer multiple peaks when the input period is short. Therefore, when the input frequency is "high", some coherent information of the input signal is smeared by the noise quickly, which makes the hump on the SNR curve unclear. In addition, the above analysis also shows us that the stochastic phase locking does not suggest the SR certainly, and the stochastic phase locking connected with SR is not confined to the 1:1 pattern, which extends the conclusion in (Shimokawa et al., 1999).

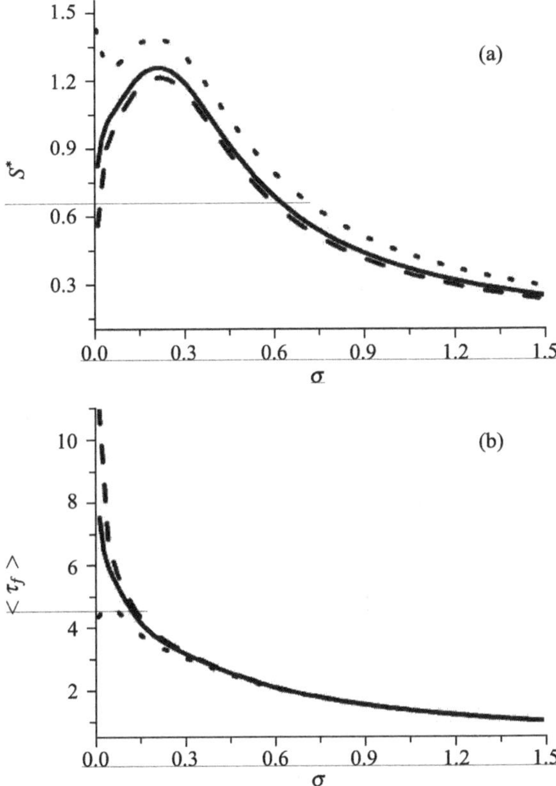

Fig. 6.31 (Figure 3 in (Kang et al., 2005a)). The evolution of (a) SNR and (b) mean ISI via noise intensity. The parameters $\tau = 1.0, S_0 = 1.0, T = 2, \mu = 0.75$ and A is 0.8242(solid), 0.8(dash), 0.9(dot).

Thus, we have drawn the conclusions about the IF neuron with the input noise as follows. Firstly, the noise intensity where the extreme of the ISI stationary probability peak $g(nT)$ is observed is greatly discordant with the optimal intensity where the SNR attains the maximum. Secondly, the double SR cannot occur in the IF, but the resonant structure similar to that of the double SR can be observed when the input frequency is low. Thirdly, for stronger subthreshold input signals, the occurrence of SR does relate to the stochastic phase locking, but the phase locking pattern can be beyond the 1:1 pattern. The conclusions might reflect some properties of real neurons, and they should be important in understanding the role of noise in neurophysicology.

6.6.6 Signal-to-noise ratio gain of a noisy neuron that transmits subthreshold periodic spike trains

We numerically investigate the transmission properties of an integrate-and-fire neuron model that transmits coherent subthreshold spike trains in a shot noise environment (Kang et al., 2005b). For very weak coherent couplings, it is shown that the input-output signal-to-noise ratio (SNR) gain is easier to exceed unity; while for stronger coherent couplings it is difficult to observe the SNR gain larger than unity at the optimal noise intensity. These observations are different from those acquired in the case of continuous noise. Our analysis further suggests that the larger SNR gain in the very weak coherent coupling case should be due to the noise induced resonance. It is also shown that there is more possibility of the SNR gain above unity for slower periodic spike trains transmitted by the model. The results may be useful in understanding the performance of real noisy neurons acting as signal-processing elements.

6.6.7 Mechanism of bifurcation-dependent coherence resonance of Morris-Lecar Model

The mechanism of bifurcation-dependent CR of excitable neuron models is related to the random transitions between attractors on two sides of bifurcation point. We examine that the relationship between the random transitions and the mechanism of bifurcation-dependent CR by use of the ML neuron model [113], and show that there exist different attractors on two sides of the Hopf bifurcation point. It follows that the neuron may transit between attractors on two sides of bifurcation point at the presence of noise. The transition frequency tends towards a certain value for a certain optimal noise intensity. Since the SNR of the neuronal response evaluated at this certain frequency is maximal at the optimal noise intensity, CR occurs.

6.7 Analysis of time series of interspike intervals

As stated previously, all neurons in the brain fire action potentials that carry information to other parts of the brain along their fibres, and neural information is embedded in the series of interspike intervals (ISIs). Consequently, the analysis methods for the series of ISIs are very important in order to extract the neural information. In what follows, we illustrate several methods from the viewpoint of nonlinear dynamics.

6.7.1 Return map

Return map is a simple method to find out the deterministic structure hidden in the series of the ISIs. In detail, return map is constructed by previous ISI versus next ISI. As we know, bifurcation and chaos phenomena of ISI are often dynamically observed along a single fiber of injured sciatic nerve in the anesthetized rat. Figure 6.32(a) exhibits the continuous bifurcation procedure of the ISI data CA381W4 that were recorded at a constant sampling interval t_s of 97.66 μs as soon as the concentration of calcium in the Krebs' solution used to perfuse the injured sciatic nerves was decreased from 5.0 mmol/L to 1.2 mmol/L (Gong et al., 1998). Therefore, the direction of the abscissa not only stands for the sampling sequence of ISI but also qualitatively represents the change in the ionic concentration of the solution with time. To reveal the mechanism underlying the irregular ISI after the period transition enlarged and shown in Fig. 6.32(b) to the left of the vertical line, the data in the point smear region from the right side of the vertical line to the end of the data file have been separated and uniformly divided into nine groups each consisting of 4000 points. Thus every group may approximately be considered as a steady-state process and return map can be applied to investigate the series of the ISIs.. Figure 6.32(c) shows the one-hump structure constructed by the ISI data of the first group using return map. This simple form is the most convincing evidence for the existence of deterministic chaotic dynamics.

6.7.2 Phase space reconstruction

Actually, phase space reconstruction is attractor reconstruction, which refers to methods for inference of geometrical and topological information about a dynamical attractor from observations (for detail see: http://www.scholarpedia.org/article/Attractor_reconstruction.). Attractor reconstruction is an important first step in the process of making predictions for nonlinear time series and in the computation of certain invariant quantities such as Lyapunov exponent used to characterize the dynamics of such series. The reliability of computed predictions and the accuracy of invariant quantities are strictly dependent on the accuracy of attractor reconstruction, which in turn is determined by the methods used in the reconstruc-

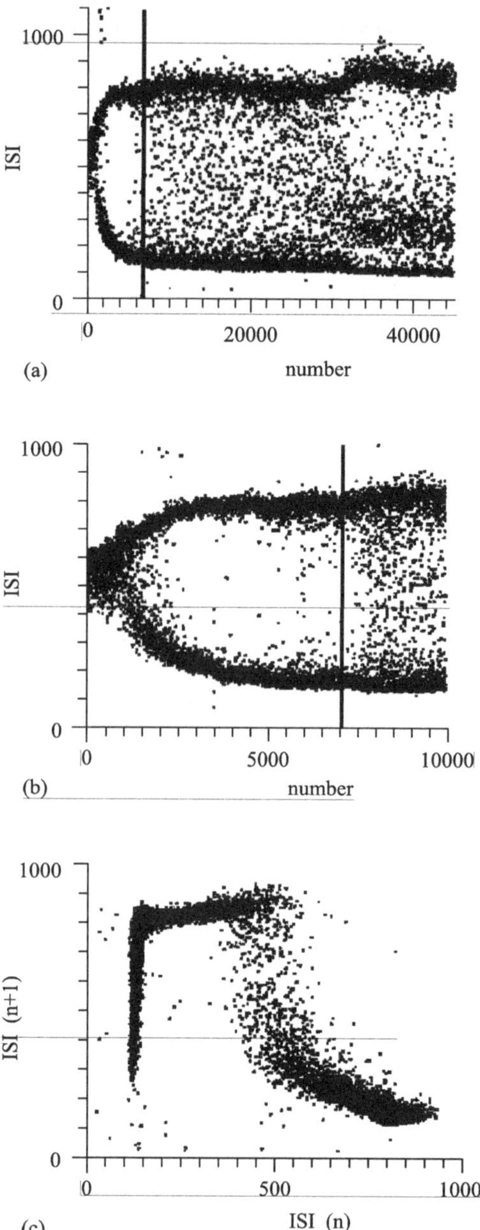

Fig. 6.32 (Figure 5 in (Gong et al., 1998)). (a) Bifurcation diagram of the ISI data CA381W4: ISI versus ISI serial number. (b) Enlargement of the period transition on the left side of the vertical line. (c) One-hump map constructed from the plot of ISI(n+1) versus ISI(n).

tion process. Usually the time delay embedding as the way to reconstruct the phase spaces of original dynamical systems. The validity of such a method is guaranteed by the Whitney embedding theorem and the Takens embedding theorem (http://www.scholarpedia.org/article/Attractor_reconstruction.). In the reconstruction process, time delay and embedding dimension are two very import quantities. Now there are various methods for determining such two quantities, autocorrelation function and mutual information for time delay, and the false near neighbor (FNN) method and the Cao method for embedding dimension, for example. Embedding ideas were later extended beyond autonomous systems with continuously-measured time series. A version was designed by Sauer for the series of ISIs. Actually, time delay for the series of ISIs can be selected as 1 because it can be considered to be a discrete dynamical system. The key problem, therefore, is to determine the embedding dimension for the series of ISIs.

6.7.3 Extraction of unstable periodic orbits

Periodic orbits play an important role in understanding the rich structures in a dynamical system (http://www.scholarpedia.org/article/Unstable_periodic_orbits). An unstable periodic orbit (UPO), however, is simply a periodic orbit which is dynamically unstable, and actually UPOs in chaotic set can be saddles as well as repellers. Since the set of UPOs is dense within the chaotic set, a typical trajectory wanders incessantly in a sequence of close approaches to these orbits. The more unstable an orbit, the less time that a trajectory spends near it. Interestingly, the set of UPOs can be considered to be the skeleton for chaotic attractors. Furthermore, many dynamical invariants, such as, natural measure, Lyapunov exponents, and fractal dimensions can by efficiently expressed in terms of a sum over UPOs. Therefore, UPOs are also important tools in affecting the behavior of dynamical systems (So et al., 1996,1997; Cvitanovic, 1998).

In most cases, time series of some variables observed in experiments are usually the only available information from a dynamical system. To further analyze the system, we firstly need to reconstruct its phase space. Recurrence method [67] is a simple one for detecting UPOs from this reconstructed data set, and the standard procedure is to look for peaks in a histogram of recurred points as a function of their recurring periods. The sensitivity of this method in finding UPOs naturally depends on the natural measure of the UPOs.

An enhancement of the standard recurrence methods was proposed later (So et al., 1996,1997). In this method, experimentally extracted linear dynamics near each state point was incorporated into a periodic-orbit transform that take experimental data into a space where the probability measure at the UPOs are enhanced and at other non-recurring points are dispersed. Similar to the previous recurrent methods, an experimenter detects UPOs by looking for peaks in the transformed space.

6.7.4 Nonlinear prediction and surrogate data methods

The nonlinear forecasting method was proposed that measures the predictability of the future state of a dynamical system (Sugihara and May, 1990). It has the feature of analyzing the experimental time series dynamically and the advantage of needing only a relatively short time series, so nonlinear forecasting quickly became one of the most important tools for the analysis of experimental data. The key of this method is that when predicting the future state from chaotic time series, the nearest-neighbor points are utilized so as to obtain the best prediction (Gong et al., 1998). Let X_i^T, $i = 1,\ d+1$ be the nearest-neighbor points that form the the smallest simplex around the point X^T, and X_{i+p}^T, $i = 1,\ d+1$ be their p steps of evolution points. Then the point X_p^T after p steps of evolution from X^T can be predicted:

$$X_p^T = \sum_{i=1}^{d+1} \exp(-\|X_i^T - X^T\|)X_{i+p}^T, \tag{6.32}$$

according to the exponential weights computed from the Euclidean distances between X^T and its nearest-neighbor points.

After taking the first difference Δ of the measured time series to reduce the effects of any short-term linear autocorrelations, the time series from observation are usually divided into two groups. One group is used as the data base for forecasting the future while the other is used for the purpose of comparison with the predicted values. Finally, Pearson's correlation coefficient is applied to evaluate the effect of prediction, i.e., the larger the values of the correlation coefficients is, the better the forecasting is.

It is known that the long term behavior of a chaotic system is almost unpredictable due to its sensitivity to the initial conditions, while the behavior of the system can still be predicted to some extent for a relatively short time. As a result, the evolution curve of the correlation coefficient first decay slowly when the prediction step is small, then quickly with the growth of the prediction step.

Nonlinear forecasting method is frequently combined with the surrogate data method (Theiler et al., 1992) together to analyse the series of ISIs. The simplest algorithm for generating surrogate data is random shuffling. According to this algorithm we will rearrange the original data in the rank order of the Gaussian white noise generated from a random number generator. This algorithm guarantees that the surrogate data is consistent with the null hypothesis of a δ-correlated random process, while exactly preserving the distribution of the original data. Another surrogate data algorithm is phase randomization, which is realized by implementing the Fourier transform for the original data set, randomizing the phases and then inverting the transform. The surrogate data generated by the algorithm of phase randomization have the same power spectrum as the original data.

Figure 6.33 shows that the ISI, i.e., the time interval between any two successive impulses, is the same as the interval between their corresponding two crosspoints of the trajectory with the Poincare section from one side. According to this, the bifurcation and chaos phenomena of ISI time series from the β-cell model can conveniently be obtained with the proper choice of Poincare section (Gong et al., 1998). As an example, for the chaotic ISI time series given in Fig. 6.34 from the β-cell model, the values of the surrogate data generated by random shuffling are the same as the original data, while the structure is identical to that of the noise. With the embedding dimension $d = 3$ that equals the number of state variables of the model and the time delay $\tau = 1$, altogether 4000 data points were used for the computation of the noisy chaotic ISI time series from the model. It can be seen in Fig. 6.35 that with the increase in the prediction step, the evolution of the correlation coefficient curve for the chaotic ISI time series decays slowly first then very quickly, which is apparently different from that of the surrogate data even in this worse SNR case. This suggests the deterministic structure in the noisy time series. Therefore, one can easily distinguish chaos from noise according to their different evolution of the correlation coefficient curves, even at a high level of noise.

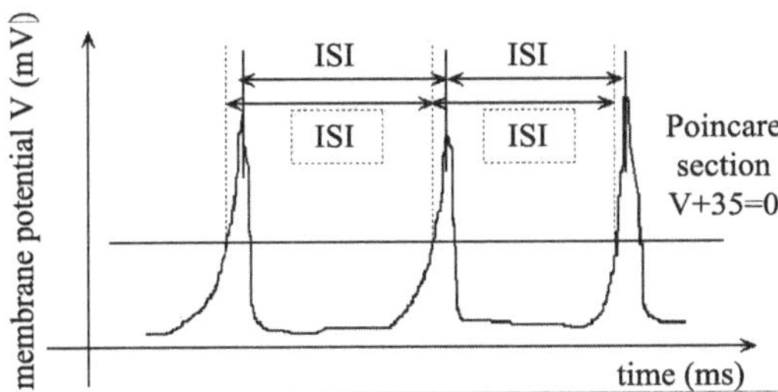

Fig. 6.33 Schematic diagram of the corresponding relationship between interspike intervals and the Poincare section (Figure 1(b) in (Gong et al., 1998)).

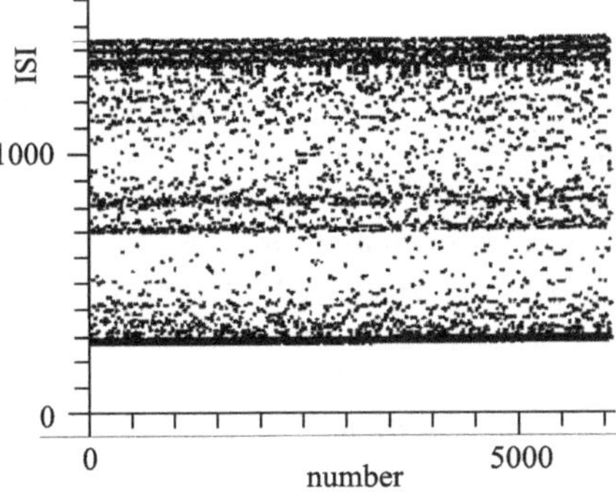

Fig. 6.34 Chaotic ISI time series obtained from the β-cell model: ISI versus ISI serial number (Figure 2(a) in (Gong et al., 1998)).

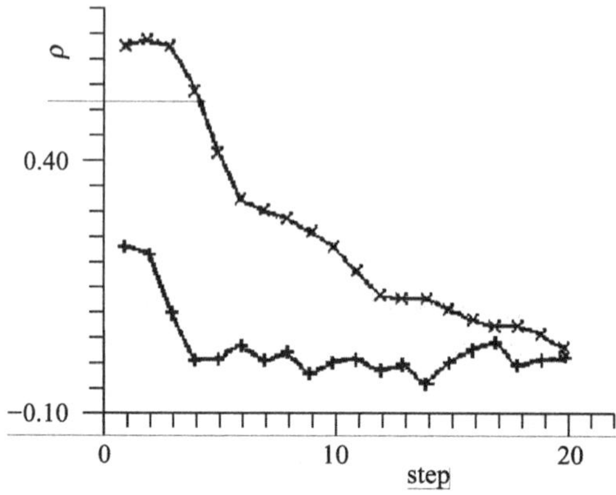

Fig. 6.35 Curves of the correlation coefficient ρ versus the prediction step. $\times$, chaotic ISI time series from the β-cell model contaminated with noise; $+$, the surrogate data (Fig. 4 in Ref. (Gong et al., 1998)).

6.7.5 Nonlinear characteristic numbers

6.7.5.1 Correlation dimension

The most widely used procedure of detecting chaos is the GP correlation dimension algorithm that quantifies the strangeness of the attractors (Grassberger and Procaccia, 1983a, b; Mizrachi et al., 1984). According to the GP algorithm, the correlation integral is first evaluated using the following expression:

$$C_d(l) = \lim_{N\to\infty} \frac{1}{N^2} \sum_{i=1}^{N} \sum_{j=1}^{N} \Theta(l - \|X_i^T - X_j^T\|). \tag{6.33}$$

Here, X_i^T and X_j^T are reconstructed vectors with the embedding dimension d. $C(l)$ is the correlation integral, $\|\cdot\|$ represents Euclidean norm, $\Theta(\cdot)$ is the Heaviside function, and l is the correlation length. Then, the correlation dimension in d-dimensional phase space is estimated form the slope of $\ln(l)$ versus $\ln C_d(l)$ plot. To identify the correlation dimension, we look for a scaling region in the correlation dimension plot. When the correlation dimension reaches a saturated state as a function of d, this value is taken as an estimate of the correlation dimension. Note that the onset of the plateau known as then scaling region in local slope curves indicates the occurrence of chaos.

Although the GP method has the advantage of simplicity, limitations still exist such as the need for a large amount of time series data (Grassberger P., 1986); also, as it is only concerned with the static values of the data, the influence of noise on its computation is almost unavoidable (Osborne and Provenzale, 1989). As an example, 4000 ISI data points are computed from the β-cell model are used to calculate the correlation dimension, and the local slope curves are given in Fig.6.36. It can be seen that the plateau of the scaling region is long and distinct.

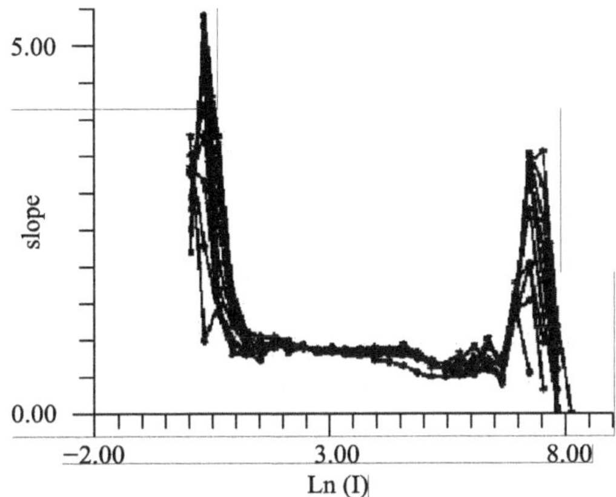

Fig. 6.36 Local slope versus logarithmic l curves. The embedding dimension d is in sequence 2; 3; 4; : : : ; 9; 10; 11 from the bottom to the top (Figure 3(a) in (Gong et al., 1998)).

6.7.5.2 Lyapunov exponent

The Lyapunov exponent or Lyapunov characteristic exponent of a dynamic system is a quantitative measure of the sensitivity of the system to the initial conditions, i.e., the Lyapunov exponent is a quantity that characterizes the rate of separation of infinitesimally close trajectories (Gong et al., 1998). Quantitatively, two trajectories in phase space with initial separation δZ_0 diverge

$$|\delta Z(t)| \approx e^{\lambda t}|\delta Z_0|, \tag{6.34}$$

where λ is the Lyapunov exponent.

The rate of separation can be different for different orientations of initial separation vector. Thus, there is a whole spectrum of Lyapunov exponents, and the number of them is equal to the number of dimensions of the phase space. On the other hand, the number of Lyapunov exponents is equal to the number of the embedding dimension of the reconstructed phase space. Usually we calculate the Largest Lyapunov exponent (LLE), because it determines the predictability of the system. A positive LLE is usually taken as an indication that the system is chaotic, and may be regarded as an estimator of the dominant chaotic behavior of a dynamical system.

The LLE can be defined as follows:

$$\lambda = \lim_{t\to\infty} \frac{1}{t} \ln\left|\frac{\delta Z(t)}{\delta Z_0}\right|. \tag{6.35}$$

Although the formula for the calculation of the Lyapunov exponents is given above, in most cases we cannot carry out analytically, and must turn to numerical techniques. One of the most used and effective numerical technique to calculate the Lyapunov spectrum for a smooth dynamical system relies on periodic Gram Schmidt orthonormalization of the Lyapunov vectors to avoid a misalignment of all the vectors along the direction of maximal expansion.

6.7.5.3 Approximate entropy

Approximate entropy has been introduced to characterize the regularity or predictability of time series (Pincus, 1991). This entropy calibrates an ensemble extent of sequential interrelations, quantifying a continuum that ranges from totally ordered to completely random. More interestingly, approximate entropy assigns a nonnegative number to a sequence or time series: smaller values of approximate entropy imply a greater likelihood that similar patterns of measurements will be followed by additional similar measurements. If the time series is highly irregular, the occurrence of similar patterns will not be predictive for the following measurements, and approximate entropy will be relatively large. Note that approximate entropy has significant weaknesses, notably its strong dependence on sequence length and its poor self-consistency. Namely, the observation that approximate entropy for one data set is larger than approximate entropy for another for a given choice of m and r should, but does not, hold true for other choices of m and r. By they way, m is the number of dimensions, and r is the tolerance. To compute approximate entropy, m and r must be fixed. The values $m = 2$ and r between 10% and 25% of the standard deviation of the data sets are recommended.

The process of calculation is as follows (Pincus, 1991; Yang et al., 2002):

First, given N data points $\{u(i)\}$, form vector sequences $x(i) = [u(i), \ldots, u(i+m-1)]$. Then define the distance $d[x(i), x(j)]$ between $x(i)$ and $x(j)$ as the maximum difference in their respective scalar components. And then use the sequence $x(1), x(2), \ldots, x(N-m+1)$ to construct, for each $i \leq N-m+1$, $C_i^m(r) = \sum_{j=1}^{N-m+1} \Theta(r - d[x(i) - d(j)])$. Here, $\Theta(\cdot)$ is the Heaviside function. And then determine $\Phi^m(r) = \dfrac{1}{N-m+1} \sum_{i=1}^{N-m+1} \ln C_i^m(r)$. Finally, we can calculate the approximate entropy as follows:

$$ApEn(m, r, N) = \Phi^m(r) - \Phi^{m+1}(r). \qquad (6.36)$$

The value of N for the computation of approximate entropy is typically between 100 and 5000; $m = 2$ and $r = 0.1 - 0.25STD$, here, STD is the standard deviation of the data set.

Figure 6.37 shows a bifurcation cascade of ISIs from the rat injured nerve with the perfusion of Ca^{2+} free solution and its approximate entropy (Han et al., 2002), and the schematic diagram of the experiment setup is shown in Fig. 6.38. The series of ISIs is transforming from period-2 bursting to period-3 bursting. We can see that the change in approximate entropy is considerably obvious at the time instant of transformation of the ISI pattern.

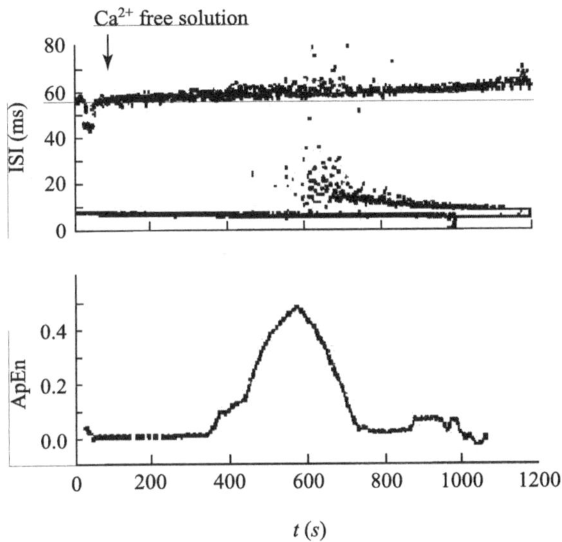

Fig. 6.37 The series of ISIs from the rat injured nerve and its approximate entropy (Figure 2 in (Han et al., 2002)).

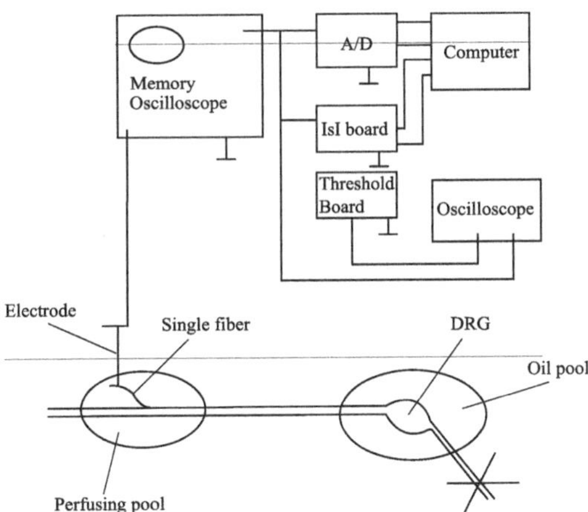

Fig. 6.38 Schematic diagram of the experiment setup about the spontaneous firings of the DRG neurons (Figure 1 in (Gong et al., 2002)).

6.7.5.4 Lempel-Ziv complexity

The Lempel-Ziv complexity algorithm can be used to quantitatively characterize the complexity of a data set (Szczepanski et al., 2003, 2004). The calculation of complexity measure consists of two steps. First, generate a binary string by comparing the original data with a threshold. Second, use the string to calculate the disorder degree of the original data based on Lempel-Ziv algorithm. In detail, in the first step, the average $\bar{x} = \frac{1}{N}\sum_{i=1}^{N} x_i$ is subtracted for every signal simple x_i, $i = 1, \ldots, N$. The sample points are then compared with the threshold $x_{th} = 0.2 \times STD(\{x_i\}_{i=1}^{N})$, where $STD(\{x_i\}_{i=1}^{N})$ is the standard deviation of the data set. If $-x_{th} < x_i < x_{th}$, then $s_i = 0$, otherwise $s_i = 1$. In this way, the set of the original data is transformed into a binary sequence $s = s_i s_2, \ldots, s_N$, where s_i is 1 or 0. In the second step, the Lempel-Ziv complexity algorithm is calculated to obtain the complexity measure of the original time series. The N-digit sequence is scanned from left to right and defined a new block of length k every time it discovers a sub-string of length k not previously encountered. After a number of operations, a decomposition of the original time series in minimal blocks is obtained. The complexity is then defined as the number of blocks in the decomposition.

6.8 Application

The dense UPO structure within a chaotic attractor and the exponential sensitivity of a chaotic system provides a very power tool for utilizing chaos for control as well as controlling chaos (http://www.scholarpedia.org/article/Unstable_periodic_orbits). Cardiac arrhythmias were successful tamed by a chaos stabilization method; and seizure activities were controlled by judiciously timed electric pulses.

SR-based techniques has been used to create a novel class of medical devices (such as vibrating insoles) for enhancing sensory and motor function in the elderly, patients with diabetic neuropathy, and patients with stroke (http://en.wikipedia.org/wiki/Stochastic_resonance). For example, a potentially very important proposed application where suprathreshold SR would be caused to occur is a cochlear implant signalling strategy. The idea is based on the fact that people requiring cochlear implants are missing the natural sensory hair cells that a functioning inner ear uses to encode sound in the auditory nerve. It is known that the stereocilia of these hair cells undergo significant Brownian motion, i.e. randomness, and the synaptic release of neurotransmitter introduces additional randomness. These sources of randomness lead to spontaneous firing in primary afferent auditory nerve fibres that is not normally present in deaf patients who benefit from cochlear implants. The hypothesis is that suprathreshold stochastic resonance induced by re-introducing this natural randomness to the encoding of sound could improve speech comprehension in patients fitted with cochlear implants (http://www.scholarpedia.org/article/Suprathreshold_stochastic_resonance).

6.9 Conclusions

The mammalian nervous system is a complex adaptive system, which is composed of a large number of neurons. In fact, a neuron can be considered to be a rather complicated nonlinear dynamical system. There are many neuron models presented for different kinds of neurons. For example, the HR model of neuronal activity is aimed to study the spiking-bursting behavior of the membrane potential of cortical neurons, and the Traub model (Traub et al., 1991) is used to characterize a CA3 hippocampal pyramidal neuron. A rich variety of dynamical behavior occurs in these neuron models, such as periodic spiking, chaotic spiking, periodic bursting and chaotic bursting. It is this point that asks us to apply the nonlinear systems theory to the investigation on the firing behavior of neuron models. And the methods or theories in nonlinear dynamics can be helpful to understand the neuronal firing patterns and the mechanisms underlying neural information encoding. Meanwhile, the complicated phenomena observed in neurophysiological experiments provide strong motivation for the development of nonlinear systems theory. Actually, the naissance of neurodynamics is just a perfect combination between neuroscience and nonlinear dynamics.

Up to now, it is very important and significant to establish corresponding mathematical models for special neurons including one-compartment model and multi-compartment model. Therefore, the numerical or theoretical investigation on new neuron models or existing neuron models with complex behavior remains very interesting, and bifurcation and chaos occurring in these models are still hot topics. As for noise in nervous systems, its role is waiting for more confirmation. As we know, the neural information transmission happens at the synapses between neurons. The collective behavior of neuronal population, therefore, should be studied from the viewpoint of coupled neuronal networks including electrical and chemical coupling. In addition, the transitory process in nervous systems deserves more attention.

Acknowledgements The research was supported by the National Natural Science Fund of China (Grant Nos. 10432010, 10502039, 10972170 and 10602041) and the President Fund of Xi'an Jiaotong University. In particular, we would like to take this opportunity to express our gratitude to Prof. Sanjue Hu, Prof. Qishao Lu, Prof. Wei Ren, Dr. Yunfan Gong, Dr. Pulin Gong, Dr. Ning Tan, Dr. Wuying Jin, Dr. Yanmei Kang, Dr. Guangjun Zhang, Dr. Ying Wu, Prof. Jun Jiang and Prof. Ling Hong for their generous support of our research over the years. We owe special thanks to Prof. Albert C.J. Luo for a careful reading of the text and valuable suggestions.

References

Abed E.H. and Fu J.H., 1986, Local feedback stabilization and bifurcation control, I. Hopf bifurcation, *Syst. Control Lett.*, **7**, 11–17.

Abed E.H., Wang H.O. and Chen R.C., 1994, Stabilization of period doubling bifurcations and implications for control of chaos, *Physica D*, **70**, 154–164.

Aihara K., and Matsumoto G., 1986, Chaotic oscillations and bifurcations in squid giant axons, In: *Chaos* (A.V. Holden, ed), Manchester and Princeton University Press, Princeton, NJ, 257–269.

Aihara K., Matsumoto G. and Ichikawa M., 1985, An alternating periodic–chaotic sequence observed in neural oscillators, *Phys. Lett. A*, **111**, 251–255.

Auerbach D., Cvitanovic P., Eckmann J.P., Gunaratne G. and Procaccia I., 1987, Exploring chaotic motion through periodic orbits, *Phys. Rev. Lett.*, **58**, 2387–2389.

Baer S.M., Rinzel J. and Carrillo H., 1995, Analysis of an autonomous phase model for neuronal parabolic bursting, *J. Math. Biol.*, **33**, 309–333.

Barbi M., Chillemi S., Garbo A.D. and Reale L., 2003, Stochastic resonance in a sinusoidally forced LIF model with noisy threshold, *BioSystems*, **71**, 23–28.

Bertram R., Butte M.J., Kiemel T. and Sherman A., 1995, Topological and phenomenological classification of bursting oscillations, *Bull. Math. Biol.*, **57**, 413–439.

Buonocore A., Nobile A.G., Ricciardi L.M., 1987, A new integral equation for the evaluation of first-passage-time probability density, *Adv. Appl. Prob.*, **19**, 784–800.

Bulsara R., Elson T.C., Doering C.R., Lowen S.B. and Lindenberg K., 1996, Coopeative behavior in periodically driven noisy integrate-and-fire models of neuronal dynamics, *Phys. Rev. E*, **53**, 3958–3969.

Braun H.A., Wissing H., Schäfer K. and Hirsch M.C., 1994, Oscillation and noise determine signal transduction in shark multimodal sensory cells, *Nature*, **367**, 270–273.

Chay T.R., Fan Y.S. and Lee Y.S., 1995, Bursting, spiking, chaos, fractals, and universality in biological rhythms, *International Journal of Bifurcattion and Chaos*, **5**, 595–635.

Chen D.S., Wang H.O. and Chen G., 2001, Anti-control of Hopf bifurcation, *IEEE Trans. Circ. Sys.-I*, **48**, 661–672.

Chialvo D.R., Longtin A. and Muller-Gerking J., 1997, Stochastic resonance in models of neuronal ensembles, *Phys. Rev. E*, **55**, 1798–1808.

Cvitanovic P., 1998, Invariant measurement of strange sets in terms of cycles, *Phys. Rev. Lett.*, **61**, 2729–2732.

Cymbalyuk G. and Shilnikov A., 2005, Coexistence of tonic spiking oscillations in a leech neuron model, *Journal of Computational Neuroscience*, **18**, 255–263.

Davidchack R.L. and Lai Y.C., 1999, Efficient algorithm for detecting unstable periodic orbits in chaotic systems, *Physical Review E*, **60**, 6172–6175.

Douglass J.K., Wilkens L., Pantazelou E. and Moss F., 1993, Noise enhancement of information transfer in crayfish mechanoreceptors by stochastic resonance, *Nature*, **365**, 337–340.

Ermentrout B., 2002,*Simulating, Analyzing, and Animating Dynamical Systems: A Guide to XPPAUT for Researchers and Students*, SIAM, Philadelphia.

Ermentrout G.B. and Kopell N., 1986, Parabolic bursting in an excitable system coupled with a slow oscillation, *SIAM J. Appl. Math.*, **46**, 233–253.

Elson R.C., Selverston A.I., Huerta R., Rulkov N.F., Rabinovich M.I. and Abarbanel H.D.I., 1998, Synchronous behavior of two coupled biological neurons, *Phys. Rev. Lett.*, **81**, 5692–5695.

Faisal A.A., Selen L.P.J. and Wolpert D.M., 2008, Noise in the nervous system, *Nat. Rev. Neurosci.*, **9**, 292–303.

Feudel U., Neiman A., Pei X., Wojtenek W. and Moss F., 2002, Homoclinic bifurcation in a thermally sensitive neuron, In: *Experimental chaos*, AIP Conference Proceedings, **622**, 139–148.

FitzHugh R., 1961, Impulses and physiological states in theoretical models of nerve membrane, *Biophys. J.*, **1**, 445–466.

Fries P., 2001, A mechanism for cognitive dynamics: neuronal communication through neuronal coherence, *TICS*, **9**, 474–480.

Gammaitoni L., Hanggi P., Jung P. and Marchesoni F., 1998, Stochastic resonance, *Rev. Mod. Phys.*, **70**, 223–287.

Gong P.L., Xu J.X., Long K.P. and Hu S.J., 2002, Chaotic interspike intervals with multipeaked histogram in neurons, *International Journal of Bifurcation and Chaos*, **12**, 319–328.

Gong Y.F., Xu J.X., Ren W., Hu S.J., and Wang F.Z., 1998, Determining the degree of chaos from analysis of ISI time series in the nervous system: a comparison between correlation dimension and nonlinear forecasting methods, *Biological Cybernetics*, **78**, 159–165.

Grassberger P., 1986, Do climatic attractors exist? *Nature*, **323**, 609–612.

Grassberger P. and Procaccia I., 1983a, Characterization of strange attractors, *Phys. Rev. Lett.*, **50**, 346–349.

Grassberger P. and Procaccia I., 1983b, Measuring the strangeness of strange attractors, *Physica D*, **9**, 189–208.

Gray C.M., Konig P., Engel A.K. and Singer W., 1989, Oscillatory responses in cat visual cortex exhibit inter-columnar synchronization which reflects global stimulus properties,*Nature*, **338**, 334–337.

Grebogi C., Ott E. and Yorke J.A., 1986, Critical exponent of chaotic transients in nonlinear dynamical systems, *Phys. Rev. Lett.*, **57**, 1284–1287.

Grebogi C., Ott E., Romeiras F. and Yorke J.A., 1987, Critical exponents for crisis-induced intermittency, *Phys. Rev. A*, **36**, 5365–80.

Gu H.G., Ren W., Lu Q.S., Wu S.G., Yang M.H. and Chen W.J., 2001, Integer multiple spiking in neuronal pacemakers without external periodic stimulation. *Physics Letter A*, **285**, 63–68.

Gu H.G., Li L., Yang M.H., Liu Z.Q., and Ren W., 2003, Integer muliple bursting generated in an experimental neural pacemaker, *Acta Biophyica Sinica*, **19**, 68–72.

Gu H., Yang M.H., Li L., Liu Z. and Ren W., 2002, Experimental observation of the stochastic bursting caused by coherence resonance in a neural pacemaker, *Neuroreport*, **13**, 1657–1660.

Guckenheimer J. and Holmes P., 1997, *Nonlinear Oscillations, Dynamical Systems, and Bifurcations of Vector Fields* 5th edition, Springer, New York,150–152.

Guckenheimer J. and Oliva R.A., 2002, Chaos in the Hodgkin-Huxley model, *SIAM J. Applied Dynamical Systems*, **1**, 105–114.

Gong Y.F., Xu J.X. and Hu S.J., 1998, Stochastic resonance: When does it not occur in neuronal models? *Phys. Lett. A*, **243**, 351–359.

Gong P.L. and Xu J.X., 2001, Globall dynamics and stochastic resonance of the forced FitaHugh-Nagumo neuron model,*Phys. Rev. E*, **63**, 031906.

Han S., Duan Y.B., Jian Z., Xie Y., Xing J.L. and Hu S.J., 2002, Calculating the degree of complexity of interspike interval with the method of approximate entropy,*Acta Biophysica Sinica*, **18**, 448–451.

Hassard B.D., Kazarinoff N.D. and Wan Y.H., 1981, Theory and application of Hopf bifurcation, in: *London Mathematical Society Lecture Note Series*, **41**, Cambridge Univ. Press, Cambridge, 86–91.

Hassouneh M.A., Lee H.C. and Abed E.H., 2004, Washout filters in feedback control: Benefits, Limitations and Extensions, In: *Proceeding of the 2004 American Control Conference*, Boston, Massachusetts, June 30-July 2, 3950–3955.

Hindmarsh J.L. and Rose R.M., 1984, A model of neuronal bursting using three coupled first order differential equations, *Proc. R. Soc. London, Ser. B*, **221**, 87–102.

Hodgkin A., and Huxley A., 1952, A quantitative description of membrane current and its application to conduction and excitation in nerve, *J. Physiol*, **117**, 500–544.

Hoffman R.E., Shi W.X. and Bunney B.S., 1995, Nonlinear sequence-dependent structure of nigral dopamine neuron interspike interval firing patterns,*Biophys. J.*, **69**, 128–137.

Hu S.J., Yang H.J., Jian Z., Long K.P., Duan Y.B., Wan Y.H., Xing J.L., Xu H., and Ju G., 2000, Adrenergic sensitivity of neurons with non-periodic firing activity in rat injured dorsal root ganglion, *Neuroscience*, **101**, 689–698.

Izhikevich E.M., 2000, Neural excitability, spiking, and bursting,*International Journal of Bifurcation and Chaos*, **10**, 1171–1266.

Izhikevich E.M., 2001, Resonate-and-fire neurons, *Neural Networks*, **14**, 883–894.

Izhikevich E.M., 2007, *Dynamical Systems in Neuroscience: The Geometry of Excitability and Bursting*, MIT, Cambridge, MA.

Jin W.Y., Xu J.X., Wu Y., Hong L. and Wei Y.B., 2006, Crisis of interspike intervals in Hodgkin-Huxley model, *Chaos, Solitons and Fractals*, **27**, 952–958.

Kang Y.M., Xu J.X. and Xie Y., 2005a, A further insight into stochastic resonance in an integrate-and-fire neuron with noisy periodic input, *Chaos, Solitons and Fractals*, **25**, 165–170.

Kang Y.M., Xu J.X. and Xie Y., 2005b, Signal-to-noise ratio gain of a noisy neuron that transmits subthreshold periodic spike trains, *Phys. Rev. E*, **72**, 021902.

Kim J.H. and Stringer J.,1992, *Appl. Chaos*, Wiley, New York, 441–455.

Kretzberg J., Warzecha A.K. and Egelhaaf M., 2001, Neural coding with graded membrane potential changes and spikes,*Journal of Computational Neuroscience*, **11**, 153–164.

Lathrop D.P. and Kostelich E.J., 1989, Characterization of an experimental strange attractor by periodic orbits, *Phys. Rev. A*, **40**, 4028–4031.

Levin J.E. and Miller J.P., 1996, Broadband neural encoding in the cricket cereal sensory system enhanced by stochastic resonance, *Nature*, **380**, 165–168.

Li L., Gu H.G., Yang M.H., Liu Z.Q. and Ren W., 2003, Bifurcation scenario rhythm in the firing pattern transition of a neural pacemaker, *Acta Biophysica Sinica*, **19**, 388–394.

Liu W.M., 1994, Criterion of Hopf bifurcation without using eigenvalues, *J. Math. Anal. Appl.*, **182**, 250–255.

Longtin A., 1993a, Nonlinear forecasting of spike trains from sensory neurons, *International Journal of Bifurcation and Chaos*, **3**, 651–661.

Longtin A., 1993b, Stochastic resonance in neuron models, *Journal of Statistical Physics*, **70**, 309–327.

Longtin A., 1995, Mechanisms of stochastic phase-locking, *Chaos*, **5**, 209–215.

Longtin A., 1997, Autonomous stochastic resonance in bursting neurons, *Phys. Rev. E*, **55**, 868–876.

Longtin A. and Bulsara A., 1991, Time-interval sequences in bistable systems and the noise-induced transmission of information by sensory neurons, *Phys. Rev. Lett.*, **67**, 656–659.

Longtin A., and Hinzer K., 1996, Encoding with bursting, subthreshold oscillations, and noise in mammalian cold receptors, *Neural Computation*, **8**, 215–255.

Mandelblat Y., Etzion Y., Grossman Y. and Golomb D., 2001, Period doubling of calcium spike firing in a model of a Purkinje cell dendrite, *Journal of Computational Neuroscience*, **11**, 43–62.

Matsumoto G., Aihara K., Hanyu Y., Takahashi N., Yoshizawa S. and Nagumo J., 1987, Chaos and phase locking in normal squid axons, *Physics Letters A*, **123**, 162–166.

Michael V.L. and Bennett D.P., 2006, Electrical synapses between neurons synchronize gamma oscillations generated during higher level processing in the nervous system, *Electroneurobiologia*, **14**, 227–250.

Mizrachi A.B., Procaccia I. and Grassberger P., 1984, The characterization of experimental (noisy) strange attractor, *Phys. Rev. A*, **29**, 975–977.

Morris C., and Lecar H., 1981, Voltage oscillations in the barnacle giant muscle fiber, *Biophys. J.*, **35**, 193–213.

Moss F., Ward L.M., and Sannita W.G., 2004, Stochastic resonance and sensory information processing: a tutorial and review of application, *Clin Neurophysiol*, **115**, 267–281.

Mpitsos G.J., Burton R.M., Creech Jr. H.C. and Soinila S.O., 1988, Evidence for chaos in spike trains of neurons that generate rhythmic motor patterns, *Brain Res. Bull.*, **21**, 529–538.

Neiman A.B. and Russell D.F., 2002, Synchronization of noise-induced bursts in noncoupled sensory neurons, *Phys. Rev. Lett.*, **88**, 138103.

Osborne A.R., and Provenzale A., 1989, Finite correlation dimension for stochastic systems with power-law spectra, *Physica D*, **35**, 357–381.

Pincus S.M., 1991, Approximate entropy as a measure of system complexity, *Proc. Natl. Acad. Sci. USA*, **88**, 2297–2304.

Pikovsky A.S. and Kurths J., 1997, Coherence resonance in a noise-driven excitable system, *Phys. Rev. Lett.*, **78**, 775–778.

Plesser H.E. and Geisel T., 1999, Markov analysis of stochastic resonance in a periodically driven integrate-and-fire neuron, *Phys. Rev. E*, **59**, 7008–7017.

Ren W., Gu H.G., Jian Z., Lu Q.S., and Yang M.H., 2001, Different classifications of UPOs in the parametrically different chaotic ISI series of neural pacemaker, *Neuroreport*, **12**, 2121–2124.

Ren W., Hu S.J., Zhang B.J., Wang F.Z., Gong Y.F. and Xu J.X., 1997 , Period-adding bifurcation with chaos in the interspike intervals generated by an experimental neural pacemaker, *International Journal of Bifurcation and Chaos*, **7**, 1867–1872.

Rinzel J., 1985, Bursting oscillations in an excitable membrane model, In: *Ordinary and partial Differential Equations Proceedings of the 8th Dundee Conference* (B.D. Sleeman and R.J. Jarvis, eds.), 304–316, Lecture Notes in Mathematics 1151, Springer, Berlin.

Rinzel J., 1987, A formal classification of bursting mechanisms in excitable systems, In: *Mathematical Topics in Population Biology; Morphogenesis; and Neurosciences* (E. Teramoto, M. Yamaguti, eds), Vol. 71 of Lecture Notes in Biomathematics, Springer, Berlin.

Rinzel J. and Ermentrout B., 1989, Analysis of neural excitability and oscillations, In: *Methods in Neuronal Modeling* (Koch C, Segev I, eds): MIT, Cambridge, MA.

Rinzel J. and Lee Y.S., 1987, Dissection of a model for neuronal parabolic bursting,*J. Math. Biol.*, **25**, 653–675.

Rozental R., Andrade-Rozental A.F., Zheng X., Urban M., Spray D.C. and Chiu F.C., 2001, Gap Junction-Mediated Bidirectional Signaling between Human Fetal Hippocampal Neurons and Astrocytes, *Develpmental Neuroscience*, **23**, 420–431.

Russell D.F., Wilkens L.A. and Moss F., 1999, Use of behavioural stochastic resonance by paddle fish for feeding, *Nature*, **402**, 291–294.

Schmelcher P., and Diakonos F.K., 1998, General approach to the localization of unstable periodic orbits in chaotic dynamical systems, *Physical Review E*, **57**, 2739–2746.

Schnitlzer A. and Gross J., 2005, Normal and pathological oscillatory communication in the brain, *Nat. Rev. Neurosci.*, **6**, 285–296.

Shimokawa T., Pakdaman K., Takahata T., Tanabe S. and Sato S., 2000, A first-passage-time analysis of the periodically forced noisy leaky integrate-and-fire model, *Biol. Cybern.*, **83**, 327–340.

Shilnikov A. and Cymbalyuk G., 2005, Transition between tonic spiking and bursting in a neuron model via the blue-sky catastrophe, *Physical Review Letters*, **94**, 048101.

Shimokawa T., Pakdaman K. and Sato S., 1999, Mean discharge frequency locking in the response of a noisy neuron model to subthreshold periodic stimulation, *Phys. Rev. E*, **60**, R33–R36.

Singer W., 1993, Synchronization of cortical activity and its putative role in information processing and learning, *Annu. Rev. Physiol.*, **55**, 349–374.

So P., Francis J.T., Netoff T.I., Gluckman B.J. and Schiff S.J., 1998, Periodic orbits: a new language for Neuronal dynamics, *Biophys. J.*, **74**, 2776–2785.

So P., Ott E., Schiff S.J., Kaplan D.T., Sauer T. and Grebogi C., 1996, Detecting unstable periodic orbits in chaotic experimental data, *Physical Review Letters*, **76**, 4705–4708.

So P., Ott E., Sauser T., Gluckman B.J., Grebogi C. and Schiff S.J., 1997, Extracting unstable orbits from chaotic time series data, *Physical Review E*, **55**, 5398–5417.

Suzuki H., Aihara K., Murakami J. and Shimozawa T., 2000, Analysis of neural spike trains with interspike interval reconstruction, *Biological Cybernetics*, **82**, 305–311.

Sugihara G., and May R.M., 1990, Nonlinear forecasting as a way of distinguishing chaos from measurement error in time series, *Nature*, **344**, 734–741.

Szczepaski J., Amigo J.M., Wajnryb E. and Sanchez-Vives M.V., 2003, Application of Lempel-Ziv complexity to the analysis of neural discharges, *Network: Computation in Neural Systems*, **14**, 335–350.

Szczepanski J., Amigo J.M., Wajnryb E. and Sanchez-Vives M.V., 2004, Characterizing spike trains with Lempel-Ziv complexity, *Neurocomputing*, 58–60, 79–84.

Takahashi N., Hanyu Y., Musha T., Kubo R. and Matsumoto G., 1990, Global bifurcation structure in periodically stimulated giant axons of squid, *Physica D*, **43**, 318–334.

Theiler J., Eubank S., Longtin A. Galdrikian B. and Farmer J.D., 1992, Testing for nonlinearity in time series: the method of surrogate data, *Physica D*, **58**, 77–94.

Traub R.D., Wong R.K.S., Miles R. and Michelson H., 1991, A model of a CA3 hippocampal pyramidal neuron incorporating voltage-clamp data on intrinsic conductances, *Journal of Neurophysiology*, **66**, 635–650.

Tsuji S., Ueta T., Kawakami H., Fujii H. and Aihara K., 2007, Bifurcations in two-dimensional Hindmarsh-Rose type model, *International Journal of Bifurcation and Chaos*, **17**, 985–998.

Wellens T., Shatokhin V. and Buchleitner A., 2004, Stochastic resonance, *Rep. Prog. Phys.*, **67**, 45–105.

Wiesenfeld K. and Moss F., 1995, Stochastic resonance and the benefits of noise: from ice ages to crayfish and Squids, *Nature*, **373**, 33–36.

Wu Y., Xu J.X. and He M., 2005, Synchronous behaviors of Hindmarsh-Rose neurons with chemical coupling, in: *Lecture in Computer Science*, 3610, 508–511.

Xie Y., Aihara K. and Kang Y.M., 2008a, Change in types of neuronal excitability via bifurcation control, *Physical Review E*, 021917.

Xie Y., Chen L., Kang Y.M. and Aihara K., 2008b, Controlling the onset of Hopf bifurcation in the Hodgkin-Huxley model, *Physical Review E*, **77**, 061921.

Xie Y., Duan Y.B., Xu J.X., Kang Y.M. and Hu S.J., 2003a, Parabolic bursting induced by Veratridine in rat injured sciatic nerves, *Acta Biochimica et Biophysica Sinica*, **35**, 806–810.

Xie Y., Duan Y.B., Xu J.X., Kang Y.M. and Hu S.J., 2003b, The interspike interval increases gradually: why? *Acta Biophysica Sinica*, **19**, 401–408.

Xie Y., Xu J.X. and Hu S.J., 2004a, A novel dynamical mechanism of neural excitability for integer multiple spiking, *Chaos, Solitons and Fractals*, **21**, 177–184.

Xie Y., Xu J.X., Hu S.J., Kang Y.M., Yang H.J. and Duan Y.B., 2004b, Dynamical mechanisms for sensitive response of aperiodic firing cells to external stimulation, *Chaos, Solitons and Fractals*, **22**, 151–160.

Xie Y., Xu J.X., Kang Y.M., Hu S.J. and Duan Y.B., 2005, Critical amplitude curves for different periodic stimuliand different dynamical mechanisms of excitability, *Communications in Nonlinear Science and Numerical Simulation*, **10**, 823–832.

Xing J.L., Hu S.J., Xu H., Han S. and Wan Y.H., 2001, Subthreshold membrane oscillations underlying integer multiples firing from injured sensory neurons, *Neuroreport*, **12**, 1311–1313.

Xu J.X., Gong Y.F., Ren W., Hu S.J., and Wang F.Z., 1997, Propagation of periodic and chaotic action potential trains along nerve fibers, *Physica D*, **100**, 212–224.

Yang H.J., Hu S.J., Jian Z., Wan Y.H., and Long K.P., 2000, Relationship between the sensitivity to tetraethylammonium and firing patters of injured dorsal root ganglion neurons, *Acta Physiologica Sinica*, **52**, 395–401.

Yang H.J., Hu S.J., Han S., Liu G.P., Xie Y. and Xu J.X., 2002, Relation between responsiveness to neurotransmitters and complexity of epileptiform activity in rat hippocampal CA1 neurons, *Epilepsia*, **43**, 1330–1336.

Yang Z.Q., Lu Q.S., Gu H.G. and Ren W., 2002, Integer multiple spiking in the stochastic Chay model and its dynamical generation mechanism, *Physics Letter A*, **299**, 499–506.

Zhang G.J., Xu J.X., Wang J., Yue Z.F., Liu C.B., Yao H. and Wang X.B., 2008, The Mechanism of Bifurcation-Dependent Coherence Resonance of Morris-Lecar Neuron Model, In: *Advances in Cognitive Neurodynamics, Proceedings*, 83–89.

Nonlinear Physical Science

(Series Editors: Albert C.J. Luo, Nail H. Ibragimov)

*__Nail. H. Ibragimov / Vladimir. F. Kovalev:__ Approximate and Renormgroup Symmetries

*__Abdul-Majid Wazwaz:__ Partial Differential Equations and Solitary Waves Theory

*__Albert C.J. Luo:__ Discontinuous Dynamical Systems on Time-varying Domains

*__Anjan Biswas / Daniela Milovic / Matthew Edwards:__ Mathematical Theory of Dispersion-Managed Optical Solitons

*__Albert C.J. Luo:__ Nonlinear Deformable-body Dynamics

*__Albert C.J. Luo / Valentin Afraimovich (Editors):__ Hamiltonian Chaos Beyond the KAM Theory

*__Albert C.J. Luo / Valentin Afraimovich (Editors):__ Long-range Interaction, Stochasticity and Fractional Dynamics

*__Vasily E. Tarasov:__ Fractional Dynamics

*__Meike Wiedemann / Florian P.M. Kohn / Harald Roesner / Wolfgang R.L. Hanke :__ Self-organization and Pattern-formation in Neuronal Systems under Conditions of Variable Gravity

*__Albert C. J. Luo / Jian-Qiao Sun (Editors):__ Complex Systems with Fractionality, Time-delay and Synchronization

Ivo Petras: Fractional Order Nonlinear Systems

Vladimir V. Uchaikin: Fractional Derivatives in Physics

Fečkan Michal: Bifurcation and Chaos in Discontinuous and Continuous Systems

Sergey N. Gurbator / Oley V. Rudenko / Alexander I. Sachev: Waves and Structures in Nonlinear Nondispersive Media

---

*published

Zeitfracht Medien GmbH
Ferdinand-Jühlke-Straße 7
99095 Erfurt, Deutschland
produktsicherheit@kolibri360.de